STP 1142

Hydraulic Conductivity and Waste Contaminant Transport in Soil

David E. Daniel and Stephen J. Trautwein, Editors

ASTM Publication Code Number (PCN)
04-011420-38

ASTM
1916 Race Street
Philadelphia, PA 19103
Printed in the U.S.A.

Library of Congress

Hydraulic conductivity and waste contaminant transport soil/David E. Daniel and Stephen J. Trautwein, editors.
    (STP; 1142)
    "Contains papers presented at the symposium of the same name held in San Antonio, TX on 21-22 January, 1993 . . . sponsored by ASTM Committee D-18 on Soil and Rock and its Subcommittee D18.04 on Hydrologic Properties of Soil and Rock"—Foreword.
    "ASTM publication code number (PCN): 04-011420-38."
    Includes bibliographical references and indexes.
    ISBN 0-8031-1442-7
    1. Soil pollution—Congresses. 2. Soil permeability—Congresses.
I. Daniel, David E. (David Edwin), 1949- . II. Trautwein, Stephen J., 1952- . III. Series: ASTM special technical publication; 1142.
    TD878.H95 1994   94-27568
    628.5'5—dc20

Copyright © 1994 AMERICAN SOCIETY FOR TESTING AND MATERIALS, Philadelphia, PA. All rights reserved. This material may not be reproduced or copied, in whole or in part, in any printed, mechanical, electronic, film, or other distribution and storage media, without the written consent of the publisher.

Photocopy Rights

Authorization to photocopy items for internal or personal use, or the internal or personal use of specific clients, is granted by the AMERICAN SOCIETY FOR TESTING AND MATERIALS for users registered with the Copyright Clearance Center (CCC) Transactional Reporting Service, provided that the base fee of $2.50 per copy, plus $0.50 per page is paid directly to CCC, 222 Rosewood Dr., Danvers, MA 01923; Phone: (508) 750-8400; Fax: (508) 750-4744. For those organizations that have been granted a photocopy license by CCC, a separate system of payment has been arranged. The fee code for users of the Transactional Reporting Service is 0-8031-1442-7/94 $2.50 + .50.

Peer Review Policy

Each paper published in this volume was evaluated by three peer reviewers. The authors addressed all of the reviewers' comments to the satisfaction of both the technical editor(s) and the ASTM Committee on Publications.

To make technical information available as quickly as possible, the peer-reviewed papers in this publication were printed "camera-ready" as submitted by the authors.

The quality of the papers in this publication reflects not only the obvious efforts of the authors and the technical editor(s), but also the work of these peer reviewers. The ASTM Committee on Publications acknowledges with appreciation their dedication and contribution to time and effort on behalf of ASTM.

Printed in Philadelphia, PA
August 1994

Foreword

This publication, *Hydraulic Conductivity and Waste Contaminant Transport in Soil*, contains papers presented at the symposium of the same name held in San Antonio, TX on 21–22 January, 1993. The symposium was sponsored by ASTM Committee D-18 on Soil and Rock and its Subcommittee D18.04 on Hydrologic Properties of Soil and Rock. David E. Daniel of University of Texas, Austin, TX and Stephen J. Trautwein of Trautwein Soils Testing Equipment Company, Houston, TX presided as symposium chairmen and are editors of the resulting publication.

Contents

Overview— vii

INVITED PAPERS

Representative Specimen Size for Hydraulic Conductivity Assessment of Compacted Soil Liners—C. H. BENSON, F. S. HARDINATO, AND E. S. MOTAN   3

State-of-the-Art: Laboratory Hydraulic Conductivity Test for Saturated Soils—D. E. DANIEL   30

Hydraulic Conductivity of Vertical Cutoff Walls—J. C. EVANS   79

Slug Tests for Determining Hydraulic Conductivity of Natural Geologic Deposits—B. L. HERZOG   95

Waste-Soil Interactions that Alter Hydraulic Conductivity—C. D. SCHACKELFORD   111

Hydraulic Conductivity Assessment of Unsaturated Soils—D. B. STEPHENS   169

In-Situ Hydraulic Conductivity Tests for Compacted Soil Liners and Caps—S. J. TRAUTWEIN AND G. P. BOUTWELL   184

OTHER PAPERS

Laboratory Testing to Evaluate Changes in Hydraulic Conductivity of Compacted Clays Caused by Freeze-Thaw: State-of-the-Art—M. A. OTHMAN, C. H. BENSON, E. J. CHAMBERLAIN, AND T. F. ZIMMIE   227

Measurement of Saturated Hydraulic Conductivity in Fine-Grain Glacial Till in Iowa: Comparison of In Situ and Laboratory Methods—D. R. BRUNER AND A. J. LUTENEGGER   255

Hydraulic Conductivity of Compacted Clayey Soils Under Distortion or Elongation Conditions—S. C. CHENG, J. L. LARRALDE, AND J. P. MARTIN   266

The Compatibility of Slurry Cutoff Wall Materials with Materials with Contaminated Groundwater—S. R. DAY   284

A Comparison Between Field and Laboratory Measurements of Hydraulic Conductivity in a Varved Clay—D. J. DeGROOT AND A. J. LUTENEGGER   300

Effects of Post Compaction Water Content Variation on Saturated Conductivity—M. A. PHIFER, E. C. DRUMM, AND G. V. WILSON   318

Lessons Learned from the Application of Standard Test Methods for Field and Laboratory Hydraulic Conductivity Measurement—R. J. DUNN AND B. S. PALMER   335

Large-Size Test for Transport of Organics Through Clay Liners—T. B. EDIL, J. K. PARK, AND D. P. HEIM   353

Field Measurement of Hydraulic Conductivity in Slowly Permeable Materials Using Early-Time Infiltration Measurements in Unsaturated Media—D. J. FALLOW, D. E. ELRICK, W. D. REYNOLDS, N. BAUMGARTNER, AND G. W. PARKIN   375

Impact of Leakage on Precision in Low Gradient Flexible Wall Permeability Testing—M. D. HAUG, W. G. BUETTNER, AND L. C. WONG   390

Influence of Polymers on the Hydraulic Conductivity of Marginal Quality Bentonite-Sand Mixtures—M. D. HAUG AND B. BOLDT-LEPPIN   407

Hydraulic Conductivity and Adsorption Parameters for Pollutant Transport Through Montmorillonite and Modified Montmorillonite Clay Liner Materials—I. M.-C. LO., H. M. LILJESTRAND, AND D. E. DANIEL   422

Hydraulic Conductivity of Borehole Sealants—A. J. LUTENEGGER AND D. J. DeGROOT   439

The Effects of Freeze/Thaw Cycles on the Permeability of Three Compacted Soils—J. J. BOWDERS, Jr. AND S. McCLELLAND   461

Volume-Controlled Hydrologic Property Measurements in Triaxial Systems—H. W. OLSEN, A. T. WILLDEN, N. J. KIUSALASS, K. R. NELSON, AND E. P. POE   482

Hydraulic Conductivity of Solidified Residue Mixtures Used as a Hydraulic Barrier—S. PAMAKCU, I. B. TOPCU, AND C. GUVEN   505

Constant-Flow and Constant-Gradient Permeability Tests on Sand-Bentonite-Fly Ash Mixtures—C. D. SHACKELFORD AND M. J. GLADE   521

A Field-Scale Study of the use of Paper Industry Sludges as Hydraulic Barriers in LANDfill Cover Systems—V. MALTBY AND L. K. EPPSTEIN   546

Two Case Histories: Field Sealed Double Ring Infiltrometer (SDRI) and Laboratory Hydraulic Conductivity Comparison Test Programs—J. F. WALLACE, R. R. SACRISON AND E. E. ROSIK   559

Effects of Electro-Kinetic Coupling on the Measurement of Conductivity—A. T. YEUNG   569

Evaluation of Attenuation Capability of a Micaceous Soil as Determined from Column Leaching Tests—R. N. YONG, B. K. TAN, AND A. M. O. MOHAMED   586

Author Index   607

Subject Index   609

Overview

There is a widespread interest among civil engineers, soil scientists, hydrologists, and geologists in the hydraulic conductivity of soils. Of the principal soil properties (strength, compressibility, and hydraulic conductivity) hydraulic conductivity is the most variable, the easiest to misjudge, and the hardest to measure accurately. Interest in hydraulic conductivity has increased substantially in recent years because of concern over ground-water contamination. Assessments of the potential for continued or future contamination at a site are only possible if accurate information is available concerning the hydraulic conductivity of subsoils. It is for these reasons that "hydraulic conductivity" and "waste contaminant transport" comprised the theme of this symposium.

This volume contains the proceedings from a specialty conference presented in January, 1993, in San Antonio, TX, on the topic of Hydraulic Conductivity and Groundwater Contaminant Transport in Soil. The symposium was sponsorerd by ASTM Subcommittee D18.04 on Hydrologic Properties of Soil and Rock, which is a subcommittee of ASTM Committee D-18 on Soil and Rock for Engineering Purposes.

This symposium is the second ASTM symposium on the subject of hydraulic conductivity and ground-water contaminant transport. The first symposium was held in 1979. The proceedings from the first symposium were published in *Permeability and Groundwater Contaminant Transport,* ASTM STP 746, T. F. Zimme and C. O. Riggs, Eds., American Society for Testing and Materials, 1981. The 1993 symposium consisted of more than twice as many papers as the 1979 symposium. In the 1993 symposium much greater emphasis was placed on testing soils of low hydraulic conductivity (primarily for waste containment applications) on field hydraulic conductivity measurements, and on the effects of chemicals upon the hydraulic conductivity of soils. A comparison of the current proceedings with the 1981 publication shows that there has been a substantial improvement in the state-of-the-art for hydraulic conductivity testing of soil.

Seven state-of-the-art papers were presented during the 1993 symposium. Daniel summarized methods for determining hydraulic conductivity of saturated soils in the laboratory. The presentation covered both fixed- and flexible-wall permeameters and described methods of permeation with both water and waste liquids. Shackelford discussed waste-soil interactions that can alter hydraulic conductivity. Methods of permeating soils in the laboratory with waste liquids were discussed in detail as were procedures for interpreting data from such tests. Stephens described the state-of-the-art for assessment of hydraulic conductivity in unsaturated soils. The presentation included a discussion of both the laboratory and field methods for evaluating the hydraulic conductivity of unsaturated soils. Trautwein and Boutwell discussed in-situ hydraulic conductivity tests for compacted soil liners and caps. The presentation focused primarily upon the sealed double-ring infiltrometer and the two-stage borehole test. Evans described hydraulic conductivity testing for vertical cutoff walls. Procedures for dealing with many potential testing errors were discussed in depth. Herzog evaluated and described methods for determining the hydraulic conductivity of natural geologic deposits. The presentation focused on four slug test procedures and presentation of data from actual tests. Benson evaluated the minimum representative elementary volume for hydraulic conductivity testing of compacted soil liners. In this presentation the question of how large a test specimen must be in order to determine a representative hydraulic conductivity was considered.

In addition to invited state-of-the-art presentations, a number of outstanding contributions were presented on various topics related to hydraulic conductivity testing. Several of the papers describe techniques for dealing with challenging hydraulic conductivity testing problems in the laboratory, including techniques for permeating with a constant rate of flow and dealing with leakage when testing materials of low hydraulic conductivity. Several papers evaluated the special problems involved in permeation of soils with chemicals and waste liquids. Various techniques for determining the hydraulic conductivity of soils in the field were discussed. Typical results obtained from a variety of field tests were presented in several papers. The effects of environmental stresses, such as freeze-thaw, were discussed in several papers. The comparison between field and laboratory tests to determine the hydraulic conductivity of soils was the topic of several papers. Finally, various papers discussed specialized problems in hydraulic conductivity testing, such as electrokinetic coupling and influence of distortion in the soil and measuring the hydraulic conductivity of bentonites.

Because of concern for the environment, the regulating community now plays a significant role in issues dealing with hydraulic conductivity. In particular, the regulating community in many cases makes the final decision on what test methods are acceptable for hydraulic conductivity measurements. For this the regulating community relies in part on ASTM standards. However, because of the rapid advancements in this field, there is a lag between the development of new and improved testing techniques and the publication of corresponding ASTM standards. It is the hope of the editors that the information presented in this symposium will serve not only to keep practitioners abreast with recent advancements, but also will provide the regulating community with reference material for updating acceptance criteria. It is also the hope of the editors that this sympoium will encourage practitioners and regulators to participate in the development of new standards for measuring hydraulic conductivity in both the laboratory and the field. In particular, there is an urgent need for the development of standard test methods to determine the effects of chemicals and waste liquids on hydraulic conductivity.

The editors wish to express their appreciation to all those who participated in the symposium. Particular thanks is extended to those who contributed papers, to the reviewers of papers, to ASTM Committee D18 on Soil and Rock for sponsoring the symposium through Subcommittee D18.04 on Hydrologic Properties of Soil and Rock, and to the editorial staff of ASTM.

David E. Daniel
University of Texas, Austin, TX;
chairman and editor.

Stephen J. Trautwein
Trautwein Soils Testing Equipment Company, Houston, TX;
chairman and editor.

Invited Papers

Craig H. Benson[1], Fransiscus S. Hardianto[2], and E. Sabri Motan[3]

REPRESENTATIVE SPECIMEN SIZE FOR HYDRAULIC CONDUCTIVITY ASSESSMENT OF COMPACTED SOIL LINERS

REFERENCE: Benson, C. H., Hardianto, F. S., and Motan E. S., "Representative Specimen Size for Hydraulic Conductivity Assessment of Compacted Soil Liners," <u>Hydraulic Conductivity and Waste Contaminant Transport in Soil, ASTM STP 1142</u>, David E. Daniel, and Stephen J. Trautwein, Eds., American Society for Testing and Materials, Philadelphia, 1994.

ABSTRACT: An alternative to field measurement of hydraulic conductivity is to conduct laboratory hydraulic conductivity tests on specimens large enough to simulate field-scale conditions. Laboratory tests can be performed rapidly using standard procedures and with accurate control of state of stress and gradient. The objective of this research program was to identify how large a specimen must be to yield field-scale hydraulic conductivity. This objective was accomplished through field testing, laboratory testing, and statistical modeling.
   Hydraulic conductivity tests were conducted on test pads at four sites that represented construction conditions ranging from poor to excellent. One test pad was deliberately constructed using poor construction methods to demonstrate "worst case" conditions. Field tests were performed with sealed double-ring infiltrometers (SDRIs) having inner rings with widths of 0.61, 0.92, 1.2, or 1.5 m. Laboratory tests were performed on block specimens with diameters ranging from 0.07 m to 0.46 m.
   For the range of construction conditions that were evaluated, the test results showed that hydraulic conductivity at or near field-scale can be measured using block specimens with a diameter of 0.30 m and a thickness of 0.15 m. A probabilistic model was designed to simulate macroscopic defects in compacted soil. Results obtained with the model supported the results of the experimental study.

KEYWORDS: representative specimen size, hydraulic conductivity, soil liner, clay liner, test pads, field-scale, in situ, sealed double-ring infiltrometer, block samples.

---

[1] Asst. Prof., Dept. of Civil and Environmental Engineering, University of Wisconsin, Madison, WI 53706.
[2] Project Engineer, Bromwell & Carrier, Inc., Lakeland, FL 33807.
[3] Senior Geotechnical Engineer, RUST Environment and Infrastructure, Naperville, IL 60563.

INTRODUCTION

Compacted fine-grained soils are widely used in liners and covers for waste containment structures. Their primary purpose is to minimize flow. Hence, low hydraulic conductivity is of utmost importance (Daniel, 1987). To ensure a liner will have sufficiently low hydraulic conductivity, measurements are performed in the laboratory on specimens removed in thin-wall sampling tubes or in the field using in situ measurement techniques.

Several studies have shown that the field-scale hydraulic conductivity of compacted soil liners may differ significantly from hydraulic conductivity measured in the laboratory on small specimens (Daniel, 1984; Day and Daniel, 1985; Elsbury et al., 1988; Benson and Boutwell, 1992). For poorly built liners, the hydraulic conductivity of small specimens (diameter~0.07m) can be several orders of magnitude lower than the hydraulic conductivity measured using large-scale field tests. These studies have shown that defects which control flow at field-scale may be inadequately represented in small specimens.

Currently, many regulatory agencies require field tests to ensure an adequate volume of soil is permeated and the field-scale hydraulic conductivity is assessed. As will be shown, however, it is not essential that these tests be performed in the field; tests only need to be conducted on a specimen that is large enough to adequately represent macroscopic defects. In fact, field tests have practical and technical problems that limit their use in quality control. From a practical perspective, field tests usually require a long time to complete and are expensive. As a result, replicate measurements are not ordinarily performed and therefore the measurements cannot be analyzed statistically. Furthermore, for most field tests, the state of stress cannot be controlled and the hydraulic gradient cannot be measured with accuracy. As a result, hydraulic conductivity is usually estimated conservatively.

A logical alternative to field testing is to conduct hydraulic conductivity tests in the laboratory on specimens large enough to simulate field-scale conditions. Laboratory tests can be performed rapidly and with accurate control of the state of stress and gradient. Furthermore, methods to perform laboratory tests in flexible-wall permeameters have been refined in recent years and an ASTM standard is now available to ensure consistency among laboratories (ASTM D5084). In addition, laboratory tests are usually less expensive than field tests. Therefore, replicate measurements can be obtained to yield statistically significant results.

Before laboratory testing can be used as an alternative for field tests, the size of specimen that is necessary to adequately represent field-scale hydraulic conductivity under a variety of construction conditions must be identified. Herein, this size is referred to as the "Representative Specimen Size" (RSS). The objective of the research program described in this paper was to determine the RSS through field testing, laboratory testing, and statistical modeling.

BACKGROUND

Discrepancies Between Small and Large-Scale Measurements

Daniel (1984) presented three case histories regarding soil liners in central Texas where hydraulic conductivities were measured in the field and laboratory. Field tests were performed using ring infiltrometers and laboratory tests were performed on small undisturbed specimens. Daniel found that the field-scale hydraulic conductivity was generally 10-1000 times larger than the hydraulic conductivity obtained from laboratory tests. Daniel also noted that the liners were thin, construction was poorly documented, and little was done to prevent desiccation.

Based on these results, Daniel (1984) concluded that laboratory hydraulic conductivity tests may yield hydraulic conductivity lower than the hydraulic conductivity at field-scale because small, non-representative specimens are used for testing. He stated that small specimens are not likely to contain a representative distribution of desiccation cracks, fissures, slickensides, or other hydraulic defects that may be present in the field.

Day and Daniel (1985) reported similar findings. They performed hydraulic conductivity tests in the laboratory and field on two "prototype" soil liners. Laboratory tests were performed on hand-carved specimens, on specimens obtained with thin-wall sampling tubes, and on laboratory-compacted specimens. Field tests were performed with infiltrometers and underdrains. Two clays, designated Clay 1 and Clay 2, were used.

The overall hydraulic conductivity of each liner was computed from the rate of outflow measured with the underdrains and was found to be 9×10^{-8} m/sec for Clay 1 and 4×10^{-8} m/sec for Clay 2. Hydraulic conductivities measured with the underdrains were assumed to be the actual hydraulic conductivities of the liners. Field hydraulic conductivity tests were also performed using single-ring and double-ring infiltrometers. The single-ring infiltrometers had a diameter of either 0.56 m or 1.12 m. The double-ring infiltrometers had inner and outer rings with diameters of 0.30 m and 0.50 m. Hydraulic conductivities of 5×10^{-8} m/sec (Clay 1) and 3×10^{-8} m/sec (Clay 2) were obtained from the infiltration tests. These hydraulic conductivities are close to the hydraulic conductivities measured with the underdrains.

Block specimens removed from the liners were trimmed to a diameter of 0.10 m or 0.064 m. Flexible-wall permeameters were used to test the 0.10-m specimens whereas consolidation cells were used to test the 0.064-m specimens. Specimens obtained with thin-wall sampling tubes were tested in flexible-wall permeameters. Results of the laboratory tests showed that the average hydraulic conductivity for Clay 1 was 1×10^{-10} m/sec whereas the average value for Clay 2 was 3×10^{-11} m/sec. Thus, the laboratory-measured hydraulic conductivities were 2 to 3 orders of magnitude lower then the field-measured values. Day and Daniel (1985) concluded that the laboratory specimens were too small to incorporate macropores controlling field-scale hydraulic conductivity, whereas the volume of soil permeated with the ring infiltration tests was large enough to be representative of the entire liner.

Elsbury et al. (1988) have also shown that field-scale hydraulic conductivity may differ substantially from hydraulic conductivity measured on small specimens in the laboratory. They constructed a test pad with a high plasticity clay. The pad was compacted dry of standard Proctor optimum water content with a lightweight padfoot compactor.

Field measurements of hydraulic conductivity were conducted using an underdrain (4.9 m x 4.9 m) and 4 sealed double-ring infiltrometers (SDRIs). Laboratory tests were conducted on small specimens removed in thin-wall sampling tubes 0.07 m in diameter and on block specimens trimmed to a diameter of 0.15 m. The field-scale measurements of hydraulic conductivity obtained with the SDRIs and the underdrain were essentially the same. However, the average hydraulic conductivity of the 0.07-m specimens was 5 orders of magnitude lower than the field-scale hydraulic conductivity. The average hydraulic conductivity of the larger block specimens (diameter=0.15 m) was approximately 2 orders of magnitude lower than the field-scale hydraulic conductivity. Elsbury et al. (1988) concluded that these discrepancies occurred because the laboratory specimens were too small to capture macropores controlling flow at field-scale.

Similar Hydraulic Conductivity at Small and Large-Scale

Lahti et al. (1987) and Reades et al. (1990) have found close agreement between hydraulic conductivity measured in the laboratory and field for a liner at the Keele Valley Landfill. The liner was constructed with glacial till placed in 0.15 m lifts and compacted to achieve a dry unit weight in excess of 95% of standard Proctor maximum dry unit weight. Water content was maintained 2 to 3% wet of optimum water content. Based on the measurements of water content and dry unit weight, the average degree of saturation was found to be approximately 95%.

Specimens for laboratory testing were obtained using thin-wall sampling tubes having a diameter of 0.07 m. The specimens were extruded and tested for hydraulic conductivity using flexible-wall permeameters. All of the tests were conducted at a hydraulic gradient of 20 and an effective stress of 165 kPa. Measurements made during the construction seasons of 1983, 1984, and 1985 showed a geometric mean hydraulic conductivity of 7.1×10^{-11}, 8.2×10^{-11}, and 7.7×10^{-11} m/sec, respectively.

Field-scale hydraulic conductivity was computed from flow rates measured in six square underdrains, each with a width of 15 m. Three underdrains were installed below the liner and three were installed within the liner. Hydraulic conductivity computed from the underdrains averaged 9×10^{-11} m/sec, which is comparable to the hydraulic conductivity measured on the small laboratory specimens.

Lahti et al. (1987) and Reades et al. (1990) concluded that the hydraulic conductivity measured in the field and laboratory was similar because proper construction techniques were employed and quality control procedures were strictly followed. Heavy rollers were used and the water content was maintained wet of optimum. As a result, the pores controlling flow through the liner were very small and were adequately represented in small specimens.

Similar agreement between laboratory and field-measured hydraulic conductivity has been observed by Johnson et al. (1990). They constructed two test pads with a moderate plasticity clay. A heavy sheepsfoot compactor was used to compact the soil in 0.15-m lifts and the degree of saturation during compaction was maintained above the degree of saturation at optimum water content. Field hydraulic conductivity tests were conducted with sealed double-ring infiltrometers, Boutwell borehole permeameters, and underdrains. Laboratory tests were conducted in flexible-wall permeameters on specimens removed in thin-wall sampling tubes (diameter=0.07 m). The field-measured hydraulic conductivity was found to range between 0.6 to 2 times the laboratory-measured hydraulic conductivity. The close agreement between the laboratory and field measurements occurred because the soil was carefully compacted and devoid of macropores. As a result, the small specimens contained pores that were representative of the pores conducting flow at field-scale.

Synthesis

The case histories show that a large discrepancy can exist between hydraulic conductivity measured in the laboratory on small specimens and in the field using large-scale tests. A large discrepancy occurs when inadequate construction techniques are employed and macropores exist in the soil. Macropores, which control flow at field-scale, are inadequately represented in small specimens (diameter~0.07m) normally tested in the laboratory. In contrast, when proper construction methods are used, a dense mass devoid of macropores is obtained. As a result, the field-scale hydraulic conductivity is controlled by very small pores that are adequately represented in small specimens traditionally used for laboratory testing.

Thus, the RSS depends on the quality of construction, which directly impacts the size of the network of pores controlling flow at field-scale. Unfortunately, the quality of construction and the size of the network of pores controlling flow at field-scale are not known a priori. Hence, an RSS needs to be identified that is applicable to a wide range of construction conditions. The aforementioned case histories illustrate that a widely applicable RSS is likely to be larger than the commonly used thin-wall sampling tube (diameter=0.07 m) and smaller than or equal in size to infiltrometers (diameter=0.5 to 1.5m).

TEST SITES

Hydraulic conductivity tests were performed on test pads at four sites. Construction methods that were used varied from poor to excellent. One test pad was deliberately constructed poorly to define "worst case" conditions that would result in an upper bound on the RSS.

Site A

Soil used to construct the test pad at Site A was a sandy clay obtained from an alluvial deposit. Properties of the soil are summarized in Table 1.

The test pad at Site A was built with 5 lifts. Soil for the pad was sieved to remove clods and rocks with a diameter greater than 0.10 m. For the upper lift, a smaller sieve was used to reduce the maximum clod size to 0.02 m. Water was added as necessary to ensure that the water content was wet of optimum based on modified Proctor effort. A Caterpillar 825C tamping foot compactor was used for compaction. The compactor weighed 320 kN and had feet 0.19 m long. A minimum of six passes were used to compact each lift. The lower 4 lifts were approximately 0.13 m thick (after compaction) and the top lift was 0.10 m thick.

There was concern after construction that the different procedure used to compact the upper lift may confound comparisons to be made between measurements of hydraulic conductivity performed in the field and laboratory. However, the thin upper lift swelled and became soft during infiltration testing (described later). Thus, the upper lift probably had little impact on the test results.

Compaction tests were conducted to determine the relationship between water content, dry unit weight, and compactive effort. Three compactive efforts were used: modified Proctor (ASTM D1557), standard Proctor (ASTM D698), and reduced Proctor. The latter effort (reduced Proctor) is used to simulate light compactive effort. The weight and drop of the hammer are the same as standard Proctor, but only 15 blows per lift are applied (Daniel and Benson, 1990). Similar procedures were used to develop compaction curves for soils from Sites B-D.

Figure 1a shows the compaction curves and measurements of water content and dry unit weight performed during construction. The designer of the test pad planned for compaction "wet of optimum" to achieve low hydraulic conductivity. Construction of the pad was performed in accord with the designer's specifications; however, 60% of the field data points fell dry of the line of optimums (Fig. 1a).

TABLE 1--Summary of soil properties.

Site	LL	PI	% Gravel	% Sand	% Fines	% Clay	USCS Class
A	24	11	3	35	62	37	CL
B	32	14	1	14	85	44	CL
C	31	15	8	18	74	26	CL
D	30	17	0	48	52	16	SC-CL

Particle Size Definitions: Gravel>4.75 mm, 4.75 mm>Sand>0.075 mm, Fines<0.075 mm, Clay<2 μm

Site B

The test pad at Site B was deliberately compacted dry of the line of optimums at low compactive effort. These conditions are conducive to the formation of macropores and hydraulic conductivity that is scale-dependent (Benson and Daniel, 1990; Benson and Boutwell, 1992). These conditions were expected to represent a "worst case" that would require the largest specimen to obtain field-scale hydraulic conductivity.

Soil used to construct the test pad at Site B was a low plasticity clay obtained from a deposit of glacial till (see Table 1 for index properties). The first two lifts were compacted to a thickness of 0.15 m using a light-weight bulldozer (weight~35 kN). Each location received about 5 passes of the dozer. After the first two lifts were completed, it was apparent that even the light bulldozer was heavy enough to remold the clods in some locations. This occurred because some of the soil was too wet because of recent rains. Hence, to reduce the compactive energy, the remaining 0.30 m of soil was placed in one lift.

No effort was made to break down the clods prior to compaction. Clods ranged in size from small particles that would pass the No. 4 sieve to large chunks with diameters of 0.15 to 0.20 m. Some of the clods were broken down as the dozer spread the soil into lifts, but many of the clods remained intact and were only pressed together by the compactive effort.

Like the test pad at Site A, there was concern after construction that the different procedure used to compact the upper lift may confound comparisons to be made between measurements of hydraulic conductivity performed in the field and laboratory. However, the macropores were so extensive in each lift of this test pad, that the different compaction

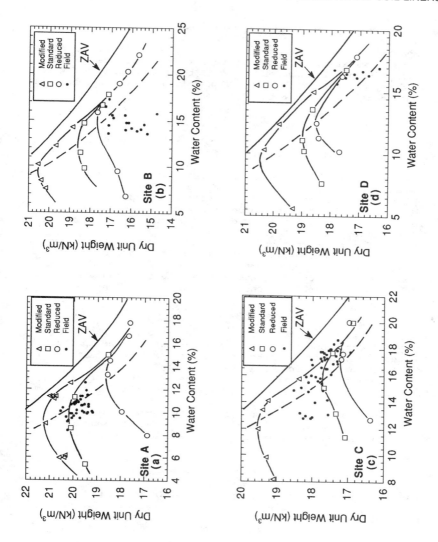

FIG. 1--Compaction curves and field data for Sites A-D.

procedures apparently had no effect on the test results (described later).
Compaction curves for Site B are shown in Fig. 1b with the measurements of water content and dry unit weight performed during construction. Figure 1b shows that 88% of the measurements of water content and dry unit weight fall dry of the line of optimums and that the compactive effort was low.

Site C

Soil used to construct the test pad at Site C was obtained from a deposit of glacial till. A summary of index properties of the soil is shown in Table 1.
The pad was constructed with six lifts each having a thickness of 0.15 m. Water was added if needed to ensure the water content remained above optimum based on modified Proctor effort. Compaction was performed using a Dynapac CA25 padfoot compactor having a weight of 90 kN and feet 0.11 m long. A minimum of four passes of the compactor were used for each lift.
Results of compaction tests performed on the soil from Site C are shown in Fig. 1c with measurements of water content and dry unit weight obtained during construction. Forty-three percent of the data points fall dry of the line of optimums even though the test pad was constructed in accord with the construction specifications.

Site D

The soil used to construct the test pad at Site D was a sandy marine clay. Index properties of the soil are summarized in Table 1.
The test pad was constructed in six lifts and each lift had a thickness of 0.15 m. A large bulldozer (weight=275 kN) was used for compaction. Compaction was controlled by ensuring that the degree of saturation at compaction exceeded the degree of saturation at optimum water content.
Compaction curves for the soil from Site D are shown in Fig. 1d with measurements of water content and dry unit weight performed during construction. Ninety-five percent of the measurements fall wet of the line of optimums.

TESTING PROCEDURES

To evaluate the relationship between hydraulic conductivity and size of specimen, experiments were conducted at various scales. Large scale tests (diameter > 0.6 m) were conducted in the field using sealed double-ring infiltrometers whereas smaller scale tests (diameter < 0.6 m) were conducted in the laboratory in a specially built large-scale flexible-wall permeameter.

Field Tests

Large-scale measurements of hydraulic conductivity were performed in the field with sealed double-ring infiltrometers (SDRIs) using the methods described in ASTM D5093. The inner and outer rings were square. For Sites A and B, the outer rings had a width of 2.45 m. The inner rings had widths of 0.61, 0.92, or 1.2 m so that different volumes of soil would be permeated. Two tests were performed for each size. At Sites C and D, only 1 SDRI was used. At both sites, the outer ring was

3.7 m wide and the inner ring was 1.5 m wide. All of the SDRIs met or exceeded dimensional requirements described in ASTM D5093.
Infiltration rate was measured with plastic bags connected to the inner ring via Tygon tubing. Double-sealing quick connects were used to join the plastic bags to the tubing. Infiltration rate (I) was determined from the change in weight of the bags (ΔW) using Eq. 1 (Daniel, 1989):

$$I = \frac{\Delta W}{\Delta t\, \gamma_w\, A} \qquad (1)$$

where Δt is the elapsed time between measurements of the bag weight, A is the horizontal cross-sectional area of the inner ring, and γ_w is the unit weight of water. The SDRI tests were deemed complete when the infiltration rate became steady.
Hydraulic conductivity (K) was computed from infiltration rate by (Daniel, 1989):

$$K = \frac{I}{i} \qquad (2)$$

In Eq. 2, i is the hydraulic gradient, which was computed using Eq. 3 (Daniel, 1989):

$$i = \frac{D_p + D_f}{D_f} \qquad (3)$$

where D_f is the depth to the wetting front and D_p is the depth of ponding.
Equation 3 ignores suction head at the wetting front, which can affect the hydraulic gradient and thus the hydraulic conductivity computed using Eq. 2. To avoid error caused by ignoring suction head, the tests at Sites A-C were continued until tensiometers installed between the inner and outer rings indicated that the wetting front passed through the test pad (Hardianto, 1992). Test pits were also excavated to confirm the depth to the wetting front. At Site D, however, the wetting front only penetrated about 0.20 m into the pad when the test was terminated. Thus, at Site D, the hydraulic conductivity (computed using Eqs. 2 and 3) may have been somewhat larger than the actual field-scale hydraulic conductivity.

Laboratory Tests

After the SDRI tests were complete, large specimens were removed as blocks from soil directly beneath each inner ring. The block specimens were shipped to the University of Wisconsin where they were trimmed and then permeated in a large-scale flexible-wall permeameter.

Sampling procedure

A trimming ring was used to carve and protect the block specimens. The ring was manufactured from PVC pipe and had an inside diameter of 0.58 m and a height of 0.3 m. A beveled cutting shoe was machined at the base of the ring. The trimming ring was similar to a consolidation ring, only much larger.
After the ring was placed on the soil, a trench was excavated surrounding the ring (Fig. 2a). Then, soil was carefully trimmed away until the ring could be moved downward over the soil with light effort

(Fig. 2b). The trimming procedure was similar to procedures used to trim specimens into rings for consolidation or direct shear testing. Trimming was continued until the upper edge of the ring was 0.05-0.10 m below the surface. The soil above the ring, which typically was very soft, was removed.

FIG. 2--Placement of ring on soil surface (a) and ring trimmed over specimen to be sampled (b).

When the trimming ring was at full depth, the specimen was separated from the underlying soil using one of the following procedures: (1) a sharpened steel plate was tapped into the soil with a hammer or (2) a flat-bladed shovel was pushed into the underlying soil at several locations. The latter method proved to be easier to implement and was less likely to cause disturbance. Afterwards, the specimen was transferred to a reinforced pallet and sealed with plastic wrap and duct tape.

Testing procedure

The rings were removed in the laboratory and soil was trimmed from the outer edge of the specimen until a diameter of 0.46 m was obtained. About 0.05 m of soil was also removed from the upper and lower surfaces. Afterwards, the upper and lower surfaces were scarified to eliminate smear.

Trimmed specimens were placed in the University of Wisconsin's large-scale flexible-wall permeameter which is similar in construction to flexible wall permeameters typically used in industry (e.g., Daniel et al., 1985), but much larger (Fig. 3). The permeameter was designed so that two specimens could be placed in the permeameter concurrently.

FIG. 3--Large-scale flexible-wall permeameter.

Hydraulic conductivity tests were conducted in accord with ASTM D5084 at an effective stress of 10 kPa and a hydraulic gradient between

3 and 5, but no backpressure was used. Permeation continued until inflow equaled outflow and the hydraulic conductivity was steady.

To evaluate the effect of specimen size, the specimens were repeatedly trimmed to smaller sizes and retested. At each size, the specimen was permeated until the hydraulic conductivity measurement became steady and inflow equaled outflow. Each block specimen was trimmed to diameters of 0.46, 0.30, 0.15, and 0.07 m with hydraulic conductivity measurements conducted at each size. Flexible-wall permeameters of various sizes were used to perform the hydraulic conductivity measurements.

Prior to performing the majority of tests, preliminary experiments were conducted to ensure that variations in aspect ratio (height/diameter) would not affect the scale-dependence of the measurements. Hydraulic conductivity tests were initially conducted on specimens having an aspect ratio of about 1.0. After equilibrium was reached, the specimens were trimmed to an aspect ratio of about 0.5 and their hydraulic conductivity was measured.

Results of the aspect ratio tests are shown in Table 2 for specimens from Site A. Specimens with an aspect ratio of 0.5 showed greater variability, but on average had similar hydraulic conductivity as the specimens having an aspect ratio of 1.0. Because aspect ratio did not have a consistent effect on hydraulic conductivity, an aspect ratio of approximately 0.5 was used for most of the laboratory tests. The exact size of each specimen is summarized in Benson and Hardianto (1992).

TABLE 2--Results of aspect ratio tests.

Diameter (m)	Initial Aspect Ratio	K (m/sec)	Shortened Aspect Ratio	K (m/sec)	K_{short}/K_{long}	Average K_{short}/K_{long}
0.15	1.13	1×10^{-9}	0.59	6.0×10^{-10}	0.6	1.3
			0.56	2.0×10^{-9}	2.0	
0.15	1.23	1×10^{-9}	0.56	2.5×10^{-9}	2.5	1.35
			0.63	2.0×10^{-10}	0.2	
0.01	1.13	1×10^{-9}	0.53	3.5×10^{-10}	0.35	0.35
					Average:	1.0

RESULTS OF FIELD AND LABORATORY TESTING

Results of the field and laboratory tests are shown in Fig. 4. Size of the specimens is described in Fig. 4 by "equivalent diameter," which is the diameter of a circle having area equal to the horizontal cross-sectional area that was permeated. Equivalent diameter was used as a common measure of specimen size to describe results obtained with SDRI tests (square cross-section) and laboratory tests (circular cross-section). Lines depicting trends in the data shown in Fig. 4 were fit by eye.

Site A

For Site A, hydraulic conductivity is related to equivalent diameter (Fig. 4a). The hydraulic conductivity of small specimens

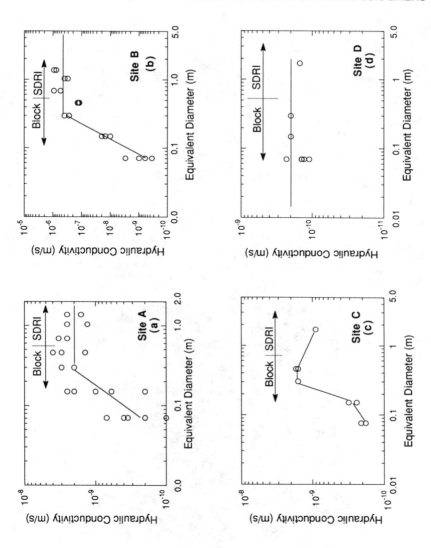

FIG. 4--Hydraulic conductivity-specimen size relationship for Sites A-D.

(diameter=0.07 m) is about one order of magnitude lower than the hydraulic conductivity measured with the SDRIs. An increase in hydraulic conductivity with increasing diameter is also apparent. However, for diameters equal to or greater than 0.30 m, hydraulic conductivity ceases to increase with further increase in diameter. To determine if the differences in hydraulic conductivity are statistically significant, a t-test was performed comparing the geometric mean hydraulic conductivity for each diameter. Results of the t-test showed that the geometric mean hydraulic conductivities are significantly different at the 5% level.

The trend of hydraulic conductivity with equivalent diameter that was observed at Site A was expected. The test pad at Site A was constructed slightly dry of the line of optimums, a condition conducive to the formation of macropores and scale-dependent hydraulic conductivities (Benson and Boutwell, 1992). Macropores existing in a specimen from Site A are evident in Fig. 5.

To confirm that macropores were carrying flow through the specimens, dye was introduced into one specimen after measuring its hydraulic conductivity at a diameter of 0.46 m. When the specimen was trimmed to smaller sizes, macropores through which the dye was flowing were evident. Trimming the specimens to smaller sizes eliminated some of these macropores and consequently the hydraulic conductivity decreased as the diameter of the specimen was reduced.

FIG. 5--Specimen from Site A showing macropores.

Site B

Results of field and laboratory tests from Site B are summarized in Fig. 4b. For Site B, hydraulic conductivity is also related to equivalent diameter. The small block specimens had hydraulic conductivity approximately 3 orders of magnitude lower than the large block specimens and the field tests.

The hydraulic conductivities of the block specimens with a diameter=0.46 m were slightly lower than the hydraulic conductivities measured with SDRIs. The lower values can be attributed in part to an artificial upper limit imposed by head losses in the permeameter. That is, the head loss in the tubes and valving was greater than the head loss in the specimen. Nevertheless, the results are similar to results obtained for Site A; for specimens having a diameter greater than 0.3 m, hydraulic conductivity near field-scale was measured.

Because Site B was deliberately constructed using procedures conducive to the formation of macroscopic defects, the large change in hydraulic conductivity with increasing diameter was expected. During installation of the SDRIs and removal of the block specimens, large macropores were observed. Dye studies showed that these macropores, which apparently controlled flow at field-scale, were inadequately represented in small (diameter < 0.30 m) specimens (Hardianto, 1992). Thus, the specimens were not representative of field-scale conditions. Trimming in the laboratory revealed that most of the macropores had an aperture width of about 1 to 3 mm. These pores were wider and longer than the pores observed in the specimens from Site A, as a result of the light compactive effort used during construction.

Site C

Results of the tests performed on specimens from Site C (Fig. 4c) show trends similar to those observed for Site A (Fig. 4a). For specimens with diameter exceeding 0.30 m, field-scale hydraulic conductivity was obtained. Like Site A, the test pad at Site C was constructed slightly dry of the line of optimums, a procedure conducive to the formation of macroscopic defects and scale-dependent hydraulic conductivities. Consequently, the small specimens (diameter=0.07 m) inadequately represented macroscopic features. Hence, their hydraulic conductivity did not represent field-scale conditions.

Figure 4c also shows that hydraulic conductivities measured on large block specimens (0.3 and 0.46 m diameter) were slightly larger than the hydraulic conductivity measured with the SDRI. The reason for this effect is not clear, but a similar trend is evident in the data from Site A and the modeling results presented later.

Site D

In contrast to the results obtained from Sites A-C, measurements of hydraulic conductivity at Site D showed no dependence on diameter. Examination of the liner during sampling and the specimens during trimming showed no presence of macroscopic features. Hence, hydraulic conductivity was not expected to depend on the diameter of specimen.

The lack of scale-dependence of hydraulic conductivity that was observed for Site D is a result of the construction methods that were employed. Compaction was achieved with a heavy compactor at water contents in excess of the line of optimums. Benson and Daniel (1990) have shown that large compactive effort, when combined with water content in excess of optimum, results in the elimination of macropores and interclod pores during compaction. As a result, the pores conducting flow are very small and are adequately represented in small and large specimens. Hence, the hydraulic conductivity does not vary with size of specimen.

Synthesis

The results of the testing program suggest that field-scale hydraulic conductivity can be assessed using laboratory tests on undisturbed block specimens having a diameter greater than 0.30 m. For the sites evaluated in this study, hydraulic conductivity measured on specimens of this size was similar to hydraulic conductivity measured using SDRIs. A variety of construction methods, ranging from very poor (Site B) to excellent (Site D), were used when compacting the test pads at these sites.

STOCHASTIC MODEL

Modeling of scale-dependent hydraulic conductivities was conducted concurrently with the experimental program. In this effort, a model of flow in compacted soil containing macropores was developed. Before modeling began, a literature review was conducted to determine if modeling techniques for flow in fractured rock could be adapted for flow in compacted soils. Two types of models were found to be widely used to simulate flow in fractured rock: dual continua models (e.g., Long et al., 1982; Andersson and Dverstorp, 1987; Berkowitz et al., 1988) and discrete fracture models (e.g., Snow, 1969; Neuzil and Tracy, 1981).

In dual continua models, the rock is assumed to consist of two continua that are joined hydraulically by a transfer function. One continuum represents the matrix and the other represents fractures. Typically, a finite difference or finite element algorithm (e.g., Long, et al., 1982) is used to link the continua together. In contrast, discrete fracture models (e.g., Neuzil and Tracy, 1981) ignore flow in the matrix and assume all flow occurs in the fractures. Laminar flow equations are used to compute flow rates in the fractures and continuity equations are written to join flows at the fracture intersections.

Approach for Compacted Soil Liners

The objective of this research program was to determine, via experiments and modeling, the size of specimen that is needed to represent field-scale hydraulic conductivity for a variety of construction conditions. In light of this objective, a simplified approach that combines the dual continua and discrete fracture methods was used. The macropores and the matrix were both treated as laminar flow media (i.e., using Darcian flow), but flow in the matrix and flow in the macropores was assumed to be uncoupled.

Figure 6 is a conceptual illustration of the model. A lift of soil is assumed to contain numerous interconnected macropores and a specimen with cross-sectional area A is sampled from the lift. The total flow rate (Q) through the specimen can be expressed as:

$$Q = Q_m + \sum_{i=1}^{N} Q_{p,i} \qquad (4)$$

where Q_m is the flow rate in the matrix, $Q_{p,i}$ is the flow rate in the ith macropore that exits the base of the specimen, and N is the number of macropores that exit the base of the specimen. The flow rate in the matrix is computed based on the hydraulic conductivity of the soil that would be achieved for wet side compaction (K_w); i.e.,

$$Q_m = K_w \, i \, A \qquad (5)$$

where i is the average hydraulic gradient and A is the cross-sectional area of flow. The cross-sectional area for matrix flow (A) was assumed equal to the gross area of the specimen because the contributions of the macropores to A are small.

The flow rate for the ith macropore ($Q_{p,i}$) is also computed using Eq. 6. However, K_w is replaced by the hydraulic conductivity of the macropore, i is the hydraulic gradient along the macropore, and A is the cross-sectional area of the macropore.

The total flow rate is evaluated in terms of equivalent hydraulic conductivity (K) by:

$$K = \frac{Q}{i \, A} = K_w + \frac{1}{i \, A} \sum_{i=1}^{N} Q_{p,i} \qquad (6)$$

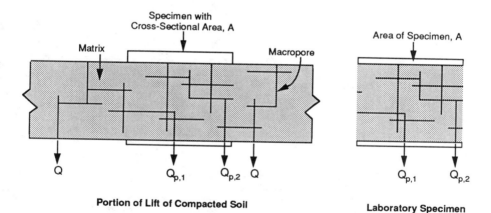

FIG. 6--Simulation of soil liner with defects.

Geometric Properties of Macropores

Little information is currently available to describe macropores in compacted soil liners. The greatest wealth of information has been collected by Elsbury et al. (1988), who describe a morphological study of a test pad that was very permeable (K~1 x 10^{-6} m/sec). Thus, their results are likely to be representative of "worst case" conditions.

The report by Elsbury et al. (1988) contains photographs of dye stained macropores they found in the test pad. These photographs were used to characterize statistics of the length and size of macropores. All macropores were assumed to be horizontally or vertically oriented and a trace was made from each photograph based on this assumption (e.g., Fig. 7). From each trace, the length, orientation, and location of each macropore was obtained.

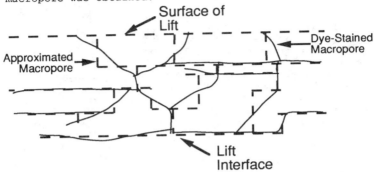

FIG. 7--Characterizing geometry of macropores from field observations.

Length of horizontal and vertical macropores

Horizontal macropores ranged in length from 0.03 m to 0.38 m with a mean 0.14 m. Vertical macropores were generally much shorter; they ranged in length from 0.02 to 0.06 m and had a mean length of 0.04 m. Aperture widths were virtually impossible to determine from the photographs; however, Elsbury et al. (1988) reported that the macropores were typically 0.001 to 0.003 m wide.

Histograms of macropore lengths were constructed to determine distributional forms that could be used to describe their variability (Fig. 8). Positive skew was evident in histograms for horizontal and vertical macropores. Hence, the lengths of the macropores were hypothesized as being log-normally distributed. To test this hypothesis, Filliben's probability plot correlation coefficient test was employed (Filliben, 1975). At a significance level of 0.05, the log-normal hypothesis was not rejected.

Location of intersections and number of vertical macropores

The photographs from Elsbury et al. (1988) were examined to develop a generating scheme based on how water infiltrating into a macropore at the surface forms a flow path through a soil liner. Figure 9a illustrates the process. Water first enters a vertical macropore at the surface of the soil and flows downward until it reaches a horizontal macropore. When water reaches the end of the vertical macropore, it stops (i.e., it reaches a dead end) or it spreads laterally in a horizontal macropore until another vertical macropore (or several vertical macropores) is found. This process is repeated until the water reaches the bottom of the lift.

To determine if probability distributions could be used to describe this process, the locations of intersections of horizontal and

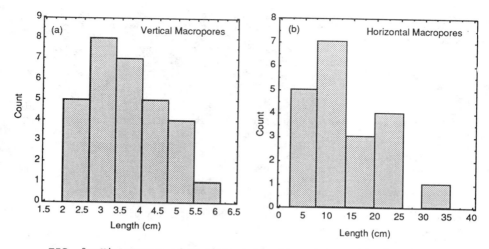

FIG. 8--Histograms of macropore lengths: (a) vertical and (b) horizontal.

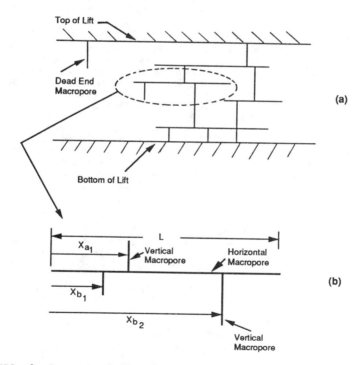

FIG. 9--Concept of flow in macropores (a) and definition of normalized location of intersections (b).

vertical macropores were evaluated and the frequency of vertical macropores stemming from horizontal macropores was determined. Locations of intersections of vertical macropores, for a given horizontal macropore, were measured relative to the length (i.e., X_a/L and X_b/L) of the horizontal macropore (Fig. 9b).

Fig. 10 shows histograms of the location of vertical macropores stemming from each horizontal macropore relative to the length of the horizontal macropore (X_a/L or X_b/L). The lack of shape present in the histograms suggests that the locations of the intersections (X_a's and X_b's) can be described by a uniform distribution. Vogel's probability plot correlation coefficient test for the uniform distribution (Vogel and Kroll, 1989) was used to test this hypothesis at a significance level of 0.05. The hypothesis was not rejected.

A similar analysis was conducted for the number of intersections of vertical and horizontal macropores (N_b). The number of vertical macropores stemming from each horizontal macropore was counted. Figure 11 is a histogram of N_b based on data collected from the macropore traces. The positive skew of N_b and the uniform distribution of the locations of the intersections suggests that a Poisson distribution can be used to describe N_b. To test this hypothesis, the chi-square test (Haan, 1977) was employed. At a significance level of 0.05, the hypothesis was not rejected.

Monte Carlo Simulation

A numerical model employing Monte Carlo simulation was developed to generate random networks of macropores. The Poisson distribution was used to generate the number of macropores and the log-normal distribution was used to specify their length. Locations of the intersections of horizontal and vertical macropores were generated from a uniform distribution.

After a network of macropores was generated, specimens of various size were "removed" from the network. Then, hydraulic heads corresponding to steady state flow throughout the network were computed and an equivalent hydraulic conductivity was determined as defined in Eq. 6. Heads were obtained by ensuring continuity at each intersection of the macropores. Details describing the simulation procedure and the method to compute heads and flow rates can be found in Benson and Hardianto (1992).

Networks of macropores were generated in a two-dimensional domain. The domain was 1 m wide and 0.15 m thick, representing one lift of compacted soil. Two different lifts were analyzed, each with different geometric properties for the macropores. Herein, the lifts are referred to as Lift 1 and Lift 2. Input data for the two lifts are summarized in Table 3. The writers admit that the input parameters were, in part, selected arbitrarily. Thus, using different parameters may yield different results.

Lift 1 was used to simulate macroscopic defects caused by poor construction practices (e.g., Site B). Statistics for the distributions used to describe macropores were obtained from the study by Elsbury et al. (1988). Eight starting points for vertical macropores were assigned on the top of Lift 1, with a spacing of 0.12 m between the starting points. Other inputs for Lift 1 included: mean number of vertical macropores (λ) of 1.31, log-mean and log-standard deviation of the length of horizontal macropores equal to $\mu=2.4$ and $\sigma=0.6$, and log-mean and log-standard deviation for length of vertical macropores of $\mu=1.9$ and $\sigma=0.3$. All lengths are in centimeters prior to logarithmic transformation. Each macropore was assigned a hydraulic conductivity of 1×10^{-4} m/sec.

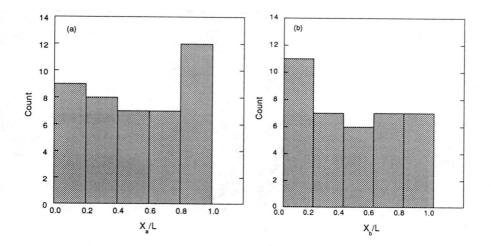

FIG. 10--Histograms of normalized locations.

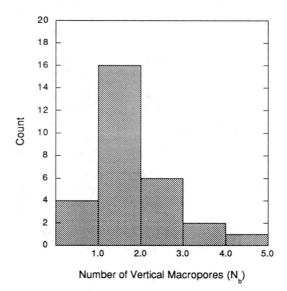

FIG. 11--Histogram of number of vertical macropores.

TABLE 3—Macropore input parameters for Lifts 1 and 2.

Parameter	Description	Lift 1	Lift 2
μ (horizontal)	log-mean length of horizontal macropores	2.4	1.2
σ (horizontal)	log-variance length of horizontal macropores	0.6	0.6
μ (vertical)	log-mean length of vertical macropores	1.9	0.95
σ (vertical)	log-variance length of vertical macropores	0.3	0.3
λ	mean of number of vertical macropores	1.31	1.31
No. starting points	number of vertical macropores at the top of the lift	8	16

Note: All units of length in centimeters prior to logarithmic conversion.

Lift 2 was used to simulate a lift with many macropores that are highly tortuous. These conditions are assumed to correspond to compaction of a pre-processed soil with heavy machinery, but slightly dry of the line of optimums (i.e., similar to conditions at Site A). Under this condition, the soil would contain numerous tortuous macropores (Benson and Daniel, 1990). The construction condition represented by Lift 2 is better than the condition represented by Lift 1, but still will result in scale-dependent hydraulic conductivity. Shorter horizontal and vertical macropores were used for Lift 2 and the number of starting points was doubled. Sixteen starting points were assigned to the top of the lift, with a 0.06 m-spacing. Log-means (μ's) of 1.2 and 0.95 were used for the length of horizontal and vertical macropores and their hydraulic conductivity was specified as 1×10^{-8} m/sec. Lower hydraulic conductivity was used to simulate greater tortuosity and smaller aperture width. The remaining parameters were the same as those used in Lift 1.

Moment sensitivity studies showed that stable estimates of the mean could be obtained with a minimum of 10 realizations (Hardianto, 1992). Because the mean hydraulic conductivity of each specimen size was the primary variable being considered, 12 realizations were used for each condition that was simulated. For each realization, a "specimen" was isolated from the simulated lift, with the center of the "specimen" always being located at the center of the lift. Each specimen had a different width; widths of 0.07, 0.15, 0.30, 0.45, 0.60, and 0.90 m were used.

Results and Discussion

Figure 12 shows results obtained with the model, experimental data obtained from Sites A and B, and data presented in Elsbury et al. (1988). Lines passing through the data points were fit by eye. For each specimen size, the geometric mean hydraulic conductivity (from test results or modeling) was computed and then normalized by dividing by the geometric mean hydraulic conductivity of the smallest specimens (diameter=0.07 m). For the modeling results, hydraulic conductivity of the matrix (K_w) was 1×10^{-10} m/sec for Sites A and B (Hardianto, 1992) and 1×10^{-11} m/sec for Elsbury et al. (1988).

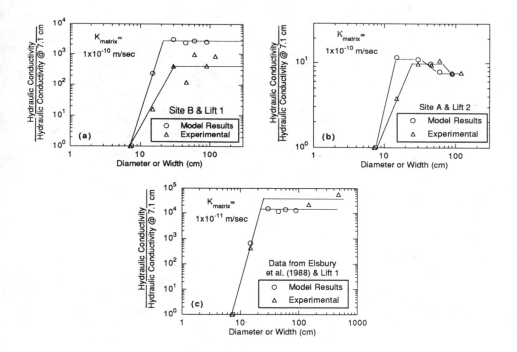

FIG. 12--Comparison of modeling results and field data: (a) Site B and Lift 1, (b) Site A and Lift 2, (c) Data from Elsbury et al., (1988) and Lift 1. Note: geometric mean hydraulic conductivities are shown.

Modeling results for Lift 1 are compared to the experimental results from the test pad at Site B in Fig. 12a. Lift 1 was selected for comparison with the experimental data because Site B had large widely spaced macropores. Figure 12a shows that trends exhibited by the experimental data and the modeling results are similar. The modeling results show a slightly larger increase in hydraulic conductivity and a slightly smaller representative specimen size than would be suggested by the experimental data. Nevertheless, both the model and the experimental data exhibit a similar trend and illustrate that hydraulic conductivity near field-scale can be measured using a specimen having a diameter greater than 0.30 m.

Fig. 12b is a comparison of experimental results obtained from Site A and modeling results for Lift 2. Results from Lift 2 were selected for comparison because macropores in the test pad at Site A were frequent and small relative to the thickness of the lifts. They also appeared highly tortuous. Thus, they were likely to have lower hydraulic conductivity than macropores for Site B.

The trends in the modeling results and experimental data shown in Fig. 12b are similar. Again, the modeling results indicate a slightly smaller specimen is required to achieve field-scale hydraulic conductivity than would be inferred from the experimental data. Nevertheless, the modeling results and the field data are similar. The modeling and experimental results in Fig. 12b also show a slight decrease in hydraulic conductivity that occurs as the specimen size is increased beyond 0.30 m. A similar decrease in hydraulic conductivity was observed in the test results for Site C.

Experimental data from Elsbury et al. (1988) and modeling results for Lift 1 are compared in Fig. 12c. The modeling results and experimental data are very similar, which is expected, because statistical data used in the model for Lift 1 was derived directly from the morphological study described in Elsbury et al. (1988). More importantly, however, the modeling results suggest that field-scale hydraulic conductivity may have been obtained if a specimen having diameter greater than 0.30 m had been tested. Unfortunately, Elsbury et al. did not perform tests on specimens of this size. Thus, this conclusion cannot be substantiated.

CONCLUSIONS AND RECOMMENDATIONS

Based on the results of this study, the following conclusions and recommendations are made:

1. The size of a representative specimen for measurement of hydraulic conductivity of compacted soil liners depends on the method and quality of construction. If the soil is compacted poorly (e.g., dry of the line of optimums or with low compactive effort), the representative specimen size is large. However, when the soil is well compacted (wet of the line of optimums or with high compactive effort), the representative size is small. The key factor controlling the representative size is the size of the network of pores conducting flow and the ability to represent the network of pores in a test specimen.

2. The experimental results suggest that field-scale hydraulic conductivity can be measured on specimens with a diameter of at least 0.30 m and a thickness of 0.15 m for a wide variety of construction conditions. Block specimens of this size are <u>recommended</u> for use in hydraulic conductivity assessment of test pads but are <u>not recommended</u> for evaluation of constructed liners because of the large holes left in the liner after sampling.

3. The modeling results suggest that specimens with a diameter of 0.30 m are adequate to represent the presence of macropores and field-scale hydraulic conductivity. Trends observed in the modeling results and experimental data were similar.

ACKNOWLEDGMENT

Financial support for the research described in this paper has been provided by Waste Management, Inc. (WMI). Philip Gagnard of RUST Environment and Infrastructure (RUST E&I) manages research and development activities for WMI. Clarke Lundell and Anthony Maxson of RUST E&I provided valuable input during the project. Mr. Norman Severson assisted in design and construction of the SDRIs and the large-scale permeameter. Xiaodong Wang, Jason Kraus, and John Trast assisted in the field and laboratory. Appreciation is also expressed to Gordon Boutwell of Soil Testing Engineers, Inc., who provided the SDRI data for Site D and assisted in sampling at the site. Others who assisted in the project include: Robert Borkenhagen, Keith Burdick, Ronald Keller, Henry Koch, Dean Mussatti, Robert Philips, Mark Schoemann, and Thomas Wimmer. However, the opinions expressed in this paper are those of the authors and do not necessarily reflect the policies or opinions of WMI, RUST E&I, or others who assisted in the project.

REFERENCES

Andersson, J. and B. Dverstorp, 1987, "Conditional Simulations of Fluid Flow in Three-Dimensional Networks of Discrete Fractures," Water Resources Research, Vol. 23, No. 10, pp. 1876-1886.

Benson, C. H. and F.S. Hardianto, 1992, "Hydraulic Conductivity Assessment of Compacted Soil Liners-Final Report: Phase I," Environmental Geotechnics Report No. 92-4, Dept. of Civil and Environmental Engineering, University of Wisconsin-Madison.

Benson, C. H. and G. P. Boutwell, 1992, "Compaction Control and Scale-Dependent Hydraulic Conductivity of Clay Liners," Proceedings, 15th Annual Madison Waste Conference, Madison, WI, pp. 62-83.

Benson, C. H. and D. E. Daniel, 1990, "Influence of Clods on Hydraulic Conductivity of Compacted Clay," Journal of Geotechnical Engineering, ASCE, Vol. 116, No. 8, pp. 1231-1248.

Berkowitz, B. et al., 1988, "Continuum Models for Contaminant Transport in Fractured Porous Formations," Water Resources Research, Vol. 24, No. 8, p. 1225-1236.

Daniel, D. E., 1984, "Predicting Hydraulic Conductivity of Clay Liners," Journal of Geotechnical Engineering, ASCE, Vol. 110, No. 2, p. 285-300.

Daniel, D. E., et al., 1985, "Fixed-Wall vs. Flexible-Wall Permeameters," Hydraulic Barriers in Soil and Rock, Special Technical Publication 867, American Society for Testing and Materials, Philadelphia, pp. 107-126.

Daniel, D. E., 1987, "Earthen Liners for Land Disposal Facilities," *Geotechnical Practice for Waste Disposal '87*, GSP No. 13, ASCE, pp. 21-39.

Daniel, D. E., 1989, "In Situ Hydraulic Conductivity Tests for Compacted Clays," *Journal of Geotechnical Engineering*, Vol. 115, No. 9, pp. 1205-1227.

Daniel, D. E., 1990, "A Note on Falling Headwater and Rising Tailwater Permeability Tests," *Geotechnical Testing Journal*, Vol. 12, No. 4, pp. 308-310.

Daniel, D. E. and C. H. Benson, 1990, "Water Content-Density Criteria for Compacted Soil Liners," *Journal of Geotechnical Engineering*, ASCE, Vol. 116, No. 12, p. 1811-1830.

Day, S. R. and D. S. Daniel, 1985, "Hydraulic Conductivity of Two Prototype Clay Liners," *Journal of Geotechnical Engineering*, ASCE, Vol. 111, No. 8, p. 957-970.

Elsbury, B.R. et al., 1988, "Field and Laboratory Testing of a Compacted Soil Liner," Report to U.S.E.P.A. for Contract No. 68-03-3250, Cincinnati, Ohio.

Filliben, J., 1975, "The Probability Plot Correlation Coefficient Test for Normality," *Technometrics*, Vol. 17, No. 1, p. 111-117.

Haan, C. T., 1977, *Statistical Methods in Hydrology*, Iowa State University Press.

Hardianto, F.S., 1992, "Representative Sample Size for Hydraulic Conductivity of Compacted Soil Liners," MSCE thesis, Dept. of Civil and Environmental Engineering, University of Wisconsin.

Johnson, G., et al., 1990, "Field Verification of Clay Liner Hydraulic Conductivity," in *Waste Containment Systems: Construction, Regulation, and Performance*, GSP. No. 26, ASCE, New York, pp. 226-245.

Lahti, L., et al., 1987, "Quality Assurance Monitoring of a Large Clay Liner," *Geotechnical Practice for Waste Disposal '87*, GSP No. 13, ASCE, pp. 640-654.

Long, J., et al., 1982, "Porous Media Equivalents for Networks of Discontinuous Fractures," *Water Resources Research*, Vol. 18, No. 3, pp. 645-658.

Mitchell, J.K., et al., 1965, "Permeability of Compacted Clay," *Journal of the Soil Mechanics and Foundations Division*, ASCE, Vol. 91, No. SM4, pp. 41-63.

Neuzil, C. and J. Tracy, 1981, "Flow Through Fractures," *Water Resources Research*, Vol. 17, No. 1, pp. 191-199.

Reades, D., et al., 1990, "Detailed History of Clay Liner Performance," in <u>Waste Containment Systems: Construction, Regulation, and Performance</u>, GSP No. 26, ASCE, New York, pp. 156-174.

Snow, D., 1969, "Anisotropic Hydraulic Conductivity of Fractured Media," <u>Water Resources Research</u>, Vol. 58, No. 6, p. 1273-1289.

Vogel, M. and C. Kroll, 1989, "Low-Flow Frequency Analysis Using Probability-Plot Correlation Coefficients," <u>Journal of Water Resources Planning and Management</u>, ASCE, Vol. 115, No. 3, pp. 338-357.

David E. Daniel[1]

STATE-OF-THE-ART: LABORATORY HYDRAULIC CONDUCTIVITY TESTS FOR SATURATED SOILS

REFERENCE: Daniel, D. E., "State-of-the-Art: Laboratory Hydraulic Conductivity Tests for Saturated Soils," Hydraulic Conductivity and Waste Contaminant Transport in Soil, ASTM STP 1142, David E. Daniel and Stephen J. Trautwein, Eds., American Society for Testing and Materials, Philadelphia, 1994.

ABSTRACT: Hydraulic conductivity of saturated soils is measured in the laboratory with either rigid- or flexible-wall permeameters. Rigid-wall cells are preferred for granular materials, and either rigid- or flexible-wall cells are preferred for low-hydraulic conductivity materials, depending on the type of test specimen and the conditions being simulated. Procedures for selection and preparation of test specimens are critically important, even for laboratory-compacted materials. Either constant- or variable-head tests can be performed; either type of test will yield satisfactory results if used properly. For compressible materials, the head cannot change much or the test specimen will change volume. Permeation with deaired water until inflow and outflow rates equilibrate, supplemented with backpressure where appropriate, are needed to saturate test specimens. Permeation should continue until chemical equilibrium is established when testing with liquids other than water, e.g., chemical wastes.

KEYWORDS: hydraulic conductivity, permeability, soil, permeameter, rigid-wall permeameter, flexible-wall permeameter, waste, leachate

Hydraulic conductivity is the coefficient of proportionality in Henry Darcy's 1856 law, which can be written as:

$$q = k i A = k \frac{\Delta H}{L} A \qquad (1)$$

where

q = rate of flow ($m^3 s^{-1}$)

k = hydraulic conductivity ($m s^{-1}$)

i = hydraulic gradient (dimensionless) = $\Delta H/L$

---

[1] Professor of Civil Engineering, University of Texas, Austin, TX 78712

ΔH = head loss across specimen (m)

L = length of specimen (m)

A = cross-sectional area of specimen perpendicular to direction of flow (m^2).

The hydraulic conductivity in Darcy's law depends not only upon the properties of the porous medium but also upon the properties of the permeating liquid (Olson and Daniel, 1981). Sand, for example, would have a lower hydraulic conductivity if permeated with motor oil than water because motor oil is more viscous than water. An alternative form of Darcy's law may be written as follows:

$$q = K \frac{\gamma}{\mu} \frac{\Delta H}{L} A \qquad (2)$$

where:

K = intrinsic permeability of the soil (m^2)

γ = unit weight of the permeant liquid (g m^{-2} s^{-2})

μ = viscosity of the permeant liquid (g m^{-1} s^{-1})

and the other terms are the same as Eq. 1. The intrinsic permeability is a function only of the properties of the porous material, not the permeating liquid. Civil engineers and ground water specialists have traditionally used hydraulic conductivity rather than intrinsic permeability because the density and viscosity of water are relatively constant (hydraulic conductivity changes about 3% for every 1°C change in temperature, but this effect is minor for ordinary ranges in temperature).

Civil engineers have traditionally called the coefficient k in Eq. 1 the *coefficient of permeability*, but soil scientists and hydrogeologists have traditionally called it *hydraulic conductivity*. Because numerous other conduction phenomena are described by an equation identical in form to Darcy's law, and because the coefficient of proportionality in these equations of conduction is usually termed the conductivity (e.g., thermal conductivity) of the conducting medium, one should call the coefficient k in Darcy's law the *hydraulic conductivity* in order to be consistent with other fields. Also, use of the term *permeability* for the coefficient k in Eq. 1 can create confusion with *intrinsic permeability*, which is the coefficient in Eq. 2. Use of *hydraulic conductivity* eliminates any confusion with *intrinsic permeability*. However, for individuals who prefer to stick with *coefficient of permeability* no matter how much confusion it might cause for others, they may wish to note that Henry Darcy himself called the parameter "un coefficient dépendant de la pérmeabilité" ("a coefficient depending on the permeability") and did not mention the term *hydraulic conductivity*.

In this paper, *hydraulic conductivity* will be used to denote the coefficient k in Eq. 1, and *intrinsic permeability* will be used for K in Eq. 2. Hydraulic conductivity has units of length per time. In the U.S., the traditional units for k have been cm/s, but centimeters are not in the SI system. Units of m/s are the SI units typically employed for hydraulic conductivity. The conversion is 1 cm/s = 0.01 m/s.

In succeeding sections, testing procedures for hydraulic conductivity are discussed. The topics to be covered include hydraulic conductivity cells, hydraulic control systems, permeant water, sample preparation procedures, control variables, special requirements for permeation with chemicals or waste liquids, and testing errors.

HYDRAULIC CONDUCTIVITY CELLS

Numerous variations of hydraulic conductivity cells are available for testing soils. The cells may be divided into two broad categories: rigid-wall and flexible-wall cells.

Rigid-Wall Permeameters

Rigid-wall permeameters consist of a rigid tube or box that contains the specimen to be permeated. The tube is almost always circular and constructed of metal (plated brass, plated steel, stainless steel, or aluminum), plastic (acrylic or polyvinyl chloride), or glass (used only for testing with chemicals or waste liquids). The permeating liquid flows along the axis of the cylindrical test specimen. Flow may be from top-to-bottom or bottom-to-top. Upward flow may help to dislodge entrapped gas, but one must be careful not to liquefy the specimen or to displace an unrestrained test specimen upward.

Four types of rigid-wall permeameters are used: (1) compaction-mold permeameter; (2) consolidation-cell permeameter; (3) sampling-tube permeameter; and (4) oversized permeameter.

Compaction-Mold Permeameter. The compaction-mold permeameter is the most commonly used type of rigid-wall permeameter. The soil to be permeated is placed in the permeameter tube, compacted in an appropriate manner, and then permeated. Sometimes the material is simply poured or rodded into the cell. The "compaction mold" doubles as a permeameter tube. Soils varying in grain-size from gravel to clay can be tested in a compaction-mold permeameter.

A schematic diagram of a simple compaction-mold permeameter is shown in Fig. 1 Porous disks are needed at the ends of the specimen to ensure that one-dimensional flow takes place within the test specimen. To accomplish this, the porous disks must have a much higher hydraulic conductivity than the test specimen. When one permeates a test specimen in any type of permeameter, it is essential that nearly all of the frictional resistance to fluid flow (i.e., head loss) occur within the test specimen and not in the porous disks, tubes leading to the permeameter, or valves. The author recommends that prior to setting up a hydraulic conductivity test, one should first set up an empty permeameter (including porous disks) and flow water through the cell at the same head loss across the cell to be employed in the hydraulic conductivity test. If the flow rate through the empty cell is at least 10 times greater than the flow rate through the cell containing a test specimen, it may be safely assumed that nearly all the head loss is within the test specimen.

The permeameter in Fig. 1 does not allow the test specimen to swell. An alternative rigid-wall, compaction-mold permeameter that

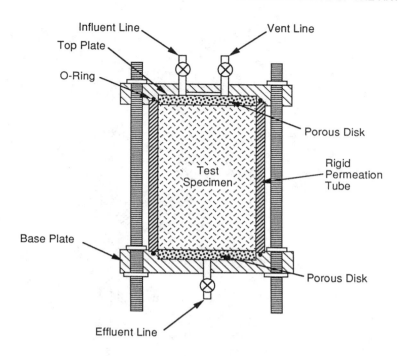

Figure 1. Rigid-Wall Permeameter that Does Not Allow Soil to Swell.

allows for free swell is shown in Fig. 2. With this type of cell, no porous disk is necessary on the upper surface, but flow is always from top-to-bottom. When materials that swell are tested, one must recognize that a change in the length (L) of the specimen will cause a change in the hydraulic gradient (Eq. 1). If soil expands into the swell ring shown in Fig. 2, the soil in the swell ring will probably have a much higher apparent hydraulic conductivity than the material contained in the permeameter tube because: (1) the effective stress acting on the soil in the swell ring is essentially zero, which leads to a high porosity, and (2) sidewall leakage is probably occurring. Accordingly, the procedure that the author recommends to deal with soil that has expanded into the swell ring is as follows:

1. The test specimen is permeated until inflow and outflow rates are reasonably equal and the computed hydraulic conductivity is steady.

2. The permeameter is disassembled, and the thickness of soil in the swell ring is measured. If the thickness of the soil layer in the swell ring is less than 5% of the thickness of the test specimen in permeation tube, no further action is needed — the test is complete. If the thickness of soil in the swell ring

is greater than 5% of the length of the permeation tube, the soil is removed from the swell ring, the test specimen is trimmed flush with the top of the permeation tube, the cell is reassembled, and permeation is continued until inflow and outflow rates are reasonably equal and hydraulic conductivity is steady. If necessary, the trimming/permeation process is repeated.

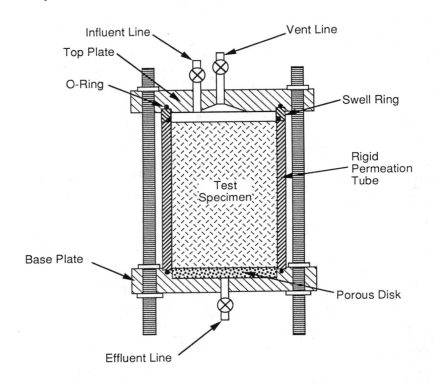

Figure 2. Rigid-Wall, Compaction Mold Permeameter that Permits the Test Specimen to Swell.

One of the potential problems with any rigid-wall permeameter is that sidewall flow can occur. The potential error is greatest for low-hydraulic-conductivity materials. The double-ring permeameter (Fig. 3) allows one to determine whether sidewall flow is occurring. Pierce et al. (1987) used a triple-ring base plate to examine the variation of hydraulic conductivity within compacted clays that were permeated in both rigid- and flexible-wall permeameters. Pierce et al. found that the flow rate in the outer ring, which represented one fifth of the total area, was 2 to 5 times larger than the flow rate in the inner two rings for both rigid- and flexible-wall permeameters. The most likely cause for the higher flow rate near the sidewall was the presence of a greater percentage of macropores near the perimeter of the test specimens.

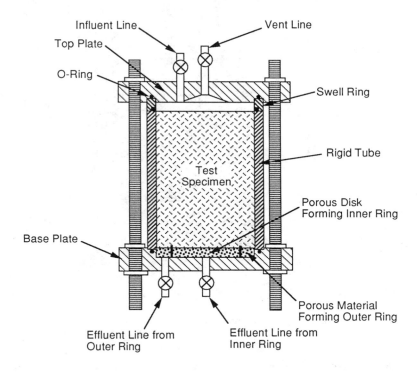

Figure 3. Double-Ring Permeameter for Identifying Sidewall Leakage.

The permeameter shown in Fig. 2 is often modified for testing granular materials. As shown in Fig. 4, screens are used rather than porous disks because porous disks are less permeable than most medium to coarse sands and most gravels. Piezometers may be inserted in the walls of the tube (Fig. 4) to measure the head loss over a known distance within the specimen. Measuring the head loss between two points in permeation tube is often the simplest way of dealing with equipment head losses in tubes, valves, fittings, end plates, and the screens. These equipment head losses serve to reduce the flow rate through the test specimen, but the flow reduction is irrelevant to determination of hydraulic conductivity if the hydraulic gradient within the test specimen is accurately determined. (If equipment head losses lower the flow rate through the specimen, the hydraulic gradient, $\Delta H/L$, will be lowered proportionally). For soils with very high hydraulic conductivity (e.g., > 1 cm/s), the main testing challenges are: (1) devising a hydraulic system to deliver a high flow rate; and (2) accurately measuring $\Delta H/L$ using piezometers as shown in Fig. 4.

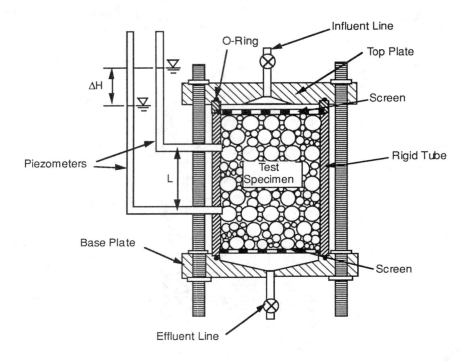

Figure 4. Rigid-Wall Permeameter for Testing Granular Materials.

The permeameter in Fig. 1 allows no swell, and the permeameter in Fig. 2 offers no resistance to swell. In many cases, an intermediate boundary condition is desired. This is achieved by controlling the vertical stress acting on the soil as shown in Fig. 5. The vertical stress simulates field conditions. The only problem is that friction between the soil and the inside of the permeameter tube will cause the applied stress to vary with depth in the permeameter. This problem is avoided by making the ratio of the length (L) to diameter (D) of the specimen small, e.g., ≤ 0.25, which is typical for consolidation testing. The compaction-mold cell with controlled vertical stress is especially useful for testing highly compressible materials such as sludges and soil-bentonite backfill for slurry walls.

<u>Consolidation-Cell Permeameter</u>. A schematic diagram of a consolidation-cell permeameter is shown in Fig. 6. This type of permeameter may be used in two ways (Olson and Daniel, 1981; and Olson, 1986): (1) the soil can be consolidated and the hydraulic conductivity computed from the rate of consolidation, or (2) the soil can be permeated directly. The problem with the first option is that errors are likely to be introduced as a result of the effects of secondary

consolidation, which are not accounted for in calculation of hydraulic conductivity and which lead to a computed hydraulic conductivity that is too low,. The error will probably be less than 50% (Olson and Daniel, 1981; Olson, 1986). The problem with the second option (direct permeation in the consolidation ring) is that sidewall leakage may occur. Sidewall leakage is rarely a problem for compressible soils that have been subjected to compressive stresses of at least 50 kPa (1,000 psf) or more, but sidewall leakage can be a serious problem when testing very stiff or hard soils or when permeating soils at a very low compressive stress.

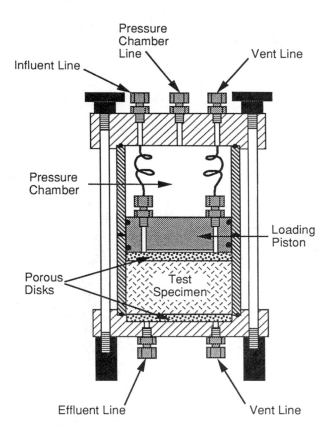

Figure 5. Rigid-Wall Permeameter with Controlled Vertical Stress (Original Drawing Courtesy of Trautwein Soil Testing Equipment Co.)

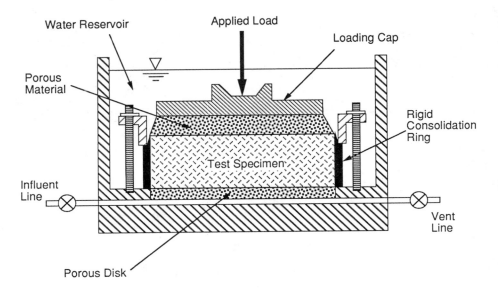

Figure 6. Consolidation-Cell Permeameter.

The consolidation-cell permeameter is useful only for clayey soils that contain no gravel or coarse sand. Because the consolidation-cell permeameter is useful only for testing a small subset of all the types of soil that require permeation, the consolidation-cell permeameter is not frequently used. In the author's experience, the use of the consolidation-cell permeameter has declined significantly in the past 10 to 15 years, mostly because other types of permeameters offer greater versatility.

Sampling-Tube Permeameter. "Undisturbed" samples of soil are frequently obtained by pushing thin-walled sampling tubes into the soil. Sometimes, the sampling tubes are cut, end plates are attached (Fig. 7), and the soil is permeated directly in the sampling tube. There is a high risk of sidewall leakage when materials containing any gravel, or very stiff or hard materials, are tested. One potential problem is damage to the thin-walled tube (such a crimping of the sampling end), which can gouge the sample, or shearing action along the sidewall during sampling, which can remold the soil. Also, many thin-walled tubes are constructed with an opening at the cutting edge that is slightly smaller than the inside diameter of the sampling tube (providing "inside clearance") — this is done to reduce friction but can promote sidewall leakage. Except for easy-to-sample soils and sampling tubes for which these problems cannot occur, this type of permeameter is not recommended.

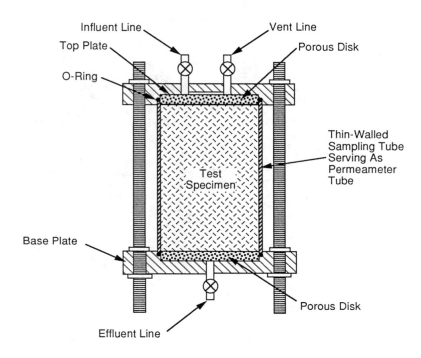

Figure 7. Rigid-Wall, Sampling-Tube Permeameter.

Oversized Permeameter. On occasion, soil samples are placed in an oversized permeameter (Fig. 8), with an annular seal provided between the test specimen and the permeameter tube. The sealing material is typically bentonite.

The author has used the oversized permeameter to permeate soil-cement samples, and results were generally satisfactory. However, problems did occasionally develop with the seal, and with each test, considerable effort was given to forming, checking, and worrying about the seal. The flexible-wall permeameter, described below, is preferred for this type of testing because it is much easier to effect a reliable seal around the perimeter of the test specimen. Because of the difficulties just described, the oversized permeameter is not recommended for general usage.

Flexible-Wall Permeameter

A schematic diagram of a flexible-wall permeameter is shown in Fig. 9. The test specimen is confined with porous disks and end caps on the top and bottom and by a latex membrane on the sides. Two drainage lines to the top and bottom end pieces are recommended for flushing air out of the lines. The cell is filled with water and pressurized to

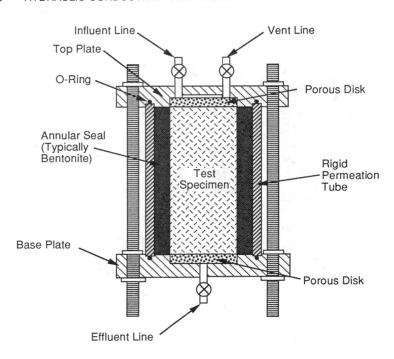

Figure 8. Rigid-Wall, Oversized Permeameter.

press the latex membrane against the test specimen and thereby minimize or eliminate sidewall leakage. The cell may be filled with some other fluid besides water for special tests. For instance, if the test specimen is to be frozen, a mixture of ethylene glycol and water (antifreeze) may be used for the cell fluid.

When the permeating liquid is a chemical or waste fluid, three potential problems must be addressed: (1) compatibility of the liquid with the components of the flexible-wall permeameter; (2) compatibility of the liquid with the latex membrane; and (3) diffusion of the permeating fluids, or solutes in the fluid, through the confining membrane and into the cell liquid. The first problem is the easiest to solve; the cell is built with materials such as stainless steel and Teflon® that are compatible with the waste liquid. The second problem can be solved in two ways: (1) another flexible membrane material besides latex (e.g., neoprene) can be used, or (2) the test specimen can first be wrapped with a thin film of Teflon® (the author uses a wide roll of Teflon® tape with an adhesive on one side), and then a latex membrane can be placed over the Teflon®. A Teflon® wrapping has worked well with soil, but some have reported problems achieving a seal with Teflon® wrapped around incompressible materials (e.g., cement-based materials). In the author's experience, the Teflon® wrapping is more practical than use of a special membrane because no type of flexible membrane is compatible with all organic solvents or hydrocarbons. The problem of diffusion through he membrane is solved by placing a diffusion barrier (Teflon® film or aluminum foil, for instance) between the membrane and test specimen or between two membranes that encase the specimen.

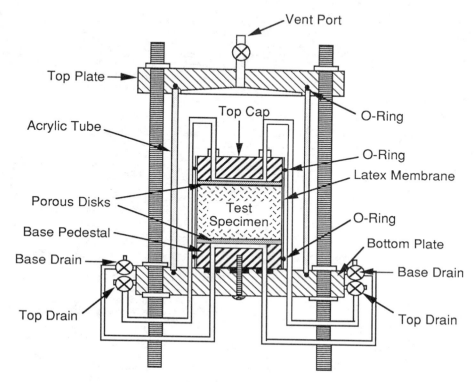

Figure 9. Flexible-Wall Permeameter.

The flexible-wall permeameter may contain a solid rod that passes through the top plate and comes into contact with the upper surface on the top cap. The purpose of this rod is to enable measurement of vertical deformation of the test specimen, which can be helpful in determining the rate of volume change and in correcting the length of the test specimen to account for swelling or consolidation. The author has found it convenient to use an optical telescope to monotor changes in the height of the specimen (Daniel et al., 1984), but optical telescope can only be used on cells equipped with a clear cell wall.

Additional information on flexible-wall cells and testing with flexible-wall permeameters is given by Zimmie (1981), Daniel et al.,

(1984), Carpenter and Stephensen (1986), Evans and Fang (1986), and ASTM D5084, "Measurement of Hydraulic Conductivity of Saturated Porous Materials Using a Flexible Wall Permeameter."

Size of Test Specimen

There is no limit on the size of a test specimen that can be used in a permeameter in terms of the physical constraints required for measurement of hydraulic conductivity. However, if the size of the test specimen is not significantly larger than the size of the particles that comprise the porous medium, the measurements cease to have any real meaning. Neither the length nor the diameter of a test specimen should be less than approximately 6 times larger than the largest particle in the test specimen per ASTM D2434, "Permeability of Granular Soils," and ASTM D5084.

Most rigid-wall, compaction-mold permeameters come in one of 3 diameters: (1) 36 mm (1.4 in.), which is the diameter of a Harvard miniature compaction mold; (2) 100 mm (4 in.), which is a typical Proctor mold diameter; and (3) 150 mm (6 in.), which is an alternative Proctor mold diameter for material with a large particle size. Compaction-mold permeameters approaching 1 m in diameter have been used, e.g., Shackelford and Javed (1991). Flexible-wall permeameters are available commercially for testing soil specimens with diameters as large as 300 mm (12 in.).

In general, it is believed that the sample size should be as large as practical in order to test the most representative specimen possible. The author typically tests specimens that are 75 to 150 mm (3 to 6 in.) in diameter but occasionally tests 300-mm-diameter specimens for special research applications. One such example of a special research application is given by Shan and Daniel (1989), who tested hydraulic contact between a geosynthetic clay liner and a defective geomembrane in a large, flexible-wall permeameter.

The ratio of length to diameter (L/D) varies considerably. Consolidation cell permeameters have the smallest L/D ratio (typically about 0.25). Many rigid-wall, compaction mold permeameters (e.g., Proctor-molds) have L/D approximately equal to unity. With rigid-wall, sampling-tube permeameters, the L/D ratio is typically 1 to 2.

There are no real limitations on the L/D ratio that relate to testing — selection is largely subjective. However, the flat, trimmed surfaces of test specimens are generally disturbed at least slightly, and may be considerably disturbed if stones have to be removed — one would not want to use too small a length because the slightly disturbed ends might dominate the hydraulic conductivity of a very thin test specimen. Decreasing the L/D ratio tends to accelerate testing times when chemicals or waste liquids are used as the permeant liquid because the time required to flush a prescribed number of pore volumes of liquid through a test specimen at a given hydraulic gradient decreases with decreasing L/D. Thus, thin (low L/D) test specimens can be advantageous when testing with chemicals and waste liquids. With flexible-wall permeameters, the larger the L/D, the greater is the difference in effective stress between influent and effluent ends of the test specimen, for a given hydraulic gradient. This can be important for compressible materials where one does not want the two ends of the specimen consolidated to significantly different void ratios. Thus, excessive L/D (i.e., L/D greater than 1 to 2) should be avoided when using flexible-wall permeameters for testing compressible materials. The author typically uses an L/D ratio of approximately unity whenever this is practical.

Comparison of Rigid- and Flexible-Wall Cells

The relative advantages of rigid and flexible-wall permeameters are discussed by Zimmie (1981) and Daniel et al. (1986). The main advantages and disadvantages are summarized in Table 1. The rigid-wall permeameter tends to be less expensive and easier to use. However, sidewall leakage is always possible, and the applied stress is not controlled with most types of rigid-wall permeameters. In principle, one could eliminate gas bubbles in a rigid-wall cell with backpressure. "Backpressure" means pressurization of both the influent and effluent liquid. Pressurizing the water in the test specimen reduces the volume of air present by compressing air bubbles and dissolving air into the pore water (backpressure dissolves air because the amount of air that can be dissolved in water increases linearly with pressure in accord with Henry's law). However, backpressure has not been found to work well in rigid-wall cells (Edil et al., 1985).

In the author's experience, sidewall leakage is not a problem for laboratory-compacted clay soils that are permeated with water (no backpressure) in a rigid-wall, compaction-mold permeameter. Swelling of the soil upon wetting seems to seal off potential sidewall leaks. In an attempt to saturate the soil as thoroughly as possible, the author takes the following steps when permeating laboratory-compacted soils in rigid-wall, compaction-mold permeameters: (1) the permeant water is thoroughly deaired so that the water is constantly dissolving and thereby removing entrapped air during permeation; (2) permeation continues until outflow and inflow rates are equal; and (3) permeation continues until hydraulic conductivity is steady. The latter two requirements usually force testing times of 1 to 8 weeks for soils having hydraulic conductivities of 1×10^{-9} m/s (1×10^{-7} cm/s) or less.

The flexible-wall permeameter is generally more versatile than the rigid-wall cell and virtually eliminates problems with sidewall leakage. Almost any soil (except granular materials, because of head losses in porous end pieces) can be permeated in a flexible-wall cell. Because the soil can be saturated fairly rapidly with backpressure in a flexible-wall cell, testing times tend to be less with flexible-wall cells. Importantly, saturation of the test specimen can be verified prior to permeation by measuring the B coefficient. The B coefficient is defined as follows:

$$B = \frac{\Delta u}{\Delta \sigma} \qquad (3)$$

where:

Δu = change in pore water pressure in the test specimen (kPa)

$\Delta \sigma$ = change in all-around confining pressure (kPa)

Table 1. Advantages and Disadvantages of Rigid- and Flexible-Wall Permeameters.

Type of Cell	Principal Advantages	Principal Disadvantages
Rigid-Wall Permeameter	• Simplicity of construction and operation of cell • Low cost of cell • Very large permeameters can be constructed fairly conveniently • Wide range of materials can be used (including chemically resistant materials) • Unrestrained vertical swelling can be allowed • Zero vertical stress can be applied, if desired	• Sidewall leakage is possible • No control over horizontal stress • If specimen shrinks, sidewall leakage is virtually guaranteed • Cannot confirm saturation via B-coefficient measurement • Cannot conveniently backpressure saturate the test specimen • Longer testing time for low-hydraulic conductivity material
Flexible-Wall Permeameter	• Can backpressure saturate the specimen • Can confirm saturation via B-coefficient measurement • Can control principal stresses • Sidewall leakage is highly unlikely, even for test specimens with rough sidewalls • Faster testing times for low-hydraulic-conductivity materials due to capability for rapid saturation via backpressure	• High equipment cost for cell • Requires 3 pressure positions (cell pressure, influent pressure, effluent pressure) • Problems with chemical compatibility of membrane with certain chemicals and waste liquids • More complicated operation of cell compared with rigid-wall cell • Difficult to perform test with extremely low compressive stress

For those not familiar with B-coefficient measurements, the following brief explanation is offered. To determine the B coefficient, the valves leading to or from the test specimen are closed to isolate the pore water in the specimen. The all-around cell pressure is changed (usually increased) by a pre-determined amount, $\Delta\sigma$. The corresponding change in pore water pressure, Δu, within the test specimen is measured with an electronic pressure transducer, and the B-coefficient is computed from Eq. 3. If the soil is completely saturated and no air is present, the pore water pressure will change by the same amount as the cell pressure and B will equal 1.0. However, for hard soils and rock, the B coefficient will be < 1.0 even if the test specimen is completely saturated (Skempton, 1954). In general, the more air present in the test specimen (or porous disks or drainage tubes), the lower will be the B coefficient.

The main problems with the flexible-wall permeameter are high equipment costs, complexity, problems with the integrity of the membrane when permeating test specimens with certain chemicals or waste liquids, and difficulty in testing at very low effective confining stress. Testing at low effective confining stress is a problem because enough confining stress must always be applied to press the membrane against the soil — it is difficult to permeate with less than about 14 kPa (2 psi) effective confining stress because sidewall leakage can develop. For testing "undisturbed" samples of soil obtained from the field and returned to the laboratory for permeation, it is the author's experience that the flexible-wall permeameter is the best type of permeameter to use in nearly all situations.

Up until the early 1980's, rigid-wall permeameters were the most commonly used type of permeameter. In the early 1980's, it was recognized that sidewall leakage may have a significant influence on tests involving clayey materials (and particularly if the clayey soil is permeated with shrinkage-producing organic liquids) that motivated development of the flexible-wall hydraulic conductivity test (ASTM D5084). Now, flexible-wall hydraulic conductivity tests are frequently specified for tests on clayey materials partly because there is an ASTM standard for flexible-wall testing (D5084) but no standard for rigid-wall testing, and also because backpressure saturation in a flexible-wall permeameter leads to shorter testing times compared to a rigid-wall permeameter. The rigid-wall cell is by far the most common type of cell to use for testing sands, gravels, and other highly-permeable granular materials (per ASTM D2434), and the rigid-wall cell is still frequently used by researchers who are studying water or chemical flow through laboratory-compacted, clayey soils.

Table 2 presents a summary of hydraulic conductivity cells used in over 50 published studies. The literature suggests a mix (roughly 50-50) of rigid- and flexible-wall permeameters. Several investigators used both types of cells.

Figure 10 presents a comparison of hydraulic conductivity determined in rigid-wall, compaction-mold permeameters to the hydraulic conductivity determined in flexible-wall permeameters. All specimens were permeated with water. Most of the tests used to prepare Figure 10 were performed by the author and his students. As mentioned earlier, when the rigid-wall permeameter is used, deaired water is employed to strip air from the specimen during permeation and the tests are continued for a sufficient period of time to ensure equal inflow and outflow as well as steady hydraulic conductivity. Under these circumstances, there is good agreement between the results of rigid-wall and flexible-wall permeameters.

HYDRAULIC CONTROL SYSTEMS

Hydraulic control systems are needed to control the delivery of fluid to a permeameter cell. Darcy's law (Eq. 1) relates the rate of flow to the hydraulic gradient. It may be assumed that in all hydraulic conductivity tests, the length (L) and cross-sectional area (A) of the test specimen are known. Thus, in order to determine the hydraulic conductivity, one must measure the flow rate (q) and head loss (ΔH) in

Table 2. Summary of Hydraulic Conductivity Studies Published Between 1980 and 1992.

Type of Soil	Reference	Rigid-Wall Permeameter	Flexible-Wall Permeameter
Native Clayey Soil	Silva et al. (1981)	X	
	Tavenas et al. (1983)	X	X
	Quigley et al. (1987)	X	
Slurry-Wall Backfill & Special Materials	D'Appolonia (1980)	X	
	Chapuis et al. (1984)		X
	Gale and Reardon (1984)		X
	Gill and Christopher (1985)	X	
	Bowders et al. (1987)	X	
	Jedele (1987)	X	
	Bodocsi and Bowers (1991)		X
	Edil et al. (1992)		X
	Estornell and Daniel (1992)		X
Compacted, Clayey Material	Garcia-Bengochea and Lovell (1981)	X	
	Daniel (1984)	X	
	Boynton and Daniel (1985)	X	X
	Day and Daniel (1985)	X	X
	Mundell and Bailey (1985)		X
	Juang and Holtz (1986)	X	
	Yong et al. (1986)	X	
	Brunnelle et al. (1987)	X	X
	Garlanger et al. (1987)		X
	Korfiatis et al. (1987)		X
	Pierce et al. (1987)	X	X
	Benson and Daniel (1990)	X	X
	Chapuis (1990)	X	X
	Daniel and Benson (1990)	X	
	Fernuik and Haug (1990)		X
	Johnson et al. (1990)	X	
	Shackelford and Javed (1991)	X	
	Chapuis et al. (1992)	X	X
	Kenney et al. (1992)	X	
	Zimmie et al. (1992)		X
Permeation with Wastes or Chemicals	Gordon and Forrest (1981)	X	
	Anderson (1982)	X	
	Crooks and Quigley (1984)	X	
	Dunn and Mitchell (1984)		X
	Acar et al. (1985)		X
	Alther et al. (1985)	X	
	Anderson et al. (1985)	X	
	Fernandez and Quigley (1985)	X	
	Peterson and Gee (1985)	X	
	Gipson (1985)	X	
	Lentz et al. (1985)		X
	Foreman and Daniel (1986)	X	X
	Pierce and Witter (1986)	X	
	Acar and D'Hollosy (1987)	X	X

Table 2. Summary of Hydraulic Conductivity Studies Published Between 1980 and 1992 (Continued).

Type of Soil	Reference	Rigid-Wall Permeameter	Flexible-Wall Permeameter
Permeation with Wastes or Chemicals	Bowders and Daniel (1987)	X	X
	Demetracopoulos and Dharmapal (1987)		X
	Ho and Pufahl (1987)	X	
	Daniel et al. (1988)	X	X
	Uppot and Stephenson (1989)		X
	Broderick and Daniel (1990)	X	
	Yanful et al. (1990)		X
	Fernandez and Quigley (1991)	X	
Granular Materials	Kenney et al. (1984)	X	
	LaFleur (1984)		X
	Scott (1984)	X	
	Poran and Faouzi (1989)	X	
	Chapuis et al. (1989a)	X	
	Chapuis et al. (1989b)	X	

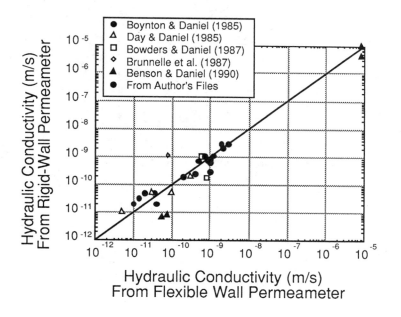

Figure 10. Comparison of Hydraulic Conductivity Measured on Laboratory-Compacted Soils Permeated with Water in Rigid-Wall, Compaction-Mold Permeameters and Flexible-Wall Permeameters.

order to determine hydraulic conductivity. Several approaches are possible:

- Constant-head test in which the head loss is kept constant and the corresponding rate of flow is measured.

- Varying-head test in which the head loss declines with time in a measured manner and the rate of flow is computed from the change in water level and the area of the tube in which the head falls.

- Constant-rate-of-flow test in which the rate of flow is kept constant and the corresponding head loss is measured.

Constant-Head Test

There are several ways of maintaining a constant head. In Fig. 11, an overflow drain in the influent reservoir maintains a constant water level in the influent reservoir, provided the supply line feeds enough water to the reservoir to supply the permeameter and drain.

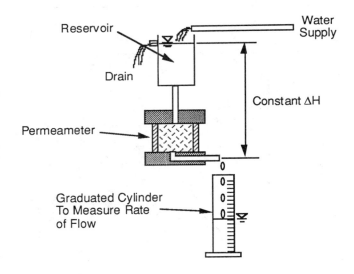

Figure 11. Constant Head Maintained with Overflow Drain.

Alternatively, one can simply use a very large reservoir of influent liquid (Fig. 12) such that the water level in the reservoir does not change significantly during the permeation stage. A small change in head occurs, but if the reservoir is sufficiently large, the change is trivial and can be ignored.

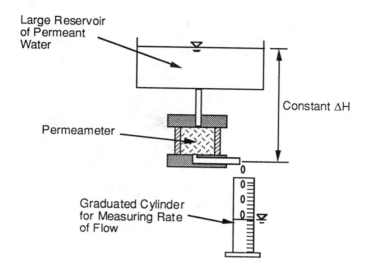

Figure 12. Constant Head Maintained with a Large Influent Reservoir.

Another method for maintaining a constant head is the Marriotte tube (Fig. 13), which has also been called a "bubble tube." The Marriotte tube works as follows. A small-diameter tube is inserted into the reservoir, and the annulus is sealed, e.g., with a rubber stopper. As water begins to flow out of the reservoir, air is drawn into the small-diameter tube. When the small-diameter tube is drained of water, air begins to bubble into the reservoir. The head at the base of the small diameter tube must be equal to the applied air pressure (atmospheric pressure if the small-diameter tube is not pressurized). As long as air is bubbling into the reservoir, the Marriotte device maintains a constant head. However, there is a small fluctuation in head of a few millimeters during the formation and release of air bubbles. This small fluctuation in head is rarely important in laboratory hydraulic conductivity testing but can be important in other types of applications, e.g., field infiltration testing.

The main advantages of a constant head test are: (1) the simplicity of calculation of hydraulic conductivity, and (2) the maintenance of constant water pressure in the test specimen. If the water pressure in the test specimen varies, air bubbles can change volume and the test specimen itself can change volume in certain types of permeameters.

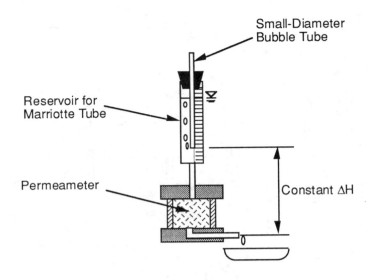

Figure 13. Constant Head Maintained with Marriotte Device.

Variable-Head Test

A variable-head test is one in which the water level in the influent or effluent reservoir (or both) changes during permeation. Two types of variable-head tests are possible: (1) a falling-headwater, constant-tailwater-pressure test (e.g., Fig. 14), and (2) a falling-headwater, rising-tailwater-pressure test (Fig. 15). The test with constant tailwater pressure tends to be more convenient for testing soils with hydraulic conductivities greater than about 1×10^{-5} m/s while the rising-tailwater-pressure test tends to be more convenient for testing under backpressure or testing materials with hydraulic conductivities less than about 1×10^{-9} m/s.

As discussed by Daniel (1989), the equations that are used for computing hydraulic conductivity are different for the conditions of a constant tailwater pressure and a rising tailwater pressure. If the influent and effluent standpipes in a rising-tailwater-pressure test have the same area, then the equations differ by a factor of 2. One must be careful to use the proper equation.

The primary advantage of a variable-head test is that equipment is simpler than for a constant head test. A slight disadvantage of the variable-head test is that the equations for computing hydraulic conductivity are more complicated. However, three potentially more significant limitations of the variable-head test are: (1) as the head falls, the pressure drops and any gas bubbles in the test specimen expand; (2) as the pressure drops, the amount of dissolved gas that can be held by the liquid decreases, which could cause release of dissolved gas from the permeant liquid and formation of air bubbles if the

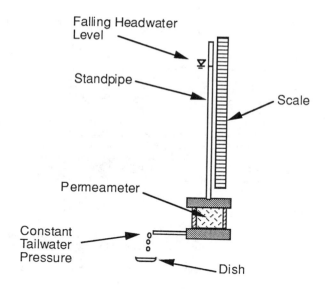

Figure 14. Variable-Head Test with Constant Tailwater Pressure.

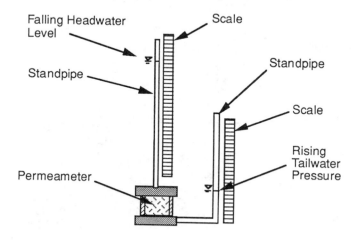

Figure 15. Variable-Head Test with Rising Tailwater Pressure.

permeant liquid is saturated with dissolved gas at the initial (maximum) pressure; and (3) in flexible-wall cells for which a constant total stress is maintained, a decline in pore water pressure causes an increase in effective stress — an increase in effective stress causes consolidation and a reduction in void ratio. The problem of consolidation can be significant for highly compressible materials. It is for this reason that ASTM D5084 limits the reduction in head loss during a variable-head test to no more than 25% of the initial head loss. The problem of gas bubbles expanding when pressure drops is eliminated if the test specimen is saturated via application of backpressure, and the problem of release of gas from solution is solved by using a permeant liquid that is not saturated with gas at the higher (influent) pressure. Also, if there are small, known head losses in tubes, valves, and fittings, these can be taken into account in constant-head tests, but not variable head tests since the head loss varies with head.

Constant-Rate-of-Flow Test

The constant-rate-of-flow test is performed by pumping permeant liquid through the test specimen at a controlled rate and measuring the pressure drop across the test specimen (Fig. 16). In the simplest arrangement, a differential electronic pressure transducer hooked up to a strip-chart recorder (Fig. 16). Once the flow rate and pressure drop become steady, the test is complete (unless a chemical or waste liquid is being used, in which case permeation continues until other termination criteria are met, as discussed later).

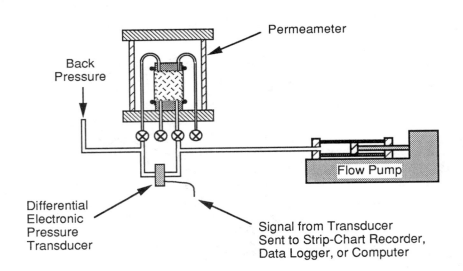

Figure 16. Constant-Rate-of-Flow Test.

One must be careful when imposing the flow rate to be sure that the resulting pressure drop across the test specimen is well below the net confining pressure to ensure good contact between the confining membrane and test specimen — this precaution can be programmed in the computer. More details about the test are provided by Olsen (this proceedings) and the references therein.

The constant-rate-of-flow test can be fully automated with computer control of all pressures and flow rates. A schematic diagram of a computer-controlled system is shown in Fig. 17.

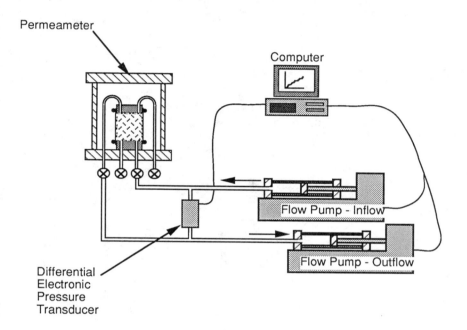

Figure 17. Computer Automated Constant-Rate-of-Flow Test (Original Drawing Courtesy of Trautwein Soil Testing Equipment Co.)

The advantages of the constant-rate-of-flow test are: (1) one can achieve equilibration very rapidly if the test specimen is saturated with water and thus minimize testing time; (2) the entire test, including control of the test, is easily automated; and (3) when the permeant liquid is a chemical or waste liquid, one can set the flow pump to deliver a known amount of liquid in a given amount of time, e.g., 2 pore volumes of liquid in 4 weeks, which can be helpful in making sure that sufficient throughput of liquid occurs within the time constraints of a project. Disadvantages of the constant-rate-of-flow test are high

equipment cost and the possibility of developing extremely large
hydraulic gradients if too large a flow rate is used.

Hydraulic Control for Constant Volume

An alternative hydraulic control system, which was developed by
Richard Ladd, based on work done at the Norwegian Geotechnical
Institute, is shown in Fig. 18. With this system, permeation is
effected with a falling headwater and rising tailwater, but the system
is designed to maintain a constant volume in the test specimen. The
closed manometer ensures that inflow and outflow are equal, which in
turn ensures that a saturated test specimen does not change volume. If
the soil does not change volume when permeation is initiated,
equilibrium is reached sooner and testing time is minimized. This type
of hydraulic control system is particularly advantageous for testing
low-hydraulic-conductivity, compacted soil liner materials. More
information about this method of hydraulic control is provided by Ladd
et al. in this proceedings.

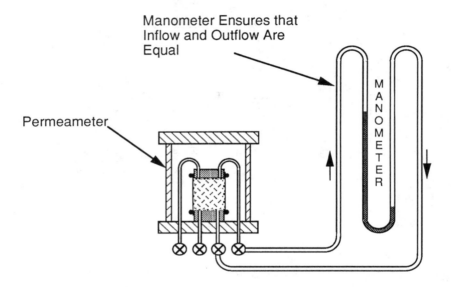

Figure 18. Hydraulic System with Closed Manometer to Maintain Constant
            Volume in the Test Specimen.

Interface between Pressure Panels and Permeameter

Most commercial hydraulic systems (except for testing granular
soils) involve use of pneumatic pressure panels. Panels are filled with
water and are pressurized with air. When a hydraulic conductivity test
is to be performed with a chemical or waste liquid, it is convenient to

provide an interface between the permeant liquid and the water in the panel board to avoid contaminating the panel board.

The interface that is commonly used is a bladder accumulator (Fig. 19). The accumulator consists of a flexible diaphragm ("bladder") contained within a chamber. Water is housed on one side of the bladder and the permeant liquid on the other. Care must be taken to make sure that the bladder and all materials on the permeameter side of the bladder are chemically compatible with the permeant liquid. Materials usually consist of stainless steel and Teflon®.

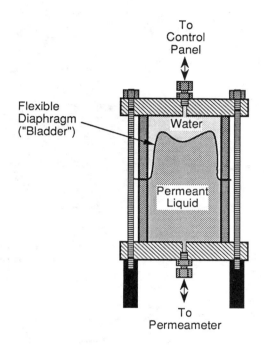

Figure 19. Bladder Accumulator.

Flexible polymer diaphragms are not completely impermeable to all organic materials. The diaphragm may absorb and allow a small amount of organic chemical to pass through it. Frequent checks of concentration of key chemicals in the permeant liquid are required to ensure that undesirable losses do not change the composition of the permeant liquid.

A sketch of a typical panel board set-up with a bladder accumulator is shown in Fig. 20. The pressure panel consists of three positions to control cell pressure, influent pressure, and effluent pressure in a flexible-wall hydraulic conductivity cell.

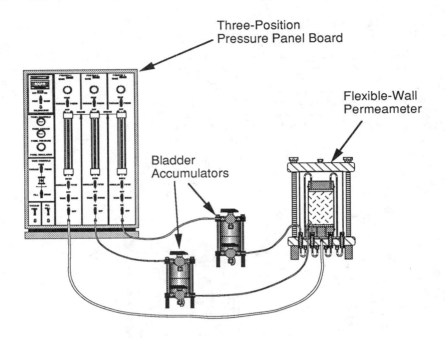

Figure 20. Pressure Panel, Bladder Accumulators, and Flexible Wall Permeameter (Courtesy of Trautwein Soil Testing Equipment Co.).

Special Requirements for Permeation of Granular Soils

The permeation of granular materials presents significant challenges. In many ways permeation of a material that has an extremely high hydraulic conductivity (e.g., > 0.01 m/s or 1 cm/s) is more difficult than permeation of a low-hydraulic-conductivity soil. Problems with permeation of granular soils include:

- Logistics of delivering large flow quantities,
- Logistics of deairing a large quantity of water,
- Head losses in tubes, valves, and fittings,
- Accurate measurement of the flow rate,

- Accurate measurement of the head loss across the test specimen,
- Deviations from Darcy's law due to turbulent flow,
- Problems with large pores at contact between particles of granular material and rigid confining walls of the permeameter.

Specially-designed devices are sometimes constructed for pumping large amounts of water through granular test specimens. Overflow systems (Fig. 11) tend to be the most convenient for maintaining a constant head with large flow quantities. Deairing the permeant water, however, can be very difficult because of the large quantities of water needed for the test. Some people use water straight out of the tap and simply ignore the issue of deairing. This may be a reasonable approach because (1) the hydraulic conductivity cannot be determined with great accuracy for granular materials, and a small amount of air in the test specimen is simply one of many limitations in determination of hydraulic conductivity, and (2) deairing a large quantity of water can be expensive. Large amounts of water can be deaired by storing the water under vacuum in a very large tank or, to some extent, by percolating the water through a sand bed. The problem with using water straight out of the tap is that the water in the water supply line is at high pressure and contains a large amount of dissolved air. When the water is depressurized, air bubbles slowly evolve as air comes out of solution. It is much better to use water that has been stored at atmospheric pressure than water out of the tap since there is a far lesser tendency to generate gas with water at atmospheric pressure.

Head losses in tubes, valves, and fittings are minimized by using very large tubes and valves. Tubing as large as 100 mm (4 in.) may be necessary, depending on the size of the test specimen. It is essential that piezometers be used to measure the head loss over a known distance for accurate determination of hydraulic gradient (Fig. 4) whenever head losses in tubes, valves, fittings, or other components of the permeameter might be significant. The rate of flow is often measured by collecting effluent liquid in a very large container and measuring the amount collected over a known period of time.

An important problem with granular materials is that the high flow rates needed to generate a measurable head loss often produce turbulent flow. Darcy's law is not valid for turbulent flow (Fig. 21). The best way to ensure that turbulent flow is not affecting results is to determine the hydraulic conductivity at three different hydraulic gradients and to ensure that approximately the same value is determined at each gradient.

PERMEANT LIQUID

Hydraulic conductivity tests are usually performed either with water or with a chemical or waste liquid as the permeant liquid. Chemicals require some special considerations in terms of equipment design.

<u>Water</u>

The most important characteristics of the permeant water are the amount of dissolved air in the water, the type and concentration of electrolytes, turbidity, nutrient content, and population of microorganisms. It is best to permeate soils with deaired water. "Deaired water" means water whose dissolved air has been substantially removed. A dissolved oxygen meter may be used to confirm proper deairing. At atmospheric pressure, water will contain approximately 8 mg/L of dissolved oxygen. Properly deaired water will contain less than 1 to 2 mg/L of dissolved oxygen. Water is deaired by boiling the water or by placing the water in a container that is subjected to vacuum. With a vacuum chamber, it is best to spray the water into the evacuated chamber in a fine mist (e.g., with a spray paint nozzle) to expose a large surface area of the water to vacuum and to facilitate rapid deairing. If the permeant water is boiled, care must be taken not to increase the salinity too much.

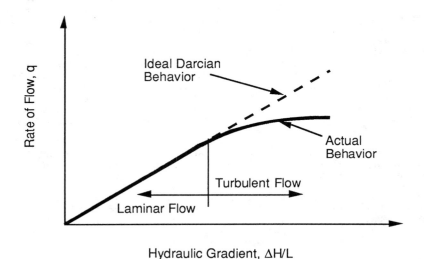

Figure 21. Effect of Turbulent Flow.

Electrolytes can influence the hydraulic conductivity of clayey soils. Most clay minerals are negatively-charged, plate-shaped particles. The soil water and cations (positively charged ions) in the soil water are attracted to the surfaces of the clay particles. The forces of attraction are so strong that the water and cations immediately adjacent to the clay particles are said to be "adsorbed" to the particle and to form what is known as a "diffuse double layer" or "electrical double layer" around the soil particle. The adsorption is of sufficient strength that both the soil particles and the double layer serve to block flow paths. As indicated by Fig. 22, the thicker the double layer, the more narrow and tortuous the flow paths through the soil and, hence, the lower the hydraulic conductivity.

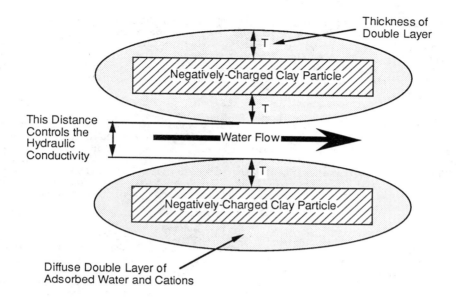

Figure 22. Diffuse Double Layer and Its Effect on Hydraulic Conductivity.

The Gouy-Chapman equation (Mitchell, 1976) may be used as a guide to the probable effects of different types of liquids on the thickness of the double layers that surround clay particles. The Gouy-Chapman equation states the following proportionality:

$$T \propto \sqrt{\frac{D}{n_o v^2}} \quad (3)$$

where

- T = thickness of electrical double layer surrounding soil particle
- D = dielectric constant of pore liquid
- n_o = electrolyte concentration (i.e., salinity)
- v = cation valence.

As suggested by Fig. 22, any increase in double layer thickness produces a decrease in hydraulic conductivity. An increase in electrolyte concentration or an increase in cation valence will tend to decrease hydraulic conductivity. Clayey soils permeated with liquids containing monovalent cations (e.g., Na^+ and K^+) tend to produce low hydraulic conductivity while those permeated with liquids containing polyvalent cations (e.g., Ca^{++} and Mg^{++}) tend to produce higher hydraulic conductivity. Strongly saline solutions (i.e., high n_o) increase the hydraulic conductivity of clayey soils. Permeation with distilled water tends to produce a very low hydraulic conductivity because $n_o = 0$.

These factors should be considered when one selects a permeant water for testing of clayey soils. Some researchers have used 0.01 N $CaSO_4$ as the permeant water (e.g., Brown and Anderson, 1983), but many laboratory technicians have resisted using standard chemical solutions because of the time and trouble of making up stock solutions. Tap water is a commonly used permeant water and is the one recommended in ASTM D5084. Except for special research applications, distilled water should never be used with clayey soils because distilled water leaches electrolytes from the soil water, which expands double layers and reduces hydraulic conductivity.

Turbidity, nutrient content, and population of microorganisms can also affect hydraulic conductivity. Turbid water is not used in hydraulic conductivity testing, unless one specifically wants to examine the effects of permeation with turbid water. Nutrients can promote growth of microorganisms, which tends to reduce hydraulic conductivity. Olson and Daniel (1981) present data on the effects of microorganisms on the hydraulic conductivity of soils. It is the author's experience in long-term hydraulic conductivity tests that the hydraulic conductivity of soils almost always decreases with time (often about one-quarter to one-half an order of magnitude per year) — growth of pore-blocking microorganisms is probably the cause for this decline. Growth of microorganisms is discouraged by use of fresh, chlorinated tap water.

Chemicals and Waste Liquids

Permeation with chemicals or waste liquids presents a number of additional challenges, which include health and safety considerations, material compatibility concerns, cross-contamination potential from one test to the next, changes in chemistry of the influent liquid, and monitoring of the chemistry of the effluent liquid.

Health and safety considerations often require one to set up the entire hydraulic conductivity system in an environmentally controlled room with separate air conditioning, constant replenishment of air, and appropriate emergency and first-aid equipment on hand. In the author's university laboratories, where a specially-equipped environmental laboratory is not available for hydraulic conductivity testing, chemical hoods in his laboratory are used to withdraw and circulate air constantly. The permeameters themselves are set up inside the hoods when necessary.

Material compatibility issues are usually addressed by using stainless steel and Teflon® for those components that will come into contact with the permeant liquid. Bladder accumulators (Fig. 19) are helpful in limiting the number of pieces of equipment that will contact the permeant liquid. Cross-contamination must be dealt with by thoroughly cleaning the equipment after each use.

It is often necessary to sample the effluent liquid for chemical analysis. The methodology and equipment depend upon the chemical constituents and type of permeameter. In a permeameter with no backpressure, the effluent liquid is discharged directly into a graduated cylinder or other device. The cylinder is covered with a thin film to minimize evaporation. Sometimes, the effluent collector is placed inside a glass beaker, which is partly filled with water and also covered with a thin film. The outer reservoir of water in the beaker helps to maintain a humid atmosphere around the effluent collector and to reduce slightly evaporative losses. In the author's experience, evaporative losses in carefully-sealed graduated cylinders are a fraction of a milliliter per week.

If volatile organic constituents are present in the permeant liquid or soil specimen, an open effluent collection system is unacceptable for chemical analysis because the volatiles will be lost due to evaporation. One approach (e.g., used by Brown and Anderson, 1983) is to refrigerate the effluent collectors to reduce the vapor pressure of the volatile organic compound. However, the author uses another scheme. A stainless steel "tee" connector is mounted to the valve on the permeameter that opens and closes the effluent line. The tee is fitted with a septum that is held in place with a drilled-out solid ferrule. A hypodermic syringe with needle is used to puncture the septum and extract a sample of liquid from the effluent line. The septum will not leak for at least 30 to 40 punctures. The same syringe is usually used to inject the effluent sample into a device for analysis of water chemistry, e.g., an ion chromatograph or gas chromatograph/mass spectrometer.

Bladder accumulators work well for effluent collectors when backpressure is used. The bladder accumulator is periodically drained to collect fluid for analysis. For volatile organics, the author usually uses the "tee" connector described above, even when a bladder accumulator is used, to minimize the possibility of losses during sampling.

Several other potential problem areas warrant mentioning:

- Precise analysis of very low levels (a few parts per million or less) of chemical contaminants is extremely difficult in permeameters. Adsorption of chemicals by components of the permeameter (including the latex membrane), volatilization, losses through tubes and fittings, and chemical adsorption by the soil make it extremely difficult to achieve a good mass balance. One would be well advised not to attempt careful tracking of chemicals at low levels of concentration unless one is prepared to go to elaborate ends to limit and account for sources and sinks.

- Effluent sampling points should be as close to the permeameter as possible — mixing occurs in the effluent tube.

- Chemical or biological degradation and alteration can occur in the influent liquid. Oxygen is often one of the main causes for these changes. One should decide at the outset whether the influent liquid will be subject to aerobic or anaerobic conditions. If anaerobic conditions are desired, nitrogen may be used to pressurize the panel system rather than air.

- It is good practice to analyze the key chemical parameters in the influent liquid on a regular basis — concentrations may change due to unforeseen chemical reactions or losses.

- The reservoir of influent liquid should be replaced periodically. Changing the liquid every one to two weeks is common practice in the author's laboratories when permeant liquids that can change concentration, e.g., due to volatilization of organics, are used.

- Many leachates tend to be rich in microorganisms. Problems have occasionally been encountered with microorganisms blocking the tubing leading from the reservoir of influent liquid to the permeameter. The influent and effluent lines should be periodically flushed or replaced, if necessary.

- Permeation with liquids that tend to increase hydraulic conductivity presents a challenge in flow control. As indicated by Fig. 23, if the liquid causes an increase in hydraulic conductivity, the increase is not seen immediately. Because the least permeable section of the test specimen controls the flow rate, the overall hydraulic conductivity does not increase until the hydraulic-conductivity-increasing permeant liquid makes its way all the way through the test specimen. The result is that flow rates can be extremely low (for fine-grained soils) for days or weeks and then rapidly increase in a matter of a few hours. Often no one is watching the test when the large increase occurs. The problem is that the reservoir of permeant liquid can easily empty when the hydraulic conductivity rapidly increases. If the influent reservoir contains pressurized air in contact with the permeant liquid, air can permeate the specimen when the reservoir of liquid is emptied, unless a mechanism is provided to automatically "shut off" or refill the reservoir of permeant liquid. The bladder accumulator is convenient for these situations because once the permeant liquid is emptied from the accumulator, no further flow occurs. Technicians can refill the bladder accumulator and continue with the test without damaging the test specimen.

Preparation of Test Specimens

So far, the focus of discussion has been on the mechanics of permeation. In fact, however, the procedures used to select or prepare a test specimen are far more important than the permeation details. Opportunities abound to alter the hydraulic conductivity of a test specimen by many orders of magnitude (particularly for fine-grained materials) simply by changing the sample selection or preparation procedure.

<u>Undisturbed Test Specimens</u>

Undisturbed test specimens are prepared from thin-walled tube samples. Care must be taken not to damage the sample during storage or transport. The main problems with preparation of undisturbed specimens are the presence of stones, trimming the ends, selection of the material for testing, and determination of specimen size.

Rocks and stones are often encountered in soils that are to be permeated. In a flexible-wall permeameter, the stones should be left in place along the side walls, if at all possible. Gouges left from stones can be filled or left open, depending on the size of the gouge. Stones can have an extremely damaging effect on the soil if the tube causes shearing or "rolling" of stones during the sampling process. The structure and hydraulic conductivity of the surrounding soil can be completely altered if stones are displaced. In the author's experience, soils that contain significant quantities of large pieces of gravel (greater than about 12 mm or 1/2 in.) are practically impossible to sample with conventional 75-mm (3-in.) diameter, thin-walled sampling tubes. Often the testing laboratory will simply discard samples until a zone of material that is relatively free of gravel is found. On some projects involving soil liner construction for landfills, 5 to 6 "undisturbed" samples have been obtained for each sample that was finally selected for testing. This process obviously introduces tremendous bias in the sampling process because the samples with the least amount of gravel are the only ones that can be made into test specimens. The author suggests avoiding laboratory hydraulic conductivity tests on undisturbed samples of soils that contain stones. If tests must be performed, larger sampling tubes (e.g., 125 mm or 5 in.) or undisturbed blocks may be helpful in minimizing damage to the interior of the sample.

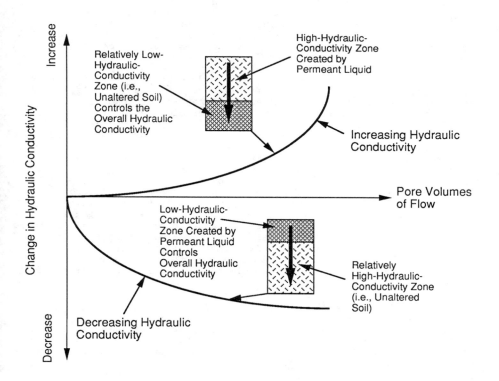

Figure 23. Change in Hydraulic Conductivity Versus Pore Volumes of Flow.

Once a cylindrical soil sample is obtained, the ends must be trimmed. When technicians trim cylindrical test specimens for strength tests, emphasis is often placed on obtaining perfectly flat ends. This same approach leads technicians to trowel the ends of the specimen and to fill in any cavities or voids with soil. It is recommended that technicians "cut" rather than "trowel" the ends of test specimens and minimize filling of voids. Soil should not be smeared across the flat surfaces of the test specimen. The ends may be lightly roughened or scarified (e.g., with a toothbrush) to ensure that smeared soil does not block preferential flow paths through the soil.

The selection of the actual material for testing is, unfortunately, not given as much thought and care as it should be. For example, on one project, a 600-mm (2-ft) thick compacted soil liner was to be sampled, and within this sample, a test specimen (nominally 100 mm in length) was to be trimmed for compliance testing. The soil was required to have a hydraulic conductivity $\leq 1 \times 10^{-9}$ m/s. The technicians at the laboratory decided to take a conservative approach and to select what appeared to be the most permeable portion of the 600-mm-long sample, but the contractor believed that this process was biased and unfair and felt that a "representative" segment should be tested instead. The process for selection of the actual test specimen was not clearly stated at the beginning of the project, and the dispute was never resolved to the satisfaction of all parties. The problem could have been avoided by stating the criteria for selection of the actual test specimen from a larger "undisturbed" sample at the beginning of the project.

In any undisturbed sample, the hydraulic conductivity is likely to vary, just as the soil itself is likely to vary. The scientists and/or engineers responsible for interpreting the hydraulic conductivity data should take an active role in determining specimen-selection criteria.

Specimen size was discussed earlier in terms of the mechanics of hydraulic conductivity testing. The larger the specimen, the more representative the specimen is likely to be. For soils with secondary structure, hydraulic conductivity tends to increase with increasing specimen size (Fig. 24). Thus, the largest practical specimen size should be selected. For undisturbed samples, conventional sampling equipment yields specimens that are approximately 75 mm (3 in.) in diameter, although equipment that will obtain 125-mm (5-in.) diameter samples is readily available and some practitioners have obtained 300-mm (12-in.) diameter samples for hydraulic conductivity testing. For laboratory-compacted materials, the diameter of the test specimen is often 100 mm (4 in.) to 150 mm (6 in.). The following simple rule is suggested: use the largest test specimen that is practical for any given project.

Compacted Test Specimens

Hydraulic conductivity testing of compacted specimens is of particular interest for low-hydraulic-conductivity compacted soil liners, e.g., for landfills. Typically, the liner is required to have a hydraulic conductivity that is less than or equal to a specified value, e.g., $\leq 1 \times 10^{-9}$ m/s. To qualify a potential borrow material for a soil liner, hydraulic conductivity tests are performed on laboratory-compacted specimens.

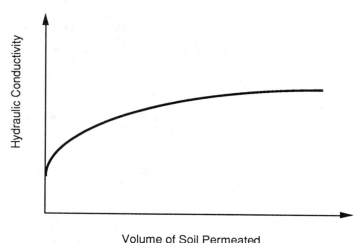

Figure 24. Effect of Sample Size on Hydraulic Conductivity for Soils that Contain Secondary Features (e.g., macropores, fissures, or slickensides).

There is no standard procedure for preparing laboratory-compacted specimens for hydraulic conductivity testing. Often, the problem is as follows. A specification writer has established a compaction criterion, e.g., the soil must be compacted to a minimum dry unit weight (γ_d) equal to 95% of the maximum dry unit weight ($\gamma_{d,max}$) obtained by standard compaction via ASTM D698, "Laboratory Compaction Characteristics of Soils Using Modified Effort (12,400 ft-lbf/ft^3 (600 kN-m/m^3))," and the water content (w) must be 0 to 4 percentage points wet of optimum water content (w_{opt}). This requirement is sketched in Fig. 25. The most critical point in terms of hydraulic conductivity is shown in Fig. 25. Soil compacted at this point (i.e., lowest acceptable dry density and lowest acceptable water content) is likely to have the greatest hydraulic conductivity of all specimens compacted within the acceptable range of water content and dry unit weight (Daniel and Benson, 1990). At the lowest water content, clods of clay are difficult to remold, and lowest dry density, the compactive energy is least likely to be large enough to remold the clods of clay into a uniform mass of low hydraulic conductivity. Given the fact that the point shown in Fig. 25 is clearly the most critical point, the key question is: how does one prepare a specimen at the desired water content and dry unit weight?

Several procedures are used. One common technique is to moisten the soil to the desired water content and weigh out the proper amount of soil to fit into a mold of known volume to produce the desired dry unit weight. The soil is compacted into the mold in lifts until the predetermined amount of soil fits exactly into the mold. A hammer or tamping rod is commonly used to pack the soil into the mold. A certain degree of art and finesse is needed to get the predetermined amount of soil to fit precisely into a given volume and to compact the soil reasonably uniformly. The problem with this technique is that there is no control over compactive energy. The energy of compaction can have a major influence on the hydraulic conductivity, independently of dry unit weight, for a given water content (Mitchell, Hooper, and Campanella, 1965; and Daniel and Benson, 1990). In some laboratories, the force applied to the compaction rod is measured and documented, which is far better than no control or documentation on compaction energy.

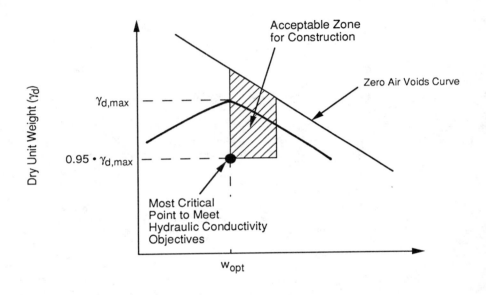

Figure 25. Typical Compaction Requirements for Soil Liners.

An alternative method, and the one used by the author, is the following. Enough soil is mixed to the desired water content to make several compacted specimens. Soil is compacted into the mold using a standard compaction procedure (e.g., ASTM D698), but with the energy of compaction modified. For instance, in the first trial, one might use 15 drops of the ram per lift rather than the usual 25. The dry unit weight is determined and compared with the target value. If the desired value is not achieved, a new specimen is prepared with the number of drops of the ram adjusted to increase or decrease the dry unit weight, as desired. The process is repeated until the desired dry unit weight is achieved. The advantage of this procedure is that a consistent and easily documented procedure is used, and the compactive energy is known and can be reported.

Other Laboratory-Prepared Test Specimens

A variety of other techniques are used for preparing laboratory test specimens. Slurry-like materials may be poured into a mold; a rod is sometimes used to make sure that the specimen uniformly fills the mold. Soil-cement specimens may be prepared using techniques that are standard for these types of materials. Other materials are prepared with various methods that should be prescribed by the individual requesting the test.

Termination Criteria

"Termination criteria" refers to criteria for terminating a hydraulic conductivity test. The criteria tend to be somewhat subjective due to the extreme variability in hydraulic conductivity (conductivity can vary over 10 orders of magnitude in soil) and the different requirements for testing such a variable material. Also, different criteria apply to tests with water and with chemicals or waste liquids.

Permeation with Water

When soils are permeated with water, the following criteria are recommended for determining when a test is complete:

- The rates of inflow and outflow should be reasonably equal. With unsaturated soils, the initial rate of inflow is always greater than the initial rate of outflow. For soils that are consolidating, the rate of outflow may exceed the rate of inflow. Eventually, equilibrium will be reached, and the rates will equilibrate. A good target value is for the ratio of outflow to inflow (q_{out}/q_{in}) to be between 0.9 and 1.1 before a test is terminated. However, with some extremely low-hydraulic-conductivity materials (i.e., hydraulic conductivity < 1×10^{-10} m/s), this goal may be very hard to achieve with testing times less than several weeks, and a ratio of q_{out}/q_{in} between 0.75 and 1.25 may be more appropriate for these types of materials.

- The hydraulic conductivity should be reasonably steady. It is good practice to plot hydraulic conductivity versus time or pore volumes of flow. (One pore volume of flow has occurred when the cumulative quantity of inflow equals the computed volume of void space in the soil.) When there is a minimal tendency for increase or decrease in hydraulic conductivity, the hydraulic conductivity is reasonably steady.

- Enough data points should be collected to ensure that the test has lasted long enough to achieve representative results. ASTM D5084 requires that 4 hydraulic conductivity determinations be made and that the highest and lowest values not be too far from the mean of the measurements (the actual acceptable range depends on the hydraulic conductivity). This type of criterion is recommended.

Permeation with Chemicals

When soils are permeated with chemicals, the same requirements for water apply, but some additional controls should be established. Because most permeant liquids are aqueous-based liquids, the density and viscosity are very similar to water, and the results are reported in terms of hydraulic conductivity. However, for non-aqueous-phase liquids, e.g., petroleum hydrocarbons, where the density or viscosity of the permeant liquid is significantly different from that of water, it may be advantageous to report intrinsic permeability rather than hydraulic conductivity.

The additional requirements that are recommended for tests with chemicals or waste liquids include the following:

- Permeation should continue until at least 2 pore volumes of flow has occurred. This will ensure that the remnant soil water has been flushed out of the test specimen.

- Permeation should continue until the chemical composition of the effluent liquid is similar to that of the influent liquid. For instance, if the permeant liquid is a strong acid, permeation should continue until the pH of the effluent liquid has dropped to a value comparable to that of the influent liquid. Permeation should continue until the key compounds that might alter hydraulic conductivity appear in the effluent liquid in concentrations that are similar to the influent liquid. However, sometimes chemical alterations occur (e.g., biotransformations), and the effluent liquid will never have the same chemistry as the influent liquid.

- A graph such as shown in Fig. 26 is recommended to illustrate the hydraulic conductivity trend and breakthrough of key ions. More than one ion may be of interest; the concentration of all critical ions in the effluent liquid should be plotted.

Other Testing Variables

Several testing variables, such as specimen size and permeant water, have been discussed. Other critical variables are reviewed in this section.

Effective Stress

The effective stress to which a soil is subjected affects the hydraulic conductivity. Highly compressible soils or soils containing secondary features such as macropores, fractures, joints, and slickensides are the most sensitive to changes in effective stress. In all cases, increasing the compressive stress reduces porosity and reduces hydraulic conductivity.

The best practice to follow is to always subject the test specimen to a compressive stress that is representative of the situation being modeled in the field. For low-hydraulic-conductivity materials (e.g., used for liner), care should be taken not to consolidate the test specimen to excessively large effective stresses because the measured hydraulic conductivity will be too low. Conversely, for granular soils (e.g., used for drainage layers), care should be taken not to consolidate the specimen to a compressive stress that is too low because the hydraulic conductivity will be overestimated.

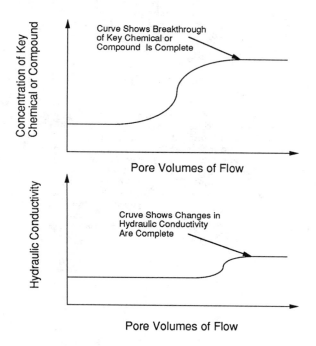

Figure 26. Recommended Presentation of Data for Hydraulic Conductivity Tests with Chemicals or Waste Liquids.

Backpressure

For flexible-wall permeameters, the level of backpressure may be selected based upon the initial degree of saturation (higher backpressure for lower initial degree of saturation) or the desired time to complete the saturation process (higher backpressure for faster saturation). In principle, the level of backpressure should not influence hydraulic conductivity — the effective confining stress and degree of saturation should control hydraulic conductivity. However, if the specimen is backpressured too quickly, the specimen is subjected to consolidation stresses during the backpressure process. There are no widely-accepted rules on how quickly to backpressure-saturate soils for hydraulic conductivity testing, but one should be aware that the rate of backpressure saturation is much more important for specimens consolidated to low effective confining stress than for specimens consolidated to high stress because consolidation to high stress masks any consolidation effects from backpressuring. In at least one commercial laboratory, the rate of backpressure application is related to the rate of water flow into the test specimen during application of backpressure; the faster the rate of water flow into the specimen, the smaller the increment of backpressure and the longer the time interval between increments. The author usually tries to increase the backpressure in 35 kPa (5 psi) increments at intervals of 1 to 4 hours. In commercial laboratories, however, where turnaround time is more critical, backpressure is usually applied more quickly.

Hydraulic Gradient

Application of an excessively large hydraulic gradient can cause several alterations in the soil:

- By necessity, the effective stress at the effluent end of the test specimen is always larger than that at the influent end, and the larger the hydraulic gradient, the larger this difference. Since the application of increasing effective stress tends to reduce void ratio and hydraulic conductivity, there is a tendency for the test specimen to have a lower hydraulic conductivity at the effluent end compared to the influent end. Thus, the larger the hydraulic gradient, the larger is the difference between hydraulic conductivity at the influent and effluent ends of the test specimen. The significance of greater effective stress at the effluent end (compared to influent end) is much greater for compressible soils than for relatively incompressible soils.

- Application of a high hydraulic gradient tends to wash fine soil particles downstream in the test specimen. Two things can happen: (1) the fine particles may become trapped in the soil near the downstream end of the test specimen (i.e., due to natural filtration), and the hydraulic conductivity of the downstream end of the test specimen is accordingly reduced; or (2) the fine particles may be washed completely out of granular soils, which would cause the hydraulic conductivity to increase.

In general one should try to use a hydraulic gradient as close as possible to the value expected in the field but in any case not to use an excessively large gradient. ASTM D5084 recommends limits on hydraulic gradient for testing low-hydraulic-conductivity soil. However, allowances for higher hydraulic gradients may have to be made for compatibility testing involving chemicals or leachate, where a minimum number of pore volumes of flow are required and where the only practical way of achieving this in a realistic time is to use elevated hydraulic gradient.

Temperature

The hydraulic conductivity of soil varies with temperature because the density and viscosity of the permeant liquid varies with temperature. With water, hydraulic conductivity changes approximately 3% for every 1°C change in temperature (Olson and Daniel, 1981). Corrections for temperature effects are easily made. Temperature fluctuations can be a major problem when testing extremely materials having an extremely low hydraulic conuctivity because expansion and contraction of the water in the hydraulic system can exceed flow through the test specimen. Temperature isolation and control may be necessary in some extreme cases.

CONCLUSIONS

The purpose of this paper was to provide a general review of the state of the art for laboratory hydraulic conductivity testing of soil. The types of cells that are used can be divided into rigid-wall and flexible-wall permeameters. Rigid-wall cells are preferred for granular materials, and either rigid- or flexible-wall cells are preferred for low-hydraulic conductivity materials, depending on the type of test specimen and the conditions being simulated. Procedures for selection and preparation of test specimens are critically important, even for laboratory-compacted materials. Either constant- or variable-head tests can be performed; either type of test will yield satisfactory results if used properly. For compressible materials, the head cannot change much or the test specimen will change volume. Permeation with deaired water until inflow and outflow rates equilibrate, supplemented with backpressure where appropriate, are needed to saturate test specimens. Chemical equilibrium must ideally be ensured when testing with liquids other than water, e.g., chemical wastes. A graphic of hydraulic conductivity and concentration of key chemicals or compounds is the best way to present the results of hydraulic conductivity tests with chemicals or waste liquids.

ACKNOWLEDGMENTS

The author would like to express his appreciation to the many graduate students whose meticulous laboratory work over the years has contributed greatly to the author's understanding of hydraulic conductivity testing. Special thanks are extented to Steve Boynton, David Foreman, Steve Day, Craig Benson, John Bowders, Greg Broderick, Chuck Shackelford, Wing Liao, and Hsin-yu Shan.

The author also acknowledges four individuals who contributed greatly to the author's understanding of hydraulic conductivity testing. Roy Olson spent many hours teaching the author how to perform laboratory tests with proper consideration to the many potential errors that can invalidate a test and has been a tremendous source of information and insight about hydraulic conductivity testing. Steve Trautwein has been generous in sharing his deep knowledge of hydraulic conductivity testing and in providing drawings, notes, and other information. In addition to the figures acknowledged directly in the text to Mr. Trautwein's company, Figs. 11-16 and Figs. 18 and 19 were adapted from Mr. Trautwein's drawings, which he was kind enough to share with the author. The process of writing and revising ASTM D5084 was a tremendous learning experience, and the author thanks Jeffrey Dunn, who drafted the original version of ASTM D5084. Finally, the author expresses his sincere appreciation to Richard Ladd, who has kindly shared information about his unique approach to laboratory testing in general and hydraulic conductivity testing in particular and whose encouragement and support over the years, within and outside of ASTM, have been very much appreciated.

REFERENCES

Acar, Y.B., Hamidon, A.B., Field, S.D., and Scott, L., 1985, "The Effect of Organic Fluids on Hydraulic Conductivity of Compacted Kaolinite," *Hydraulic Barriers in Soil and Rock, ASTM STP 874*, A.I. Johnson, R.K. Frobel, N.J. Cavalli, and C.B. Pettersson, Eds., American Society for Testing and Materials, Philadelphia, pp. 171-187.

Acar, Y.B., and D'Hollosy, E., 1987, "Assessment of Pore Fluid Effects Using Flexible Wall and Consolidation Permeameters," *Geotechnical Practice for Waste Disposal '87*, R.D. Woods, Ed., American Society of Civil Engineers, New York, pp. 231-245.

Alther, G., Evans, J.C., Fang, H.-Y., and Witmer, K., 1985, "Influence of Inorganic Permeants upon the Permeability of Bentonite," *Hydraulic Barriers in Soil and Rock, ASTM STP 874*, A.I. Johnson, R.K. Frobel, N.J. Cavalli, and C.B. Pettersson, Eds., American Society for Testing and Materials, Philadelphia, pp. 64-73.

Anderson, D.C., 1982, "Does Landfill Leachate Make Clay Liners More Permeable?", *Civil Engineering*, Vol. 52, No. 9, pp. 66-69.

Anderson, D.C., Crawley, W., and Zabcik, J.D., 1985, "Effects of Various Liquids on Clay Soil: Bentonite Slurry Mixtures," *Hydraulic Barriers in Soil and Rock, ASTM STP 874*, A.I. Johnson, R.K. Frobel, N.J. Cavalli, and C.B. Pettersson, Eds., American Society for Testing and Materials, Philadelphia, pp. 93-103.

Benson, C.B., and Daniel, D.E., 1990, "Influence of Clods on Hydraulic Conductivity of Compacted Clay," *Journal of Geotechnical Engineering*, Vol. 116, No. 8, pp. 1231-1248.

Bodocsi, A., and Bowers, M.T., 1991, "Permeability of Acrylate-, Urethane-, and Silicate-Grouted Sands with Chemicals," *Journal of Geotechnical Engineering*, Vo. 117, No. 8, pp. 1227-1244.

Bowders, J.J., and Daniel, D.E., 1987, "Hydraulic Conductivity of Compacted Clay to Dilute Organic Chemicals," *Journal of Geotechnical Engineering*, Vo. 113, No. 12, pp. 1432-1448.

Bowders, J.J., Usmen, M.A., and Gidley, J.S., 1987, Stabilized Fly Ash for Use as Low-Permeability Barriers," *Geotechnical Practice for Waste Disposal '87*, R.D. Woods, Ed., American Society of Civil Engineers, New York, pp. 320-333.

Boynton, S.S., and Daniel, D.E., 1985, "Hydraulic Conductivity Tests on Compacted Clay," *Journal of Geotechnical Engineering*, Vol. 111, No4, pp. 465-478.

Broderick, G.P., and Daniel, D.E., 1990, "Stabilizing Compacted Clay Against Chemical Attack," *Journal of Geotechnical Engineering*, Vol. 116, No. 10, pp. 1549-1568.

Brown, K.W., and Anderson, D.C., 1983, "Effects of Organic Solvents on the Permeability of Clay Soils," U.S. EPA, EPA-600/2-83-016, Cincinnati, Ohio, 153 p.

Brunelle, T.M., Dell, L.R., and Meyer, C.J., 1987, "Effect of Permeameter and Leachate on a Clay Liner," *Geotechnical Practice for Waste Disposal '87*, R.D. Woods, Ed., American Society of Civil Engineers, New York, pp. 347-361.

Carpenter, G.W., and R.W. Stephenson, 1986, "Permeability Testing in the Triaxial Cell," *Geotechnical Testing Journal*, Vol. 9, No. 1, pp. 3-9.

Chapuis, R.P., 1990, "Sand-Bentonite Liners: Predicting Permeability from Laboratory Tests," *Canadian Geotechnical Journal*, Vol. 27, No. 1, pp. 47-57.

Chapuis, R.P., Paré, J.J., and Loiselle, A.A., 1984, "Laboratory Test Results on Self-Hardening Grouts for Flexible Cutoffs, *Canadian Geotechnical Journal*, Vol. 21, No. 1, pp. 185-191.

Chapuis, R.P., Baass, K., and Davenne, L., 1989a, "Granular Soils in Rigid-Wall Permeameters: Methods for Determining the Degree of Saturation," *Canadian Geotechnical Journal*, Vol. 26, No. 1, pp. 71-79.

Chapuis, R.P., Gill, D.E., and Baass, K., 1989b, "Laboratory Permeability Tests on Sand: Influence of the Compaction Method on Anisotropy," *Canadian Geotechnical Journal*, Vol. 26, No. 4, pp. 614-622.

Chapuis, R.P., Lavoie, J., and Girard, D., 1992, "Design, Construction, Performance, and Repair of the Soil-Bentonite Liners of Two Lagoons," *Canadian Geotechnical Journal*, Vo. 29, No. 4, pp. 638-649.

Crooks, V.E., and Quigley, R.M., 1984, "Saline Leachate Migration through Clay: a Comparative Laboratory and Field Investigation," *Canadian Geotechnical Journal*, Vol. 21, No. 2, pp. 349-362.

Daniel, D.E., 1984, "Predicting Hydraulic Conductivity of Clay Liners," *Journal of Geotechnical Engineering*, Vol. 110, No. 2, pp. 385-300.

Daniel, D.E., Trautwein, S.J., Boynton, S.S., and Foreman, D.E., 1984, "Permeability Testing with Flexible-Wall Permeameters," *Geotechnical Testing Journal*, Vol. 7, No. 3, pp. 113-122.

Daniel, D.E., Anderson, D.C., and Boynton, S.S., 1986, "Fixed-Wall Versus Flexible-Wall Permeameters," *Hydraulic Barriers in Soil and Rock, ASTM STP 874*, A.I. Johnson, R. K. Frobel, N.J. Cavalli, and C.B. Pettersson, Eds., American Society for Testing and Materials, Philadelphia, pp. 107-126.

Daniel, D.E., Liljestrand, H.M., Broderick, G.P., and Bowders, J.J., 1988, "Interaction of Earthen Liner Materials with Industrial Waste Leachate," *Hazardous Waste & Hazardous Materials*, Vol. 5, No. 2, pp. 93-108.

Daniel, D.E., and Benson, C.H., 1990, "Water Content-Density Criteria for Compacted Soil Liners," *Journal of Geotechnical Engineering*, Vol. 116, No. 12, pp. 1811-1830.

D'Appolonia, D.J., 1980, "Soil-Bentonite Slurry Trench Cutoffs," *Journal of the Geotechnical Engineering Division*, ASCE, Vol. 106, No. GT 4, pp. 399-417.

Day, S.R., and Daniel, D.E., 1985, "Hydraulic Conductivity of Two Prototype Clay Liners," *Journal of Geotechnical Engineering*, Vol. 111, No. 8, pp. 957-970.

Demetracopoulos, A.C., and Dharmapal, A.P., 1987, "Flow and Mass Transport for Hazardous Waste Liners," *Geotechnical Practice for Waste Disposal '87*, R.D. Woods, Ed., American Society of Civil Engineers, New York, pp. 392-405.

Dunn, R.J., and Mitchell, J.K., 1984, "Fluid Conductivity Testing of Fine-Grained Soils," *Journal of Geotechnical Engineering*, Vol. 110. No. 11, pp. 1665.

Edil, T.B., and Erickson, A.E., 1985, "Procedure and Equipment Factors Affecting Permeability Testing of a Bentonite-Sand Liner Material," *Hydraulic Barriers in Soil and Rock, ASTM STP 874*, A.I. Johnson, R.K. Frobel, N.J. Cavalli, and C.B. Pettersson, Eds., American Society for Testing and Materials, Philadelphia, pp. 155-170.

Edil, T.B., Sandstrom, L.K., and Berthouex, P.M., 1992, "Interaction of Inorganic Leachate with Compacted Pozzolanic Fly Ash," *Journal of Geotechnical Engineering*, Vol. 118, No. 9, pp. 1410-1430.

Evans, J.C., and Fang, H.Y., 1986, "Triaxial Equipment for Permeability Testing with Hazardous and Toxic Permeants," *Geotechnical Testing Journal*, Vol. 9, No. 3, pp. 126-132.

Estornell, P.M., and Daniel, D.E., 1992, "Hydraulic Conductivity of Three Geosynthetic Clay Liners," *Journal of Geotechnical Engineering*, Vol. 118, No. 10, pp. 1592-1606.

Fernandez, F., and Quigley, R.M., 1985, "Hydraulic Conductivity of Natural Clays Permeated with Simple Liquid Hydrocarbons," *Canadian Geotechnical Journal*, Vol. 22, No. 1, pp. 205-214.

Fernandez, F., and Quigley, R.M., 1991, "Controlling the Destructive Effects of Clay-Organic Liquid Interactions by Application of Effective Stresses," *Canadian Geotechnical Journal*, Vol. 28, No. 3, pp. 388-398.

Fernuik, N., and M. Haug, 1990, "Evaluation of In Situ Permeability Testing Methods," *Journal of Geotechnical Engineering*, Vol. 116, No. 2, pp. 297-311.

Foreman, D.E., and D.E. Daniel, 1986, "Permeation of Compacted Clay with Organic Chemicals," *Journal of Geotechnical Engineering*, Vol. 12, No. 7, pp. 669-681.

Gale, J.E., and E.J. Reardon, 1984, "Effect of Groundwater Geochemistry on the Permeability of Grouted Fractures," *Canadian Geotechnical Journal*, Vol. 21, No. 1, pp. 8-20.

Garcia-Bengochea, I., and Lovell, C.W., 1981, "Correlative Measurements of Pore Size Distribution and Permeability in Soils," *Permeability and Groundwater Contaminant Transport, ASTM STP 746*, T.F. Zimmie and C.O. Riggs, Eds., American Society for Testing and Materials, Philadelphia, pp. 137-150.

Garlanger, J.E., Cheung, F.K., and Tannous, B.S., 1987, "Quality Control Testing for a Sand-Bentonite Liner," *Geotechnical Practice for Waste Disposal '87*, R.D. Woods, Ed., American Society of Civil Engineers, New York, pp. 488-499.

Gipson, A.H., 1985, "Permeability Testing on Clayey Soil and Silty Sand-Bentonite Mixture Using Acid Liquor," *Hydraulic Barriers in Soil and Rock, ASTM STP 874*, A.I. Johnson, R.K. Frobel, N.J. Cavalli, and C.B. Pettersson, Eds., American Society for Testing and Materials, Philadelphia, pp. 140-154.

Gill, S.A., and Christopher, B.R., 1985, "Laboratory Testing of Cement-Bentonite Mix for Proposed Plastic Diaphragm Wall for Complexe LaGrande Reservoir, Caniapiscau, James Bay, Canada," *Hydraulic Barriers in Soil and Rock, ASTM STP 874*, A.I. Johnson, R.K. Frobel, N.J. Cavalli, and C.B. Pettersson, Eds., American Society for Testing and Materials, Philadelphia, pp. 75-92.

Gordon, B.B., and Forrest, M., 1981, "Permeability of Soils Using Contaminated Permeant," *Permeability and Groundwater Contaminant Transport, ASTM STP 746*, T.F. Zimmie and C.O. Riggs, Eds., American Society for Testing and Materials, Philadelphia, pp. 101-120.

Ho, Y.A., and Pufahl, D.E., 1987, "The Effects of Brine Contamination on the Properties of Fine Grained Soils, *Geotechnical Practice for Waste Disposal '87*, R.D. Woods, Ed., American Society of Civil Engineers, New York, pp. 547-561.

Jedele, L.P., 1987, "Evaluation of Compacted Inert Paper Solids as a Cover Material," *Geotechnical Practice for Waste Disposal '87*, R.D. Woods, Ed., American Society of Civil Engineers, New York, pp. 562-577.

Johnson, G.W., Crumley, W.S., and Boutwell, G.P., 1990, "Field Verification of Clay Liner Hydraulic Conductivity," *Waste Containment Systems: Construction, Regulation, and Performance*, R. Bonaparte, Ed., American society of Civil Engineers, new York, pp. 226-245.

Juang, C.H., and Holtz, R.D., 1986, "Fabric, Pore Size Distribution, and Permeability of Sandy Soils, *Journal of Geotechnical Engineering*, Vol. 112, No. 9, pp. 855-868.

Kenney, T.C., Lau, D., and Ofoegbu, G.I., 1984, "Permeability of Compacted Granular Materials," *Canadian Geotechnical Journal*, Vol. 21, No. 4, pp. 726-729.

Kenney, T.C., van Veen, W.A., Swallow, M.A., and Sungaila, M.A., 1992, "Hydraulic Conductivity of Compacted Bentonite-Sand Mixtures," *Canadian Geotechnical Journal*, Vol. 29, No. 3, pp. 364-374.

Kim, W.H., and Daniel, D.E., 1992, "Effects of Freezing on Hydraulic Conductivity of Compacted Clay," *Journal of Geotechnical Engineering*, Vo. 118, No. 7, pp. 1083-1097.

Korfiatis, G.P., Rabah, N., and Lekmine, D., 1987, "Permeability of Compacted Clay Liners in Laboratory Scale Models," *Geotechnical Practice for Waste Disposal '87*, R.D. Woods, Ed., American Society of Civil Engineers, New York, pp. 611-624.

LaFleur, J., 1984, "Filter Testing of Broadly Graded Cohesionless Tills," *Canadian Geotechnical Journal*, Vol. 21, No. 3, pp. 634-643.

Lentz, R.W., Horst, W.D., and Uppot, J.O., 1985, "The Permeability of Clay to Acidic and Caustic Permeants," *Hydraulic Barriers in Soil and Rock, ASTM STP 874*, A.I. Johnson, R.K. Frobel, N.J. Cavalli, and C.B. Pettersson, Eds., American Society for Testing and Materials, Philadelphia, pp. 127-139.

Mitchell, J.K., 1976, *Fundamentals of Soil Behavior*, John Wiley & Sons, New York, 422 p.

Mitchell, J.K., Hooper, D.R., and R.G. Campanella, 1965, "Permeability of Compacted Clay, *Journal of the Soil Mechanics and Foundations Division*, ASCE, Vol. 91, No. SM4, pp. 41-65.

Mundell, J.A., and Bailey, B., 1985, "The Design and Testing of a Compacted Clay Barrier Layer to Limit Percolation through Landfill Covers," *Hydraulic Barriers in Soil and Rock, ASTM STP 874*, A.I. Johnson, R.K. Frobel, N.J. Cavalli, and C.B. Pettersson, Eds., American Society for Testing and Materials, Philadelphia, pp. 246-262.

Olson, R.E., and Daniel, D.E., 1981, "Measurement of the Hydraulic Conductivity of Fine-Grained Soils," *Permeability and Groundwater Contaminant Transport, ASTM STP 746*, T.F. Zimmie and C.O. Riggs, Eds., American Society for Testing and Materials, Philadelphia, pp. 18-64.

Olson, R.E., 1986, "State of the Art: Consolidation Testing," *Consolidation of Soils: Testing and Evaluation, ASTM STP 892*, R.N. Yong and F.C. Townsend, Eds., American Society for Testing and Materials, Philadelphia, pp. 7-70.

Peterson, S.R., and Gee, G.W., 1985, "Interactions Between Acidic Solutions and Clay Liners: Permeability and Neutralization," *Hydraulic Barriers in Soil and Rock, ASTM STP 874*, A.I. Johnson, R.K. Frobel, N.J. Cavalli, and C.B. Pettersson, Eds., American Society for Testing and Materials, Philadelphia, pp. 229-245.

Pierce, J.J., and Witter, K.A., 1986, "Termination Criteria for Clay Permeability Testing," *Journal of Geotechnical Engineering*, Vol. 112, No. 9, pp. 841-856.

Pierce, J.J., Salfors, G., and Ford, K., 1987, "Differential Flow Patterns through Compacted Clays," *Geotechnical Testing Journal*, Vol. 10, No. 4, pp. 218-222.

Poran, C.J., and Faouzi, A.A., 1989, "Properties of Solid Waste Incinerator Fly Ash," *Journal of Geotechnical Engineering*, Vol. 115, No. 8, pp. 1119-1133.

Quigley, R.M., Fernandez, F., Yanful, E., Helgason, T., and Margaritis, A., 1987, "Hydraulic Conductivity of Contaminated Natural Clay Directly Below a Domestic Landfill," *Canadian Geotechnical Journal*, Vol. 24, No. 3, pp. 377-383.

Scott, J.C., 1984, "Effects of Dust Suppressants on Tailing Sand Permeability," *Geotechnical Testing Journal*, Vol. 7, No. 1, pp. 41-44.

Shackelford, C.C., and Javed, F., 1991, "Large-Scale Laboratory Permeability Testing of a Compacted Clay Soil," *Geotechnical Testing Journal*, Vol. 14, No. 2, pp. 171-179.

Shan, H.Y., and D. E. Daniel, 1989, "Results of Laboratory Tests on a Geotextile/Bentonite Liner Material," *Proceedings, Geosynthetics '91*, Industrial Fabrics Association International, St. Paul, Minnesota, Vol. 2, pp. 517-535.

Silva, A.J., Hetherman, J.R., and Calnan, D.I., 1981, "Low-Gradient Permeability Testing of Fine-Grained Marine Sediments," *Permeability and Groundwater Contaminant Transport, ASTM STP 746*, T.F. Zimmie and C.O. Riggs, Eds., American Society for Testing and Materials, Philadelphia, pp. 121-136.

Skempton, A.W., 1954, "The Pore Pressure Coefficients A and B," *Geotechnique*, Vol. 4, No. 4, pp. 143-147.

Tavenas, F., Leblond, P., Jean, P., and Leroueil, S., 1983, "The Permeability of Natural Soft Clays. Part I: Methods of Laboratory Measurement," *Canadian Geotechnical Journal*, Vol. 20, No. 4, pp. 629-644.

Uppot, J.O., and Stephenson, R.W., 1989, "Permeability of Clays under Organic Permeants," *Journal of Geotechnical Engineering*, Vol. 115, No. 1, pp. 115-131.

Vesperman, K.D., Edil, T.B., and Berthouex, P.M., 1985, "Permeability of Fly Ash and Fly Ash-Sand Mixtures," *Hydraulic Barriers in Soil and Rock, ASTM STP 874*, A.I. Johnson, R.K. Frobel, N.J. Cavalli, and C.B. Pettersson, Eds., American Society for Testing and Materials, Philadelphia, pp. 289-298.

Yanful, E.K., Haug, M.D., and Wong, L.C., 1990, "The Impact of Synthetic Leachate on the Hydraulic Conductivity of a Smectitic Till Underlying a Landfill Near Saskatoon, Saskatchewan," *Canadian Geotechnical Journal*, Vol. 27, No. 4, pp. 507-519.

Yong, R.N., Boonnsinsuk, P., and Wong, G., 1986, "Formulation of Backfill Material for a Nuclear Fuel Waste Disposal Vault," *Canadian Geotechnical Journal*, Vol. 23, No. 1, pp. 216-228.

Zimmie, T.F., 1981, "Geotechnical Testing Considerations in the Determination of Laboratory Permeability for Hazardous Waste Disposal Siting," *Hazardous Solid Waste Testing: First Conference, ASTM STP 760*, R.A. Conway and B.C. Malloy, Eds., American Society for Testing and Materials, Philadelphia, pp. 293-304.

Zimmie, T.F., LaPlante, C.M., and Bronson, D., 1992, "The Effects of Freezing and Thawing on the Permeability of Compacted Clay Landfill Covers and Liners," *Environmental Geotechnology*, M.A. Usmen and Y.B. Acar, Eds., Balkema, Rotterdam, pp. 213-217.

Jeffrey C. Evans[1]

HYDRAULIC CONDUCTIVITY OF VERTICAL CUTOFF WALLS

REFERENCE: Evans, J. C., "**Hydraulic Conductivity of Vertical Cutoff Walls,**" Hydraulic Conductivity and Waste Contaminant Transport in Soil, ASTM STP 1142, David E. Daniel and Stephen J. Trautwein, Eds., American Society for Testing and Materials, Philadelphia, 1994.

Abstract: Vertical cutoff walls have been used to control the movement of contaminants and contaminated groundwater since the remediation of contaminated sites began. There are, however, significant hydraulic conductivity differences between soil-bentonite, cement-bentonite, plastic concrete, and in situ mixed cutoff walls. The results of laboratory and field studies were assessed to show the influence of material properties, confining stress, permeameter type, water table position, and state of stress, on the hydraulic conductivity of vertical cutoffs.
 The results of these studies show the range of hydraulic conductivity expected for each of the cutoff wall types. Increasing confining stress markedly decreases the hydraulic conductivity of soil-bentonite and has a measurable but reduced impact on stronger backfill materials. Studies on soil-bentonite cutoff walls show that the stress at depth is less than predicted using the effective weight of the overlying materials. This reduction in stress is a result of soil-bentonite materials "hanging-up" on the side walls of the trench. Thus, applying the effective stress calculated from the effective weight of the overlying backfill overestimates the stress to be used in the laboratory tests and results in unconservative measures of hydraulic conductivity. Field data also reveals that, with time, the hydraulic conductivity of soil-bentonite above the water table may increase substantially. Further, the hydraulic conductivity does not significantly decrease upon re-saturation.

Keywords: slurry wall, hydraulic conductivity, soil-bentonite, cement-bentonite, plastic concrete

---

[1]Associate Professor of Civil Engineering, Bucknell University, Department of Civil Engineering, Lewisburg, PA 17837

INTRODUCTION

Vertical barriers are widely employed in the subsurface to control the flow of ground water and to reduce the rate of contaminant transport. Vertical cutoff walls have been used to control the movement of contaminants and contaminated groundwater since the remediation of contaminated sites began; one of first superfund sites where remedial technologies were implemented employed a slurry trench cutoff wall (Salvesen 1983). The principal factor in the performance of vertical barrier systems is the hydraulic conductivity. Like other geotechnical materials, there is no unique value of hydraulic conductivity. In cases where the ubiquitous value of 1×10^{-7} cm/s is specified, it is necessary to identify additional parameters which control this value in the laboratory and in the field. These parameters include the material composition, effective stress, field environment and cutoff wall defects. This paper will address the hydraulic conductivity of vertical cutoff walls with particular emphasis on soil-bentonite cutoff walls but including cement-bentonite, and plastic concrete slurry walls as well as in situ mixed walls (also known as deep soil mixed, auger mixed, soil mixed walls).

Soil-Bentonite Slurry Trench Cutoff Walls

The construction methods of soil-bentonite slurry trench cutoff walls are well-established (Spooner et al., 1984, Ryan 1987, Evans, 1993). A narrow (typically 0.5 to 1 m), slurry filled trench is first excavated in the subsurface. The slurry, comprised of a mixture of about 5% bentonite and 95% water by weight, is employed to maintain trench stability as the excavation proceeds downward from the ground surface. As the excavation proceeds longitudinally, the trench is backfilled by displacing the slurry with a mixture of soil, bentonite-water slurry, and occasionally dry bentonite. The soil used in the backfill may be soil excavated from the trench, borrow soil imported from offsite, or a mixture of both, depending upon grain size characteristics, the presence/absence of contamination and project hydraulic conductivity requirements. The hydraulic conductivity of soil-bentonite is typically between 1×10^{-7} cm/s and 1×10^{-8} cm/s. The excavation, backfill mixing, and backfill placement are shown schematically on Fig. 1.

Cement-Bentonite Slurry Trench Cutoff Walls

The construction methods of cement-bentonite slurry trench cutoff walls are also well-established (Spooner et al., 1984, Ryan 1987, Evans, 1993). A narrow (typically 0.5 to 1 m), slurry filled trench is excavated in the subsurface as with the soil-bentonite technique. The slurry in this case is comprised of a mixture of about 5% bentonite, 10% to 20% cement, and 75% to 85% water by weight. Cement-bentonite mixes have also incorporated fly ash as cement replacement (Carr 1990). In Europe, slag is commonly incorporated in the mix (Jefferis, 1981b). The slurry is employed to maintain trench stability and is left to harden in place to form the completed cutoff wall. Cement-bentonite may be the cutoff wall of choice where strength considerations indicate the need for a material stronger than soil-bentonite. The hydraulic conductivity of cement-bentonite is typically between 1×10^{-5} cm/s and 1×10^{-6} cm/s and occasionally lower.

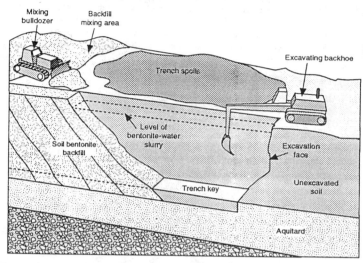

Fig. 1 Construction of a Soil-Bentonite Slurry Trench Cutoff Wall (from LaGrega et al. 1994)

Plastic Concrete Slurry Trench Cutoff Walls

Plastic concrete is a mixture of cement, aggregate, bentonite and water resulting in a material that is relatively strong with a relatively low hydraulic conductivity (Evans et al. 1987) Plastic concrete cutoff walls are usually constructed using the panel method of slurry trench construction. In this method of construction, the trench is excavated in panels using bentonite water slurry to maintain trench stability. The excavated panel is backfilled using plastic concrete placed using a tremie method of concrete placement. This panel excavation and backfill technique is similar that used for diaphragm walls (Xanthakos 1974). Although plastic concrete has been used in several applications, the higher cost when compared to soil-bentonite cutoff walls has limited its use. The hydraulic conductivity of plastic concrete barrier walls is typically between 1×10^{-7} cm/s and 1×10^{-8} cm/s.

In Situ Mixed Vertical Barriers

Using specially designed and fabricated augers, vertical barriers can be mixed in place. In-situ mixing is often called deep soil mixing (DSM) or a soil-mixed wall (SMW) process. Regardless of the name the process is similar; a special auger mixing shaft is rotated into the ground while simultaneously adding bentonite-water slurry or cement-bentonite-water slurry. The construction sequence shown in Fig. 2 results in a column of blended soil when multiple mixing shafts are employed. If additional strength is needed reinforcing can be added to the treated soil columns. The resulting wall is typically from 0.5 to 0.8 m wide. The bentonite-water slurry normally contains about 5% bentonite and 95% water. Mixing this slurry with the soil typically results in a wall with a bentonite content of about 1%. Since the wall is constructed as a mixture of the in situ soils, variability in the soil properties both along the wall alignment and with depth results in variability in the hydraulic conductivity of the completed wall. The hydraulic conductivity of in situ soil mixed walls is typically between 1×10^{-6} cm/s and 1×10^{-7} cm/s.

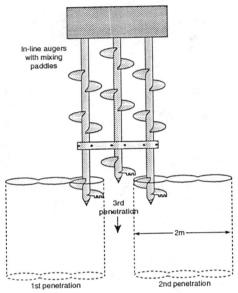

Fig. 2   Construction of an In Situ Mixed Cutoff Wall

PARAMETERS AFFECTING HYDRAULIC CONDUCTIVITY OF CUTOFF WALLS

What is the "true" hydraulic conductivity of a completed vertical barrier wall? How does the hydraulic conductivity of the wall relate to the hydraulic conductivity measured in the laboratory or in the field on samples of the wall? What are the factors which influence the hydraulic conductivity of the completed wall? What are the factors which influence the measurement of the hydraulic conductivity of the cutoff wall material? Without attempting to revisit all the factors involved in hydraulic conductivity testing, the remainder of this paper will focus on several parameters which influence the hydraulic conductivity of the vertical barrier walls described above.

Parameters which influence the hydraulic conductivity and our measures of hydraulic conductivity include:

1) grain size distribution
2) bentonite content, type and gradation

3) effective consolidation pressure in the laboratory
4) field state of stress
4) homogeneity of the cutoff wall
5) hydraulic fracturing
6) permeameter type
7) location of the water table
8) variability
9) nature of the pore fluid and permeant
10) potential for defects

Effect of Grain Size Distribution

It has long been established that the type and nature of the fines fraction influences the hydraulic conductivity of the soil-bentonite backfill (D'Appolonia 1980) In general, as the fraction of the soil finer than the No. 200 sieve increases, the hydraulic conductivity decreases. Shown on Fig. 3 is the relationship between the hydraulic conductivity and the fines content for the soils of a specific project. The data demonstrate the importance of "adequate" natural fines in achieving a low hydraulic conductivity. The low hydraulic conductivity is achieved without enriching the mix with additional dry bentonite. For this particular study, the addition of 20% plastic fines from a clayey borrow source to the base soil of about 20% gravel, 70% sand, and 10% silt, lowered the hydraulic conductivity to 5×10^{-8} cm/s, below the project target of 1×10^{-7} cm/s. The mixture using virtually 100% plastic fines from the borrow source resulted in a hydraulic conductivity of 3×10^{-8} cm/s, not significantly lower that for the mix containing only 20% natural fines.

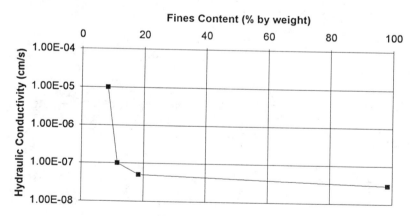

Fig. 3 Site Specific Relationship between Fines Content and Hydraulic Conductivity

The results shown in Fig. 3 were determined for specific soils from a specific site. Although there is clearly a relationship between fines content and hydraulic conductivity for these materials, that is not to say the relationship may be generalized. In fact, the data presented by Ryan (1987) in updating a relationship published earlier by D'Appolonia dismissed the notion that one can achieve a certain hydraulic conductivity by simply choosing a fines content.

These data are presented to illustrate the approach to determining the desired optimum soil-bentonite backfill mixture. A well-graded soil, consisting of a blend of gravel, sand, silt and clay results in a backfill of low hydraulic conductivity, low compressibility, and as discussed later in this paper, greater resistance to degradation by contaminants than a backfill containing a very high percentage of fines in the mixture. This approach is shown schematically on Fig. 4. The Figure shows the arrangement of progressively finer particles plugging progressively finer voids, leaving only the smallest voids to be filled with the clayey fines and the bentonite which is added via the slurry. The natural analogy to this approach is glacial till, generally well-graded and having a low hydraulic conductivity. Although segregation of the larger particles is theoretically possible, grain size distribution data indicates that the gravel remains well-distributed throughout the backfill.

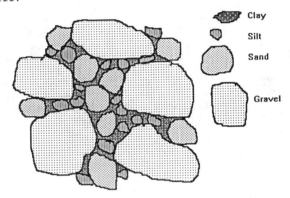

Fig. 4   Schematic of Well-Graded Soil

An added benefit results from a well-graded backfill when contamination resistance is taken into account. This will be discussed in more detail later in this paper.

Effect of Bentonite Content

Generally speaking, increasing the bentonite content in a vertical barrier will decrease the hydraulic conductivity in soil-bentonite and in situ soil mixed walls; there may , however, be an optimum. Shown on Fig. 5 is a relationship between hydraulic conductivity and the bentonite content. The data reveal that, for this particular mix, the minimum hydraulic conductivity was found at a bentonite content of about 3%. Although such correlations may be developed for site specific use, when data from about thirty soil-bentonite projects were combined, little correlation of hydraulic conductivity to bentonite was found (Ryan 1987). The same conclusion can be reached for cement-bentonite cutoff walls.

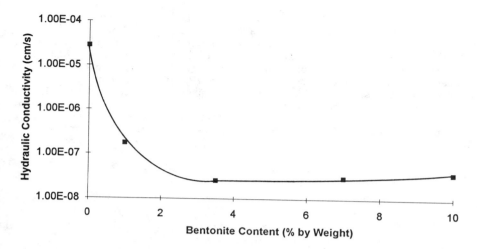

Fig. 5 Effect of Bentonite Content on Hydraulic Conductivity of a Soil Bentonite backfill (after Barvenik 1992)

Effect of Consolidation Pressure

For any give sample of vertical barrier material, the hydraulic conductivity decreases as the effective consolidation pressure increases. This trend is predictable on a theoretical basis from considerations of decreasing void ratio with increasing effective stress. Shown on Fig. 6 are relationships between effective consolidation pressure and hydraulic conductivity showing significant decreases in the hydraulic conductivity of soil-bentonite backfill as the effective consolidation pressure is increased.

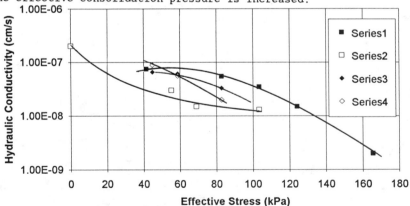

Fig. 6 Effect of Consolidation Pressure on Hydraulic Conductivity of a Soil-Bentonite Backfill (data from McCandless and Bodocsi 1988; Day 1992)

The impact of confining pressure in a laboratory permeability test may go beyond that expected from void ratio considerations. Shown in Fig. 7 are the results of a series of laboratory tests on molded samples of an in situ mixed wall of soil, bentonite and cement. The trend is evident, as the confining pressure increases, the hydraulic conductivity decreases. The authors conclude that, as a result of the rough surface of the cemented samples, a high confining pressure is needed to maintain contact between the membrane and the sample to prevent sidewall leakage.

Using the authors' data, the relationship between the confining pressure and void ratio is shown on Fig. 8. As shown, the decrease in void ratio due to the increasing confining pressure is quite small.

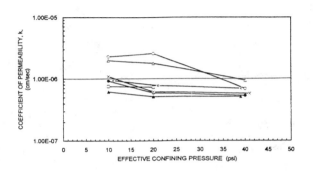

Fig. 7 Effect of Confining Pressure on the Hydraulic Conductivity of an In Situ Mixed Wall (after Yang 1993)

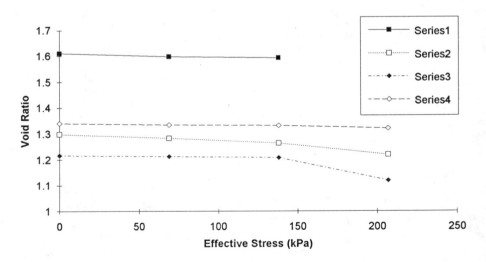

Fig. 8 Effect of Confining Pressure on the Void Ratio of an In Situ Mixed Wall (data from Yang, 1993)

Field Stress Conditions

The study of the influence of effective consolidation pressure on hydraulic conductivity is more than academic. Unless the state of stress in the field is known, the hydraulic conductivity remains uncertain. Laboratory model and field data obtained to date on soil-bentonite slurry trench cutoff walls indicated that the stress does not increase hydrostatically with depth (McCandless and Bodocsi 1987; Cooley 1991). Analytical approaches reveal similar findings (Sweidan, 1990). Data on a fully instrumented soil-bentonite cutoff wall measuring total and effective stress with depth was not found in the published literature. The soil-bentonite backfill is quite compressible compared with the relatively rigid trench sidewalls; as a result the consolidation of the backfill is limited by the friction at the trench/backfill interface, termed arching by some authors (Millet et al. 1992). Based upon the information available to date for soil-bentonite walls, the effective stress distribution with depth depends upon:
1) the wall thickness (or thickness/depth ratio),
2) the backfill compressibility,
3) the backfill/trench wall interface friction, and
4) the backfill density
5) poisson's ratio

The arching or reduction in effective stress at depth can be minimized by increasing the wall thickness or reducing the backfill compressibility.

Hydraulic Fracturing Effects on Hydraulic Conductivity

The nature and potential for hydraulic fracturing in soil-bentonite slurry trench cutoff walls is often misunderstood. Handbook guidance quotes a rule of thumb of 1 psi (of excess hydraulic pressure) per foot of wall depth as "safe against hydraulic fracturing" (USEPA 1984). That is, at a depth of 20 feet (6.1 m), the wall can withstand an excess hydraulic head of 20 psi (138 kPa). Alternatively, a width of 15 to 23 cm (0.5 to 0.75 ft.) per 3 m (10 ft.) of hydrostatic head is cited (Case 1980). For slurry wall use in waste containment applications, the guidance has been incorrectly interpreted to calculate the maximum drawdown from within the barrier. The guidance was originally developed for pore water pressure in excess of the original pore water pressure (i.e. pressure above hydrostatic) as in the case hydrofracturing rock to enhance oil recovery by pumping fluid into the formation at pressures large enough to reduce the effective stress to zero and "lift" the rock. This guidance is also applicable to the case of a slurry wall beneath the core of a dam where the upstream reservoir induces high hydraulic head. In such an application, the excess pore water pressure could exceed the minor principal total stress within the cutoff wall, reducing the minor principal effective stress to zero and resulting in hydraulic fracturing. Dewatering from within a cutoff wall lowers the phreatic surface relative to the original phreatic surface and results in an increase in effective stress within the wall. Since dewatering from within the cutoff wall cannot cause a reduction in minor principal effective stress, hydraulic fracturing can not result from dewatering.

Laboratory Permeameters and Their Influence on Hydraulic Conductivity

Much has been written regarding laboratory test methods and equipment and their effect on hydraulic conductivity values (Olson and Daniel 1981). The discussion here is limited to those unique equipment considerations that influence the hydraulic conductivity of vertical barriers. In particular, a fixed wall API Filter Press (API 1984) has been used to conduct rapid evaluations of hydraulic conductivity in the field as the construction progresses. Traditional fixed wall permeability tests have also been used. The data shown on Fig. 9 indicate a some correlation between the API filter press fixed wall test method and the triaxial test methods for two particular projects. These data show the need for site specific correlations if fixed wall permeability tests are to be used for field quality control tests.

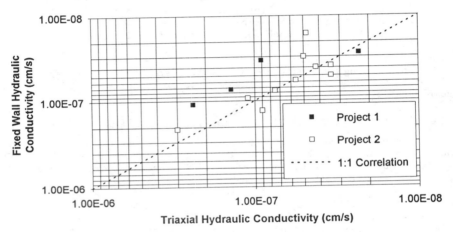

Fig. 9 Permeameter Test Results (data from Barvenik 1992 and Day 1992)

Influence of Fluctuating Water Table

The principal purpose of a cutoff wall is typically to impede the horizontal flow of ground water (and the associated transport of contaminants in many environmental applications). Considerable effort is made to hydrate the bentonite in the cutoff wall in an effort to minimize the hydraulic conductivity. Further, it is common for the wall to have a portion that is expected to remain permanently below the water table, another portion permanently above the water table and a portion which may be in the range of a fluctuating water table. Limited information on the long-term performance of cutoff walls is available, however, in one recent study measurements of permeability were made for a soil-bentonite cutoff wall that was constructed in 1981 and another constructed in 1987 (Cooley 1991). The cutoff walls joined to form a vertical cutoff surrounding a wet ash handling facility that maintained essentially constant water levels year-round. The investigator obtained thin-walled tube samples above and below the water table for each of the different age cutoffs. The results are as shown in Table 1.

Table 1 - Effect of Water Table and Wall Age on Hydraulic Conductivity of Soil-Bentonite

Construction Date	Position w.r.t water table	Hydraulic Conductivity (cm/s)
1981	above	1×10^{-3}
1981	below	1×10^{-8}
1987	above	6×10^{-6}
1987	below	1×10^{-7}

To further investigate the potential for "rehydration" the permeability tests were repeated after 16 days of backpressure saturation with no change in the results. To examine the phenomena further red dye was introduced to see if the increased hydraulic conductivity could be attributed to defects in the sample. After permeation the samples were cut apart and examined; no dye paths were observed and the samples was noted to be uniform in cross-sectional appearance. Although these data are limited and the time spent rehydrating the clay was limited to 16 days, they give rise to concern that if soil-bentonite materials are not kept saturated, the hydraulic conductivity may increase and such increases are not reversible.

Variability of Cutoff Walls

As described above, the construction of these cutoff walls typically employs the in-place mixing of natural soils with bentonite, bentonite water slurry, and/or cement-bentonite slurry. It has also been established that the hydraulic conductivity is a function of the properties of the base soil to be blended (i.e., grain size-distribution, plasticity, water content, fines content). As a result, it is expected that the variability in the hydraulic conductivity of the completed soil-bentonite or in situ mixed barrier would be a function of the variability of the soils along the cutoff wall alignment. Thus, it is important to fully characterize the distribution of materials with depth and along the trench alignment in order to properly predict the range of hydraulic conductivity to be expected.

Perhaps expectedly, the test values of hydraulic conductivity of the completed cutoff wall depends on the method of sampling and testing. For one study of an in situ mixed soil, bentonite and cement barrier (Yang et al. 1993), the data ranged from a low of about 1×10^{-8} cm/s to a high of 1×10^{-4} cm/s. About 45% of the data meet the project requirements of 1×10^{-6} cm/s. A number of parameters were found to affect the laboratory test results. Thin wall samples were affected by damage to sampling tubes including cutting edges both during and after sampling. Soil-cement samples were observed to exhibit rough and loose surface zones and cracking. In essence, the scatter in permeability test data is attributed to the inferior quality of bulk samples and sample disturbance of core samples. In contrast to the laboratory data on both field and laboratory prepared and obtained samples, all of the data obtained from in situ permeability tests met the project requirements of 1×10^{-6} cm/s. The case study just described (Yang et al. 1993) suggests a need to develop more reliable methods for sampling in situ mixed soil/cement/bentonite materials and for determining the hydraulic conductivity testing of these materials as part of the construction quality control process.

<u>Effect of Pore Fluid and Permeant</u>

The hydraulic conductivity of soil-bentonite backfill can be altered as a result of permeation with permeants having a different chemistry than the original pore fluid. The practical questions are two; will the hydraulic conductivity increase or decrease and what will the magnitude of the change be? The nature of clay-pore fluid interactions has been well studied (Mitchell 1976; Evans et al. 1985; Brown and Anderson 1983). It is generally considered that the behavior of soils in the presence of contaminants can be modeled by the clay-water-electrolyte model as developed for colloidal suspensions (Mitchell and Madsen, 1987). In general, little effect is observed for clays permeated with chemicals at low concentrations. In contrast, permeation with concentrated organics may result in significant increases in hydraulic conductivity. Thus, to minimize detrimental clay/contaminant interactions it is important to minimize 1) the activity of the clay fraction, and 2) the amount of the clay fraction.

To meet these goals it is necessary to include only enough low plasticity clay to reduce the hydraulic conductivity to the desired level and to include only the quantity of bentonite which is mixed in by virtue of the addition of bentonite-water slurry for workability. Thus, for the schematic shown in Fig. 4, the gravel, sand, and silt components are virtually non-reactive and the slightly reactive low-plasticity clay is present in the minimum quantity necessary to achieve the desired hydraulic conductivity. In this way, the potential for major changes in the hydraulic conductivity due to incompatibility with the surrounding ground water environment are minimized.

Indicator tests such as sedimentation tests, cracking pattern tests, and/or Atterberg limits may be used to initially evaluate the potential for long term compatibility problems or short term construction problems (Alther et al. 1988). Compatibility testing should ultimately include a long-term triaxial permeability test using the expected leachates/permeants (Evans and Fang 1988). Although passing of two to three pore volumes of the contaminant is usually considered sufficient to investigate compatibility, recent research has shown that the permeant volume needed is dependent upon the contaminant mass needed to complete the reaction (Jefferis 1992).

Limited data indicate that plastic concrete may be less susceptible to changes in hydraulic conductivity when permeated with contaminated permeants than soil-bentonite (Evans et al. 1987).

Based upon the research to date, the presence of non-aqueous phase liquids may pose the greatest risk to the degradation of vertical cutoff walls. For additional detail regarding the compatibility of slurry cutoff wall materials the reader is referred to Day in this same proceedings Day 1993).

Potential for Defects in Vertical Cutoff Walls

No discussion of the hydraulic conductivity of vertical barriers would be complete without mention of the potential for defects, i.e. areas of high hydraulic conductivity. A defect is defined as that portion of the cutoff wall where the hydraulic conductivity is beyond the limits of that expected due to the statistical variability of the cutoff wall materials. The potential defects in slurry trench cutoff walls are many and have been described elsewhere (Evans 1993; Evans 1990; McCandless et al. 1993). The probability that any given defect will be detected in any given verification testing program is small. Most testing programs use laboratory tests of field prepared samples to verify the hydraulic conductivity of the cutoff. Even where field tests are used, it may not be economically feasible to conduct enough in situ permeability tests to reduce the probability of missing a defect to a reasonably small number. Non-destructive geophysical techniques have also been considered (Barvenik and Ayers 1987). Pumping tests may be used but in situ heterogeneity often precludes definitive conclusions regarding the integrity of the completed barrier. Recent studies show that the use of standpipes along the wall alignment may provide useful information if properly spaced (Bodocsi et al. 1993). Further research in this area is needed to better verify the as-constructed condition of vertical barrier walls.

THE FUTURE OF BARRIER TECHNOLOGIES

There is little doubt that advances will be and are being made along several fronts. These include construction techniques, design and analysis methods, laboratory and insitu testing methods, and in the philosophy of application. It is this last topic that perhaps offers the most promise. Historically, barriers have been constructed as the title of this paper reflects, as hydraulic barriers. However, it is understood and recognized that the ultimate goal may be more precisely stated as contaminant transport barriers. Thus there is a need to develop barrier techniques that are improved methods of reducing contaminant transport. This can be done by either further reducing the hydraulic conductivity or increasing the attenuation capacity of the barrier. Thus, HDPE membranes are being placed in cutoffs to achieve low hydraulic conductivity. The use of attenuating materials in the barrier system is also under study (Evans et al, 1990; Mott and Weber 1989, 1991).

REFERENCES

Alther, G. R., Evans, J. C. and Andrews, E., "Organic Fluid Effects Upon Bentonite" Proceedings of the 5th National Conference on Hazardous Wastes and Hazardous Materials, Las Vegas, Nevada, April, 1988, pp. 210-214.

American Petroleum Institute. Standard Procedure for Field Testing Drilling Fluids. API Recommended Practice, API RP 138, 10th Ed. Dallas, Texas, June 1, 1984.

Barvenik, M. J., and Ayers, J. E., "Construction Quality Control and Post-Construction Performance Verification for the Gilson Road Hazardous Waste Site Cutoff Wall," EPA/600/2-87/065, 1987.

Barvenik, M. J., "Design Options Using Vertical Barriers Systems," presented at ASCE International Convention and Exposition, Environmental Geotech Symposium, New York, New York, Sept., 1992.

Bodocsi, A., McCandless, R. M. and Ling, F. K., "Detection of Hydraulic Defects in Soil-Bentonite Cutoff Walls," in preparation, 1993.

Brown, K. W., and D. C. Anderson. "Effects of Organic Solvents on the Permeability of Clay Soils". EPA 600/2-83-016. U.S. Environmental Protection Agency, Cincinnati, Ohio, 1983.

Carr, F. H., "Fly Ash Added to Cut Losses in Deep Bentonite Slurry Wall" ENR, January, 1990, pp. 21-22.

Case International Co. Slurry Cut-off Trenches and Walls, Case International Co., Roselle, IL, 1980

Cooley, B. "Behavior of Soil-Bentonite Slurry Trench Cutoff Walls," Honors Thesis, Bucknell University, Dec. 1991.

Day, S. GeoCon, Inc. project data, personal communication, 1992.

Day, S. "Compatibility of Slurry Cutoff Wall Materials with Contaminated Groundwater", Hydraulic Conductivity and Waste Contaminant Transport in Soils, ASTM STP 1142, David E. Daniel and Stephen J. Trautwein, Eds., American Society for Testing and Materials, Philadelphia, 1993.

D'Appolonia, D. "Soil-Bentonite Slurry Trench Cutoffs", Journal of the Geotechnical Engineering Division, ASCE, Vol. 106, No. 4, 1980, pp. 399-418.

Evans, J. C., "Geotechnics of Hazardous Waste Control Systems," Chapter 20, Foundation Engineering Handbook, 2nd ed., Ed. H. Y. Fang, Von Nostrand Reinhold Company, New York, NY, 1991.

Evans, J. C., "Vertical Cutoff Walls,", Geotechnical Practice for Waste Disposal, Ed. D. E. Daniel, Chapman and Hall, 1993.

Evans, J. C., Stahl, E. D. and Droof, E., "Plastic Concrete Cutoff Walls," Geotechnical Practice for Waste Disposal '87, ASCE Geotechnical Special Publication No. 13, June, 1987, pp. 462-472.

Evans, J. C., Fang, H. Y., and Kugelman, I. J., "Organic Fluid Effects on the Permeability of Soil-Bentonite Slurry Walls," Proceedings of the National Conference on Hazardous Wastes and Environmental Emergencies, Cincinnati, OH, May, 1985, pp. 267-271.

Evans, J. C., Fang, H. Y., and Kugelman, I. J., "Containment of Hazardous Materials with Soil-Bentonite Slurry Walls," Proceedings of the 6th National Conference on the Management of Uncontrolled Hazardous Waste Sites, Washington, D. C., November, 1985, pp. 249-252.

Evans, J. C. and Fang, H. Y., "Triaxial Permeability and Strength Testing of Contaminated Soils," Advanced Triaxial Testing of Soil and Rock, ASTM STP 977, ASTM STP 977, American Society for Testing and Materials, Philadelphia, 1988. pp. 387-404.

Evans, J. C., Sambasivam, Y. and Zarlinski, S. J., "Attenuating Materials in Composite Liners," Waste Containment Systems: Construction, Regulation, & Performance, ASCE Geotechnical Special Publication No. 26, November, 1990, pp. 246-263.

Jefferis, S. A. "Bentonite-Cement Slurries for Hydraulic Cut Offs," Proceedings of the Tenth International Conference on Soil Mechanics and Foundation Engineering, Stockholm, June 15-19, 1981b, A. A. Balkema, Rotterdam, pp. 435-440.

Jefferis, S. A., "Contaminant-Grout Interaction," in-preparation, 1992.

LaGrega, M. L., Buckingham, P. and Evans, J. C. Hazardous Waste Management, McGraw-Hill, NY, NY, (in press)

Ryan, C. R. "Vertical Barriers in Soil for Pollution Control", Geotechnical Practice for Waste Disposal '87, Geotechnical Special Publication No. 13, ASCE, June, 1987, pp. 182-204.

McCandless, R. M., and Bodocsi, A., "Investigation of Slurry Cutoff Wall Design and Construction Methods for Containing Hazardous Wastes." EPA 600-S2-87/063, U. S. Environmental Protection Agency, Cincinnati, Ohio, November 1987.

McCandless, R. M., and Bodocsi, A., "Hydraulic Characteristics of Model Soil-Bentonite Slurry Cutoff Walls", Proceedings of the 5th National Conference on Hazardous Wastes and Hazardous Materials, Las Vegas, Nevada, April, 1988, pp. 198-201.

McCandless, R. M., Bodocsi, A. and Ling, F. K., "Development of Hydraulic Defects in Model Soil-Bentonite Cutoff Walls," in preparation, 1993.

Millet, R. A., Perez, J.-Y., Davidson, R. R., "USA Practice Slurry Wall Specifications 10 years Later," Slurry Walls: Design, Construction and Quality Control, ASTM STP 1129, David B. Paul, Richard R. Davidson and Nicholas J. Cavalli, Eds., American Society of Testing and Materials, Philadelphia, 1992.

Mitchell, J. K. Fundamentals of Soil Behavior, John Wiley and Sons, Inc., New York, New York, 1976.

Mitchell and Madsen "Chemical Effects on Clay Hydraulic Conductivity", Geotechnical Practice for Waste Disposal '87, Geotechnical Special Publication No. 13, ASCE, June, 1987, pp. 87-116.

Mott, H. V. and Weber, W. J., "Diffusion of Organic Contaminants Through Soil-Bentonite Cut-Off Barriers," Research Journal WPCF, Vol. 63, No. 2, March/April, 1991, pp. 166-176.

Mott, H. V. and Weber, W. J., "Solute Migration in Soil-Bentonite Containment Barriers," Proceedings of the 10th National Conference on the Management of Uncontrolled Hazardous Waste Sites, HMCRI, November , 1989, Washington, D.C., pp. 526-533.

Olson, R. E., and Daniel, D. E., "Measurement of the Hydraulic Conductivity of Fine-Grained Soils," ASTM STP 746, pp. 18-64, 1981.

Salvesen, R. H. "Downtown Carcinogens - A Gaslight Legacy," Proceedings of the 5th National Conference on the Management of Uncontrolled Hazardous Waste Sites, HMCRI, November , 1984, Washington, D.C., pp. 11-15.

Sweidan, R. "Analysis of Soil-Bentonite Slurry Trench Cutoff Wall", Bucknell University, May 15, 1990.

USEPA   Slurry Trench Construction for Pollution Migration Control. EPA-540/2-84-001, U.S. Environmental Protection Agency, Cincinnati, Ohio, February, 1984.

Xanthakos, P. P. Slurry Walls, McGraw-Hill Book Company, New York. 1974

Yang, D. S., Luscher, U., Kimoto, I., and Takeshima, S., SMW Wall for Seepage Control in Levee Reconstruction, Proceedings of the 3rd International Conference on Case Histories in Geotechnical Engineering, June, 1993.

Beverly L. Herzog[1]

SLUG TESTS FOR DETERMINING HYDRAULIC CONDUCTIVITY OF NATURAL GEOLOGIC DEPOSITS

REFERENCE: Herzog, B. L., "Slug Tests for Determining Hydraulic Conductivity of Natural Geologic Deposits," *Hydraulic Conductivity and Waste Contaminant Transport in Soil, ASTM STP 1142*, David E. Daniel and Stephen J. Trautwein, Eds., American Society for Testing and Materials, Philadelphia, 1994.

ABSTRACT: This paper reviews four methods for analyzing slug test data to determine hydraulic conductivity and examines the effects of slug size on slug test results. Data from more than 100 slug tests, including tests for low permeability (glacial tills and fractured rock) deposits, coarse-grained (sand) deposits, and simulated tests were analyzed. Analysis of the data showed that all four analytical methods can produce similar values of hydraulic conductivity for low permeability deposits. Values determined for coarser deposits were highly dependent on method; values commonly spanned up to two orders of magnitude. The method of Cooper et al. (1967) generally produced the highest values, followed by the methods of Bouwer and Rice (1976), Hvorslev (1951), and Nguyen and Pinder (1984).

For sandy materials, tests were repeated on each well using several different slug sizes, ranging from 0.5 to 6 meters. Several of the glacial till materials were tested using two different slug sizes. Calculated hydraulic conductivity values were independent of these slug sizes.

KEYWORDS: hydraulic conductivity; slug tests; Cooper, Bredehoeft, and Papadopulos method; Bouwer and Rice method; Hvorslev method; Nguyen and Pinder method.

Because large differences have been found between values for hydraulic conductivity determined in the field and laboratory (Olson and Daniel 1981, Herzog and Morse 1986, and Muldoon et al. 1987), regulatory agencies are increasingly requiring field tests, which are believed to produce more accurate values. The most commonly used field test is the instantaneous head change test conducted in a piezometer, frequently called the slug test. When the water level in a piezometer is instantaneously lowered, the test is sometimes called a bail test. In a slug test, the hydraulic head in a piezometer is instantaneously raised or lowered by the addition or removal of water or a solid cylinder. Water-level measurements are collected until the water level in the well returns to the level it was before the slug was added or removed. This method is popular because of its simplicity and speed, especially for coarse-grained deposits. Hydraulic conductivity is calculated from time-versus-head data using one of several methods.

This paper examines both the field procedures and analytical methods used for slug tests. In particular, two questions are addressed.

---

[1]Senior Hydrogeologist, Illinois State Geological Survey, 615 E. Peabody Drive, Champaign, IL 61820.

First, how dependent are hydraulic conductivity values on the method of analysis? The methods of analysis examined are those by Hvorslev (1951), Cooper et al. (1967), Bouwer and Rice (1976), and Nguyen and Pinder (1984). Each has different assumptions about the hydraulics of the slug test, such as whether the aquifer is confined, whether the well fully penetrates the aquifer, and whether groundwater flow is horizontal or 3-dimensional.

Second, are slug tests results dependent on the size of the slug? Most procedures require that a slug be of sufficient size to induce a significant head change in the piezometer. This is similar to the philosophy for pumping tests. Significant head change has not been quantified, either in terms of head change or effect on test duration. Bradbury and Muldoon (1990) demonstrated that hydraulic conductivity values are dependent upon the scale of the test by looking at scales ranging from laboratory tests to regional flow models.

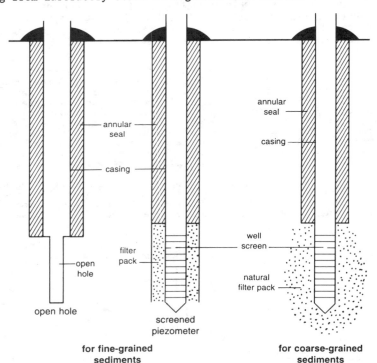

FIG. 1--Construction details for piezometers.

PROCEDURE

For this project, data were analyzed from 73 slug tests conducted in 40 piezometers located in Illinois. Three types of piezometer design were involved (Fig. 1). These included open-hole piezometers in glacial tills, screened piezometers with added filter packs in glacial tills, and screened piezometers with natural filter packs in sand and gravel deposits. Some results from the open-hole piezometers have been reported previously (Herzog and Morse 1990).

For the open-hole piezometers, a hollow-stem auger was used to bore to the desired depth, and a split spoon sample was collected to verify the geologic horizon. A 6.4-cm inner-diameter (ID) PVC casing was lowered through the hollow-stem auger to the bottom of the hole. The augers were then withdrawn from the hole and the annular space was sealed with expanding cement or a bentonite slurry. This assured that the tested interval was isolated to avoid the potential water level measurement problems, such as those noted by Chapuis and Sabourin (1989). After the surface seal had set, a 5-cm Shelby tube was lowered through the casing and pushed 0.6 m below the casing to create an open hole for the slug test. The hole and casing were then immediately filled with water, initiating the slug test and applying outward seepage forces to keep the hole open. This piezometer design minimizes the influences of side-wall smearing, the well screen, the filter pack and seepage from above on water levels. The open-hole design, however, can only be used in cohesive soils. Tests were conducted on 29 open-hole piezometers, 9 of which were constructed at a 45° angle to examine the effects of vertical fractures (Herzog and Morse 1990). In 8 of the 29 piezometers a second test was conducted using a smaller slug volume.

Screened piezometers were designed to serve as monitoring wells (Fig. 1). These were also drilled with hollow-stem auger and sealed from the top of the screen to ground level to assure that the tested interval was isolated. They were constructed of 5-cm ID PVC casing with a continuous-slot well screen. Screens were 0.4 to 0.8 meters long. Five tests were conducted in 4 screened piezometers completed in glacial tills. Seven screened piezometers were finished in sand deposits and a total of 36 slug tests were conducted using screened piezometers.

To supplement these data sets, 21 additional data sets were analyzed by the four methods. These included four data sets from Hsieh et al. (1983) for fractured rocks, 15 from Oertel (1993) for glacial deposits, and one each from Ferris and Knowles (1954), Lohman (1972), and Nguyen and Pinder (1984).

Parameters measured for analysis of slug test data are presented in Figure 2. These include radius or diameter of casing and well, length of screen or open hole, water level in piezometer throughout the test, aquifer thickness, and depth to piezometer bottom from water table. Not all parameters are needed for all methods of analysis. For example, only the Bouwer and Rice (1976) method requires the depth of the piezometer from the water table and aquifer thickness.

Slug sizes were arbitrary for tests in piezometers finished in glacial till deposits; generally the first slug was the volume of water required to fill the casing to the top. Because of the long time required for these tests, slugs for the second tests in these piezometers generally were less than 1 meter. Water levels were measured using a steel tape, an electronic water level meter or a transducer connected to a data logger.

Slug sizes in piezometers finished in sand deposits ranged from 0.5 to 2 meters in a medium sand to 1 to 6 meters in a coarser sand. Water levels were measured using a pressure transducer connected to a data logger capable of taking readings at 2-second intervals.

METHODS OF DATA ANALYSIS

Cooper, Bredehoeft, and Papadopulos Method

The first method used to analyze the results of the slug tests was the curve-matching method of Cooper et al. (1967). They produced a series of type curves by solving the partial differential equation representing flow in a confined aquifer. Additional type curves were later published by Papadopulos, Bredehoeft, and Cooper (1973). The analysis assumes an infinite, horizontal confined aquifer and that the open hole completely penetrates the aquifer to be tested. Formation and water compressibility are included in the analysis. The piezometers used

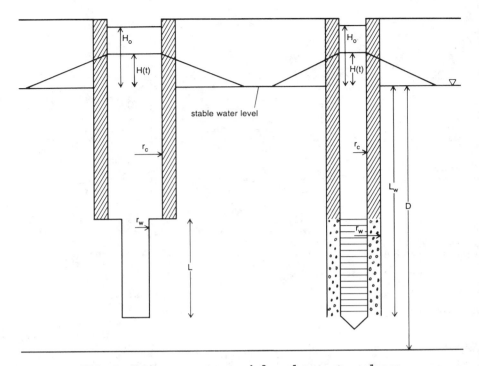

FIG. 2--Parameters measured for slug test analyses.

in this study did not completely penetrate the unit tested. However, because the length of the screen (or open hole) was much greater than its radius, horizontal flow was assumed for the vertical installations. Angle holes were expected to provide an average of the vertical and horizontal hydraulic conductivity.

During a slug test, water levels in the piezometer are measured until they stabilize. The stabilized (or initial) water level is used as the datum from which the heads in the piezometer are measured. The ratio of the measured head at any given time to the slug height ($H(t)/H_0$) is plotted on a logarithmic time scale as shown by the dots on Figure 3. This curve is then matched to one of the type curves presented by Papadopulos et al. (1973), (Fig. 3). Each type curve corresponds to a value of storativity for a particular radius of casing. From this curve match, the time (t) is selected such that $Tt/r_c^2 = 1$, where T is the transmissivity (L^2/t) of the tested material units and r_c is the radius of the casing, in which the head change occurs. The transmissivity is then determined from the equation

$$T = 1.0 \frac{r_c^2}{t} \tag{1}$$

As pointed out by Cooper et al. (1967), matching of the data to the type curves depends primarily on the slopes of the curves, which differ only slightly between adjacent curves. The determination of storativity by curve fitting is, therefore, of questionable reliability. In

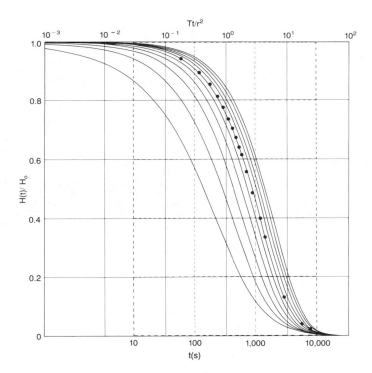

FIG. 3--Semilog plot of head ratio versus time for Illinois slug test data overlain on type curves for method Cooper et al. (1967). (Type curves from Papadopulos et al. 1973.)

contrast, the determination of transmissivity is relatively insensitive to the choice of type curves to be matched.

Because this method assumes full aquifer penetration, the length of vertical penetration of the open hole was used as the aquifer thickness. Hydraulic conductivity was then calculated by dividing the transmissivity obtained in equation (1) by the length of vertical penetration. These calculations were done using a computer program called AQUITEST which employed a nonlinear least squares technique to determine the optimal match from an infinite number of possible type curves. Of the 94 sets of field data, five could not be reasonably analyzed by this method because they produced curves that did not match any of the type curves.

Hvorslev Method

Hvorslev's (1951) basic time-lag method is based on the assumption that inflow at any time is proportional to the hydraulic conductivity and to the unrecovered head difference. His assumptions include a homogeneous, isotropic medium and incompressibility of both formation and water. The geologic unit may be confined or unconfined; if unconfined, the water table must be stable during the test. These assumptions lead to the differential equation

$$q(t) = \pi r_c^2 \frac{\partial H}{\partial t} = FK(H_0 - H(t)) \qquad (2)$$

where q is the rate of flow into the formation, F is a factor which depends on the shape and size of the intake area, K is the hydraulic conductivity and $H(t)$ is the head at any time, using the stabilized or initial head, H_0, as a datum.

Hvorslev also defined a basic time lag, T_0, as

$$T_0 = \frac{\pi r_c^2}{FK} \qquad (3)$$

For hydrostatic pressure changes, the time lag was defined as "the time for water to flow into or from the device until a desired degree of pressure equalization is obtained." The basic time lag, T_0, is the time that would be required for a complete pressure equalization if the original flow rate were maintained. When equation (3) is substituted into equation (2), and $H(t) = H_0$ is used as the initial condition, the solution to equation (2) is

$$\frac{H(t)}{H_0} = e^{-\frac{t}{T_0}} \qquad (4)$$

When the left hand of the equation is plotted against time, as shown in Figure 4, a straight line should result. The basic time lag is the time at which $t = T_0$ or when $H(t)/H_0 = 0.37$ (i.e. ln $(H(t)/H_0) = -1$). The

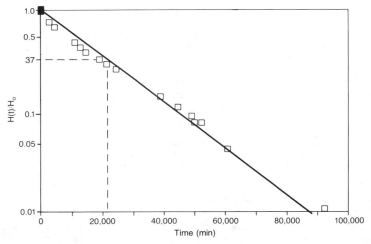

FIG. 4--Semilog plot of read ratio versus time for analysis by the method of Hvorslev (1951). Data set is the same as used in Figure 3.

hydraulic conductivity is then determined from equation (3). Hvorslev applied this equation for a laboratory permeameter and four field geometries. These four field situations are shown in Figure 5. The 94 slug tests in this study were analyzed using cases c and d, depending on whether the piezometer was at the top or middle of a unit. These analyses assumed an open hole or slotted pipe below the bottom of the casing. They differ only as to whether the open area is at the top or in the middle of the zone to be tested. Horizontal hydraulic conductivity for case c may be calculated by

$$K_h = \frac{d^2 \ln\left[\frac{2mL}{D} + \sqrt{1 + \left[\frac{2mL}{D}\right]^2}\right]}{8 L T} \tag{5}$$

where d = standpipe diameter (cm)
       m = transformation ratio, K_h/K_v
       L = intake length (cm)
       D = intake diameter (cm)
       T = basic time lag (s)
       K_h = hydraulic conductivity in the horizontal direction (cm/s)
       K_v = hydraulic conductivity in the vertical direction (cm/s).
If $(2mL)/D > 4$, equation (5) can be simplified to

$$K_h \approx \frac{d^2 \ln\left[\frac{4mL}{D}\right]}{8 L T} \tag{6}$$

Similarly, the horizontal hydraulic conductivity for case d may be calculated by

$$K_h = \frac{d^2 \ln\left[\frac{mL}{D} + \sqrt{1 + \left[\frac{mL}{D}\right]^2}\right]}{8 L T} \tag{7}$$

where the terms are defined as for equation (5). If $mL/D > 4$, equation (7) can be simplified to

$$K_h \approx \frac{d^2 \ln\left[\frac{2mL}{D}\right]}{8 L T} \tag{8}$$

A difficulty arises from using cases c and d in that the equation requires that the ratio of vertical to horizontal conductivity be known. For these analyses, the ratio was estimated from the values determined for paired vertical and angle holes using the Cooper et al. (1967) method. For formations without these data pairs, a value of 1 was assumed. Since equations (5) through (8) are not very sensitive to this ratio, this estimation technique should not be a significant problem. Of the 94 field data sets, 2 could not reasonably be described by a straight line. A least squares technique was used for the linear regression to determine K.

Alternatively, vertical hydraulic conductivities values for cases c and d can be determined using a two-stage test (Boutwell 1992). In this test, a slug test is conducted in a cased borehole, such as shown in Figure 5, cases a and b. After the first test is completed, the borehole is advanced through the casing, producing an open-hole piezometer, and a second slug test is performed.

Bouwer and Rice Method

The Bouwer and Rice (1976) method is based on the Thiem (1906) equation for steady flow to a well and was designed for the analysis of bail test data. This method is primarily applicable to fully or partially penetrating piezometers in homogeneous, isotropic, unconfined aquifers. It may also be applied to a confined aquifer if the aquifer receives water from the upper confining layer through leakage or

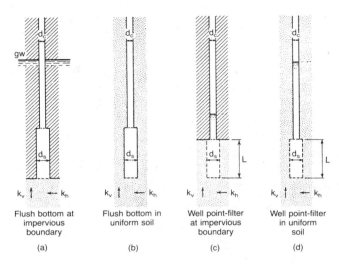

FIG. 5--Geometric configuration for use with the method of Hvorslev (after Hvorslev 1951).

compression. Assumptions include a stable water level around the piezometer, no flow above the water table, and negligible head losses as the water enters the piezometer. Although slug and bail tests are mathematically identical, the assumption of no flow above the water table may create problems when using this method to interpret slug test data.

Bouwer and Rice (1976) developed a pair of empirical equations for relating hole geometry to the effective radius of the bail test (R_e) using an electrical resistance network analog. For partially penetrating wells

$$\ln \frac{R_e}{r_w} = \left[\frac{1.1}{\ln(L_w/r_w)} + \frac{A + B\ln[(D-L_w)/r_w]}{L/r_w} \right]^{-1} \quad (9)$$

where L_w = depth from the water table to the bottom of the hole (cm)
r_w = radius of the hole or filter pack (cm)
D = aquifer thickness (cm)
L = length of open hole or screen (cm).
A and B = dimensionless coefficients which are functions of L/r_w.
Values for A and B are selected from Figure 6.

For fully penetrating wells, equation 9 becomes

$$\ln \frac{R_e}{r_w} = \left[\frac{1.1}{\ln(L_w/r_w)} + \frac{C}{L/r_w} \right]^{-1} \quad (10)$$

where C = dimensionless function of L/r_w selected from Figure 6.

From equations 9 or 10, hydraulic conductivity is calculated as

$$K = \frac{r_c^2 \ln(R_e/r_w)}{2L} \frac{1}{t} \ln \frac{H_0}{H(t)} \quad (11)$$

where all terms have been previously defined. Because the parameters to the left of $1/t$ are constant, $1/t \ln(H_0/H(t))$ must also be constant.

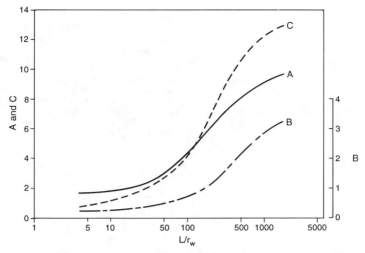

FIG. 6--Curves relating A, B, and C to L/r_w (from Bouwer and Rice 1976).

Thus when H(t) is plotted against t on semilog paper, as in Figure 7, a straight line should result. The slope of this line was determined by least squares linear regression to calculate K. Five of the 94 field data sets could not be analyzed by this method.

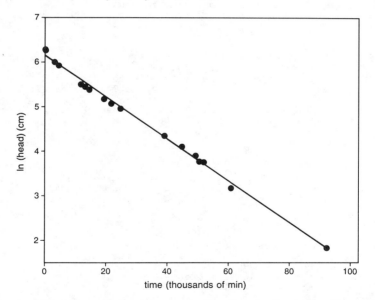

FIG. 7--Semilog plot of head versus time for analysis by the method of Bouwer and Rice (1976). Data set is the same as plotted in Figure 3.

Nguyen and Pinder Method

The method of Nguyen and Pinder (1984) was designed to consider problems involving partially penetrating wells in infinite aquifers in which the short term effects of a water table or leakage from a confining layer can be ignored. To this end, they solved the 3-dimensional axisymmetric form of the continuity equation. Energy loss due to friction or other sources of energy dissipation are ignored in this discussion, but well-bore storage is included. The parameters to be measured are the same as in the Cooper et al. (1967) method (Fig. 2). Nguyen and Pinder thought their method was most appropriate for slug tests in materials of moderate to low hydraulic conductivity.

The method of Nguyen and Pinder (1984) is illustrated in Figures 8 and 9. Figure 8 is a log-log plot of head, H(t), versus time, t. The

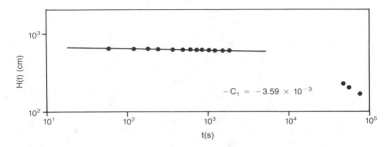

FIG. 8--Log-log plot of head versus time for analysis by method of Nguyen and Pinder (1984). The data set plotted is the same that was plotted in Figure 3.

slope of this line (C_1) is used to calculate both the storage coefficient (S) and the hydraulic conductivity (K). When the data begin to deviate from a straight line on the log-log plot, the boundary conditions are no longer being met and the test is over. Figure 9 is a semi-log plot of $1/t$ versus $\ln(-dh(t)/dt)$. The slope of this line (C_2) is then used in the calculation of K. Both slopes were determined by linear regression. The aquifer parameters, K and S are calculated as follows:

$$K = \frac{r_c^2 C_1}{4 C_2 L} \qquad (12)$$

$$S = \frac{r_c^2 C_1}{r_s^2 L} \qquad (13)$$

where:
- r_c = radius of the well casing (cm)
- r_s = radius of the hole or filter pack (cm)
- L = length of the screen or open hole (cm)
- C_1 = slope from log-log plot
- C_2 = slope from semilog plot (sec).

The method is appropriate until a finite "effective time", t_e, is reached. Until t_e, the assumptions for this method remain satisfied. From a practical standpoint, t_e appears to coincide with the time for which the computation of $\Delta H/\Delta t$ from the data exhibits questionable

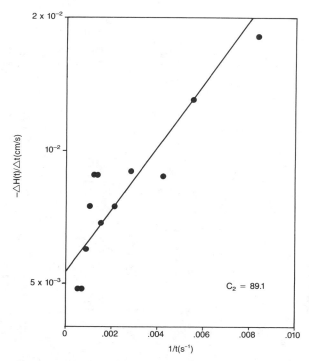

FIG. 9--Semilog plot of 1/t versus $H(t)/\Delta t$ analysis by the method of Nguyen and Pinder (1984). The data set plotted is the same that was plotted in Figure 3.

accuracy. After t_e, the plots are no longer linear. For our calculations, the t_e was determined from the graphs for each test and the slopes were obtained by linear regression. Because this method solves the governing equation and results in a straight-line plot, significant deviation from a line indicates either inaccurate data or an inappropriate mathematical model.

Five of the 94 sets of field data could not be analyzed by the Nguyen and Pinder (1984) method because the data did not produce a straight line on both plots.

RESULTS

Values for hydraulic conductivity, calculated by the four methods ranged from near exact agreement to differences of more than two orders of magnitude. Within each type of material tested, the range of hydraulic conductivities calculated was much greater. These results are summarized in Table 1. Where multiple tests were performed on a lithologic unit, field data are presented as the geometric means because hydraulic conductivity values in field soils are generally acknowledged to be log-normally distributed (Nielson et al. 1973; Rogowski 1972); therefore, arithmetic averaging would not be appropriate (Keisling et al. 1977). For this computation, only one value from each boring was used; if multiple tests were made of the same boring, a mean was taken of these values and that value was used to calculate the geometric mean for the

Table 1--Summary of Hydraulic Conductivity Data.

Material & Orientation[1]	Cooper, Bredehoeft & Papadopulos Method		Hvorslev Method		Bouwer & Rice Method		Nguyen & Pinder Method	
	Geometric Mean (cm/sec)	Log Standard Deviation (cm/sec)	Geometric Mean (cm/sec)	Log Standard Deviation (cm/sec)	Geometric Mean (cm/sec)	Log Standard Deviation (cm/sec)	Geometric Mean (cm/sec)	Log Standard Deviation (cm/sec)
ISGS tests								
Till 1 V	1×10^{-7}	0.72	1×10^{-7}	0.02	2×10^{-8}	0.67	1×10^{-7}	...
Till 1 A	1×10^{-6}	...	3×10^{-7}	...	2×10^{-7}	...	4×10^{-7}	...
Till 2 V	3×10^{-5}	1.13	1×10^{-5}	1.12	1×10^{-5}	0.88	2×10^{-5}	0.88
Till 2 A	2×10^{-5}	1.41	8×10^{-6}	1.10	2×10^{-5}	0.54	2×10^{-5}	0.62
Till 3 V	2×10^{-6}	1.26	8×10^{-7}	1.19	3×10^{-6}	1.39	2×10^{-6}	1.38
Till 3 A	2×10^{-6}	0.35	2×10^{-6}	0.35	5×10^{-6}	0.88	8×10^{-6}	0.33
Till 4 V	4×10^{-8}	0.74	4×10^{-8}	1.13	2×10^{-7}	0.84	7×10^{-8}	0.76
Till 4 A	1×10^{-6}	0.57	6×10^{-7}	0.89	4×10^{-6}	0.70	7×10^{-7}	0.32
Till 5 V	7×10^{-6}	0.94	3×10^{-6}	1.06	5×10^{-6}	0.98	3×10^{-6}	0.86
Sand 1 V	1×10^{-1}	0.41	3×10^{-2}	0.22	9×10^{-2}	0.25	2×10^{-3}	0.41
Sand 2 V	2×10^{-2}	0.15	4×10^{-3}	0.08	3×10^{-3}	0.09	6×10^{-4}	0.09
Sand 3 V	2×10^{-2}	0.25	5×10^{-3}	0.11	3×10^{-3}	0.06	1×10^{-3}	0.42
Sand 4 V	6×10^{-1}	...	2×10^{-2}	...	2×10^{-2}	...	...	...
Other data sets								
Ferris & Knowles (1954)	2×10^{-1}	...	7×10^{-2}	...	2×10^{-2}	...	...	...
Lohman (1972)	5×10^{-4}	...	9×10^{-4}	...	7×10^{-4}	...	1×10^{-4}	...
Hsieh et al. (1983)	2×10^{-6}	0.26	2×10^{-6}	0.08	2×10^{-6}	0.05	4×10^{-7}	0.21
Nguyen & Pinder (1984)	4×10^{-2}	...	2×10^{-2}	...	3×10^{-2}	...	2×10^{-3}	0.86
Oertel (1993)	8×10^{-5}	1.02	1×10^{-4}	1.17	2×10^{-4}	0.61	2×10^{-5}	0.33

[1] V = denotes vertical piezometer
    A = denotes piezometer installed at 45° angle

unit. This procedure was followed to give equal weight to all borings. For some glacial till formations, the data had a range of about three orders of magnitude. Because of the heterogeneity of glacial tills, this variability was not unexpected. However, the geometric means of hydraulic conductivity values calculated by the four methods for all low permeability units were within one order of magnitude for all lithologic units tested and well within one standard deviation for all glacial till units. Similarly, Welby (1992) compared Bouwer and Rice's (1976) method with Hvorslev's (1951) method for saprolitic deposits and found that neither could be judged better.

Agreement among the geometric means of hydraulic conductivity for the coarse-grained deposits was much worse than for the fine-grained deposits. Values produced by Cooper et al. (1967) were the highest. These are probably inaccurate because these tests did not meet the assumptions of a confined aquifer and fully penetrating piezometer. The Nguyen and Pinder (1984) method produced the lowest values. The methods of Hvorslev (1964) and Bouwer and Rice (1976) produced values between the two other methods and also produced the most consistent values, as shown by the low standard deviation values. Since these two methods are based on the same equation, they should be expected to produce similar results. Although it is quite clear that the Cooper et al. (1967) method is not applicable to these unconfined aquifers, it is not clear which of the remaining three methods yields the most accurate values.

To address this problem, nine several theoretical data sets were developed using the Cooper et al. (1967) equation and analyzed by the four methods. This method was selected because its assumptions can be met by all four methods. The theoretical data sets, which covered a wide range of simulated hydraulic conductivity and storage coefficient values, eliminated any possible problem with measurement error. The difference in calculated hydraulic conductivity remained systematic for hydraulic conductivity values greater than approximately 10^{-4} cm/s.

The differences between calculated hydraulic conductivity values can explained examining the 3-dimensional equation used by Nguyen and Pinder (1984) and the 2-dimensional equation used by Cooper et al. (1967). The former includes the term $\partial^2 h/\partial z^2$, which is missing in the latter. If all other factors are equal, $\partial^2 h/\partial z^2 \propto 1/K$. Thus, the 2-dimensional equation places an upper bound on the value of hydraulic conductivity. The other two methods become 3-dimensional by incorporating shape factors. These quasi 3-dimensional equations should produce values between the fully 2- and 3- dimensional equations. Thus, for coarse-grained deposits, one must estimate the importance of vertical flow to select the correct analysis method.

The effect of slug size on hydraulic conductivity was also examined. Slugs in coarse-grained sediments ranged in volume from 1 to 12 liters; these should correspond to head changes of 0.5 to 6 meters. Slugs in fine-grained deposits ranged from 0.5 to 2 meters. Figure 10 is based on the 27 tests performed in Sand 1. It is evident from the graph that hydraulic conductivity was independent over the range of slug sizes tested. Similarly, no correlation between slug size and hydraulic conductivity was discovered for any of the other sand or till units tested.

For coarse-grained sediments, the time required to inject the slug becomes important because of the assumption of an instantaneous head change for all methods of analysis. From a practical standpoint, very small slugs are dissipated before many measurements can be taken. In one test using a 0.1 meter slug in a highly conductive sand, the entire recovery curve was missed with by the first 2-second reading. Conversely, the largest slugs sometimes required 10 to 14 seconds to inject, eliminating the early part of the recovery curve. Despite the large difference in slug sizes, tests with slugs of 0.5 to 6 meters of head change produced a maximum of 30 seconds of usable data. For slugs of at least 0.5 meters, the size of the slug should be governed by the speed at which it can be introduced and the amount of data produced.

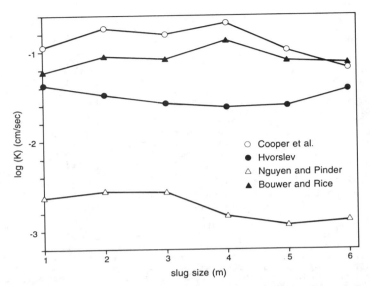

FIG. 10--Effect of slug size on calculated hydraulic conductivity for Sand 1.

CONCLUSIONS

This paper examined the effects of slug size and method of analysis on hydraulic conductivity values. Calculated values of geometric means from tests in fine-grained deposits were independent of calculation method, despite the different assumptions of each method. These are deposits that are not normally considered aquifers, leading to the violation of such assumptions as confined or unconfined conditions and fully penetrating wells. While some data sets did not match the type curves for the Cooper et al. (1967) method or produce straight lines for the other three methods, all data sets appeared to be suitable for at least one of the calculation methods. However, because values of hydraulic conductivity can vary significantly for a single piezometer depending on the calculation method, one should pick the method that best reflects the physical setting of the piezometer. However, if one is unsure of the physical setting, it may be advisable to calculate hydraulic conductivity by several of the methods.

Calculated values for hydraulic conductivity from tests in material with hydraulic conductivity of greater than approximately 10^{-4} cm/s were highly dependent on the method of calculation. The method of Cooper et al. (1967) generally produced the highest values, whereas the method of Nguyen and Pinder (1984) produced the lowest. The methods of Hvorslev (1954) and Bouwer and Rice (1976) produced similar results. Values of hydraulic conductivity varied up to two orders or magnitude. This pattern held for both real and simulated data sets. An examination of the analytical equations solved explains that a difference should be expected. Thus, for highly permeable deposits, selection of analysis method is extremely important.

Slug test results were found to be independent of slug size for slug of 0.5 to 6 meters. In coarse-grained deposits, slug size should be a function of the speed of introduction of the slug and amount of data produced. As a practical consideration in fine-grained deposits, relatively small slugs are preferred because they reduce the time required to conduct the test.

ACKNOWLEDGMENTS

Many of the piezometers used in this study were installed under research contracts sponsored by the U.S. Environmental Protection Agency, Land Pollution Control Division, Hazardous Waste Engineering Research Laboratory in Cincinnati, Ohio and the Illinois Department of Agriculture in Springfield, Illinois. Although these agencies sponsored the piezometer installations as part of larger projects, this research was not subjected to these agencies for review and, therefore, does not necessarily reflect the views of either agency.

Colleagues at the ISGS assisted in this research. Walter Morse and Michael Barnhardt assisted in the field work for the slug tests. Manoutcher Heidari provided the PC program he developed, called AQUITEST, for slug test calculations by the Cooper et al. (1967) method and the Bouwer and Rice (1976) method. Allan Oertel of Crawford, Murphy and Tilley supplied 15 data sets.

REFERENCES

Boutwell, G. P., 1992, "The STEI Two-Stage Borehole Field Permeability Test", <u>Containment Liner Technology and Subtitle D</u>, seminar sponsored by the Geotechnical Committee, Houston Branch, American Society of Civil Engineers, Houston, TX, March 12, 1992.

Bouwer, H. and Rice, R. C., 1976, "A Slug Test for Determining Hydraulic Conductivity of Unconfined Aquifers with Completely Penetrating Wells," <u>Water Resources Research</u>, V. 12, No. 3, pp. 428-428.

Bradbury, K. R. and Muldoon, M. A., 1990, "Hydraulic Conductivity Determinations in Unlithified Glacial and Fluvial Materials," <u>Ground Water and Vadose Zone Monitoring</u>, <u>ASTM STP 1053</u>, D. M. Nielsen and A. I. Johnson, Eds., American Society for Testing and Materials, pp. 138-151.

Chapuis, R. P. and Sabourin L, 1989, "Effects of Installation of Piezometers and Wells on Groundwater Characteristics and Measurements," <u>Canadian Geotechnical Journal</u>, V. 26, No. 4. pp 604-613.

Cooper, H. H., Bredehoeft, J. D. and Papadopulos, I. S., 1967, "Response of a Finite-Diameter Well to an Instantaneous Change of Water," <u>Water Resources Research</u>, V. 3, No. 1, pp. 263-269.

Ferris, J. G. and Knowles, D. B., 1954, <u>The Slug Test for Estimating Transmissibility</u>, U. S. Geological Survey Ground Water Notes 26, 7 pp.

Herzog, B. L., and Morse, W. J., 1986, "Hydraulic Conductivity at a Hazardous Waste Disposal Site: Comparison and Laboratory- and Field-Determined Values," <u>Waste Management and Research</u>, V. 4, No. 2, pp. 177-187.

Herzog, B. L., and Morse, W. J., 1990, "Comparison of Slug Test Methodologies for Determination of Hydraulic Conductivity in Fine-Grained Sediments," <u>Ground Water and Vadose Zone Monitoring</u>, <u>ASTM STP 1053</u>, D. M. Nielsen and A. I. Johnson, Eds., American Society for Testing and Materials, pp. 152-164.

Hsieh, P. A., Neuman, S. P., and Simpson, E. S., 1983, <u>Pressure Testing of Fractured Rocks: a Methodology Employing Three-Dimensional Cross-Hole Tests</u>. U. S. Nuclear Regulatory Commission, Washington D. C., NUREG/CR-3213, 176 pp.

Hvorslev, M.J, 1951, Time Lag and Soil Permeability in Ground-water Observation, Bulletin 36, Waterways Experiment Station, U.S. Army Corps of Engineers, Vicksburg, MS.

Keisling, T. C., Davidson, J. M., Weeks, D. L. and Morrison, R. D., 1977, " Precision with Which Selected Soil Parameters Can be Estimated," Soil Science, V. 124, 4, pp. 241-248.

Lohman, S. W., 1972, Ground-Water Hydraulics, U.S. Geological Survey Professional Paper 708, 70 pp.

Muldoon, M. A., Attig, J. W., Bradbury, K. R., and Mickelson, D. M., 1987, "Hydraulic Conductivity and Pre-Late-Wisconsin Till Units in Central Wisconsin," Abstracts with Programs, Geological Society of America, V. 19, No. 4, p. 235.

Nielson, D. R., Biggar, J. W., and Ehr, K. R., 1973, "Spatial Variability in Field-Measured Soil-Water Properties," Hilgardia, Vol. 42, pp. 215-259.

Nguyen, V. and Pinder, G.F, 1984, "Direct Calculation of Aquifer Parameters in Slug Test Analysis," Groundwater Hydraulics, Water Resources Monograph 9, J. Rosensheim and G. D. Bennet, Eds., American Geophysical Union, p. 222-239.

Oertel, A, 1993, personal communication.

Olson, R. E. and Daniel, D. E., 1981, "Measurement of Hydraulic Conductivity of Fine-Grained Soils," Permeability and Groundwater Contaminant Transport, ASTM STP 746, T. F. Zimmie and C. O. Riggs, Eds., American Society for Testing and Materials, Philadelphia, pp. 18-64.

Papadopulos, S. S., Bredehoeft, J. D., and Cooper, H. H., 1973, "On the Analysis of Slug Test Data," Water Resources Research, V. 9, No. 4, pp. 1087-1089.

Rogowski, A. S., 1972, "Watershed Physics: Soil Variability," Water Resources Research, V. 8, No. 4, pp. 1015-1023.

Thiem, G. Hydrologische Methoden, Leipzig: J. M. Gebhardt Verlag, 1906, 56 p.

Welby, C. W., 1992, "Field Investigation of Hydraulic Conductivity in Saprolitic materials - Comparison of Methods and Techniques," Bulletin of the Association of Engineering Geologists, v. 29, no. 2, pp. 119-130.

Charles D. Shackelford[1]

WASTE-SOIL INTERACTIONS THAT ALTER HYDRAULIC CONDUCTIVITY

REFERENCE: Shackelford, C. D., "Waste-Soil Interactions that Alter Hydraulic Conductivity," Hydraulic Conductivity and Waste Contaminant Transport in Soil, ASTM STP 1142, David E. Daniel and Stephen J. Trautwein, Eds., American Society for Testing and Materials, Philadelphia, 1994.

ABSTRACT: Liquid-soil compatibility of soil hydraulic conductivity is reviewed with respect to the materials and methods used in evaluating waste-soil interactions, factors influencing interpretation of test results, and the interactions that are thought to alter significantly the hydraulic conductivity of clay soils. Significant increases in hydraulic conductivity may result from flocculation of clay particles due to interactions with electrolyte solutions, shrinkage of the soil matrix in the presence of concentrated organic solvents, and acid-base dissolution of the soil. Observed effects typically are greater in rigid-wall permeameters than flexible-wall permeameters.
 Considerable evidence supports the use of the Gouy-Chapman theory for describing the influence of aqueous solutions on the fabric and, therefore, the hydraulic conductivity of clay soils. However, swelling test results suggest that the Gouy-Chapman theory does not account properly for shrinkage effects which have been observed to result in large increases in hydraulic conductivity upon permeation with concentrated organic solvents.
 Three mechanisms may contribute to an increase in the hydraulic conductivity of clay soils upon permeation with acid permeants: (1) flocculation of the clay, (2) dissolution of the clay minerals (aluminosilicates), and (3) dissolution of other minerals (e.g., $CaCO_3$) in the clay soil. Dissolution and piping of clay minerals leads to increases in hydraulic conductivity. Dissolution of carbonates initially leads to buffering, re-precipitation, pore clogging, and a decrease in hydraulic conductivity. Depletion of the buffering capacity leads to a decrease in pH, dissolution of constituents, and a possible increase in hydraulic conductivity.
 The measured hydraulic conductivity of a compacted sand-bentonite mixture is shown to be significantly affected by the sequence of permeation to a saturated calcium solution. The effect, termed "first exposure", has important implications with respect to the application of laboratory test results for evaluation of the suitability of a material for a waste containment barrier.

KEYWORDS: permeability, hydraulic conductivity, compatibility, laboratory testing, waste disposal, clay liners, soil fabric, diffuse double layer, volume change, inorganic chemicals, organic chemicals

---

[1]Associate Professor, Department of Civil Engineering, Colorado State University, Fort Collins, CO 80523.

INTRODUCTION

The results of numerous studies have shown that interactions between soil particles and the pore liquid between the soil particles may alter significantly the hydraulic conductivity of clay soil (e.g., see Anderson and Jones 1983; Griffin and Roy 1985; Madsen and Mitchell 1987; Mitchell and Madsen 1987; and Goldman et al. 1988). These results have significant ramifications with respect to the use of clay soils for containment of wastes since an adverse interaction may result in unexpectedly large release rates of contaminants from disposal facilities into the surrounding environment.

The primary focus of this paper is to describe the properties of the soils and liquids typically associated with waste containment barriers which affect liquid-soil interactions, the influence of the methods and procedures for evaluating the effect of waste-soil interactions on hydraulic conductivity, and the results of studies indicating the influence of various waste-soil interactions in altering hydraulic conductivity.

HYDRAULIC CONDUCTIVITY

One-dimensional flow of liquids in and through soil may be described by Darcy's law, or

$$q = \frac{Q}{A} = ki = -k\frac{\partial h}{\partial l} \qquad (1)$$

where q is the liquid flux (length/time, LT^{-1}), Q is the volumetric flow rate (L^3T^{-1}), A is the total (solids plus voids) cross-sectional area of the soil perpendicular to the direction of flow (L^2), k is the hydraulic conductivity or coefficient of permeability of the soil (LT^{-1}), i is the hydraulic gradient (dimensionless), h is the total head (L), and l is the direction of flow (L). The proportionality constant, k, is a function of the properties of both the soil and the liquid in accordance with the following equation:

$$k = K\frac{\gamma}{\mu} = K\frac{\rho g}{\mu} = K\frac{g}{\upsilon} \qquad (2)$$

where K is the intrinsic, absolute, or specific permeability of the soil (L^2), γ is the unit weight of the liquid (mass/length^2/time, $ML^{-2}T^{-1}$), μ is the absolute or dynamic viscosity of the liquid ($ML^{-1}T^{-1}$), ρ is the mass density of the liquid (ML^{-3}), g is acceleration due to gravity (LT^{-2}), and υ is the kinematic viscosity of the liquid (L^2T^{-1}). The intrinsic permeability, K, represents the influence of the soil structure on the hydraulic conductivity. In some disciplines, q is referred to as the Darcian velocity, v, and the symbol k is used to represent intrinsic permeability whereas the symbol K is used to represent hydraulic conductivity (e.g., see Freeze and Cherry, 1979).

Based on Eq. (2), differences in hydraulic conductivity, k, can result from differences in the intrinsic permeability, K, or from differences in the ratio of unit weight to viscosity, γ/μ, of the liquid. For example, a decrease in K and, therefore, k, may occur from physical forces, such as externally applied loads, resulting in consolidation or compression of the soil with a subsequent decrease in the pore space available for flow. Differences in k also may result even in the absence of changes in intrinsic permeability when the liquid permeating a soil (e.g., water) is changed to a liquid with a significantly different ratio of unit weight to viscosity (e.g., oil). In addition, internal interactions between the liquid in the pore spaces of the soil and the solid soil particles may result in structural changes in the soil which alter the intrinsic permeability of the soil. Such structural changes typically occur when a soil is permeated with liquids with significantly different chemical properties, such as in the case of clay soils used for waste containment barriers. In the case where the ratio γ/μ is not significantly different for different liquids, the effect of these liquid-soil interactions also is reflected directly by changes in hydraulic conductivity, k.

Based on the above considerations, it is important to distinguish between changes in k due to changes in the value of γ/μ versus changes in k resulting from liquid-soil interactions. In addition, an evaluation of the potential effects of liquid-soil interactions on the hydraulic conductivity of a soil requires consideration of not only the properties of the soil and the liquid which might influence such interactions but also the influence of the test conditions, such as the stress conditions on the soil.

MATERIALS AND METHODS

Waste Liquids

Waste liquids in the environment may result from several sources, e.g., uncontrolled dumping of pure solvents, spills, or infiltration of water through solid waste resulting in leachate. As indicated in Fig.1, waste liquids may be categorized as aqueous liquids or solutions containing contaminants which are miscible in water (also known as hydrophilic or "water-loving"), non-aqueous liquids consisting of organic compounds immiscible in water (also known as hydrophobic or "water-hating"), or mixtures of both aqueous and non-aqueous liquids resulting in the formation of two separate liquid phases. The migration of mixtures containing two separate liquid phases in liquid saturated soil, referred to as "two-phase flow", requires special consideration for the interaction between the two liquid phases. This interaction is particularly important in the case of permeation of a water saturated soil by a non-polar organic liquid. Due to surface and interfacial tension effects, the non-polar liquid (a "non-wetting fluid") can displace the free water (or "wetting fluid") in the pore space of the soil only after a minimum pressure, known as the "entry pressure", for the non-wetting fluid is reached. In the case of air flow through soil, this minimum pressure is known as the "air entry pressure".

Aqueous liquids contain inorganic chemicals (acids, bases, salts) and/or hydrophilic organic compounds. Hydrophilic organic compounds are distinguished from hydrophobic organic compounds based on the concept of "like dissolves like", i.e., polar organic compounds usually will readily dissolve in water, a polar molecule, whereas nonpolar organic compounds are repelled by water. A hydrophobic compound also is further separated into either a LNAPL (Light Non-Aqueous Phase Liquid) or DNAPL (Dense Non-Aqueous Phase Liquid) based on whether the density of the compound is lower or greater than water, respectively.

In some cases, the actual waste liquid is either unknown or is not readily available for use in laboratory tests. In such cases, a synthetic liquid made to simulate the properties expected for the actual waste liquid typically is used. The chemical properties of the synthetic

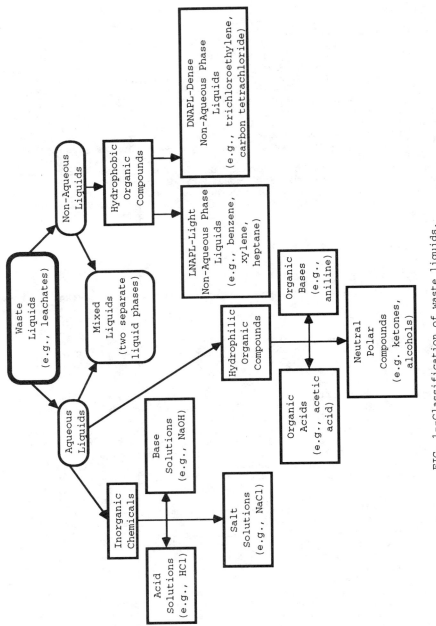

FIG. 1--Classification of waste liquids.

waste liquid which typically are measured or controlled include (1) the concentration and types of chemicals present in the liquid, (2) the density and viscosity of the liquid, (3) the pH and specific (or electrical) conductance (EC) of aqueous liquids, and (4) the polarity, solubility, and dielectric constant (ε) of organic compounds. Other parameters, such as the electron potential (Eh), total dissolved solids (TDS), total organic carbon (TOC), chemical oxygen demand (COD), and biological oxygen demand (BOD), also may be important.

The concentration and type of solutes in the pore fluid play an important role in determining the fabric of soil. In addition, available evidence indicates that only pure organic solvents significantly alter soil hydraulic conductivity (e.g., Daniel and Liljestrand 1984; Bowders 1985; Bowders and Daniel 1986; Daniel et al. 1988; and Ilgenfritz et al. 1988). The significance of the density and viscosity of the liquid already has been mentioned in terms of Eq. 2 as well as in the migration of DNAPLs and LNAPLs.

The pH of solutions ranges from 0 to 14. Solutions which are highly reactive with soils include strong acids (pH ≤ 2) and strong bases (pH ≥ 12). Neutral solutions are characterized by pH values between about 5 and 9. The pH of a solution not only characterizes chemical reactions (e.g., dissolution/ precipitation) which are important in terms of soil hydraulic conductivity but also can affect significantly soil fabric.

The specific or electrical conductance (EC) of a solution is a measure of the ability of a solution to carry an electrical current, and varies directly with the number and type of ions in solution. As a result, the electrical conductance is an indirect indication of the strength of the solution and, therefore, has been correlated with the ionic strength of inorganic solutions as follows (Griffin and Jurinak, 1973):

$$I = 0.013 EC \qquad (3)$$

where EC is in milliSiemens/cm (mS/cm) at 25°C, and I is the ionic strength (moles/liter). The ionic strength is based on the charges and concentrations of ions in solution as follows:

$$I = \frac{1}{2} \sum_{i=1}^{j} c_i z_i^2 \qquad (4)$$

where c_i and z_i are the molar concentration and charge, respectively, of the ionic species i, and j is the number of all ionic species in the solution. The empirical relationship given by Eq. 3 has a high correlation (r=0.996) based on 27 soil extracts and 124 river water samples, and is valid for ionic strengths ≤ 0.46M.

Electrical conductance (EC) provides an indirect measure of the potential for an electrolyte solution to affect the fabric of a clay soil. In addition, EC is particularly useful for indicating the relative strength of a solution when a detailed characterization of the concentration and types of chemical constituents is not possible.

The magnitudes of the significant properties of some organic compounds representing several different classes of organic chemicals are summarized in Table 1. Water also is included in Table 1 for comparison. A brief discussion of some of these properties is warranted.

Dipole moments occur in molecules containing an uneven distribution of electron charge due to an asymmetric molecular structure. The dipole moment of a molecule is a measure of the polarity

Table 1 - Properties of water and organic compounds used in compatibility testing of clay soils[1] (compiled from Bowders 1984; Griffin and Roy 1985; CRC Handbook of Chemistry and Physics 1985; Madsen and Mitchell 1987; and Acar and Olivieri 1989).

Compound Name	Compound Formula	Absolute Viscosity ($\times 10^{-3}$ N-s/m^2)	Density (kg/m^3)	Dielectric Constant	Dipole Moment (debyes)	Molar Volume ($\times 10^{-4}$ m^3/mole)	Water Solubility (kg/m^3)
Acetic Acid	$C_2H_4O_2$	1.16	1049.2	(6.15)	1.04;1.74	0.572	∞
Acetone	C_3H_6O	0.316	(789.9)	20.7	1.66; 2.9	(0.735)	∞
Aniline	$C_6H_5NH_2$	3.71(4.40)	(1021.7)	(6.89)	1.55	(0.911)	36
Benzene	C_6H_6	(0.65)	(876.5)	2.274	0	(1.02)	0.7
Carbontetra-chloride	CCl_4	(0.969)	1594	2.228	0	0.965	0.8
Cyclohexane	C_6H_{12}	1.02	(778.5)	2.015	0	(1.08)	<0.3
1,4-dioxane	$C_4H_8O_2$	(1.44)	(1033.7)	2.209	0.45	(0.852)	∞
Ethanol	C_2H_5OH	(0.2)	(789.3)	24.30	1.69	(0.584)	∞
Ethylene Glycol	$C_2H_6O_2$	(21)	(1108.8)	(38.66)	2.28	(0.560)	∞
Heptane	C_7H_{16}	0.386(0.409)	(683.7)	(1.0)	0	(1.08)	<0.3
Methanol	CH_3OH	0.547 (0.597)	(791.4)	32.63	1.70	(0.405)	∞
Nitrobenzene	$C_6H_5NO_2$	(2.03)	1204	34.82	4.22	1.02	2
Phenol	C_6H_5OH	(12.7)	(157.6)	(13.13)	1.45	(0.890)	86
Tetrachloro-ethylene	C_2Cl_4	(0.72)	1620	(2.30)	0	1.02	1.0
Toluene	C_7H_8	(0.590)	(866.9)	2.379	No Data	(1.06)	0.5
Trichloro-ethylene	C_2HCl_3	(0.58)	1464	3.42	0; 0.90	(0.897)	1.1
Water	H_2O	1.01	997.07 (998.23)	78.5 (80.4)	1.83	0.181	997.07 (998.23)
m-Xylene	C_8H_{10}	(0.620)	(864.2)	(2.374)	No Data	(0.123)	0.2
o-Xylene	C_8H_{10}	(0.810)	(880.2)	(2.568)	0.62	(0.121)	0.2
p-Xylene	C_8H_{10}	(0.648)	(861.1)	(2.270)	0	(0.123)	0.2

[1] Values in ( ) are at 20°C; otherwise values are at 25°C.

of the molecule, i.e., the greater the magnitude of the dipole moment, the greater the polarity of the molecule. For example, the polar water molecule, H_2O, is asymmetrical and possesses a dipole moment of about 1.83 debyes (1 debye = 3.34 x 10^{-30} coulomb-meters, C-m). Symmetrical molecules, such as hexane (C_6H_{12}), typically have zero dipole moments and, therefore, are nonpolar. Based on the premise that "like dissolves like", a knowledge of the dipole moment of a molecule provides an indication of the hydrophilic or hydrophobic nature of the compound; i.e., the greater the dipole moment, the more polar the molecule and the more likely the compound will be miscible in water. This trend is evident from the data in Table 1.

The dielectric constant, ε, is a particularly important property since the magnitude of the dielectric constant of a pore liquid can have a profound effect on the fabric and hydraulic conductivity of clay soils. The dimensionless dielectric constant of any material (not just liquids) is a measure of the ability of the material to conduct an electric field relative to vacuum, and is defined as the ratio of the capacitance with a dielectric material between two charged plates to the capacitance with a vacuum between the same two charged plates. The dielectric constant also affects the magnitude of the force, F (N), between two point charges, q_1 and q_2 (C), separated a distance r (m) in accordance with Coulomb's law, as follows:

$$F = \frac{q_1 q_2}{\varepsilon \varepsilon_o 4 \pi r^2} \quad (5)$$

where ε_o is the permitivity constant (=8.85418 x 10^{-12} $C^2N^{-1}m^{-2}$). Therefore, the greater the dielectric constant of the medium, the smaller the force of attraction between opposite point charges (or repulsion between the same point charges).

The molar or molecular volume is the ratio of molecular weight to the density of a compound which provides an indication of the size of the compound. A knowledge of the molar volume may aid in interpretation of test results. For example, Anderson and Jones (1983) describe test results in which an initial decrease in hydraulic conductivity followed by a large increase was observed when a compacted clay soil was permeated with acetone. This trend was attributed to the offsetting influences of the larger dipole moment and molar volume of acetone relative to water initially present in the soil. Initial permeation of the clay resulted in a dilution of the acetone, a slightly greater dipole moment for the pore liquid, swelling of the clay soil, and a slight decrease in hydraulic conductivity. However, the acetone eventually replaced most of the pore water after further permeation resulting in a large increase in hydraulic conductivity. Anderson and Jones (1983) attributed this increase in hydraulic conductivity to shrinkage of the clay soil due to the greater molar volume of acetone relative to water.

Soils

In most cases, relatively low ($\leq$ 1 x 10^{-9} m/s) hydraulic conductivity values are required for soils used for waste containment barriers. Since low hydraulic conductivity values are associated only with clay soils, only soils with a significant clay content are considered for use as waste containment barriers. However, the same properties which make clay soils desirable from the standpoint of the ability to achieve relatively low hydraulic conductivity values also

typically make the same clay soils more susceptible to adverse interactions with waste liquids. Therefore, evaluations of the effects of waste liquids on soil hydraulic conductivity usually are made with respect to clay soils and, in particular, the clay content of clay soils.
 The clay soils considered for use as waste containment barriers typically include *in situ* and recompacted natural clays, compacted mixtures of natural or processed (mined and treated or purified) clays (e.g., bentonite) with other soils (e.g., sand) or materials (e.g., cement), and clay slurry suspensions with backfill material. The principal clay minerals encountered in these soils are the layered silicate minerals (aluminosilicates), primarily illite, kaolinite, and montmorillonite (a smectite), and the chain structure mineral, attapulgite (e.g., see Mitchell 1993).
 Illite occurs naturally in many clay soils, some of which have been evaluated for the effects of liquid-soil interactions on hydraulic conductivity (e.g., Hardcastle and Mitchell 1974; Bowders 1985; and Foreman and Daniel 1986). Processed clay soils containing primarily kaolinite, also known as kaolins, have been used extensively in the study of liquid-soil interactions (e.g., Bowders 1985; Bowders and Daniel 1986; Bowders et al. 1986; and Acar and Olivieri 1989). Montmorillonites typically are the major clay mineral constituents in bentonite clay soils used frequently as soil admixtures in both compacted clay liners and geosynthetic clay liners (GCLs) used to impound solid and liquid wastes as well as an additive for soil slurries in the construction of vertical cutoff walls (i.e., slurry walls) used for waste containment barriers (Alther 1987; and Evans 1993). The amount of bentonite used in clay soil liners typically ranges between 6 and 15 percent by dry weight (Alther, 1987). The most common bentonite clays are sodium bentonite or calcium bentonite depending on whether the principal cation on the exchange complex of the clay is sodium (Na-montmorillonite) or calcium (Ca-montmorillonite), respectively. Processed clays containing attapulgite, sometimes referred to as attapulgus, are used extensively for gelling drilling fluids in salt water environments (Sawyer 1983), and have been used in slurries in the construction of vertical cutoff walls (Tobin and Wild 1986; and Ryan 1987). In addition to these clay minerals, other layered silicate minerals, such as vermiculite and chlorite, also have been encountered in soils evaluated with respect to waste-soil interactions affecting hydraulic conductivity (e.g., Bowders 1985; and Daniel et al. 1988).
 The clay content of clay soils is evaluated in terms of particle size and mineralogy. Atterberg limits (ASTM Test for Liquid Limit, Plastic Limit, and Plasticity Index of Soils, D 4318) are correlated with the type of clay mineral present in the clay fraction of the soil and usually are used *in lieu* of a more direct analysis of the clay mineral content based on x-ray diffraction, scanning electron microscopy, differential thermal analysis, or other more sophisticated techniques (e.g., see Mitchell 1993). Activity, A, combines the influence of the plasticity of the soil and the clay fraction of the soil, or:

$$A = \frac{PI(\%)}{\% < 2\mu m} \tag{6}$$

where PI is the plasticity index, i.e., the difference between the liquid limit (LL) and the plastic limit (PL). In general, the greater the activity of the soil, the more reactive the soil. For example, montmorillonite with a relatively large activity tends to be more reactive (e.g., more susceptible to swelling and/or shrinking) than kaolinite with a lesser activity, and vice versa. A comparison of the activities and other properties of three processed clay soils typically considered for use in waste containment barriers is provided in Table 2.

Other compositional factors of clay soils which can play a significant role in waste-soil interactions include the pH and EC of the soil (typically determined on saturated extracts of the soil), carbonate content of the soil (e.g., calcium carbonate, $CaCO_3$, and dolomite, $CaMg(CO_3)_2$), organic matter content, metal oxide content (e.g., Fe_2O_3), cation exchange capacity (CEC), exchangeable cations, soluble salts, sodium adsorption ratio (SAR), and specific surface area. Detailed descriptions regarding the measurement of these parameters are provided by Mitchell (1993) and Page et al. (1982). Measured values for some of these other compositional factors for the three processed clay soils presented in Table 2 are summarized in Table 3.

Methods

Modified Atterberg Limits Tests--Modified Atterberg limits tests using various liquids other than water have been used to provide a qualitative indication of the effect of the liquid on the hydraulic conductivity of the soil. The results of several studies involving modified Atterberg limits tests are provided in Table 4.

TABLE 2--*Physical properties of three processed clay soils typically evaluated for use as waste containment barriers.*

Property	Reference	Processed Clay Soils		
		Attapulgus	Bentonite	Kaolin
Source	Product Information	Floridin Co., Quincy, FL	Eisenman Chemical Co., Greeley, CO	Georgia Kaolin Co., Union, N.J.
Trade Name	Product Information	"Florigel H-Y"	"ECCO Gel"	"Standard Air Float"
Principal Mineral	Product Information	Attapulgite	Na-Montmorillonite	Kaolinite
Water Content	ASTM D 4959	8.62	4.18	1.11
Liquid Limit	ASTM D 4318	353%	461%	41.0%
Plasticity Index	ASTM D 4318	238%	427%	17.0%
% < 2µm	ASTM D 422	76	84	61
Specific Gravity, G_s	ASTM D 854	2.58	2.82	2.65
Activity, A		3.1	5.1	0.28
Classification	ASTM D 2487	CH	CH	CL

TABLE 3--Chemical properties of three processed clay soils typically evaluated for use as waste containment barriers.

Property	Processed Clay Soils		
	Attapulgus	Bentonite	Kaolin
Saturated Extracts:			
% Saturation (g solution/g soil)	515	620	125
pH @ 25°C	9.8	8.7	4.8
Electrical Conductance @ 25°C (μS/cm)	500	300	100
Soluble Salts (meq/liter):			
Ca^{2+}	0.4	0.8	0.2
Mg^{2+}	3.5	0.2	0.1
Na^+	1.1	47.8	0.7
K^+	1.0	0.2	<0.1
Sum	5.0	49.0	<1.1
Sodium Adsorption Ratio, SAR[1]	0.8	68	1.8
Exchangeable Cations (meq/100 g):			
Ca^{2+}	22.4	33.5	4.9
Mg^{2+}	36.1	5.1	0.3
Na^+	0.7	55.0	0.2
K^+	1.4	1.0	0.0
Si^{4+}	2.0	3.8	1.1
Fe^{3+}	0.1	0.2	0.1
Al^{3+}	0.0	0.1	0.1
Sum	62.7	98.7	6.7

[1] SAR = $\{Na^+\}/[(\{Ca^{2+}\}+\{Mg^{2+}\})/2]^{0.5}$, where {} are concentrations in meq/L.

For example, Bowders (1985), Bowders and Daniel (1986), and Bowders et al. (1986) performed Atterberg limits tests and hydraulic conductivity tests with various aqueous concentrations of methanol (0, 20, 40, 60, 80, and 100% methanol, by volume) and a kaolinite soil, and found that increases in both the liquid limit, plastic limit, and hydraulic conductivity all occurred only after the methanol concentration was greater than 80 percent. Daniel et al. (1988) also found no significant differences in the Atterberg limits or hydraulic conductivity of three natural soils when dilute organic liquids were used in the tests.

Acar et al. (1985) and Acar and Olivieri (1989) performed liquid limit and plastic limit tests with three soils, a kaolinite (A=0.32), Na-montmorillonite (A=4.5), and Ca-montmorillonite (A=2.8), and 11 pure organic compounds with various dielectric constants and water. The response in the liquid limit of the kaolinite soil as a function of dielectric constant was variable. However, the response of the montmorillonite soils indicated an increase in liquid limit with an increase in dielectric constant, a trend opposite to that observed by

TABLE 4--Results of some modified Atterberg Limits tests.

Soil Type	Liquid Type	Concentration	Atterberg Limts (%) LL	PL	PI	References
Kaolinite	Water	100%	58	34	24	Bowders (1985); Bowders et al. (1986); Daniel and Bowders (1986)
	Methanol	20%	60	32	28	
		40%	60	31	29	
		60%	59	32	27	
		80%	58	33	25	
		100%	74	45	29	
	Acetic Acid	20%	62	34	28	
		40%	66	36	30	
		60%	73	41	32	
		80%	80	44	36	
		100%	87	56	31	
	Heptane	53 mg/L	56	32	34	
		100%	82	NP	NP	
	Trichloro-ethylene	1100 mg/L	57	31	26	
		100%	158	NP	NP	
Illite-Chlorite	Water	100%	33	18	15	Bowders (1985); Bowders et al. (1986); Daniel and Bowders (1986)
	Methanol	20%	35	16	19	
		40%	32	16	16	
		60%	33	17	16	
		80%	30	19	11	
		100%	27	22	5	
	Acetic Acid	20%	30	15	15	
		40%	30	17	13	
		60%	30	20	10	
		80%	34	20	14	
		100%	37	28	9	
	Heptane	53 mg/L	35	16	19	
		100%	24	NP	NP	
	Trichloro-ethylene	1100 mg/L	36	16	20	
		100%	47	NP	NP	

TABLE 4--Results of some modified Atterberg Limits tests (continued).

Soil Type	Liquid Type	Concentration	LL	Atterberg Limts (%) PL	PI	References
Lufkin Clay (smectite)	Water	100%	48	19	29	Foreman and Daniel (1984)
	Methanol	5%	52	21	31	Daniel and Liljestrand (1984)
		100%	33	NP	NP	Foreman and Daniel (1984)
	Xylene	196 mg/L	53	20	33	Daniel and Liljestrand (1984)
		100%	28	NP	NP	"
Hoytville Clay (illite)	Methanol	0%	48	19	29	Foreman and Daniel (1984)
		100%	35	29	6	"
Soil S1 (chlorite)	Water	100%	52	20	32	Daniel et al. (1988);
	Methanol	5%	50	25	25	Daniel and Liljestrand (1984)
		100%	34	NP	NP	
	Xylene	196 mg/L	51	24	27	
		100%	29	NP	NP	
	Leachate L1	ε=80;TOC=1440 mg/L; pH=6.9; EC=23,700 μS/cm	53	23	31	
	Leachate L2	ε=83;TOC=13 mg/L; pH=7.1; EC=865 μS/cm	54	23	31	
	Leachate L3	ε=83;TOC=82 mg/L; pH=1.5; EC=4,620 μS/cm	52	21	31	
	Spiked L1	L1 w/200 mg/L TCE and 200 mg/L chloroform	61	23	38	
Soil S2 (smectitie)	Water	100%	87	28	59	
	Methanol	5%	79	44	35	
		100%	53	NP	NP	
	Xylene	196 mg/L	81	36	45	
		100%	38	NP	NP	
	Leachate L1	ε=80;TOC=1440 mg/L; pH=6.9; EC=23,700 μS/cm	85	33	52	
	Leachate L2	ε=83;TOC=13 mg/L; pH=7.1; EC=865 μS/cm	88	32	56	
	Leachate L3	ε=83;TOC=82 mg/L; pH=1.5; EC=4,620 μS/cm	82	32	50	
	Spiked L1	L1 w/200 mg/L TCE and 200 mg/L chloroform	93	30	63	

TABLE 4--Results of some modified Atterberg Limits tests (continued).

Soil Type	Liquid Type	Concen-tration	Atterberg Limts (%) LL	PL	PI	References
Soil S3 (illite)	Water	100%	36	20	16	Daniel et al. (1988); Daniel and Liljestrand (1984)
	Methanol	5%	36	21	15	
	Xylene	100%	30	NP	NP	
		196 mg/L	36	20	16	
	Leachate L1	100%	32	NP	NP	
		ε=80;TOC=1440 mg/L; pH=6.9; EC=23,700 μS/cm	34	20	14	
	Leachate L2	ε=83;TOC=13 mg/L; pH=7.1; EC=865 μS/cm	34	18	16	
	Leachate L3	ε=83;TOC=82 mg/L; pH=1.5; EC=4,620 μS/cm	34	19	15	
	Spiked L1	L1 w/200 mg/L TCE and 200 mg/L chloroform	33	21	12	

Bowders (1985), Bowders and Daniel (1986), and Bowders et al. (1986) for their kaolinite soil, but consistent with the trends observed by Foreman and Daniel (1984) for the liquid limits of smectitic and illitic soils with methanol relative to water (see Table 4). Acar et al. (1985) concluded that changes in liquid limit of kaolinite were consistent with the small changes in the intrinsic permeability of the kaolinite. However, Acar and Olivieri (1989) concluded that significant decrease in the liquid limit of a high activity soil (e.g., montmorillonite) resulting from the use of an organic compound with a low dielectric constant (i.e., relative to water) provides an indirect and conservative indication that the soil would be expected to shrink with a concomitant increase in hydraulic conductivity upon permeation with the organic liquid.

"Standard" Compatibility Test Procedure--There currently is no official standard test procedure for evaluating the effect of waste-soil interactions on hydraulic conductivity. However, the unofficial "standard" laboratory test procedure typically is performed by permeating a soil specimen with water to establish a baseline value for the hydraulic conductivity followed by permeation of the same specimen with the waste liquid to determine the change in hydraulic conductivity. These tests frequently are referred to as "compatibility tests" since the purpose of the test often is to determine whether or not the soil and the waste permeant are compatible with respect to hydraulic conductivity. As a result of this test procedure, the final hydraulic conductivity value after permeation with the waste liquid, k_f, is measured relative to an initial hydraulic conductivity value, k_i, established by permeating the soil with water. A large change in hydraulic conductivity ($k_f/k_i \gg 1$) is an indication that the waste permeant and soil are not compatible. If there is a significant difference between the ratio of γ/μ for the waste permeant and water, the evaluation of compatibility should be made in terms of the ratio of intrinsic permeability values, K_f/K_i.

Compatibility Test Equipment

Permeameters--Several studies have shown that large differences in measured hydraulic conductivity values can occur depending on the type of permeameter used in the test (e.g., Foreman and Daniel 1984; and Acar et al. 1985). Permeameters (permeability cells) used to evaluate the effects of waste-soil interactions on the hydraulic conductivity of soils usually can be separated into three categories (e.g., see Zimmie et al. 1981; Daniel et al. 1984, 1985; Bowders et al. 1986; Carpenter and Stephenson 1986; Evans and Fang 1986; Madsen and Mitchell 1987; and Mitchell and Madsen 1987): (1) rigid-wall or fixed-wall permeameters, (2) flexible-wall or triaxial-cell permeameters, and (3) consolidation-cell (consolidometer or oedometer type) permeameters. Advantages and disadvantages of each type of permeameter are listed in Table 5. Additional comparisons can be found in Daniel et al. (1985).

For testing relatively undisturbed samples of natural clay, rigid-wall permeameters may consist simply of a Shelby tube with attached top and bottom caps (Fig. 2a). For compacted specimens, the rigid-wall cell typically consists of a standard Proctor mold (ASTM Standard Test Methods for Moisture-Density Relations of Soils and Soil-Aggregate Mixtures Using 5.5-lb (2.49-kg) Rammer and 12-in. (305-mm) Drop D 698) fitted with top and bottom caps (Fig. 2b). In this case, the cells frequently are referred to as compaction-mold permeameters. Although not

TABLE 5--Advantages and disadvantages of permeameters for studying waste-soil compatibility (modified after Madsen and Mitchell, 1987).

Permeameter Type	Advantages	Disadvantages
Rigid-Wall (also referred to as Fixed-Wall or Compaction-Mold)	Low cost Simplicity Useful for compacted soil No confining pressures required Side-wall leakage may provide rapid indication of incompatibility between soil and permeant	Difficult to saturate (back-pressure not recommended) Volume and/or deformation changes cannot be measured or controlled Stresses on specimen are unknown and uncontrollable Side-wall leakage may occur even if soil and waste are compatible Large hydraulic gradients may result in hydraulic fracturing of specimen or piping.
Flexible-Wall or Triaxial Cell	Specimens can be back-pressure saturated Irregular specimen surfaces can be accommodated easily Side-wall leakage prevented or minimized Stresses on specimen can be controlled Volume changes and/or deformations can be measured Useful for undisturbed or compacted specimens	Membrane may be incompatible with chemical permeant Chemicals in permeant may diffuse through membrane Shrinkage and cracking of specimen may not be detected Large hydraulic gradients may result in unreasonable effective stresses in specimen Higher cost than rigid-wall permeameters
Consolidation-Cell (Consolidometer or Oedometer)	Vertical pressure in field can be simulated Vertical deformations can be measured A range of vertical stresses can be tested on one specimen Useful for undisturbed or compacted specimens Short testing time with thin specimens Cost effective for measuring hydraulic conductivity over a range of specimen states Upward flow aids saturation	Thin specimens may not be representative Potential for side-wall leakage Some samples may be difficult to trim into consolidation ring resulting in specimen disturbance Higher cost than rigid-wall permeameters Horizontal stresses on specimen are unknown (K_o is unknown) Back-pressure saturation not recommended

shown in Fig. 2b, a liquid reservoir may be placed between the cell and the top cap (e.g. see Daniel et al. 1985). The possibility that the waste-soil interaction will result in shrinkage of the soil specimen and significant leakage between the specimen and the inner wall of the permeameter, known as "side-wall leakage", is greater in rigid-wall cells than in either consolidation or flexible-wall cells.

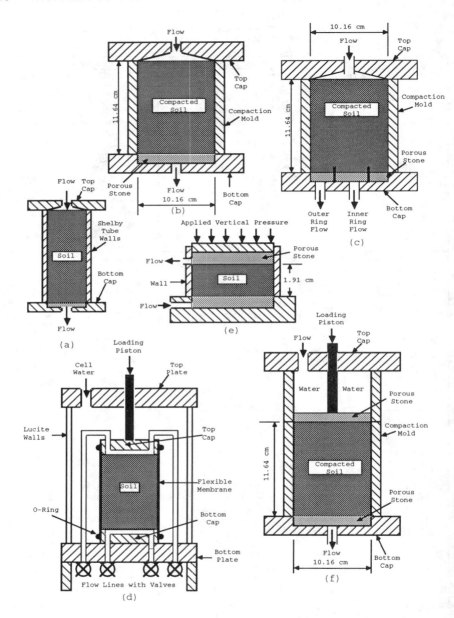

FIG. 2--Permeability cells: (a) Shelby tube; (b) compaction mold; (c) double-ring; (d) triaxial cell; (e) consolidation cell; and (f) oedometer-type compaction mold.

Anderson et al. (1985a) developed a double-ring rigid-wall permeameter to account for (not prevent) side-wall leakage by inserting a ring into the base of the permeameter to separate the outflow into inner and outer portions. Pierce et al. (1987) describe a triple-ring rigid-wall permeameter developed to study differential flow patterns through compacted clay specimens. A schematic of a double-ring permeameter is shown in Fig. 2c; similar double-ring permeameters have been used by McNeal and Coleman (1966) and Sai and Anderson (1991). In this case, the measured hydraulic conductivity is calculated to be a weighted-average value, as follows:

$$k = \frac{A_i k_i + A_o k_o}{A_i + A_o} \tag{7}$$

where A_i and A_o are the inner and outer areas of the base of the permeameter, and k_i and k_o are the hydraulic conductivity values determined with respect to the flow occurring through the inner and the outer base areas, respectively. Side-wall leakage is evident when $k_o \gg k_i$.

A triaxial-cell permeameter is depicted schematically in Fig. 2d. The cell may accommodate either a compacted soil specimen or an extruded soil sample recovered from a field boring. The specimen is covered by a flexible membrane which is sealed to bottom and top caps using O-rings to isolate the specimen from the annulus area between the membrane and the cell walls. A fluid, usually water, is placed in the annulus area and pressurized to apply normal stresses (cell pressures) to the outside of the specimen. Water pressure may be applied through the flow lines in the top and bottom caps to back-pressure saturate the specimen and/or to induce flow. In addition, a vertical load may be applied through the loading piston to vary the state-of-stress on the specimen. Therefore, the entire state-of-stress on the specimen can be controlled. In addition, the ability of the membrane to deform with the test specimen can prevent side-wall leakage, but also tends to underestimate the effect of volume changes resulting from waste-soil interactions (e.g., shrinkage of the test specimen upon exposure to a waste liquid). The term flexible-wall permeameter is more general than triaxial-cell permeameter in that the vertical loading piston is not necessary to perform a flexible-wall permeability test.

Consolidation-cell (oedometer-type) permeameters are rigid-wall permeameters that have been adapted to allow for a vertically applied load on the test specimen. Consolidation-cell (oedometer-type) permeameters traditionally have been used to measure hydraulic conductivity values between incremental loadings of a standard consolidation test for comparison with back-calculated hydraulic conductivity values based on Terzaghi's theory for one-dimensional consolidation (e.g., see Mesri and Olson 1971). A schematic diagram of a consolidation cell for permeability testing is shown in Fig. 2e. The major advantage of this type of permeameter is that the applied vertical stress on the soil specimen increases and/or maintains the lateral stress of the soil against the inner walls of the cell thereby reducing the likelihood of significant, if any, side-wall leakage. In addition, the application of vertical stresses to test specimens has been justified on the basis that the field compacted clay liners will be subject to significant overburden pressures from the overlying waste. However, it may take several years before the stress conditions assumed in the laboratory tests are realized in the field and, therefore, low stress conditions in laboratory tests may be more appropriate, at least with respect to the earlier years of operation of the facility. Also, volume changes resulting from waste-soil interactions during permeation may not be detected due to application of the vertical stress and may result in unconservative results (i.e., hydraulic conductivity values which are too low). As shown in Fig. 2f, this concept also can be applied to compaction-mold permeameters.

Other Equipment--Additional equipment required for laboratory compatibility testing includes a pressure control panel for controlling applied pressures (e.g., cell pressure, back pressure, and pressure gradient), and cell water, headwater, and tailwater accumulators (or reservoirs) for providing air-water interfaces and for allowing for flow measurements in the case of the constant-head test method and hydraulic head measurements in the case of the falling-head or falling headwater-rising tailwater test methods (see ASTM Standard Test Method for the Measurement of Hydraulic Conductivity of Saturated Porous Materials Using a Flexible-Wall Permeameter D 5084; and Daniel 1989). Daniel et al. (1984) and Evans and Fang (1986) describe pressure control panels and Daniel et al. (1984) describe in detail several different types of accumulators which can be used in permeability/compatibility tests. In addition, a flow pump with a differential pressure transducer can be used to perform constant flow tests (Olsen 1966; ASTM D 5084).

A bladder accumulator may be used to separate physically the waste liquid from the pressurized water to minimize human exposure to the waste liquid, as shown in Fig. 3. The flexible bladder, usually rubber, allows transmission of applied pressure yet prevents physical contact between the water and the waste liquid. The bladder also may help to reduce the dissolved air content in the waste liquid relative to the pressurized water. An elevated dissolved air content in the waste liquid may result in non-representative oxidation-reduction reactions in the test specimen during permeation.

In the case of permeation of volatile chemicals, such as methanol, care must be taken in sampling the effluent to minimize losses of the chemical to the environment. In these cases, a sampling system closed to the environment must be used. For example, Bowders et al. (1986) describe a system in which a syringe is used to obtain liquid samples on the order of 25µL through a septum fitted to a tee connection attached to the effluent line of the permeameter.

Equipment Materials--In most cases, compatibility test equipment must be made of non-corrosive, non-reactive materials. Permeameters, accumulators, and fittings typically are made of high grade stainless steel. In some cases, glass permeameters have been used to reduce the potential for adverse reactions between the permeameter material and organic solvents (e.g., Green et al., 1981; Ilgenfritz et al. 1988; and Abdul et al. 1990). Plastic tubing also must be non-reactive to the waste liquid. Teflon is used in most cases when concentrated organic solvents are the waste liquids.

Compatibility Test Considerations

In the "standard" compatibility test procedure, the test may be performed on soil specimens under a variety of test conditions using several different permeameters. In order to provide for an accurate evaluation of the test results, some consideration of these test conditions and permeameters is required. An extensive discussion of test conditions affecting the measurement of the hydraulic conductivity of fine-grained soil is provided by Olson and Daniel (1981).

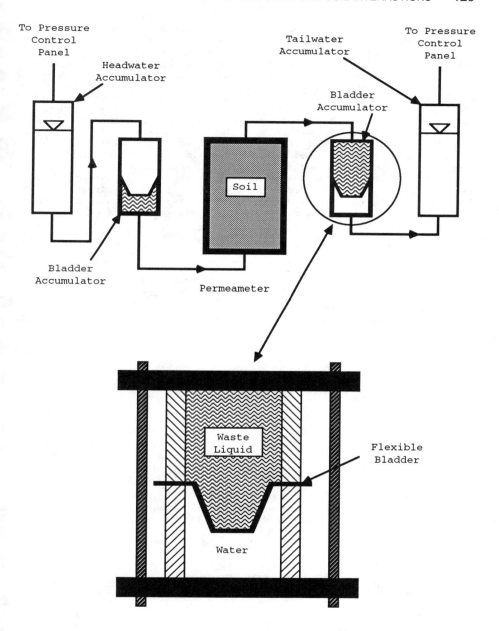

FIG. 3--Compatibility test apparatus with bladder accumulators.

Permeant Water--As shown in Table 6, different types of permeant water can be used to establish a baseline hydraulic conductivity of the test specimen before permeation with the waste liquid. However, the permeant water that should be used is the same water that is expected in the field application. In many cases, access to this water is either inconvenient or impossible, and a substitute water must be used. The use of a permeant water with low ionic strength, such as distilled water (DW) or deionized-distilled water (DDW), is not recommended because the resulting baseline hydraulic conductivity will be unrepresentative, i.e., too low (Olson and Daniel 1981; Dunn and Mitchell 1984; Evans and Fang 1988; and ASTM D 5084). A "standard water" consisting of a 0.01N $CaSO_4$ solution has been used in many cases in order to facilitate a comparison between the results of different studies and to increase the ionic strength of DDW to a value more representative of natural waters (e.g., Olson and Daniel 1981; Foreman and Daniel 1984; Acar et al. 1985; Bowders 1985; Fernandez and Quigley 1985; Bowders and Daniel 1986; and Bowders et al. 1986). In addition, ASTM D 5084 recommends using de-aired

TABLE 6--Types of permeant water based on electrical conductance and ionic strength at 25°C.

Water Type	Symbol[1]	Reference	Maximum Electrical Conductance[2], EC (μS/cm)	Maximum Ionic Strength[2], I (moles/liter)
Distilled	DW	Standard Methods (1985)	1.0	0.000013[3]
Deionized-Distilled	DDW	Standard Methods (1985)	0.2	0.0000026[3]
Type I Reagent	T1RW	ASTM D 1193 Standard Methods (1985)	0.06 0.1	0.00000078[3] 0.0000013[3]
Type II Reagent	T2RW	ASTM D 1193 Standard Methods (1985)	1.0 1.0	0.000013[3] 0.000013[3]
Type III Reagent	T3RW	ASTM D 1193 Standard Methods (1985)	1.0 10.0	0.000013[3] 0.00013[3]
"Standard": 0.01N $CaSO_4$	SW1	Olson and Daniel (1981)	1538[3]	0.02
0.005N $CaSO_4$	SW2	ASTM D 5084	769[3]	0.01
Tap Water	TW	NA	Variable	Variable

[1] Symbol designation based on this paper.
[2] Maximum values are based on assumption of complete dissociation.
[3] By correlation using I(moles/liter) = 0.013 x EC(mS/cm), where 1 mS/cm = mmho/cm (after Griffin and Jurinak, 1973).

water to maximize the potential for measurement of a saturated hydraulic conductivity. However, the use of de-aired water may result in the artificial inducement of anaerobic conditions within the test specimen and influence the effects, if any, of chemical reactions (e.g., oxidation-reduction) on the hydraulic conductivity.

In some cases, a "standard" permeant solution which has the same chemical properties as the analyzed soil pore water may be used. For example, Uppot and Stephenson (1989) specified a 0.004N magnesium sulfate heptahydrate (Epsom salt) solution for compatibility tests with a magnesium montmorillonite soil. When a permeant water is not specified, ASTM D 5084 recommends the use of tap water. Since tap water can vary widely in chemical properties based on location, a complete characterization of the chemical properties (e.g., soluble salts, pH, EC, etc.) of the tap water should be performed and reported with the test results. When tap water is not used, as well as in the case of extremely brackish water, ASTM D 5084 recommends the use of a 0.005N $CaSO_4$ solution. This "modified standard water" solution has been used in several recent studies (e.g., Abdul et al., 1990; Sai and Anderson 1991).

Membrane Compatibility--Organic solvents, such as xylene, are particularly destructive when placed in contact with rubber or latex (Daniel et al. 1984; Rad and Acar 1984; and Acar et al. 1985). Daniel et al. (1984) suggest wrapping the test specimen in two or three revolutions of Teflon tape (0.13 mm) with a width greater than the height of the test specimen before placing and sealing the membrane over the tape and test specimen. Although Teflon tape possesses some rigidity, Daniel et al. (1984) found that comparative tests on three compacted soils performed with and without Teflon tape using water as the permeant resulted in only minor differences in the measured hydraulic conductivity.

Rad and Acar (1984) determined the effect of eight different organic liquids on the deformation behavior of latex, polypropylene, and Teflon membranes. They found that latex membranes were incompatible with organic liquids with dielectric constants less than about seven. However, wrapping the test specimens in two layers of a Teflon sheet (0.03 mm) effectively remedied the problem. In addition, coating the membrane with silicon-based grease may be required to minimize the potential for chemical diffusion of the waste liquid through the membrane (Acar et al. 1985; and Acar and Olivieri 1989).

Back-Pressure Saturation--Current regulations typically require a maximum value for the saturated hydraulic conductivity of a soil used for waste disposal liners (e.g., $k \leq 1 \times 10^{-9}$ m/s). Therefore, test specimens in flexible-wall permeameters typically are back-pressure saturated before permeation. Back pressure applied equally to both ends of the specimen results in dissolving of air into the pore liquid of the specimen, due to an increase in the partial pressure of the air, resulting in an increase in the degree of saturation, S_r.

Although it is physically possible to back pressure rigid-wall specimens, it is not recommended (Edil and Erickson 1985). Since the effective stresses in rigid-wall permeameters are unknown and cannot be controlled, application of an excessive back pressure may result in expansion of the permeameter, side-wall leakage, and hydraulic fracturing. Hydraulic fracturing occurs when the effective stress in the test specimen is reduced to zero resulting in the formation of fractures or large channels. For example, Edil and Erickson (1985) found that back-pressuring compacted sand-bentonite specimens in rigid-wall permeameters resulted in measured hydraulic conductivity values which were as much as 1000 times greater than hydraulic conductivity values measured on rigid-wall specimens not back pressured as well as on flexible-wall specimens with or without back pressure. Edil and Erickson (1985) also noticed upon permeation of the test specimens with red dye that the back pressure apparently had resulted in the formation of channels between the specimen and the compaction mold resulting in sidewall leakage.

Since clays being evaluated as liner materials typically are compacted at relatively high initial degrees of saturation ($S_r \geq 80\%$) such that the final S_r for the specimen after permeation is close to 100 percent, the measured k value and the saturated k value typically are not significantly different and, therefore, back-pressure saturation may not be necessary, particularly when the measured k value is much less than the regulatory limit. However, this may not be the case for undisturbed specimens (Zimmie et al. 1981). Nevertheless, compacted clay specimens which are not back pressured more realistically represent field liner conditions.

Hydraulic Gradient and Stress Conditions--Elevated hydraulic gradients (e.g., i $\geq$ 100) typically are applied in laboratory hydraulic conductivity tests to speed the testing times. A number of studies have investigated the effect of applied hydraulic gradient and the associated stress levels on the measurement of the hydraulic conductivity of soils (e.g., Mitchell and Younger 1967; Olson and Daniel 1981; Edil and Erickson 1985; Boynton and Daniel 1985; Carpenter and Stephenson 1986; and Shackelford and Glade 1993). Most of these studies pertain to the use of flexible-wall tests since the effective stress conditions in rigid-wall and consolidometer-type hydraulic conductivity tests are unknown.

Based on these and other studies, the influence of hydraulic gradient on the flexible-wall test results is well known. In general, the hydraulic gradient in a flexible-wall test can be applied by either increasing the headwater pressure and/or decreasing the tailwater pressure (e.g., see Shackelford and Glade 1993). A decrease in the tailwater pressure to induce a hydraulic gradient is limited by the amount of back pressure applied to the test specimen and can result in release of dissolved air. In addition, a decrease in tailwater pressure will result in higher effective stresses and consolidation at the effluent end of the test specimen, and lower hydraulic conductivity values (e.g., Carpenter and Stephenson 1986). An increase in the headwater pressure is limited by the requirement to keep the cell pressure greater than the sum of the back pressure and the headwater pressure to maintain contact between the flexible membrane and the test specimen. In addition, excessively large hydraulic gradients may result in piping and migration of soil particles which, in turn, may result in clogging of the soil pores and a decrease in the hydraulic conductivity of the test specimen (Olson and Daniel 1981). Since relatively low hydraulic gradients are expected in the field, the effect of elevated hydraulic gradients in the laboratory is unconservative; i.e., flexible-wall tests tend to underestimate hydraulic conductivity.

In an effort to reduce these effects, ASTM D 5084 recommends maximum hydraulic gradients in flexible-wall hydraulic conductivity tests based on the hydraulic conductivity of the soil. For example, the maximum recommended hydraulic gradient for a soil with k < 1 x 10^{-9} m/s is 30. Carpenter and Stephenson (1986) recommend limiting the hydraulic gradient to a value which would prevent the effective stress in the soil specimen from exceeding the maximum previous consolidation stress, thus minimizing large volumetric strains with corresponding large decreases in hydraulic conductivity.

Although it is generally accepted that rigid-wall tests tend to overestimate hydraulic conductivity values for various reasons, principally side-wall leakage (e.g., Acar et al. 1985; Foreman and Daniel 1984, 1986; and Mitchell and Madsen 1987), less information on the effects of hydraulic gradient exists for rigid-wall tests. In terms of hydraulic gradient, as the headwater pressure at the inflow end of a rigid-wall test specimen is increased to apply a greater hydraulic gradient, a decrease in effective stress with a corresponding increase in void ratio at the inflow end of the specimen is expected resulting in an overall increase in measured hydraulic conductivity. This trend is evident in the data presented in Fig. 4 in which compacted specimens of fly ash, sand, bentonite, and/or cement were permeated with water under constant-head conditions in rigid-wall permeameters (Creek and Shackelford 1992). In all cases, the hydraulic conductivity increases as the hydraulic gradient increases from about 20 to about 100. After a gradient of about 100, there is only a slight change in k. In these tests, the tailwater pressure at the outflow end of the permeameter was essentially at atmospheric pressure and, therefore, the effective stress at the outflow end should have been greater than zero in all cases (see Creek and Shackelford 1992). As a result, the expected trend is opposite to that of a flexible-wall test; i.e., a larger hydraulic gradient in a rigid-wall test results in a higher measured hydraulic conductivity. However, excessively large hydraulic gradients may result in conditions which favor hydraulic fracturing within the specimen and/or side-wall leakage.

Based on consideration of the expected trends in the measurement of hydraulic conductivity using flexible-wall or rigid-wall hydraulic conductivity tests, Mitchell and Madsen (1987) recommend use of the consolidation-cell permeameter. The vertically applied stress conditions help to minimize side-wall leakage and allow for measurement of k under a range of stress conditions. While this capability for the consolidation cell permeameter is recognized, the question as to which permeameter is "best" depends on the intended purpose. For example, a rigid-wall test may be the best permeameter to indicate the potential for excessive shrinkage of the specimen upon exposure to the waste liquid. For measurement of saturated hydraulic conductivity at a prescribed effective stress, the flexible-wall permeameter is preferred since the test specimen can be back-pressure saturated relatively easily and the stresses can be controlled.

Hydraulic Gradient and Hydrophobic Compounds--The permeation of immiscible, non-polar liquids may require excessively large entry pressures to achieve permeation. Large entry pressures correlate to large hydraulic gradients in hydraulic conductivity tests. For example, Foreman and Daniel (1986) found that the entry pressure for permeating water saturated kaolin with non-polar heptane to be greater than 324 kPa (47 psi) for flexible-wall tests, corresponding to hydraulic gradients of more than 300. For similar tests in compaction-mold permeameters, Foreman and Daniel (1986) found that the heptane would enter the water saturated kaolin at an entry pressure of only 55 kPa (8 psi), or a hydraulic gradient of 50. The difference was attributed to a few relatively larger pores in the compaction-mold permeameter relative to the flexible-wall permeameter, presumably due to the absence of confinement in the rigid-wall permeameter.

Fernandez and Quigley (1985) describe constant-flow hydraulic conductivity tests in which two pore volumes of hydrophobic organic

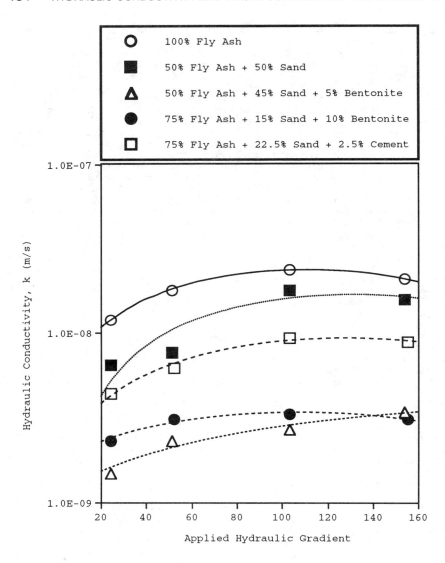

FIG. 4--Influence of hydraulic gradient on hydraulic conductivity of compacted fly ash-soil specimens in rigid-wall permeameters (data from Creek and Shackelford 1992).

compounds (benzene, xylene, and cyclohexane) were forced through saturated compacted natural soils over a period of 4.2 days. The flow rates resulted in induced hydraulic gradient of about 500, apparently due to the high entry pressures of the non-polar organics. Acar and Seals (1984) report that benzene could not be permeated through a water saturated Ca-montmorillonite at a hydraulic gradient of 150.

Since the large hydraulic gradients typically required to permeate water saturated clay soils with non-polar, hydrophobic organic compounds are unrealistic from a practical standpoint, the practical usefulness of evaluating waste-soil interactions of hydrophobic compounds under these laboratory conditions is questionable.

<u>Retention Time</u>--Another potential deleterious aspect of the application of excessively large hydraulic gradients is the generation of excessively large flow rates. Excessive flow rates in the laboratory may result in short retention (residence) times and inadequate exposure of the waste liquid to the soil, conditions which likely are not representative of the field conditions. Although little, if any, study of this effect has been made, such conditions would seem to be conducive to unconservative evaluations of waste-soil compatibility (e.g., see Bowders 1991).

<u>Termination Criteria</u>--The criteria required for termination of compatibility tests for soil hydraulic conductivity are subject to opinion. Pierce and Witter (1987) suggest that clay permeability tests be continued until (1) at least one pore volume of permeant is passed through the clay specimen and (2) the slope of hydraulic conductivity versus the number of cumulative pore volumes does not differ significantly from zero. A pore volume of flow (PVF) is the cumulative quantity of flow relative to the volume of void space conducting flow in the soil specimen, or:

$$PVF = \frac{Qt}{n_e AL} = \frac{kit}{n_e L} = \frac{vt}{L} \tag{8}$$

where L is the length of the test specimen [L], t is the cumulative time [T], v is the seepage velocity [LT^{-1}], and n_e is the effective porosity representing the pore space in the soil available for flow relative to the total volume of the test specimen. An effective porosity might exist, for example, when the soil specimen is characterized by dead-end pore space or as a result of resistance of highly viscous, adsorbed water in tightly packed clay soils (e.g., see Bear, 1972). However, in most cases, the effective porosity is assumed to be equal to the total porosity, n. In the case of unsaturated flow conditions, a clear distinction should be made between the volumetric water content, θ, representing the volume of water per unit volume of soil pertinent for liquid flow through unsaturated soil and the effective porosity, n_e.

Bowders et al. (1986) suggest that compatibility tests should continue until (1) at least two pore volumes of flow are achieved to ensure that the original soil water in the test specimen is replaced by the waste liquid (or leachate), and (2) the effluent concentration ratio (i.e., the effluent concentration, c, of each principal chemical species relative to the influent concentration, c_o, of the same chemical species) approaches unity when c_o is maintained constant. In addition, Bowders (1988) noted that the termination criteria proposed by Pierce and Witter (1986) were sound when the permeant is water, but suggested that the criteria are unacceptable in terms of compatibility tests because the criteria account only for hydraulic conductivity equilibrium, not chemical equilibrium. Bowders (1988) presented data which indicated that a significant increase (17X) in the hydraulic conductivity of a compacted specimen of a kaolin to acetic acid occurred suddenly and only after more than six pore volumes of flow. In addition, the hydraulic conductivity had stabilized essentially after six pore volumes of flow. Based on this example, both criteria proposed by Pierce and Witter (1986) as well as the criterion proposed by Bowders et al. (1985) requiring the passage of a minimum of two pore volumes of flow would not have been adequate. Bowders (1988) suggested that a number of chemical parameters, besides the relative effluent concentration of principal solutes, could be monitored, such as the pH of acid solutions and the total organic carbon (TOC) content of organic liquids. However, TOC is not always a reliable indicator of the potential impacts of organic liquids since two organic liquids with the same TOC can have different impacts depending on the composition of the organic liquids.

Based on these considerations, the criteria proposed by Pierce and Witter (1986) seem acceptable for the initial stage of the "standard" compatibility test procedure, i.e., establishing the initial or baseline value of hydraulic conductivity, k_i, to water. However, with respect to the second stage of the "standard" test procedure involving the use of a different permeant and the measurement of k_f, a criterion based on an arbitrary minimum number of pore volumes of flow does not appear suitable. Therefore, the primary criterion for termination of compatibility tests should be establishment of chemical equilibrium between the effluent and the influent of the test. The establishment of hydraulic conductivity equilibrium for k_f should be considered only after the chemical equilibrium criterion has been achieved.

<u>Measurement of pH and Electrical Conductance</u>--Two parameters which always should be measured are the electrical conductance(EC) and the pH. Both of these parameters are relatively easy to measure using electrodes and provide fairly reproducible results. However, in the case of pH, care should be taken to measure the effluent pH soon after the liquid sample is recovered because release of aqueous phase carbon dioxide ($CO_{2(aq)}$) into the atmosphere will result in a continual increase in the measured pH with time in accordance with the following equilibrium reactions:

$$\left. \begin{array}{l} CO_{2(aq)} \leftrightarrow CO_{2(g)} \\ H_2CO_3 \leftrightarrow CO_{2(aq)} + H_2O \\ H_2CO_3 \leftrightarrow HCO_3^- + H^+ \\ HCO_3^- \leftrightarrow CO_3^{2-} + H^+ \\ pH = -\log(H^+) \end{array} \right\} \tag{9}$$

Due to respiration by organisms, the carbon dioxide concentration ($CO_{2(aq)}$) in the pore liquid of a soil in a relatively closed system (e.g., in groundwater) may be several hundred times greater than it would be when exposed to the atmosphere (Stumm and Morgan 1981). As a result, when a groundwater sample is brought to the surface, $CO_{2(aq)}$, H_2CO_3, and H^+ all decrease in accordance with the reactions in Eq. 9 to re-establish equilibrium with the atmosphere and, therefore, the pH increases. This same sequence of events will occur in specimens being permeated with waste liquids, particularly when long-term tests in which the test specimen is under high back-pressures are performed.

For example, consider the effluent data in Fig. 5 for a flexible-wall hydraulic conductivity test performed on a compacted test specimen of the processed kaolin soil characterized in Tables 2 and 3. The test specimen was subjected to a back pressure of 517 kPa (75 psi) before

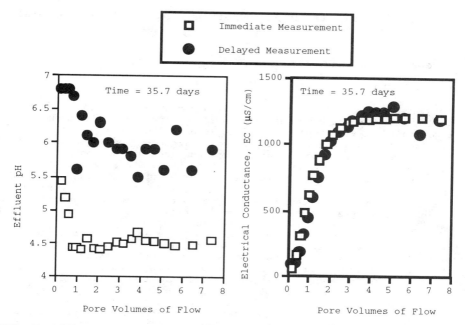

FIG. 5--Influence of delay in measurement of pH and electrical conductance (EC) on a back-pressure saturated specimen of compacted kaolin permeated in a flexible-wall permeameter with 0.01M NaCl solution.

permeation with a 0.01M NaCl solution over a period of 35.7 days. Effluent samples were recovered for pH and electrical conductance (EC) analysis. The pH and EC of the samples were either measured within minutes after collection (designated as "immediate" in Fig. 5), or placed in "air-tight" containers for pH and EC analysis by an independent laboratory at a later date (designated as "delayed" in Fig. 5). Although care was taken to minimize sample losses (e.g., $CO_{2(g)}$), it is evident that the "delayed" pH values are consistently higher than the "immediate" pH values by about 1.5 pH units. However, the EC measurements are very similar regardless of the elapsed time before measurement. Therefore, in order to measure an accurate pH of the pore liquid in the test specimen, the pH of the effluent sample should be measured as rapidly as possible after collection. Otherwise, elevated pH values should be expected resulting in improper interpretation of the test results, such as in the case of exposure of soils to acidic permeants.

Biological Activity--Biological reactions within the permeameter can affect the test results. This is particularly true in the case of organic compounds subject to biodegradation, as well as for tests performed over long periods. In some cases, permeants have been spiked with from 10 ppm to 500 ppm of mercuric chloride ($HgCl_2$) solutions to inhibit biodegradation of chemicals (e.g., McNeal and Coleman 1966; Ilgenfritz et al. 1988; and Klecka et al. 1990). Biodegradation may result in measured effluent concentrations which never reach chemical equilibrium relative to the influent (i.e., $c/c_o=1$).

WASTE-SOIL INTERACTIONS

Soil Fabric

Soil fabric refers to the the arrangement of soil particles in a soil matrix. The influence of soil fabric on the hydraulic conductivity of compacted clays is well recognized (e.g., Mitchell et al. 1965). Based on this knowledge, compacted clay soils used for waste containment liners usually are compacted wet of optimum moisture content to provide the lowest possible hydraulic conductivity. Since these compaction water contents typically are associated with degrees of saturation, S_r, ≥ 80 percent, a significant change in the fabric and, therefore, hydraulic conductivity of the compacted clay may result upon permeation of the compacted clay by a waste liquid with properties significantly different than those of the water used in compaction.

Mitchell and Madsen (1987) indicate three levels of soil fabric that are important for describing the flow of waste liquids through clay soils. The microfabric, or smallest level of fabric, refers to the arrangement of individual particles into small aggregations. Lambe (1958) further defined three types of microfabric, as illustrated in Fig. 6. Flocculated microfabrics are characterized by large micropores relative to dispersed microfabrics and, therefore, relatively larger hydraulic conductivities are associated with flocculated microfabrics. The minifabric is an intermediate scale of fabric representing the arrangement of several microfabric assemblages. The pores between microfabric assemblages, or inter-assemblage pores, are larger than micropores and, therefore, allow for much greater flow rates. The largest scale of fabric is the macrofabric consisting of large assemblages of both microfabric and minifabric assemblages, sometimes referred to as "clods" of soil. The transassemblage pores between clods, often referred to as "macropores", are significantly greater than interassemblage pores in minifabric, as illustrated in Fig. 6. Since liquids follow paths of least resistance, the flow rate through clay soils is dominated by the existence of macropores. Therefore, the objective in construction of clay soil barriers for waste containment is to destroy the clods and the associated macropores, thereby minimizing the hydraulic conductivity of the clay soil (Benson and Daniel 1990; and Shackelford and Javed 1991). However, even if the clods are destroyed during construction, post-construction effects (e.g., desiccation, freeze-thaw) can result in the formation of large macropores, such as cracks and fissures, and failure of the barrier material. Based on these considerations, two aspects of soil fabric with respect to the influence of waste-soil interactions are of prime importance: (1) the influence of the waste liquid on the microfabric of clay soils; and (2) the influence of the waste liquid on determining the volume changes of the soil matrix.

Influence of Waste Liquids on Soil Microfabric--In terms of microfabric, the principles of colloidal chemistry generally are used to describe the influence of liquids on clay soils (e.g., see Lambe 1958, van Olphen 1963; and Mitchell 1993). The application of colloidal science to clay particles seems appropriate based on size since colloidal particles range from 1 nm to 1 μm and clays typically include particles $\leq$ 2 μm.

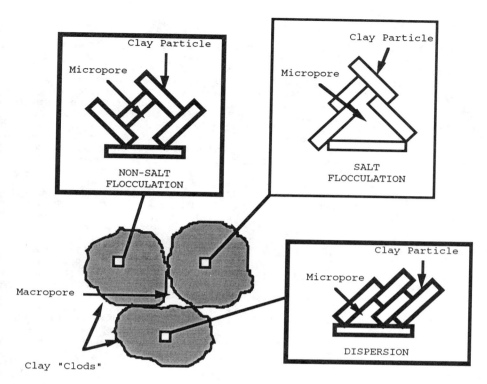

FIG. 6--Schematic illustration of microfabric and macrofabric of clay soils (after Lambe 1958; and Mitchell and Madsen 1987).

Based on the Gouy-Chapman theory from colloidal chemistry, the colloidal particles are assumed to be negatively charged, infinitely long plates, surrounded by a layer of electrolyte liquid. Due to the requirement for electroneutrality, the liquid layer surrounding the particle contains a net positive charge equal to the negative charge of the colloidal plate. The net positive charge in the liquid layer results from the difference between an excess of hydrated cations over oppositely charged hydrated anions which are attracted to the cations. The high concentration of cations near the particle surface relative to the "free water" (i.e., water outside the realm of influence of the particle) results in a concentration gradient and a tendency for diffusion of the cations away from the particle surface. These counter influences of electrostatic attraction and diffusion result in a distribution of cations referred to as an electric double layer or a diffuse double layer (DDL).

The thickness, D, of the DDL may be estimated from Gouy-Chapman theory in accordance with the following equation (van Olphen 1963; and Mitchell 1993):

$$D = \sqrt{\frac{\varepsilon \kappa T}{8\pi \eta e^2 z^2}} \qquad (10)$$

where κ is Boltzmann's constant, T is absolute temperature, η is the concentration of the cation (number of ions per cm^3), e is a unit electronic charge, and z is the valence of the cation. Based on the Gouy-Chapman theory, factors which cause a decrease in thickness, D, promote flocculation, whereas an increase in D promotes dispersion. Therefore, an increase in cation concentration and valence and a decrease in dielectric constant and temperature results in flocculation. In addition to the parameters appearing in Eq. 10, other factors, namely cation size, pH, and anion adsorption can affect the thickness of the DDL and the fabric of a clay soils (Mitchell 1993).

The influence of the major pore fluid parameters on the fabric and, therefore, hydraulic conductivity of clay soils based on the Gouy-Chapman theory is summarized in Table 7. The results of several studies have indicated that these factors generally apply when clay soils are permeated with salt (electrolyte) solutions (e.g., Quirk and Schofield 1955; McNeal and Coleman 1966; McNeal et al. 1968; Hardcastle and Mitchell 1974; Dunn and Mitchell 1984; and Alther et al. 1985). However, the correlation between the Gouy-Chapman theory and the hydraulic conductivity of clay soils permeated with electrolyte solutions tends to be more apparent for high activity clays, such as clays with appreciable amounts of montmorillonite (e.g., see McNeal and Coleman 1966). Also, the Gouy-Chapman theory tends to be more applicable for clay soils under low confining stresses, such as in clay slurry filter cakes formed in the construction of slurry walls, in which the colloidal suspension model is approximated more closely (e.g., see Alther et al. 1985).

Influence of Clay Particle Shape--The assumption of clay colloidal particles which are infinitely long plates is critical for the application of the Gouy-Chapman theory for description of the

TABLE 7--Effect of relative changes in major pore fluid parameters on hydraulic conductivity of clay soils based on microfabric (after Evans et al. 1985b).

Pore Fluid Parameter	Effect of Relative Change in Parameter	
	Decrease in DDL thickness, flocculated fabric, and an increase in hydraulic conductivity	Increase in DDL thickness, dispersed fabric, and a decrease in hydraulic conductivity
Electrolyte Concentration	Increase	Decrease
Cation Valence	Increase	Decrease
Dielectric Constant	Decrease	Increase
pH	Decrease	Increase
Cation Size	Decrease	Increase
Anion Adsorption	Decrease	Increase

DDL = diffuse double layer

microfabric of clay particles. For example, Tobin and Wild (1986) provided a qualitative comparison of the properties of three processed clay soils with different particle shapes, as shown in Table 8. Their comparison indicates that the needle-shaped particles associated with the attapulgite clay are affected only slightly in the presence of electrolytes, whereas the kaolin and bentonite soils with the plate-like particle shapes tend to flocculate in the presence of electrolytes. The implications of the study by Tobin and Wild (1986) are that attapulgite-dominated clay soils may be useful alternatives to other commonly used clay soils in terms of the relative stability of the attapulgite particles in the presence of waste liquids.

For example, Ryan (1987) reports results of a case study in which a natural leachate with several organic solvents (e.g., phenols, acetone, benzene, toluene, xylene, and gasoline) was used in sedimentation tests, compatibility tests, and cracking tests on attapulgite, bentonite, and treated bentonite backfills. In the sedimentation tests, the attapulgite clay remained stable in the leachate whereas both bentonite soils settled out of solution rapidly, presumably due to flocculation in the presence of the low dielectric constant organic leachate. The compatibility test results, reproduced in Fig. 7, indicated that the leachate adversely affected the flow rate through the bentonite and treated bentonite backfill materials, but not the attapulgite backfill material. In the cracking tests, the leachate was mixed with the slurries of the three backfill materials and allowed to air dry. Whereas the bentonite slurries cracked excessively, the attapulgite slurry remained stable. The cracking of the bentonite resulted from shrinkage of the slurries in the presence of the leachate, the subject of the next topic. Similar results were reported by Broderick and Daniel (1990) who measured the hydraulic conductivities of compacted attapulgite clay to methanol and heptane and concluded that the attapulgite clay was far less susceptible to chemical attack than the more common plate-like clays.

<u>Volume Change Considerations</u>--Considerable evidence exists to suggest that physical or bulk swelling of clay soils will be much less in organic liquids than in water (e.g., see Murray and Quirk 1982; Green et al. 1981; Acar et al. 1985; Brown and Daniel 1988; and Acar and Olivieri 1989). For example, Acar et al. (1985) and Acar and Olivieri (1989) performed free-swell tests using three clay soils (kaolinite, Na-montmorillonite, and Ca-montmorillonite) with 11 organic liquids and

TABLE 8--Qualitative comparison of properties of processed clay soils (after Tobin and Wild, 1986).

Property	Processed Clay Soils		
	Attapulgus	Bentonite	Kaolin
Principal mineral	attapulgite	montmorillonite	kaolinite
Crystalline structure	chain	three-layer sheet	two-layer sheet
Particle shape	needle	flake	plate
Surface area	high	medium	low
Swell potential	low	high	low
Cation exchange capacity	low	high	low
Effects of electrolytes	slight	flocculates	flocculates
Sorptivity	high	medium	low

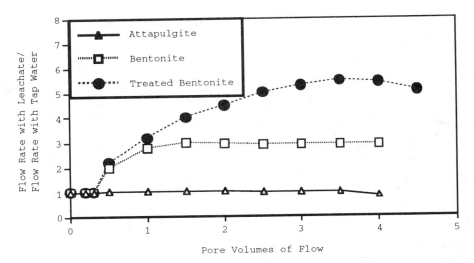

FIG. 7--Compatibility of three backfill materials to a natural organic leachate (data from Ryan 1987).

water. The free-swell tests were performed by pouring 10 cm³ of dry soil into 100 cm³ graduated cylinders, filling the cylinders with the liquid, and measuring the volume of the soil after 24 hours of exposure. The free swell is calculated as the ratio of the free swell volume to the original volume of the dry soil. The results of these tests, shown in Fig. 8, are a combined function of pH and dielectric constant.

A comparison of the data in Fig. 8 indicates that the montmorillonite soils clearly display an increasing trend in free swell with an increase in dielectric constant, with a more noticeable trend for the more active (A=4.5) Na-montmorillonite versus the less active (A=2.8) Ca-montmorillonite. The free swell response of the low activity kaolinite clay (A=0.32) is more variable and less drastic, but appears to be opposite of the free swell response for the montmorillonite soils; i.e., a decrease in dielectric constant results in a slight increase in the free swell of the kaolinite. In addition, the opposite trend in free swell between the relatively low activity kaolinite and high activity Na-montmorillonite clays correlates well with the opposite trend in modified Atterberg limits test results previously described and summarized in Table 4. Based on the free swell test results of Acar et al. (1985) and Acar and Olivieri (1989), a major factor controlling the swelling (or shrinkage) of soils is the dielectric constant of the pore liquid, and the effect tends to be more noticeable as the activity of the clay increases.

Similar behavior was observed by Acar et al. (1985) when permeability tests were performed on compacted specimens of kaolin (A=0.32) in consolidometers with volume change measurements. The specimens first were permeated with 0.01N $CaSO_4$ solution followed by permeation with acetone. The specimen indicated a slight shrinkage after permeation with the acetone. Acar et al. (1985) compared their results with other tests on Na-montmorillonite and Ca-montmorillonite as well as tests performed by Green et al. (1981) on three low activity natural

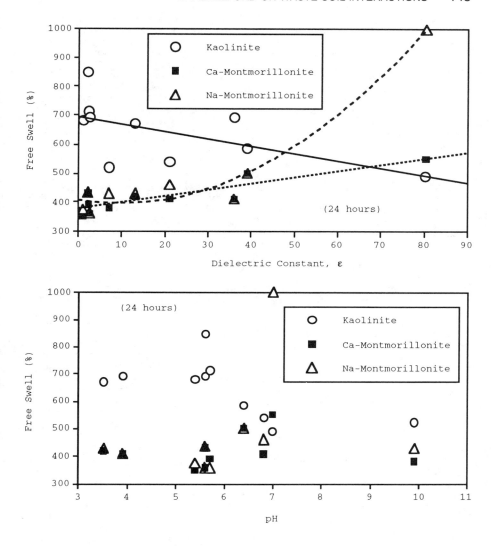

FIG. 8--Effect of dielectric constant (e) and pH of liquid on the free swell of three clay soils (data from Acar et al. 1985; and Acar and Olivieri 1989).

clay soils (A = 0.2, 0.23, and 0.25). Although the trends for the low activity natural clays were similar, a much more significant trend towards increasing shrinkage with decreasing dielectric constant was observed with the higher activity montmorillonite clays.

Murray and Quirk (1982) measured the swelling of illitic clays to 18 different organic liquids and water varying in dielectric constants from 1.9 to 187 over a period of 6 to 8 months of exposure. Although the activity values for the two illitic soils were not reported, the CEC (33 meq/100 g and 41 meq/100 g) and surface area (110 m^2/g and 188 m^2/g) suggest that the activity of the two soils was intermediate between a kaolinite and a montmorillonite. Unlike many other studies in which slurry solutions are used to measure swelling behavior, Murray and Quirk (1982) performed their investigation on compacted, thin (1-mm) discs of soil initially saturated with water and then slowly air-dried before exposure to the organic liquids. Based on their data, Murray and Quirk (1982) proposed the following linear correlation between the dielectric constant of the liquid and the swelling of the soil (also see Griffin and Roy 1985):

$$d = \frac{ze}{2\pi R \Gamma}(\varepsilon - 1) \qquad (11)$$

where d is the distance between parallel silicate sheets in the clay mineral, Γ is the surface charge density of the clay particle, and R is the cation radius. Based on Eq. 11, as the dielectric constant decreases, the distance between parallel silicate sheets decreases resulting in shrinkage of the the clay soil. Murray and Quirk (1982) suggest that the swelling behavior they observed does not correspond with Gouy-Chapman (diffuse double layer) theory but, rather, is more in accord with electric field theory involving parallel plates, or capacitors.

For example, consider the effects of liquid properties on clay suspensions illustrated in Fig. 9 (Lambe 1958; and Lambe and Whitman 1969). Based on the trends in Fig. 9, factors which cause a flocculated microfabric (e.g., lower dielectric constant) also result in a larger volume of sediment, a trend which is opposite to that observed by both Murray and Quirk (1982), Acar et al. (1985),and Acar and Olivieri (1989) for the relatively high activity montmorillonite soils. However, the low activity kaolinite soil tested by Acar et al. (1985) and Acar and Olivieri (1989) illustrated a trend opposite to that of the montmorillonite soils and, therefore, tends to correspond to the behavior illustrated in Fig. 9.

As a result of the above considerations, it appears that one possible factor not accounted for in the trends illustrated in Fig. 9 is the activity of the clay, i.e., a low activity clay, such as kaolinite, tends to correspond to the trends evident in Fig. 9 whereas a high activity clay, such as montmorillonite, does not. However, van Olphen (1963) presents results of sedimentation tests using Na-montmorillonite supporting the trends in Fig. 9. In van Olphen's tests, specially prepared sols of Na-montmorillonite were suspended for two months in solutions containing different concentrations of NaCl. A clay sol, or colloidal solution, is a special type of clay suspension in which all of the particles are $\leq$ 1 µm. Therefore, since the soils used by Murray and Quirk (1982) and Acar and Olivieri (1989) were not prepared specially to contain only particles $\leq$ 1 µm, a comparison of the their test results with those of van Olphen (1963) may not be valid. In addition, there is extensive evidence to suggest that compacted test specimens of high activity clay soils tend to shrink upon exposure to full strength organic liquids with low dielectric constants resulting in formation of cracks, macropores, and large increases in hydraulic conductivity (e.g., Anderson et al. 1982, 1985a,1985b; Brown and Thomas 1984; Acar and Ghosn 1986; Brown and Daniel 1988). The implications of these, and other, studies are significant in that high activity clay soils usually are used for waste containment barriers in order to achieve low hydraulic conductivity values. Also, compacted clay soils, initially compacted with water, would be expected to shrink upon exposure to organic waste

Chemical Solution Parameter	Value		
Electrolyte Concentration	0.5N CaCl$_2$	1.0N CaCl$_2$	2.0N CaCl$_2$
Cation Charge	1N NaCl (Na$^+$)	1N CaCl$_2$ (Ca^{2+})	1N FeCl$_3$ (Fe^{3+})
Dielectric Constant (ε)	Water (ε=80)	Alcohol (ε=25)	Benzene (ε=2)
Temperature (°C)	1N CaCl$_2$ (5°C)	1N CaCl$_2$ (20°C)	1N CaCl$_2$ (95°C)
Hydrated Ion Radius (R)	1N LiCl (R_{Li+}=9.9Å)	1N NaCl (R_{Na+}=7.3Å)	1N KCl (R_{K+}=5.3Å)
pH	1N NaOH (pH=14)	1N NaCl (pH=7)	1N HCl (pH=0)

FIG. 9--Effects of liquid properties on clay suspensions according to Lambe (1958) and Lambe and Whitman (1969).

liquids with lower dielectric constants resulting in the formation of cracks or macropores, as found by Ryan (1987).

Influence of Concentration Level--As shown in Fig. 10, results of compatibility tests in which organic solvents of varying concentrations were permeated through compacted clay soils tend to indicate that dilute concentrations (< 80% by volume) do not alter significantly the hydraulic conductivity of compacted clay soils relative to water (e.g., Bowders 1985; Evans et al. 1985a, 1985b; Bowders and Daniel 1986; and Brown and Daniel 1988). The test results with methanol tend to confirm the observations of Anderson and Jones (1983) in that the hydraulic conductivity initially decreases slightly before increasing significantly as the methanol concentration increases from zero to 100 percent. These test results have been confirmed recently by Sai and Anderson (1991) who found no substantial increase in the hydraulic conductivity of undisturbed samples of compacted clay liners permeated with leachate over a two-year period. In addition, larger increases in

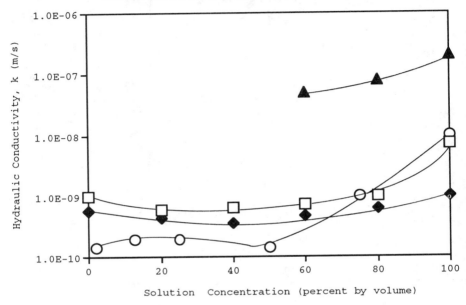

FIG. 10--Effect of concentration of organic liquids on the hydraulic conductivity of compacted clay soils (data from Bowders 1985; Brown and Daniel 1988).

hydraulic conductivity tend to occur in rigid-wall permeameters relative to flexible-wall permeameters when high concentrations of organic solvents are used as the permeant (see Fig. 10). These trends suggest that the primary interaction affecting large increases in hydraulic conductivity is shrinkage of the compacted clay upon exposure to relatively pure organic solvents with dielectric constants significantly lower than water.

Sedimentation Tests--In some cases, sedimentation tests have been performed to discern the effects of waste liquids on the fabric of a clay soil and, therefore, the potential for the waste liquid to adversely affect the hydraulic conductivity of the soil (e.g., Dunn and Mitchell 1984; Bowders 1985; Bowders and Daniel 1986; and Bowders et al. 1986). These tests typically are similar to hydrometer tests (ASTM D 422) except the waste liquid is used instead of water and a dispersing agent may or may not be used. The settling behavior can be analyzed using a hydrometer (e.g., Dunn and Mitchell 1984) to determine the particle-size distribution, or the settling rate can be measured by determining the depositional height of the suspension versus time (e.g., Bowders 1985; Bowders et al. 1985; and Bowders and Daniel 1986). Particle distributions indicating larger sizes and/or relatively rapid settling rates are indicative of a flocculated fabric. However, extrapolation of test results to describe field behavior should be made with caution, particularly in the case where a dispersing agent is used in the test, since the particle sizes being tested necessarily will not be the same as the particle sizes under field conditions.

In an attempt to discern further the relationships between fabric and depositional volume, sedimentation tests using the three processed clay soils characterized in Tables 2 and 3 and the solutions described in Table 9 were performed. The solution compositions were chosen to correspond with some of the parameters in Fig. 9. The sedimentation tests were performed by mixing 50 grams of dry soil in one liter cylinders used for hydrometer tests, and measuring the depositional height of the suspension from the bottom of the cylinders versus time. After shaking the suspensions for approximately five minutes, the suspensions were allowed to settle for periods which varied with concentration of the three solutions and soil type, as shown in Table 10.

Plots of time versus depositional height (height from base of cylinder to solution-suspension interface) are shown in Figs. 11, 12, and 13 for attapulgus, bentonite, and kaolin clays, respectively. These plots indicate that equilibrium conditions were achieved in the given time frames in all tests except for the kaolin test with 0.01M NaOH and bentonite tests with 0.01M solutions and with 0.1M HCl and 0.1M NaCl solutions. The low pH of the HCL solutions, particularly the 1M HCl solution, probably caused some dissolution of the clays, particularly the bentonite. An attempt to verify dissolution of the clay slurries at low pH by measuring the Al and Si concentrations in the solutions before and after exposure with the soils was met with mixed results due to suspended solids interferences with the chemical analysis.

The final depositional heights for all clays are compared in Fig. 14. The Gouy-Chapman theory is followed in all cases for bentonite soil (A=5.1), i.e., low pH (HCl) or higher concentrations of Na^+ (NaCl and NaOH) tend to result in flocculation and a more rapid settling of the

TABLE 9--Solutions used for sedimentation studies.

Solution	Concentration	$pH^{(1)}$	Electrical Conductance, $EC^{(1)}$ (μS/cm)
Hydrochloric Acid (HCl)	1M	0.01	262,300
	0.1M	1.0	38,780
	0.01M	2.0	3,770
Sodium Chloride (NaCl)	1M	5.9	66,280
	0.1M	5.8	10,450
	0.01M	5.4	1,200
Sodium Hydroxide (NaOH)	1M	13.3	130,400
	0.1M	12.8	20,900
	0.01M	12.0	2,260

(1) At 20.5°C to 22°C.

TABLE 10--Sedimentation test durations.

Solution Concentration[1]	Test Duration (days)		
	Attapulgus	Bentonite	Kaolin
1M	6.8	4.4	4.7
0.1M	5.0	51.7	5.1
0.01M	30.8	44.8	6.9

[1]HCl, NaCl, or NaOH; temperatures from 20.5°C to 22°C.

clay. Also, it appears that the volume of the dispersed bentonite is greater than the flocculated bentonite, which is opposite to the accepted trend shown in Fig. 9. However, since the bentonite had not reached an equilibrium depositional height in all cases (see Fig. 12), no conclusion in this regard can be made.

For a given solution, the attapulgus (A=3.1) tends to be relatively unaffected in terms of equilibrium height. Also, in extremely low pH environments (pH=0.01 or pH=1) or extremely high pH environments (pH=13.3), the attapulgus tends to be more stable than the other two clay soils, a trend which follows previous observations.

The trends for the kaolin (A=0.28) are less apparent. Relative to the bentonite, the kaolin test durations were short, yet it appears that all kaolin fabrics are flocculated except for 0.01M NaOH solution. The pH of all of the HCl solutions probably was sufficiently low to result in sufficient protonation of exposed hydroxyl ($-OH^-$) sites on the particle edges, edge-to-face (non-salt) flocculation, a rapid settling of the particles (Bohn et al. 1985). The pH of the NaCl solutions are sufficiently high such that protonation probably is not favored; yet, Na^+ concentration also does not appear to have had an effect. Since the particle sizes for kaolinite are larger than for montmorillonite, the test durations may have been longer than required for the dispersed kaolin particles to stay in suspension. However, this hypothesis is not supported by the depositional height versus time plots nor by the result of the test with 0.01M NaOH. This latter test corresponds to deprotonation of exposed hydroxyl sites resulting in negatively-charged particle edges, particle repulsion and dispersion. The higher pH values of the 1M and 0.1M NaOH solutions would be expected to result in the same behavior, but appear to be flocculated. One possible, but questionable, explanation is that adsorption of Na^+ at particle edges could have resulted in a link between exposed hydroxyls at the particle edges and the negatively charged particle surfaces of adjacent particles.

These sedimentation test results tend to corroborate accepted behavior regarding the effect of electrolyte solutions on the fabric of clay suspensions, but do not contribute to an indication of the effect of activity on the depositional volume of clay suspensions. In this regard, further studies appear necessary.

Acid-Base Reactions

Acids and bases are known to dissolve clay minerals. The solubility of a clay mineral depends on the nature and concentration of the acid (or base), the ratio of the amount of acid (or base) to clay, and the temperature and duration of the treatment (Grim 1953). The relative dissolution of clay minerals to acids and bases typically is based on measurement of aluminum and silica, respectively, before and after treatment with the acid or base.

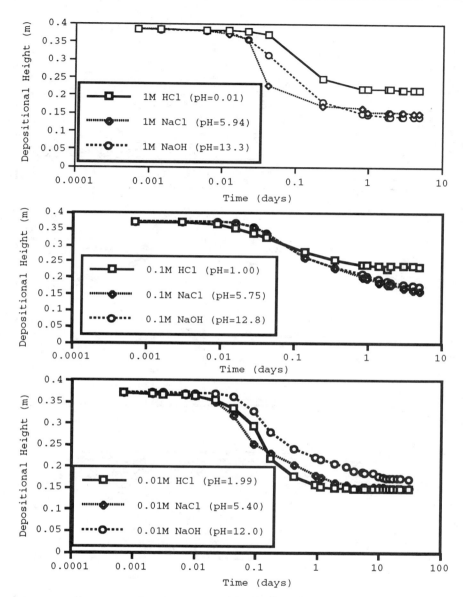

FIG. 11--Time-settlement plots for sedimentation tests with attapulgus clay.

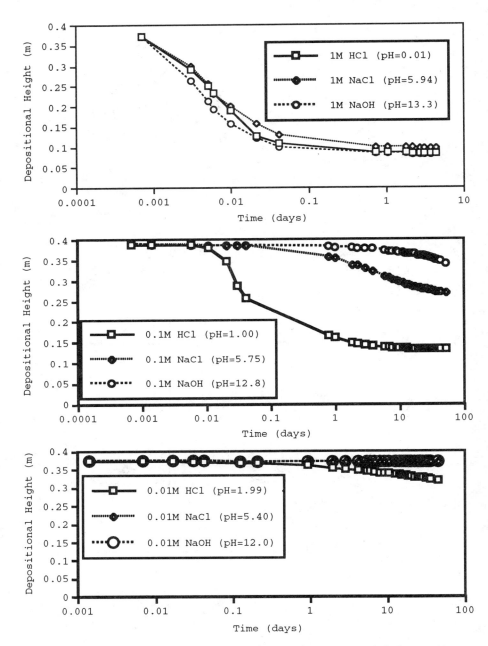

FIG. 12--Time-settlement plots for sedimentation tests with bentonite clay.

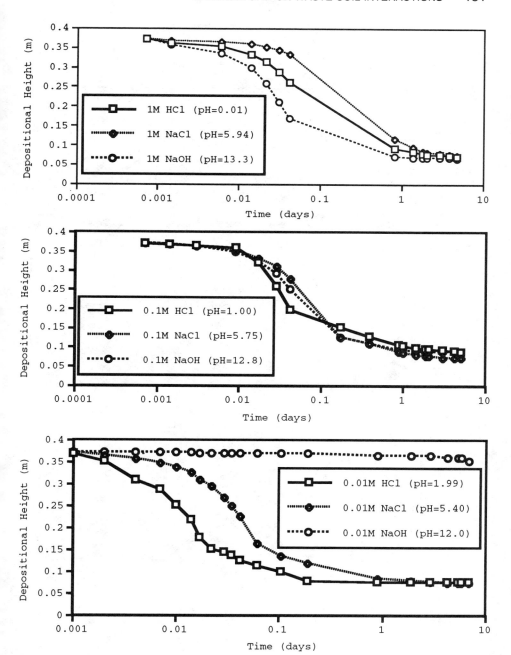

FIG. 13--Time-settlement plots for sedimentation plots with kaolin clay.

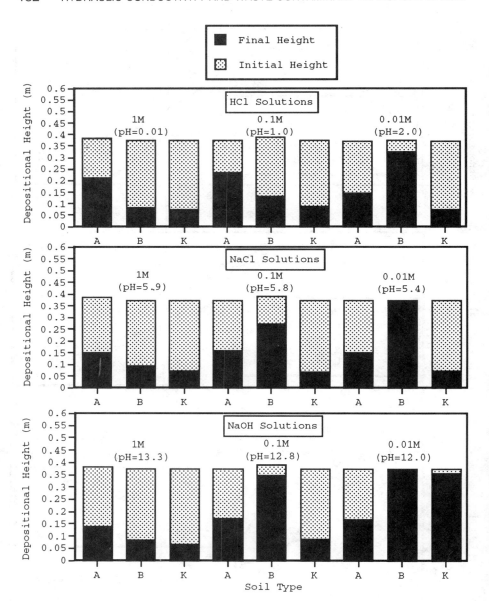

FIG. 14—Final depositional heights for sedimentation tests with attapulgus (A), bentonite (B), and kaolin (K) clays.

The solubility relationship between clay minerals tends to increase, in part, with a decrease in particle size (increase in exposed surface area). Grim (1953) reports results of a study in which 3, 11, and from 33 to 89 percent of the aluminum present in kaolinite, illite, and montmorillonite, respectively, was released when the clays were immersed in sulfuric acid (H_2SO_4). For bases, Grim (1953) reports results in which various amounts of silica (SiO_2) were released from montmorillonite and attapulgite clays after exposure to sodium carbonate (Na_2CO_3) concentrations ranging from 0 to 1 percent (by weight).

In addition to dissolution of aluminosilicate minerals, dissolution of carbonates ($CaCO_3$, $CaMg(CO_3)_2$), and sulfates (e.g., $CaSO_4 \cdot 2H_2O$) in the presence of acid (low pH) environments is well recognized (e.g., Freeze and Cherry 1979). Dissolution of these minerals may be represented by the following reactions:

$$\left.\begin{array}{l} CaCO_{3(s)} \rightarrow Ca^{2+} + CO_3^{2-} \\ CaMg(CO_3)_{2(s)} \rightarrow Ca^{2+} + Mg^{2+} + 2CO_3^{2-} \\ CaSO_4 \cdot 2H_2O_{(s)} \rightarrow Ca^{2+} + SO_4^{2-} + 2H_2O \end{array}\right\} \quad (12)$$

Several studies have evaluated the influence of acidic and/or basic permeants on the hydraulic conductivity of clay soils (e.g., D'Appolonia 1980; Gordon and Forrest 1981; Nasiastka et al. 1981; Anderson et al. 1982, 1985; Relyea and Martin 1982; Gipson 1985; Lentz et al. 1985; Peterson and Gee 1985). Interactions with inorganic acidic permeants typically are explained in three ways. First, as previously discussed, a low pH environment promotes flocculation and, therefore, an increase in hydraulic conductivity. For example, Gipson (1985) reports that the hydraulic conductivity of mixtures of silty sand with 7.5, 10, and 15 percent bentonite increased by 7X, 13X, and 41X, respectively, after permeation with an acid liquor (pH=2.2). Gipson (1985) noted that the permeant was initially high in metal cations (e.g., Ca^{2+}, Mg^{2+}, etc.). Since a low pH also promotes high solubility of metal cations which, in turn, promotes flocculation of clay particles, the fabric of the bentonite clay may have played a dominant role in affecting the change in the hydraulic conductivity of the sand-bentonite mixtures. Although the increase was not specifically attributed to fabric effects, the progressively larger increases in hydraulic conductivity with increase in bentonite content also lends credence to the influence of fabric on the hydraulic conductivity of the sand-bentonite mixtures.

D'Appolonia (1980) notes that in sand-bentonite mixtures with low bentonite contents (≤3%), a low permeant pH may lead to "smaller" clay particles which are then flushed or piped from the soil resulting in larger pore channels and an increase in hydraulic conductivity. However, a dispersion of clay particles would occur only in the case where a tendency towards flocculation due to a low pH is offset by a greater tendency towards dispersion due to high concentrations of monovalent cations (e.g., Na^+) in solution. Also, Relyea and Martin (1982) describe this same process in terms of a decrease in hydraulic conductivity resulting from swelling of the clay particles in a soil-bentonite mixture. However, the soil used by Relyea and Martin (1982) was a natural clay soil, and the bentonite content was 10 percent, much greater than the three percent limit proposed by D'Appolonia (1980).

A significant change in hydraulic conductivity also may result when a strong acid or base disintegrates the soil into small fragments or dissolves the clay mineral into solution resulting in larger pore channels (e.g., D'Appolonia 1980). If the fragments are flushed or piped from the system (e.g., permeameter), the hydraulic conductivity of the soil increases. Anderson et al. (1982, 1985) used this explanation for observed increases in hydraulic conductivity of a non-calcareous soil and an illite soil upon permeation with acetic acid. However, if the particles are restrained from piping, due to a sieving effect (e.g., filter paper or porous stone), the hydraulic conductivity may decrease. This interaction appears to be significant particularly in the case of bentonite filter cakes and soil-bentonite (SB) backfill materials when exposed to extremely acidic or basic permeants (2>pH>11) (D'Appolonia 1980). Also, D'Appolonia (1980) noted that extremely basic solutions usually produce a greater increase in hydraulic conductivity relative to extremely acid solutions since amorphous silica is quite soluble in highly alkaline solutions as previously noted.

A third possible mechanism for acid-soil interactions affecting soil hydraulic conductivity results from initial dissolution of carbonates and oxides in the soil followed by neutralization (buffering) of the pH of the liquid. As the pH increases, the dissolved constituents in the liquid re-precipitate and the clay particles disperse (higher pH) resulting in an initial decrease in the hydraulic conductivity (e.g., see Gordon and Forrest 1981; Nasiastka et al. 1981; Relyea and Martin 1982; Gipson 1985; Lentz et al. 1985; and Peterson and Gee 1985). For example, consider the neutralization process for calcite ($CaCO_{3(s)}$). Initial dissolution of calcite in accordance with the equilibrium reaction in Eq. 12 results in aqueous phase distribution of Ca^{2+}, CO_3^{2-}, and $CaCO_3$ in accordance with the following equilibrium reaction:

$$Ca^{2+} + CO_3^{2-} \leftrightarrow CaCO_3 \qquad (13)$$

From Eq. 9, the distribution of H_2CO_3, HCO_3^-, and CO_3^{2-} depends on pH. For pH < 6, H_2CO_3 is dominant, for pH > 11, CO_3^{2-} is dominant, and for 7 < pH < 10, HCO_3^- is dominant, with higher pH values favoring re-precipitation of dissolved constituents. In the case of low pH, a reaction between the $CaCO_3$ and protons (H^+) is favored, as follows:

$$2CaCO_3 + 3H^+ \rightarrow 2Ca^{2+} + HCO_3^- + H_2CO_3 \qquad (14)$$

The reaction in Eq. 14 represents a neutralization reaction (e.g., see Cherry et al. 1984; and Peterson and Gee 1985); i.e., the aqueous phase $CaCO_3$ consumes protons (H^+) resulting in an increase in pH and a dissociation of $CaCO_3$ into Ca^{2+}, HCO_3^-, and H_2CO_3. As the amount of $CaCO_{3(s)}$ available for dissolution decreases upon continued exposure to fresh acid, the buffering capacity of the system becomes exhausted, and the pH of the pore liquid approaches the pH of the acid influent. This continued permeation of the soil by fresh acid eventually will result in dissolution and flushing of the re-precipitated solids, larger pore channels, and an increase in hydraulic conductivity (e.g., see Bowders 1991). The full effect of this cycle of reactions usually can be observed only after several pore volumes of flow have been forced through the soil and chemical equilibrium between effluent and influent pH is achieved. These conditions may be unreasonable relative to the time frames and conditions prevalent in the field.

For example, Gordon and Forrest (1985) and Relyea and Martin (1982) describe neutralization reactions occurring during permeation of natural clay soils with acidic permeants in which the protons in the reaction in Eq. 14 are derived from sulfuric acid (H_2SO_4), or:

$$H_2SO_4 \leftrightarrow H^+ + SO_4^{2-} \tag{15}$$

As the pH rises due to the neutralization reaction (Eq. 14), the free sulfate (SO_4^{2-}) reacts with the free calcium (Ca^{2+}) and sulfuric acid reacts with calcium carbonate to precipitate as calcium sulfate ($CaSO_{4(s)}$), or:

$$\left. \begin{array}{l} Ca^{2+} + SO_4^{2-} \rightarrow CaSO_{4(s)} \\ CaCO_3 + H_2SO_4 \rightarrow CaSO_{4(s)} + CO_{2(g)} + H_2O \end{array} \right\} \tag{16}$$

Gordon and Forrest (1981) believed the precipitation of $CaSO_{4(s)}$ was indicated by (1) a slight weight gain in the soil, (2) decrease in effluent electrical conductance (EC), and (3) an increase in effluent pH. In addition, Gordon and Forrest (1981) attributed an increase in volumetric outflow relative to volumetric inflow in the hydraulic conductivity tests to the production of water related to the second reaction in Eq. 16. Gipson (1985) describes similar waste-soil interactions in which phosphoric acid (H_3PO_4) is the source of protons for the buffering of acid permeants in the presence of carbonates in clay soils.

Another source of proton supply for neutralization reactions is the precipitation of amorphous hydroxides (e.g., Nasiatka et al. 1981; Relyea and Martin 1982; and Peterson and Gee 1985). For example, as the acid permeant is neutralized in accordance with the reaction in Eq. 14, iron (Fe^{3+}) and aluminum (Al^{3+}) cations in the permeant will precipitate from solution as ferric hydroxide ($Fe(OH)_{3(s)}$) and aluminum hydroxide ($Al(OH)_{3(s)}$), as follows:

$$\left. \begin{array}{l} Fe^{3+} + 3H_2O \rightarrow Fe(OH)_{3(s)} + 3H^+ \\ Al^{3+} + 3H_2O \rightarrow Al(OH)_{3(s)} + 3H^+ \end{array} \right\} \tag{17}$$

resulting in a decrease in pH (more protons) which dissolves more calcite. Once the calcite is completely dissolved, the carbonate buffering capacity is exhausted and the pH of the pore liquid will decrease. However, in this case, the solid ferric and aluminum hydroxides offer a residual buffering capacity, grabbing up free protons in accordance with the reverse of the reactions in Eq. 17. As a result, the influent pH is buffered, the precipitates continue to clog the pore channels, and the hydraulic conductivity is maintained at a lower value for a longer period. This residual buffering capacity was considered to be the reason the hydraulic conductivity of the soil tested by Peterson and Gee (1985) did not increase after as much as 30 pore volumes of flow corresponding to permeation periods greater than three years.

There have been only a few studies on the effects of bases on the hydraulic conductivity of clay soils. Gordon and Forrest (1981) permeated sedimented mud slimes with sodium carbonate (Na_2CO_3; pH=8 to 9) solutions and noted that the stronger solution (14% Na_2CO_3) resulted in a smaller hydraulic conductivity than the weaker solution (1% Na_2CO_3). Therefore, they attributed the difference to chemical reactions in the pore liquid instead of soil structure effects. Lentz et al. (1985) attributed a 13X reduction in the hydraulic conductivity of a magnesium montmorillonite upon permeation with a sodium hydroxide solution (pH=13) to precipitation of magnesium hydroxide in much the same way as described above for the acid permeants. Anderson et al. (1982, 1985) studied the effect of the organic base, aniline, on the hydraulic conductivity of four soils and observed hydraulic conductivities of one to two orders of magnitude greater than measured with water. However, the increases in hydraulic conductivity to this organic base were attributed to structural changes in the soil, not to the causes described above for inorganic bases.

Oxidation-Reduction Reactions

Oxidation-reduction (redox) reactions also play a significant role in determining the pH of a pore liquid and, therefore, potentially can affect the hydraulic conductivity of the soil in much the same way as acid-base reactions. For example, in groundwater systems, organic matter (CH_2O) is oxidized in the presence of bacteria and oxygen as follows (Freeze and Cherry 1979):

$$O_{2(g)} + CH_2O \leftrightarrow CO_{2(g)} + H_2O \tag{18}$$

As previously shown (Eq. 9), an increase in $CO_{2(g)}$ results in an increase in $CO_{2(aq)}$, and $CO_{2(aq)}$ combined with H_2O forms carbonic acid, or

$$CO_{2(aq)} + H_2O \leftrightarrow H_2CO_3 \tag{19}$$

which is an acid of considerable strength in soils.

The solubility of O_2 in water under atmospheric pressure is low and replenishment of O_2 in groundwater systems is limited except near the ground surface (Freeze and Cherry 1979). However, in flexible-wall hydraulic conductivity tests in which a significant back-pressure is applied, the dissolved oxygen content is increased artificially and, therefore, a significant increase in carbonic acid production with a corresponding decrease in pH is possible. The low effluent pH values in Fig. 5 may be thought to be a result of this back-pressure effect; however, a bladder accumulator system (Fig. 3) with a flexible-wall permeameter was used in this test, so the dissolved oxygen content in the permeant probably was limited by the flexible bladder in the accumulator.

A CASE OF "FIRST EXPOSURE"

The author was hired to evaluate the hydraulic conductivity of compacted sand-clay mixtures for potential use as a liner material for the containment of a mine tailings solution at a site in Washington. Due to the client's desire for confidentiality, both the permeant and the compacted test specimens were shipped to the laboratory, and no additional characterization of either the permeant or the soil was performed by the author. However, the author was informed that the permeant was a calcium saturated solution, and a white precipitant was observed in the bottom of the permeant containers. Also, it was indicated that both the sand and the clay were from borrow pits at the mining site, and that the clay was a bentonite. The test specimens consisting of mixtures of the sand and the bentonite were compacted according to ASTM D 698 at moisture contents wet of optimum in accordance with standard practice for compacted clay liners.

Initial hydraulic conductivity tests using compaction-mold permeameters resulted in the measurement of excessively high hydraulic conductivity values ($k \geq 10^{-7}$ m/s) which was attributed to side-wall leakage, possibly resulting from disturbance to the test specimens during shipping. As a result, flexible-wall (FW) permeameters without back pressure were used in all subsequent hydraulic conductivity tests. In addition, FW tests performed on test specimens consisting of 10 and 13 percent bentonite contents resulted in unacceptably high hydraulic conductivity values ($k \geq 10^{-9}$ m/s), so two additional FW tests were performed on test specimens containing 16 percent bentonite. All of the flexible-wall tests were performed at a constant hydraulic gradient of 48 with the outflow end of the specimen essentially at atmospheric pressure. The average effective stress on the specimens (i.e., based on the assumption of saturated conditions) was 41.4 kPa (6 psi).

In the first test (FW1), "standard" compatibility test conditions were used in that the test specimen initially was permeated with tap water followed by permeation with the waste liquid. In the second test (FW2), the test specimen was permeated only with the waste liquid. The test results are shown in Fig. 15 and summarized in Table 11. The initial and final compaction properties of the test specimens are presented in Table 12. As shown in Table 13, the final water contents were based on the average of three samples taken from different segments of the test specimens immediately after breakdown of the tests, and

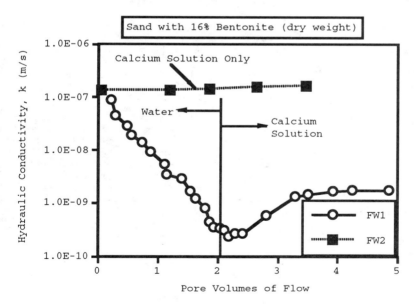

FIG. 15--Hydraulic conductivity test results for compacted sand-bentonite specimens tested in flexible-wall permeameters without backpressure.

TABLE 11--Hydraulic conductivity results for flexible-wall permeability tests on compacted specimens of 16 percent bentonite with sand[1].

Test No.	Hydraulic Conductivity (m/s)				k_f/k_i	Total Pore Volumes of Flow
	Water		Calcium Solution			
	k_i	PVF	k_f	PVF		
FW1	3.4×10^{-10}	2.05	1.7×10^{-9}	2.81	5.0	4.86
FW2	NA	NA	2.3×10^{-7}	3.47	NA	3.47

[1] PVF = pore volumes of flow; NA = not applicable; hydraulic gradient = 48.

indicate a relatively uniform water content throughout both test specimens.

As shown in Fig. 15, the hydraulic conductivity using tap water (see Creek and Shackelford 1992) as the permeant for test FW1 decreases until equilibrium is reached after a little more than two pore volumes of flow, and then increases to a value of 1.7×10^{-9} m/s upon subsequent permeation by the waste liquid. These trends are consistent with swelling of the bentonite clay upon exposure to water resulting in the establishment of a relatively "tight" soil matrix followed by flocculation of the clay due to exposure of the high concentration of calcium in the waste liquid. However, the trends apparent in test FW1 are not evident in test specimen FW2 which was exposed only to the calcium solution. The final hydraulic conductivity of test FW2 is approximately 135 times greater than that of test FW1 due apparently to the simultaneous contrasting effects of swelling and flocculation.

These test results should be viewed with caution due to inadequate characterization of the waste liquid and soils as well as lack of duplication of the tests. Also, the potential for a significant first exposure effect, if any, probably is limited to highly swelling clay soils and/or high activity clays, such as bentonite or other natural soils with significant montmorillonite content. Nonetheless, the test results indicate a potential for a significant effect due to first exposure; i.e., a test specimen exposed first to a waste liquid may not result in the same hydraulic conductivity as the same test specimen exposed first to water before permeation with the waste liquid. In addition, Lo et al. (1994) report similar results in that the hydraulic conductivity of a manufactured bentonite liner material permeated first with pure methanol is significantly greater than the same material permeated first with water followed by permeation with methanol. Similar

TABLE 12--Dry unit weights and water contents for compacted specimens of 16 percent bentonite with sand[1].

Test No.	As-Compacted			Final (After Testing)		
	γ_d (kN/m^3)	w (%)	S_r (%)	γ_d (kN/m^3)	w (%)	S_r (%)
FW1	15.77	20.7	84.8	16.03	22.0	93.7
FW2	15.79	20.9	85.7	16.04	22.6	96.3

[1] γ_d = dry unit weight; w = water content; S_r = degree of saturation

TABLE 13--Final water contents for compacted specimens of 16 percent bentonite with sand.

Test No.	Final Water Content, w (%)			
	Top	Middle	Bottom	Average
FW1	21.9	21.7	22.3	22.0
FW2	22.7	21.9	23.3	22.6

test results may have appeared in the literature, but there has been no comparison of "first exposure" test results with those corresponding to the "standard" test procedure for compatibility testing with respect to hydraulic conductivity.

Since it is unlikely that the clay soil barrier in the field will be exposed to several pore volumes of water before being exposed to the waste liquid, the potential ramifications of a first exposure effect from a practical standpoint can be severe. In the above example, the test results corresponding to "standard" test procedures (FW1) might be acceptable from a regulatory standpoint, but would be extremely unconservative from the standpoint of the field situation. Therefore, further study of the potential consequences of a "first exposure" effect is warranted.

SUMMARY AND CONCLUSIONS

Liquid-soil interactions which potentially can alter hydraulic conductivity of soil are reviewed. The review is based upon the consideration of clay soils for use as waste containment barriers. A knowledge of several material properties is important for an evaluation of the effects of waste-soil interactions on hydraulic conductivity. The most significant properties of the soils and liquids were identified in this paper.

Modified Atterberg limits tests, modified hydrometer (sedimentation) tests, and modified hydraulic conductivity (compatibility) tests typically are used to study the effect of waste-soil interactions on soil hydraulic conductivity. The procedures for these tests are essentially the same as the corresponding tests with water (e.g., ASTM D 4318, ASTM D 422, and ASTM D 5084, respectively), except the actual waste liquid is used in lieu of water.

The results of several studies in which modified Atterberg limit tests were performed indicate that significant changes in the Atterberg limits of clay soils are to be expected only in the case where pure organic solvents are used in the test. In addition, test results indicate that only pure organic solvents tend to decrease significantly the liquid limit and the plasticity of highly active clays, such as montmorillonite clays, but do not affect the Atterberg limits to the same extent for less active clays, such as kaolins. These trends tend to be reflected by the results of compatibility tests; i.e., large increases in hydraulic conductivity of clay soils typically are observed only in the case of high activity clay soils permeated with pure organic solvents. As a result, modified Atterberg limits tests provide an indirect correlation of the effects of the waste liquid on the hydraulic conductivity of the clay soil.

The advantages and disadvantages of rigid-wall, flexible-wall, and consolidation-cell permeameters were reviewed from the standpoint of compatibility tests. In addition, the effects of the permeant water, membrane compatibility, back-pressure saturation, hydraulic gradient, stress conditions, retention time, termination criteria, effluent pH and electrical conductance measurements, and biological activity were presented from the standpoint of compatibility test procedure. A thorough understanding of these aspects of compatibility testing is essential for a correct interpretation of the test results.

Sedimentation test results help to illustrate the influence of several liquid properties on the microfabric of clay soil in accordance with the Gouy-Chapman theory for behavior of colloidal particles in suspensions. In addition, extensive evidence corroborates the use of the Gouy-Chapman theory to explain the increases in hydraulic conductivity observed in clay soils. Liquid properties which tend to flocculate soil result in an increase in the hydraulic conductivity of the clay soil, whereas the opposite trend is evident when the liquid properties result in a dispersed soil microfabric. These trends are more significant in clay slurry backfill mixtures relative to compacted clay, relatively high activity clays, and for tests performed in rigid-wall permeameters relative to either flexible-wall permeameters or consolidation-cell permeameters.

Volume changes observed in sedimentation tests of clay slurry suspensions do not correlate necessarily with those observed for clay soils in free swell tests or for compacted specimens of clay upon exposure to concentrated organic liquids with dielectric constants much lower than water. As a result, the Gouy-Chapman theory for suspended colloids does not describe necessarily the shrinkage effects of compacted clays which have resulted in large increases in hydraulic conductivity when permeated with concentrated organic chemicals. In addition, the application of the Gouy-Chapman theory and Eq. 10 is restricted to miscible liquids, in general, and electrolyte solutions, in particular. Therefore, prediction of volume changes resulting from permeation of concentrated organic liquids, especially non-aqueous phase solutions or non-polar organic compounds, should not be expected necessarily to follow the volume change behavior described by the Gouy-Chapman theory.

Three mechanisms may contribute to an increase in the hydraulic conductivity of clay soils upon permeation with acid permeants: (1) flocculation of the clay, (2) dissolution of the clay minerals (aluminosilicates), and (3) dissolution of other minerals (e.g., $CaCO_3$) in the clay soil. Dissolution and piping of clay minerals leads to increases in hydraulic conductivity. Dissolution of carbonates initially leads to buffering, re-precipitation, pore clogging, and a decrease in hydraulic conductivity. Depletion of the buffering capacity leads to a decrease in pH, dissolution of constituents, and a possible increase in hydraulic conductivity. However, considerable time typically is required to exhaust the buffering capacity of natural soils.

The influence of oxidation-reduction reactions in laboratory compatibility tests is seldom, if ever, considered. Laboratory test specimens which are back-pressure saturated may result in artificially induced low pH values due to an increase in dissolved oxygen, dissolved carbon dioxide, and carbonic acid production. In addition, failure to measure the solution pH immediately after recovery of the effluent sample may result in relatively high measured pH values masking these effects and leading potentially to a misinterpretation of test results.

The measured hydraulic conductivity of a compacted sand-bentonite mixture is shown to be significantly affected by the sequence of permeation to a saturated calcium solution. The effect, termed "first exposure", probably is only relatively important in the case of clay soils with a large swelling potential, such as bentonite. Nevertheless, the "first exposure" effect observed in the test results has important implications for evaluating the suitability of a clay material for use as a waste containment barrier.

ACKNOWLEDGMENTS

The sedimentation test results reported in this paper were performed by Mary DiMartini and Todd Cotten, and the data for Fig. 5 are from tests performed by Pat Redmond. Their assistance is appreciated. The critical review of a draft of the manuscript for this paper by Dr. John Bowders is appreciated. Partial financial support for this material is based upon work supported by the National Science Foundation in the form of Research Experience for Undergraduate (REU) supplements to Grant No. MSS-8908201. The Government has certain rights in this material. Any opinions, findings, and conclusions or recommendations expressed in this material are those of the author and do not necessarily reflect the views of the National Science Foundation. The support of Dr. Mehmet T. Tumay, Director, Geomechanics, Geotechnical, and Geo-Environmental Systems Program at NSF, is appreciated.

REFERENCES

Abdul, A. S., Gibson, T. L., and Rai, D. N., 1990, "Laboratory Studies of the Flow of Some Organic Solvents and Their Aqueous Solutions Through Bentonite and Kaolin Clays," Ground Water, Vol. 28, No. 4, pp. 524-533.

Acar, Y. B. and Ghosn, A. A., 1986, "Role of Activity in Hydraulic Conductivity of Compacted Soils Permeated with Acetone," Proceedings, International Symposium on Environmental Geotechnology, Vol. 1, H.-Y. Fang, Ed., Envo Publishing Co., Inc., Bethlehem, Pennsylvania, pp. 403-412.

Acar, Y. B. and Olivieri, I., 1989, "Pore Fluid Effects on the Fabric and Hydraulic Conductivity of Laboratory-Compacted Clay," Transportation Research Record 1219, Geotechnical Engineering 1989, Transportation Research Board, National Research Council, Washington, D. C., pp. 144-159.

Acar, Y. B. and Seals, R. K., 1984, "Clay Barrier Technology for Shallow Land Waste Disposal Facilities," Hazardous Waste, Vol. 1, No. 2, pp. 167-181.

Acar, Y. B., Hamidon, A., Field, S. and Scott, L., 1985, "The Effect of Organic Fluids on Hydraulic Conductivity of Compacted Kaolinite," Hydraulic Barriers in Soil and Rock, ASTM STP 874, A. I. Johnson, R. K. Frobel, N. J. Cavalli, and C. B. Pettersson, Eds., American Society for Testing and Materials, Philadelphia, pp. 171-187.

Alther, G. R., 1987, "The Qualifications of Bentonite as a Soil Sealant," Engineering Geology, Vol. 23, pp. 177-191.

Alther, G. R., Evans, J. C., Fang, H-Y, and Witmer, K., 1985, "Influence of Organic Permeants upon the Permeability of Bentonite," Hydraulic Barriers in Soil and Rock, ASTM STP 874, A. I. Johnson, R. K. Frobel, N. J. Cavalli, and C. B. Pettersson, Eds., American Society for Testing and Materials, Philadelphia, pp. 64-73.

Anderson, D. C., Brown, K. W., and Green, J. W., 1982, "Effect of Organic Fluids on the Permeability of Clay Soil Liners," Proceedings, Eighth Annual Research Symposium on Land Disposal of Hazardous Waste, EPA-600/9-82-002, Ft. Mitchell, Kentucky, March 8-10, pp. 179-190.

Anderson, D. C., Brown, K. W. and Thomas, J. C., 1985a, "Conductivity of Compacted Clay Soils to Water and Organic Liquids," Waste Management and Research," Vol. 3, No. 4, pp. 339-349.

Anderson, D. C., Crawley, W., and Zabcik, J. D., 1985b, "Effects of Various Liquids on Clay Soil: Bentonite Slurry Mixtures," Hydraulic Barriers in Soil and Rock, ASTM STP 874, A. I. Johnson, R. K. Frobel, N. J. Cavalli, and C. B. Pettersson, Eds., American Society for Testing and Materials, Philadelphia, pp. 93-103.

Anderson, D. C., and S. G. Jones, 1983, "Clay Barrier-Leachate Interaction," National Conference on Management of Uncontrolled Hazardous Waste Sites, October 31-November 2, 1983, Washington, D.C., Hazardous Materials Control Research Institute, Silver Spring, Maryland, pp. 154-160.

Bear, J, 1972, Dynamics of Fluids in Porous Media, Elsevier Publ., Inc., New York, 764 pp.

Benson, C. H. and Daniel, D. E., 1990, "Influence of Clods on Hydraulic Conductivity of Compacted Clay," Journal of Geotechnical Engineering, ASCE, Vol. 116, No. 8, pp. 1231-1248.

Bohn, H., McNeal, B., and O'Connor, G, 1985, Soil Chemistry, 2nd Ed., John Wiley and Sons, Inc., New York, 341 pp.

Bowders, J. J., Jr., 1985, "The Influence of Various Concentrations of Organic Liquids on the Hydraulic Conductivity of Compacted Clay," Geotechnical Engineering Dissertation GT85-2, Geotechnical Engineering Center, Civil Engineering department, University of Texas, Austin, Texas, 219 pp.

Bowders, J. J., Jr., 1988, Discussion of "Termination Criteria for Clay Permeability Testing," Journal of Geotechnical Engineering, ASCE, Vol. 114, No. 8, pp. 947-949.

Bowders, J. J., Jr., 1991, Discussion of "Permeability of Clays Under Organic Permeants," Journal of Geotechnical Engineering, ASCE, Vol. 117, No. 8, pp. 1278-1280.

Bowders, J. J., Jr. and Daniel, D. E., 1986, "Hydraulic Conductivity of Compacted Clay to Dilute Organic Compounds," Journal of Geotechnical Engineering, ASCE, Vol. 113, No. 12, pp. 1432-1448.

Bowders, J. J., Jr., Daniel, D. E., Broderick, G. P., and Liljestrand, H. M., 1986, "Methods for Testing the Compatibility of Clay Liners with Landfill Leachate," *Hazardous and Industrial Solid Waste Testing: Fourth Symposium, ASTM STP 886*, J. K. Petros, Jr., W. J. Lacy, and R. A Conway, Eds., American Society for Testing and Materials, Philadelphia, pp. 233-250.

Boynton, S. S. and Daniel, D. E., 1985, "Hydraulic Conductivity Tests on Compacted Clay," *Journal of Geotechnical Engineering*, ASCE, Vol. 111, No. 4, pp. 465-478.

Broderick, G. P., and Daniel, D. E., 1990, "Stabilizing Compacted Clay Against Chemical Attack," *Journal of Geotechnical Engineering*, ASCE, Vol. 116, No. 10, pp. 1549-1567.

Brown, K. W. and Daniel, D. E., 1988, "Influence of Organic Liquids on the Hydraulic Conductivity of Soils," Chapter 4.3, *Land Disposal of Hazardous Waste: Engineering and Environmental Issues*, J. R. Gronow, A. N. Schofield, and R. K. Jain, Eds., Ellis Horwood Ltd., Chichester, England, pp. 199-216.

Brown, K. W., and Thomas, J. C., 1984, "Conductivity of Three Commercially Available Clays to Petroleum Products and Organic Solvents," *Hazardous Waste*, Vol. 1, No.4, pp. 545-553.

Carpenter, G. W. and Stephenson, R. W., 1986, "Permeability Testing in the Triaxial Cell," *Geotechnical Testing Journal*, ASTM, Vol. 9, No. 1, pp. 3-9.

Cherry, J. A., Gillham, R. W., and Barker, J. F., 1984, "Contaminants in Groundwater: Chemical Processes," *Studies in Geophysics, Groundwater Contamination*, National Academy Press, Washington, D. C. pp. 46-64.

Creek, D. N., and Shackelford, C. D., 1992, "Permeability and Leaching Characteristics of Fly Ash Liner Materials," *Transportation Research Record No. 1345*, Transportation Research Board, National Academy Press, Washington, D. C., pp. 74-83.

Daniel, D. E., 1989, "A Note on Falling Headwater and Rising Tailwater Permeability Tests," *Geotechnical Testing Journal*, ASTM, Vol. 12, No. 4, pp. 308-310.

Daniel, D. E., Anderson, D. C., and Boynton, S. S., 1985, "Fixed-Wall Versus Flexible Wall Permeameters," *Hydraulic Barriers in Soil and Rock, ASTM STP 874*, A. I. Johnson, R. K. Frobel, N. J. Cavalli, and C. B. Pettersson, Eds., American Society for Testing and Materials, Philadelphia, pp. 107-126.

Daniel, D. E., and Liljestrand, H. M., 1984, "Effects of Landfill Leachates on Natural Liner Systems," *A Report to Chemical Manufacturers Association*, Department of Civil Engineering, University of Texas, Austin, 86 pp.

Daniel, D. E., Liljestrand, H. M., Broderick, G. P., and Bowders, J. J., Jr., 1988, "Interaction of Earthen Liner Materials with Industrial waste Leachate," Hazardous Waste and Hazardous Materials, Vol. 5, No. 2, pp. 93-107.

Daniel, D. E., Trautwein, S. J., Boynton, S. S., and Foreman, D. E., 1984, "Permeability Testing with Flexible-Wall Permeameters," Geotechnical Testing Journal, ASTM, Vol. 7, No. 3, pp. 113-122.

D'Appolonia, D., 1980, "Soil-Bentonite Slurry Trench Cutoffs," Journal of Geotechnical Engineering, ASCE, Vol. 106, No. GT4, pp. 399-417.

Dunn, R. J., and Mitchell, J. K., 1984, "Fluid Conductivity Testing of Fine-Grained Soils," Journal of Geotechnical Engineering, ASCE, Vol. 110, No. GT11, pp. 1648-1665.

Edil, T. B. and Erickson, A. E., 1985, "Procedure and Equipment Factors Affecting Permeability Testing of a Bentonite-Sand Liner Material," Hydraulic Barriers in Soil and Rock, ASTM STP 874, A. I. Johnson, R. K. Frobel, N. J. Cavalli, and C. B. Pettersson, Eds., American Society for Testing and Materials, Philadelphia, pp. 155-170.

Evans, J. C., 1993, "Vertical Cutoff Walls", Chapter 17, Geotechnical Practice for Waste Disposal, D. E. Daniel, Ed., Chapman and Hall, London, pp. 430-454.

Evans, J. C. and H.-Y. Fang, 1986, "Triaxial Equipment for Permeability Testing with Hazardous and Toxic Permeants," Geotechnical Testing Journal, ASTM, Vol. 9, No. 3, pp. 126-132.

Evans, J. C. and H.-Y. Fang, 1988, "Triaxial Permeability and Strength Testing of Contaminated Soils," Advanced Triaxial Testing of Soil and Rock, ASTM STP 977, Robert T. Donaghe, Ronald C. Chaney, and Marshall L. Silver, Eds., American Society for Testing and Materials, Philadelphia, pp. 387-404.

Evans, J. C., Fang, H-Y, and Kugelman, I. J, 1985a, "Organic Fluid Effects on the Permeability of Soil-Bentonite Slurry Walls," Proc. National Conf. Hazardous Waste and Environmental Emergencies, May 14-16, 1985, Cincinnati, Hazardous Materials Control Research Institute, Silver Spring, Maryland, pp. 267-271.

Evans, J. C., Kugelman, and Fang, H.-Y., 1985b, "Organic Fluids Effects on Strength, Deformation and Permeability of Soil-Bentonite Slurry Walls," Proceedings, 17th Mid-Atlantic Industrial Waste Conference, Toxic and Hazardous Wastes, I. J. Kugelman, Ed., Technomic Publ. Co., Inc., Lancaster, Pennsylvania, pp. 275-291.

Fernandez, F. and Quigley, R. M., 1985, "Hydraulic Conductivity of Natural Clays Permeated with Simple Liquid Hydrocarbons," Canadian Geotechnical Journal, Vol. 22, pp. 205-214.

Foreman, D. E., and Daniel, D. E., 1984, "Effects of Hydraulic Gradient and Method of Testing on the Hydraulic Conductivity of Compacted Clay to Water, Methanol, and Heptane," Proceedings of the Tenth Annual Research Symposium on Land Disposal of Hazardous Waste, EPA-600/9-84-007, Ft. Mitchell, Kentucky, April 3-5, 1984, pp. 138-144.

Foreman, D. E., and Daniel, D. E., 1986, "Permeation of Compacted Clay with Organic Chemicals," Journal of Geotechnical Engineering, ASCE, Vol. 112, No. 7, pp. 669-681.

Freeze, R. A. and Cherry, J. A., 1979, Groundwater, Prentice-Hall, Inc., Englewood Cliffs, New Jersey, 604 pp.

Gibson, A. H., Jr., 1985, "Permeability Testing on Clayey Soil and Silty Sand-Bentonite Mixture Using Acid Liquor," Hydraulic Barriers in Soil and Rock, ASTM STP 874, A. I. Johnson, R. K. Frobel, N. J. Cavalli, and C. B. Pettersson, Eds., American Society for Testing and Materials, Philadelphia, pp. 140-154.

Goldman, L. J., Greenfield, L. I., Damle, A. S., Kingsbury, G. L., Northeim, C. M., and Truesdale, R. S., 1988, "Design, Construction, and Evaluation of Clay Liners for Waste Management Facilities," U. S. Environmental Protection Agency, Washington, D. C., EPA/530/SW-86/007F.

Gordon, B. B. and Forrest, M., 1981, "Permeability of Soils Using Contaminated Permeant," Permeability and Groundwater Contaminant Transport, ASTM STP 746, T. F. Zimmie and C. O Riggs, Eds., American Society for Testing and Materials, Philadelphia, pp. 101-120.

Green, W. J., Lee, G. F. and Jones, R. A.., 1981, "Clay-Soils Permeability and Hazardous Waste Storage," Journal Water Pollution Control Federation, Vol. 53, No. 8, pp. 1347-1354.

Griffin, R. A., and Jurinak, J. J., 1973, "Estimation of Activity Coefficient from Electrical Conductivity of Natural Aquatic Systems and Soil Extracts," Soil Science, Vol. 116, No. 1, pp. 26-30.

Griffin, R. A., and Roy, W. R., 1985, "Interaction of Organic Solvents with Saturated Soil-Water Systems," Open File Report No. 3, Environmental Institute for Waste Management Studies, University of Alabama, Tuscaloosa, Alabama, 86 pp.

Grimm, R. E., 1953, Clay Mineralogy, McGraw-Hill Book Co., Inc., New York, 384 pp.

Hardcastle, J. H. and Mitchell, J. K., 1974, "Electrolyte Concentration-Permeability Relationships in Sodium Illite-Silt Mixtures," Clays and Clay Minerals, Vol. 22, pp. 143-154.

Ilgenfritz, Balanchard, F. A., Masselink, R. L., and Panigraphi, B. K., 1988, "Mobility and Effects in Liner Clay of Fluorobenzene Tracer and Leachate," Ground Water, Vol. 26, No. 1, pp. 22-30.

Klecka, G. M., Davis, J. W., Gray, D. R., and Madsen, S. S., 1990, "Natural Bioremediation of Organic Contaminants in Ground Water: Cliffs-Dow Superfund Site," *Ground Water*, Vol. 28, No. 4, pp. 534-543.

Knight, M. J., 1988, "Reactivity of Aluminum Potline Waste Components with Laterized Clay and Geotechnical Significance for a Landfill at Wallaroo, New South Wales, Australia," *Bulletin of the International Association of Engineering Geology*, April, No. 37, Paris, pp. 49-60.

Lambe, T. W., 1958, "The Structure of Compacted Clay," *Journal of the Soil Mechanics and Foundations Division*, ASCE, Vol. 84, No. SM2, pp. 1654-1 -1654-34

Lambe, T. W. and Whitman, 1969, *Soil Mechanics*, John Wiley and Sons, Inc., New York, 553 pp.

Lentz, R. W., Horst, W. D., and Uppot, J. O., 1985, "The Permeability of Clay to Acidic and Caustic Permeants," *Hydraulic Barriers in Soil and Rock*, ASTM STP 874, A. I. Johnson, R. K. Frobel, N. J. Cavalli, and C. B. Pettersson, Eds., American Society for Testing and Materials, Philadelphia, pp. 127-139.

Lo, I. M.-C., Liljestrand, H. M., and Daniel, D. E., 1994, "Hydraulic Conductivity and Pollutant Transport Through Modified Montmorillonite Clay Liner Materials," *Hydraulic Conductivity and Waste Contaminant Transport in Soils*, ASTM STP 1142, David E. Daniel and Stephen J. Trautwein, Eds., American Society for Testing and Materials, Philadelphia.

Madsen, F. T. and Mitchell, J. K., 1987, "Chemical Effects on Clay Hydraulic Conductivity and Their Determination," *Open File Report No. 13*, Environmental Institute for Waste Management Studies, The University of Alabama, Tuscaloosa, Alabama, 70 pp.

McNeal, B. L. and Coleman, N. T., 1966, "Effect of Solution Composition on Soil Hydraulic Conductivity," *Proceedings, Soil Science Society of America*, Vol. 30, pp. 308-312.

McNeal, B. L., Layfield, D. A., Norvell, W. A., and Rhoades, J. D., 1968, "Factors Influencing the Hydraulic Conductivity of Soils in the Presence of Mixed-Salt Solutions," *Proceedings, Soil Science Society of America*, Vol. 32, pp. 187-190.

Mesri, G. and Olson, R. E., 1971, "Mechanisms Controlling the Permeability of Clays," *Clays and Clay Minerals*, Vol. 19, pp. 151-158.

Mitchell, J. K., 1993, *Fundamentals of Soil Behavior*, 2nd Edition, John Wiley and Sons, Inc., New York, 437 pp.

Mitchell, J. K. and Madsen, F. T., 1987, "Chemical Effects on Clay Hydraulic Conductivity," *Proceedings*, Geotechnical Practice for Waste Disposal '87, R. D. Woods, Ed., Geotechnical Special Publication No. 13, ASCE, pp. 87-116.

Mitchell, J. K., and Younger, J.S., 1967, "Abnormalities in Hydraulic Flow Through Fine-Grained Soils," *Permeability and Capillarity of Soils, ASTM STP 417*, American Society for Testing and Materials, Philadelphia, pp. 106-139.

Mitchell, J. K., Hooper, D. R., and Campanella, R. G., 1965, "Permeability of Compacted Clay," *Journal of the Soil Mechanics and Foundation Division*, ASCE, Vol. 91, No. SM4, pp. 41-66.

Murray, R. S. and Quirk, J. P., 1982, "The Physical Swelling of Clays in Solvents," *Soil Science Society of America, Journal*, Vol. 46, pp. 865-868.

Nasiatka, D. M., Shepherd, T. A., and Nelson, J. D., 1981, "Clay Liner Permeability in Low pH Environments," *Proceedings*, Fourth Symposium on Uranium Mill Tailings Management, Geotechnical Engineering Program, Department of Civil Engineering, Colorado State University, Fort Collins, Colorado, October 26-27, pp. 627-645.

Olson, R. E. and Daniel, D. E., 1981, "Measurement of the Hydraulic Conductivity of Fine-Grained Soils," *Permeability and Groundwater Contaminant Transport, ASTM STP 746*, T. F. Zimmie and C. O Riggs, Eds., American Society for Testing and Materials, Philadelphia, pp. 18-64.

Olsen, H. W., 1966, "Darcy's Law in Saturated Kaolinite," *Water Resources Research*, Vol. 2, No. 2, pp. 287-295.

Page, A. L., Miller, R. H., and Keeney, D. R., 1982, *Methods of Soil Analysis, Part 2, Chemical and Microbiological Properties*, 2nd Edition, American Society of Agronomy, Inc., Soil Science society of America, Inc., Madison, Wisconsin, 1159 pp.

Peirce, J. J. and Witter, K. A., 1986, "Termination Criteria for Clay Permeability Testing," *Journal of Geotechnical Engineering*, ASCE, Vol. 112, No. 9, pp. 841-854.

Pierce, J. J., Salfours, G., and Ford, K., 1987, "Differential Flow Patterns Through Compacted Clays," *Geotechnical Testing Journal*, ASTM, Vol. 10, No. 4, pp. 218-222.

Peterson, S. R. and Gee, G. W., 1985, "Interactions Between Acidic Solutions and Clay Liners: Permeability and Neutralization," *Hydraulic Barriers in Soil and Rock, ASTM STP 874*, A. I. Johnson, R. K. Frobel, N. J. Cavalli, and C. B. Pettersson, Eds., American Society for Testing and Materials, Philadelphia, pp. 229-243.

Quirk, J. P. and Schofield, R. K., 1955, "The Effect of Electrolyte Concentration on Soil Permeability," *Journal of Soil Science*, Vol. 6, No. 2, pp. 163-178.

Rad, N. S., and Acar, Y. B., 1984, "A Study on Membrane-Permeant Compatibility," *Geotechnical Testing Journal*, ASTM, Vol. 7, No. 2, pp. 104-106.

Relyea, J. F. and Martin, W. J., 1982, "Evaluation of Inactive Uranium Mill Tailings Sites for Liner Requirements: Characterization and Interaction of Tailings, Soil and Liner Materials," Proceedings, Fifth Symposium on Uranium Mill Tailings Management, Geotechnical Engineering Program, Department of Civil Engineering, Colorado State University, Fort Collins, Colorado, December 9-10, pp. 507-519.

Ryan, C. R., 1987, "Vertical Barriers in Soil for Pollution Containment," Proceedings, Geotechnical Practice for Waste Disposal '87, R. D. Woods, Ed., Geotechnical Special Publication No. 13, ASCE, pp. 182-204.

Sai, J. O. and Anderson, D. C., 1991, "Long-Term Effect of an Aqueous Leachate on the Permeability of a Compacted Clay Liner," Hazardous Waste and Hazardous Materials, Vol. 8, No. 4, pp. 303-312

Sawyer, E. W., 1986, "The Characterization and Use of Clays for Gelling Salt Water Fluids," Preprints Society of Mining Engineers Annual Meeting, AIME, Atlanta, March 6-10.

Shackelford, C. D., and Glade, M. J., 1994, "Constant-Flow and Constant-Gradient Permeability Tests on Sand-Bentonite-Fly Ash Mixtures," Hydraulic Conductivity and Waste Contaminant Transport in Soils, ASTM STP 1142, David E. Daniel and Stephen J. Trautwein, Eds., American Society for Testing and Materials, Philadelphia.

Shackelford, C.D. and Javed, F., 1991, "Large-Scale Laboratory Permeability Testing of a Compacted Clay Soil," Geotechnical Testing Journal, ASTM, Vol. 14, No. 2, pp. 171-179.

Standard Methods for the Examination of Water and Wastewater, 1985, 16th Edition, Arnold E. Greenberg, R. Rhodes Trussell, and Lenore S. Clesceri, Eds., American Public Health Association, Washington, D.C.

Stumm, W. and Morgan, J. J., 1981, Aquatic Chemistry, 2nd Ed., John Wiley and Sons, Inc., New York, 780 pp.

Tobin, W. R. and Wild, P. R., 1986, "Attapulgite: A Clay Liner Solution?," Civil Engineering, ASCE, Vol. 56, No. 2, pp. 56-58.

Uppot, J. O. and Stephenson, R. W., 1989, "Permeability of Clays Under Organic Permeants," Journal of Geotechnical Engineering, ASCE, Vol. 115, No. 1, pp. 115-131.

van Olphen, H., 1963, An Introduction to Clay Colloid Chemistry, John Wiley and Sons, Inc., New York, 301 pp.

Zimmie, T. R., Doynow, J. S., and Wardell, J. T., 1981, "Permeability Testing of Soils for Hazardous Waste Disposal Sites," Proceedings, Tenth International Conference on Soil Mechanics and Foundation Testing, Vol. 2, Stockholm, pp. 403-406.

Daniel B. Stephens[1]

HYDRAULIC CONDUCTIVITY ASSESSMENT OF UNSATURATED SOILS

REFERENCE: Stephens, D. B., "**Hydraulic Conductivity Assessment of Unsaturated Soils**," Hydraulic Conductivity and Waste Contaminant Transport in Soil, ASTM STP 1142, David E. Daniel and Stephen J. Trautwein, Eds., American Society for Testing and Materials, Philadelphia, 1994.

ABSTRACT: Predicting the direction and rate of contaminant transport in the vadose zone requires quantification of soil hydraulic properties. For many problems, the most appropriate approach is to assume that the contamination is moving with the water phase. Therefore, the unsaturated hydraulic conductivity is the most important property governing water movement in soils. There is a wide variety of methods to determine this important property including in situ methods (e.g., instantaneous profile method, crust method), laboratory methods (e.g., one-step outflow, pressure plate, long column methods), and calculations from other data (e.g., Brooks and Corey, van Genuchten models). In this paper we present a summary and brief critique of the current methods for determining unsaturated hydraulic conductivity.

KEYWORDS: unsaturated hydraulic conductivity, vadose zone, seepage

The movement of water in soils often occurs under unsaturated conditions. This is true even for many problems of seepage from surface impoundments and channels, as well as for problems of infiltration and drainage on hillslopes, watersheds, and landscapes. It is becoming more common in performance assessments at hazardous waste sites and solid waste disposal facilities to recognize that flow occurs under partly saturated conditions. Analytical and numerical models used for engineering analysis of these types of problems are often based on Richard's equation:

$$\nabla \cdot K(\psi) \nabla(\psi + z) = C(\psi) \frac{\partial \psi}{\partial t} \qquad (1)$$

where

K is hydraulic conductivity (L/T),

ψ is the pressure head (negative of soil tension head, L),

C is the specific moisture capacity (L^{-1}),

z is the vertical coordinate measured positive upward, and

---

[1]President, Daniel B. Stephens & Associates, Inc., 6020 Academy Road NE, Suite 100, Albuquerque, NM 87109.

$$C(\psi) = \frac{d\theta}{d\psi} \qquad (2)$$

where θ is soil moisture content (L^3/L^3). The relationship between θ and ψ is referred to as the moisture retention curve, or soil-water characteristic curve. This relationship is hysteretic, in that there are usually different values of water content for a given pressure head, depending upon whether the soil is drying or wetting. At a given pressure head or tension, a soil will retain more water during drainage than during wetting.

Hydraulic conductivity is a maximum at saturation, but it decreases dramatically with decreasing water content. It would not be unexpected to find the decrease in hydraulic conductivity from saturation to that at ambient field moisture to be roughly nine orders of magnitude for sands and five orders of magnitude for clays. Additionally, the relationship between unsaturated hydraulic conductivity and pressure (K-ψ) is hysteretic, like moisture retention (θ-ψ). Such a wide variation in this coefficient makes modeling of variably saturated flow problems much more complex than most saturated flow problems. Furthermore, determining this soil property in the field and laboratory is often one of the most difficult tasks in site characterization.

The purpose of this paper is to review some of the methods used to determine unsaturated hydraulic conductivity in the field and laboratory. Methods to estimate unsaturated hydraulic conductivity are also presented. These are invaluable when the task at hand does not require direct measurement of soil properties or when other data are available which can be used to calculate the hydraulic conductivity.

FIELD METHODS

Field methods are often preferred over laboratory methods because they typically are more representative of bulk or average properties in a heterogeneous soil. Additionally, field methods seem to offer greater reliability because laboratory samples are sometimes disturbed to the extent that the value of conductivity is not representative of the field soil sampled. Included in this summary of field methods are the instantaneous profile method, flux control methods, and the flow net method.

Instantaneous Profile Method

The instantaneous profile method is based on the transient drainage of a soil (Watson 1966). A square plot, approximately 3 to 10 m on a side, is prepared at a level site and a berm is made on the perimeter. Duplicate tensiometers and a neutron probe access tube are emplaced inside the plot, near the center where the flow field is likely to be unaffected by lateral flow beyond the perimeter. The plot is filled with water until the profile is saturated to the depth of interest. Usually this is within 3 m of the surface, but for clay soils the practical depth of testing may be much less. Either the infiltration rate at constant ponding when the soil is saturated, or the rate of decline in ponded depth upon cessation of water application, may be used to estimate the field saturated hydraulic conductivity if lateral seepage is negligible; otherwise, a ring can be placed in the center of the berm and the infiltration rate inside it can be used for determining the saturated hydraulic conductivity, provided that the head inside the ring is equal to that outside. Tensiometers are used to calculate the hydraulic gradient, and the neutron probe is used to monitor soil moisture content.

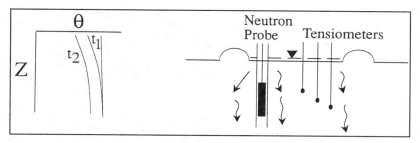

FIG. 1--Instantaneous profile test for field analysis of unsaturated hydraulic conductivity.

To obtain the unsaturated hydraulic conductivity, the water supply is shut off, the plot is covered to prevent evaporation, and pressure head and water content are measured as the profile drains (Figure 1). The measurements are most frequent immediately after infiltration stops and become less frequent with time. The drainage monitoring usually continues until the rate of decrease in moisture content is insignificant. For sandy soils, the test may require only several days to complete. For finer textured materials, tests may require weeks or months to wet and drain the profile. The transient data are used to calculate unsaturated hydraulic conductivity at some depth below land surface, L, according to the following equation:

$$K(\overline{\theta}) = \frac{\int_0^L \left(\frac{\partial \theta}{\partial t}\right) dz}{dH/dz} \qquad (3)$$

Covering the plot to prevent evaporation after ponding creates a no-flow boundary condition; this enables one to quantify the water flux as shown in the numerator of Eq. 3. At discrete depths, simultaneously, the hydraulic gradient is calculated from tensiometric data, and the rate of change in moisture content is calculated from the slope of the moisture content versus time plot. The hydraulic conductivity calculated from Eq. 3 is associated with the mean water content or pressure head at a particular depth. The analysis progresses from wet to dry conditions to obtain discrete values of hydraulic conductivity over a range of saturations. If the soil is layered, there may be significant differences in conductivity calculated at each depth. In layered soils, care must be taken in plot preparation to assure that flow is one-dimensional across the layers. If tensiometers are used to compute the gradient, the practical lower limit of hydraulic conductivity will correspond to about -850 cm of water pressure head. However, in practice the hydraulic gradient is usually near one so that the determination of pressure head is not always critical to the analysis, at least in relatively uniform soils. In which case, Eq. 3 becomes:

$$K(\overline{\theta}) = \int_0^L \frac{\partial \theta}{\partial t} dz \qquad (4)$$

Consequently, the hydraulic conductivity often can be estimated from moisture content data alone.

The instantaneous profile method is probably the most widely used of the field techniques. Its disadvantages are the greater amount of time required to complete the tests, the practical limitation to relatively shallow depths and permeable soils, the potential for channeling outside the instrumentation tubing, and the inherent analytical and measurement inaccuracies and uncertainties (e.g., Flühler et al. 1976). For further details on this procedure, refer to Hillel (1980).

Constant Flux Methods

There are two general types of field tests to determine unsaturated hydraulic conductivity which are accomplished by applying water to soil at a constant flux until the profile is at steady-state. The rate of application for both tests is less than that required to cause ponding. The smaller the application rate, the smaller the hydraulic conductivity. To calculate hydraulic conductivity, one divides the flux by the hydraulic gradient. The gradient is determined by tensiometers or is estimated to be unity. In contrast to the instantaneous profile method, the constant flux methods determine the hydraulic conductivity during wetting rather than drying conditions.

Crust method--The first constant flux method is referred to in Bouma et al. (1974). Here, a pedestal of soil is carved with a nominal diameter and depth of about 0.3 m. The side walls are sealed with a metal cylinder or other impermeable material such as a fiberglass coating. The top surface of the soil pedestal is prepared with a porous material, such as a plaster of Paris and sand slurry, which is less permeable than the underlying soil. This impeding layer forms a crust which, when a constant head of water exists above it, allows water to infiltrate the soil pedestal at a constant rate. The infiltration rate can be controlled by mixing different proportions of plaster of Paris and sand.

The soil-water pressure head is monitored by installing a tensiometer in the soil beneath the crust. The vertical hydraulic conductivity at the measured pressure head is simply equal to the steady infiltration rate, inasmuch as the hydraulic gradient is assumed to be 1. To obtain a range of values, the test must be run at different rates of infiltration that can be controlled by varying either the crust composition or thickness.

The crust method is not widely used in practice. To get each value of K-θ requires a new crust and test. There is concern that the flow field may be complex due to the potential for unstable flow created by the low-permeable crust (e.g., fingering). For layered soils, multiple tensiometers are required to determine both the hydraulic gradient across the layers and average pressure head in the layers. This method is most useful for the wet range, due mostly to the long time to reach steady-state at low flows.

Sprinkler method--Sprinklers are also used to control water flux to soil. Hydraulic conductivity is determined by the same analysis as described above for the crust method (e.g., Youngs 1964). The sprinkler infiltrometer is commonly used for agricultural and rangeland studies. It or any similar device can be utilized to apply water uniformly over the soil at a constant rate. The area of application is roughly 3 m^2 prepared directly on the soil surface, with an outer buffer area beneath the sprinkler to inhibit lateral flow. Tensiometers can be used to determine the gradient and mean pressure head. This approach overcomes some of the logistical and operational difficulties and perhaps the unstable flow issue which affects the crust method, but some of the same limitations remain.

Jeppson et al. (1975) developed a numerical inverse procedure to determine unsaturated hydraulic conductivity from transient sprinkler infiltration data. Their method allows the flow field to be axisymmetric, that is, unconstrained with respect to horizontal flow. However, the porous media must be homogeneous. Instrumentation requirements include tensiometers for measuring pressure head and either neutron probe or gamma ray methods for measuring water content in situ. Except for tests in the original article, in which vegetated sites were tested to the 0.3 m depth. There does not appear to be other examples of applications of this method.

Flow Net Method

A relatively recent method was proposed to determine the unsaturated hydraulic conductivity by mapping the hydraulic head fields near constant head sources (Stephens 1985). These sources could be surface impoundments, ditches, or even water-filled boreholes which produce steady-state flow fields above a water table. The principal of the method is based on observations and theory of multi-dimensional flow in which the water content decreases with increasing distance of flow along a stream tube emanating from the water source. The hydraulic head field is mapped by installing pressure head sensors such as tensiometers at a sufficient number of locations to contour the measurements in a vertical slice through the flow field. Several flow tubes, arbitrarily selected, are divided into segments which coincide with the equipotential surfaces. The segment of the stream tube closest to the source must be saturated. The hydraulic conductivity of soil at other saturations along the stream tube can be calculated from:

$$K_i = \frac{K_{S,1} J_1 A_1}{J_i A_i} \quad (5)$$

where the subscript "1" refers to the stream tube segment closest to the course where saturation occurs, subscript "i" indicates stream tube segments between equipotential lines, J is hydraulic gradient, and A is cross-sectional area of the stream tube in the center of the segment where K_i is calculated (e.g., Figure 2). In this method, only the relative hydraulic conductivity (ratio of unsaturated to saturated hydraulic

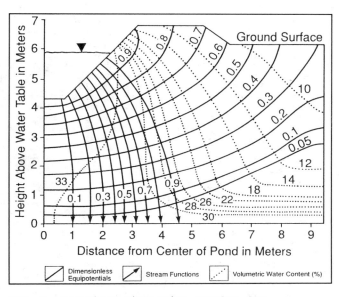

FIG. 2--Numerical simulation results for axisymmetrical flow from an impoundment in gravelly sand G.E.9 [modified from Riesenauer (1963)].

conductivity, K_i/K_s) can be obtained, if the saturated hydraulic conductivity is not determined independently (such as with a borehole permeameter).

The method has been applied to sandy and loamy soils with results which compare favorably with other methods (e.g., Larson and Stephens 1985). The method, which is most useful in the wet range of conductivities, may produce conductivities which are influenced by anisotropy where the flow lines are not in the principal directions. Uncertainties in the graphical flow net procedure can lead to significant uncertainties in the calculated conductivity. Additionally, installation of perhaps more than two dozen tensiometers can be quite tedious. Although this method shows considerable promise, it requires further testing before it can be recommended for standard practice.

LABORATORY METHODS

In comparison to the field methods, there are a larger number of laboratory methods for determining unsaturated hydraulic conductivity. These have been grouped into two general categories: steady-state flow methods and transient flow methods.

Steady-State Methods

Steady flow can be introduced in horizontal or vertical laboratory columns under constant head or constant flux conditions. By what ever method this is achieved, the hydraulic conductivity is calculated from Darcy's equation as simply the ratio of steady flow rate per unit cross-sectional area to the hydraulic gradient. The hydraulic conductivity is associated with the mean water content and pressure head established in the column. Separate tests must be run to obtain conductivities over a range of water contents or potentials.

A constant head of water at pressure less than atmospheric can be applied to the top and bottom of a vertical soil column when the ends are fitted with porous plates (Figure 3). These plates fit tightly against the soil and have high air-entry pressures so that the plate remains saturated as water flows through it under a tension. To supply the water to and receive water from the soil column under a constant tension, Mariotte siphons are commonly used. The flow rate can be calculated from transient water level measurements in a calibrated reservoir or burette. Tensiometers are often installed through the wall of the soil column at two different positions to measure soil-water pressure head for computing both the hydraulic gradient and the mean pressure head at steady-state. Usually, a series of steady-state tests are run beginning with a nearly saturated column. Therefore, it is important to recognize that the unsaturated hydraulic conductivity is associated with pressure head on the drainage cycle, owing to hysteresis in the K-ψ relationship.

Steady-state can also be achieved by applying water at a constant rate until the inflow and outflow, as well as the pressure in the soil column, are constant. The easiest means to apply water at a constant rate is with a peristaltic pump. The water is diffused uniformly across the soil surface through a porous end-plate or a network of small tubes. Water is allowed to freely drain to the atmosphere at the lower end of the column. Under this condition, and if the column length is much greater than the height of the capillary fringe for the soil, the hydraulic gradient will be close to unity in the upper part of the column. Such a test is called the long column method. If the gradient is unity, then the hydraulic conductivity is equal to the steady applied flux. At this hydraulic conductivity, either a tensiometer can be used to measure pressure, or a gamma ray device can be used to measure *in*

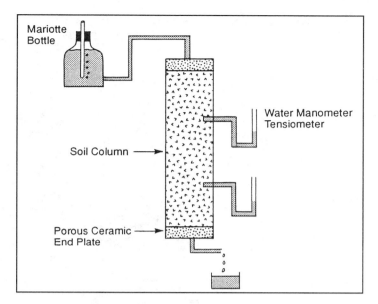

FIG. 3--Column method for establishing steady infiltration using a Mariotte siphon.

situ water content. As with the constant head methods, a series of tests must be conducted to calculate conductivity over a range of moisture conditions.

Although the conductivity analysis is very simple, the steady-state methods are tedious to conduct. The entire process is very slow to complete, especially for low water content conditions, due to the long time requirements to reach steady-state. During this time, care must be taken to overcome potential problems such as bacterial growth, air entrapment, dissolution, or shrinking and swelling.

Transient Methods

To reduce the time to obtain useful unsaturated hydraulic conductivity data, a variety of transient methods have been proposed. In general, the time savings is at the expense of increased experimental data requirements and more complex mathematical analysis. Here, five transient techniques will be discussed: instantaneous profile method, Bruce-Klute method, pressure plate method, one-step outflow method, and ultracentrifuge.

Instantaneous profile method--This method for laboratory analyses of unsaturated hydraulic conductivity is virtually the same as described previously for the field. The length of the soil column must be significantly greater than the height of the capillary fringe, so the method is best applied to sandy soils. The soil column is saturated, the top is loosely capped to prevent evaporation, and the profile is allowed to drain. During drainage, water content is measured, usually by the gamma attenuation method. However, time domain reflectometry may be an alternative method to measure moisture content which is less expensive and environmentally safe. Also, pressure head is measured with tensiometers located at several locations along the column. The mathematical analysis of data is identical to that described above for the field technique. This method is frequently used by researchers interested in relatively large laboratory scale results. The hydraulic conductivities are typically representative of the wet range of the imbibition cycle.

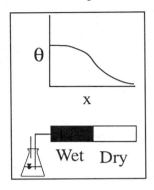

FIG. 4--Bruce-Klute apparatus for measuring soil water diffusivity.

Bruce-Klute method--This method utilizes soil packed into a thin column which is oriented horizontally (Bruce and Klute 1956). Water is introduced to one end at a small but constant tension using, for example, a Mariotte siphon (Figure 4). The only data requirements are water content measurements over time, either at one location or at many locations along the column at about the same time. Both approaches produce a set of $\theta(x,t)$ data. Water content distributions at an instant in time are best obtained by using a specially designed column composed of segments such that, at the desired time, the column is instantly sectioned and water content can be determined gravimetrically from the soil in each segment. A non-destructive method utilizes the gamma beam attenuation method to measure moisture content.

The mathematical model is based on the diffusivity form of the flow equation:

$$\nabla \cdot D(\theta) \nabla \theta = \frac{\partial \theta}{\partial t} \quad (6)$$

where

$$D(\theta) = K(\psi)/C(\psi) = diffusivity \quad (7)$$

The two independent variables measured during the test (x,t) are used to generate the Boltzman variable $\lambda(\theta)$:

$$\lambda = \frac{x}{\sqrt{t}} \quad (8)$$

where x is the horizontal distance from the source to a particular value of θ at time t. The Boltzman variable $\lambda(\theta)$ is used to linearize Eq. 6 to give the solution for diffusivity at water content θ' from the following equation:

$$D(\theta') = -\frac{1}{2}\left(\frac{d\lambda}{d\theta}\right)_{\theta=\theta'} \int_{\theta_1}^{\theta'} \lambda(\theta)\, d\theta \quad (9)$$

The terms in Eq. 9 can be determined graphically from λ versus θ plots or by new techniques which are more accurate (e.g., Shan and Stephens 1993). To compute the hydraulic conductivity, the specific moisture capacity needs to be determined from an independent analysis of the moisture retention curve, and then Eq. 7 is applied to calculate conductivity.

The Bruce-Klute method is a valuable tool for obtaining hydraulic conductivity under wetting conditions. In almost all published applications, the tested soils are repacked, and therefore disturbed. However, there is nothing to preclude the extension to undisturbed samples if the moisture content is determined nondestructively. The graphical method of analysis produces results which are less reliable near saturation, because of the large slope of the λ-θ curve there; however, the method is generally most useful for obtaining conductivities at relatively low water content. Bear in mind that this method does not produce unsaturated hydraulic conductivity directly. Consequently, the method has a disadvantage because of the extra effort required to characterize moisture retention. Also, there is a question about bias introduced from the possible variability in properties between the sample used for the imbibition experiment and the sample used for moisture retention analysis.

<u>Pressure plate method</u>--This is a third method to obtain conductivity in the laboratory (e.g., Gardner 1955). The principle here is to force water out of a soil sample under pressure, and examine the rate of water outflow over time. The testing assembly consists of a tightly sealed metal chamber which contains inside near its base a saturated porous ceramic plate with a high air entry value (Figure 5). A thin soil sample is firmly placed on the ceramic plate to ensure good contact. The top of the chamber is connected to a pressurized source of nitrogen gas. Usually the test is initiated with the sample near saturation. A small increment of gas pressure in excess of atmospheric pressure is applied and maintained constant, as the volume of water forced from the sample chamber is measured over time. The rate of outflow is conveniently measured in a horizontal glass tube and burette assembly. When the outflow from that pressure increment ceases, another small increment of pressure is applied, and so on, until the water content range of interest is covered. With proper equipment design, the apparatus can be used to rewet the soil, and therefore examine hysteretic behavior.

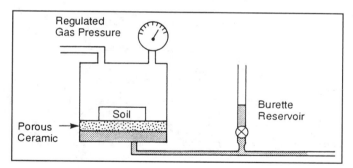

FIG. 5--Volumetric pressure plate extractor. Rate of outflow (or inflow) from sample due to an applied pressure change is measured by the advance (or retreat) of the meniscus in the thin-horizontal glass tube.

Although the pressure plate method is relatively easy to apply, even in the dry range, it does not seem to be widely used in practice. This stems from several problems related to the theoretical development and logistics. For example, the mathematical model is based upon a form of Eq. 6 which assumes that during a pressure increment, the diffusivity is constant even though the moisture content is actually changing. At high permeability, flow impedance through the ceramic plate can affect results unless this is taken into account (e.g., Kunze and Kirkham 1962). Furthermore, shrinkage of the soil during drainage may cause loss of contact with the plate.

One-step outflow method--This currently popular variation on the pressure outflow method was proposed by Passioura (1976). His approach forces water from the soil in response to a large step increase of known applied pressure and then examines the rate of outflow in a graphical procedure. The analysis produces soil-water diffusivity as in Eq. 7, but it is assumed that diffusivity is an exponential function of water content, instead of a constant over the increment of applied pressure. One-step outflow data can be collected by draining the sample under a tension using a hanging water column apparatus or by applying a positive pressure in a pressure cell. Constantz and Herkelrath (1984) designed an outflow cell and procedure for conducting tests under controlled temperature conditions in order to analyze the effect of high temperatures on unsaturated hydraulic conductivity. Kool et al. (1985) developed a convenient computer code to analyze the outflow data using Richards equation and a nonlinear least squares technique to adjust unsaturated flow parameters to obtain a best fit to measured outflow. Unfortunately, this approach may lead to nonunique results (e.g., Toorman et al. 1990). There are presently a number of researchers attempting to overcome this limitation of the one-step outflow method which, in virtually all other respects, is a very practical method.

Centrifuge technique--This last method is rather new (Nimmo et al. 1987). The principal is to establish a steady-state flow through a sample while it is in a spinning centrifuge and to compute hydraulic conductivity using a form of Darcy's equation which takes into account both matric potential and centrifugal forces. At sufficiently high rotational speeds, the gradients due to matric potential are negligible compared to those imposed by the rotation. Therefore, hydraulic conductivity can be determined from the following equation:

$$q = K(\psi)\rho\omega^2 r \qquad (10)$$

where ρ is the fluid density, ω is the rotation speed, and r is the radius from the axis of rotation. Conca and Wright (1990) have developed an ultracentrifuge apparatus that is capable of measuring conductivities at moisture contents of less than a few percent. Not only does the centrifuge method provide conductivities in the very low water content range, but a steady flow field is achieved in a matter of hours, at an extraordinary savings of time. Problems with the technique are few and lie primarily with issues centering on compression and change of pore structures during centrifuging. Additionally, at this time the application of the method is limited because of the expense of the ultracentrifuge equipment.

ESTIMATING UNSATURATED HYDRAULIC CONDUCTIVITY

It should be obvious now that most of the field and laboratory methods to characterize unsaturated hydraulic conductivity are either tedious and time-consuming, or expensive. This has led to considerable research in developing ways to estimate unsaturated hydraulic conductivity. There are two general approaches for estimation. The first is an empirical approach and the other is to calculate conductivity from moisture retention data.

Empirical Approach

Unsaturated hydraulic conductivity can be estimated by assuming that the soil follows a particular model or relationship between conductivity and pressure or moisture content. For example, an equation commonly used, especially in conjunction with analytical solutions, is that developed by Gardner:

$$K(\psi) = K_0 \exp(\alpha\psi) \qquad (11)$$

where K_0 is the conductivity extrapolated to zero pressure head. For most soils, α ranges from about 0.1 cm^{-1} for sands to 0.005 cm^{-1} for clays. Brooks and Corey (1964) proposed another commonly used equation:

$$K_r = (\psi/\psi_{cr}); \qquad (\psi \leq \psi_{cr}) \qquad (12)$$

where ψ_{cr} is the critical pressure or bubbling pressure, and λ is the pore size distribution index. These two parameters can be calculated from moisture retention curves, or they can be estimated from textural characteristics (e.g., McCuen et al. 1981).

Clapp and Hornberger (1978) proposed a power function model:

$$K_r = S^{2b+3} \qquad (13)$$

where $S = \theta/\theta_{sat}$, and b is an empirical parameter determined for each distinct soil textural classification. Representative values of b were derived by statistical analysis of many data sets of texture and measured moisture retention. To estimate conductivity, only an analysis of soil texture is needed. A similar power function model was proposed by Mualem (1978), based in part on statistical analyses from many soils:

$$K_r(S_e) = S_e^{3.0 + 0.015w} \qquad (14)$$

where w is evaluated from the moisture retention curve for the soil. In Eq. 14, S_e is the effective saturation defined according to:

$$S_e = \frac{\theta - \theta_r}{\theta_s - \theta_r} \qquad (15)$$

where θ_r is the residual water content and θ_s is porosity. These methods are frequently used in lieu of field or laboratory testing, chiefly because they utilize common and easily obtainable parameters.

Calculation From Water Retention Data

Calculating hydraulic conductivity from moisture retention curves is perhaps the most popular of all means in current use. The popularity stems from the ease of computation, but the method is also attractive because it is semi-quantitative in that measured properties (porosity, saturated hydraulic conductivity, and moisture retention) are used in the calculation. Statistical models of the pore-size distribution have been applied to develop a mathematical tool for calculating conductivity from moisture retention data (e.g., Childs and Collis-George 1950, Burdine 1953, Mualem 1976). van Genuchten (1980) developed a closed-form analytical solution for conductivity which is based on the following functional form for the moisture retention curve:

$$S_e = [1 - (\alpha\psi)^n]^{-m} \qquad (16)$$

where α and n are fitting parameters. For the Burdine model, $m = 1 - 2/n$ and for the Mualem model $m = 1 - 1/n$. van Genuchten uses a nonlinear least-squares routine to determine the best fit of Eq. 15 to observed moisture retention (θ-ψ) data. When residual moisture content is not measured (Eq. 15), it can be obtained by the fitting routine. These fitting parameters are used to compute unsaturated hydraulic conductivity from the following equation for Mualem's model:

$$K(\psi) = \frac{\{1 - (\alpha\psi)^{n-1}[1 - \alpha\psi)^n]^{-m}\}^2}{[1 - \alpha\psi^n]^{m/2}} \qquad (17)$$

The methods to estimate unsaturated hydraulic conductivity generally have produced good results in non-structured soils, especially in the wet range (e.g., Stephens and Rehfeldt 1985). However, there is little experience to establish the reliability of the predictions in the dry range. This includes the pressure head range below about -800 cm where the LaPlace capillary model is not valid because water in this range is held mostly on the particle surfaces rather than in the pore throats.

The moisture retention curve, which is one of the key data needs for estimating hydraulic conductivity, is most commonly obtained by laboratory tests. At pressure heads greater than about -200 cm of water, the hanging water column method is the most reliable method to determine the moisture retention curve. The apparatus consists of a Büchner funnel (a beaker having a fritted glass bottom plate with glassware that necks down to about a 2 cm diameter) connected by thin, water-filled plastic tubing to a burette. The test begins with the soil sample near saturation on the fritted glass plate and the water level in the burette at the base of the sample. The burette is quickly lowered to a new position. As water drains from the sample, the water level in the burette rises to a new position of equilibrium which is lower in elevation than the base of the sample. The elevation difference between the burette water level and the center of the sample represents the soil-water tension associated with the retained water in the soil sample. Testing proceeds sequentially to lower or higher tensions. By measuring the initial or final water content, one can determine paired water content-pressure head values for a sufficient number of points to define the moisture retention curve. For pressure heads less than about -200 cm, positive pressure methods, such as those illustrated in Figure 5, are more appropriate. Here, one assumes that the water retained in the soil sample inside the chamber is in equilibrium with the applied pressure. For additional information on measuring soil-water retention, refer to Klute (1986)

CONCLUSION

There are a variety of field and laboratory methods to characterize the unsaturated hydraulic conductivity of soil. Except for measurements on cores evaluated for petroleum recovery and tests conducted as part of agricultural studies, these methods are not widely utilized in the geotechnical and hydrogeological fields. Nevertheless, there is a growing need to be aware of the available techniques, particularly for applications to environmental problems.

Most of the field methods are applicable to the wet range of moisture conditions. Laboratory methods, such as the Bruce-Klute and centrifuge methods, are more applicable to the dry range. Estimation procedures are the most convenient methods to determine unsaturated hydraulic conductivity; however, these require measuring other properties of the soil. There remains a great need for developing reliable and practical tools for determining conductivity especially in the field at a scale relevant to the problems of interest to geotechnical and environmental engineers.

REFERENCES

Bouma, J., Baker, F. G., and Veneman, P. L. M., 1974, "Measurement of Water Movement in Soil Pedons Above the Water Table," Information Circular No. 27, University of Wisconsin, Madison.

Brooks, R. H., and Corey, A. T., 1964, "Properties of Porous Media Affecting Fluid Flow, Proceedings of the American Society of Civil Engineers, Journal of the Irrigation and Drainage Division, Vol. IR2, pp. 61-88.

Bruce, R. R., and Klute, A., 1956, "The Measurement of Soil-Water Diffusivity," Soil Science Society of America Proceedings, Vol. 20i, pp. 458-562.

Burdine, N. T., 1953, "Relative Permeability Calculation From Pore Size Distribution Data," Petroleum Transactions, American Institute of Mining and Metallurgical Engineers, Vol. 198, pp. 71-77.

Childs, E. C., and Collis-George, N., 1950, "The Permeability of Porous Materials," Proceedings of the Royal Society of London, Vol. 201A, pp. 392-405

Clapp, R. B., and Hornberger, G. M., 1978, "Empirical Properties for Some Soil Hydraulic Properties," Water Resources Research, Vol. 14, No. 4, pp. 601-604.

Conca, J. L., and Wright, J., 1990, "Diffusion Coefficients in Gravel Under Unsaturated Conditions," Water Resources Research, Vol. 26, No. 5, pp. 1055-1066.

Constantz, J., and Herkelrath, W. N., 1984, "Submersible Pressure Outflow Cell for Measurement of Soil Water Retention and Diffusivity from 5 to 95°C," Soil Science Society of America Journal, Vol. 48, No. 1, pp. 7-10.

Flühler, H., Ardakani, M.S., and Stolzy, L. H., 1976, "Error Propagation in Determining Hydraulic Conductivities From Successive Water Content and Pressure Head Profiles," Soil Science Society of America Journal, Vol. 40, pp. 830-836.

Gardner, R., 1955, "Relation of Temperature to Moisture Tension of Soil," Soil Science, Vol. 79, pp. 257-265.

Hillel, D., 1980, Fundamentals of Soil Physics, Academic Press, Inc., New York.

Jeppson, R. W., Rawls, W. J., Hamon, W. R., and Schreber, D.L., 1975, "Use of Axisymmetric Infiltration Model and Field Data to

Determine Hydraulic Properties of Soils," *Water Resources Research*, Vol. 11, No. 1, pp. 127-138.

Klute, A., 1986, *Methods of Soil Analysis*, Part 1: Physical and Mineralogical Methods, 2nd ed., American Society of Agronomy-Soil Science Society of America, Madison, WI.

Kool, J. B., Parker, J.C., van Genuchten, M. Th., 1985, "Determining Soil Hydraulic Properties Form One-Step Outflow Experiments by Parameter Estimation, I. Theory and Numerical Studies, *Soil Science Society of America Journal*, Vol. 49, No. 6, pp. 1348-1354.

Kunze, R. J., and Kirkham, D., 1962, "Simplified Accounting for Membrane Impedance in Capillary Conductivity Determinations," *Soil Science Society of America Proceedings*, Vol. 26, pp. 421-426.

Larson, M. B., and Stephens, D. B., 1985, "A Comparison of Methods to Characterize Unsaturated Hydraulic Properties of Mill Tailings," Seventh Symposium on Management of Uranium Mill Tailings, Low-Level Waste and Hazardous Waste, Ft. Collins, Colorado, Colorado State University, Ft. Collins, February 6-8, 1985.

McCuen, R. H., Rawls, W. H., and Brakensisk, D. L., 1981, "Statistical Analysis of the Brooks-Corey and the Green-Ampt Parameters Across Soil Textures," *Water Resources Research*, Vol. 17, No. 4, pp. 1005-1013.

Mualem, Y., 1976, "A New Model for Predicting the Hydraulic Conductivity of Unsaturated Porous Media," *Water Resources Research*, Vol. 12, No. 3, pp. 513-522.

Mualem, Y., 1978, "Hydraulic Conductivity of Unsaturated Porous Media: Generalized Macroscopic Approach," *Water Resources Research*, Vol. 14, No. 2, pp. 325-334.

Nimmo, J. R., Rubin, J., Hammermeister, D. P., 1987, "Unsaturated flow in a centrifugal field: Measurement of Hydraulic Conductivity and Testing of Darcy's Law," *Water Resources Research*, Vol. 12, pp. 513-522.

Passioura, J. E., 1976, "Determining Soil Water Diffusivities From One-Sided Outflow Experiments," *Australian Journal of Soil Research*, Vol. 15, pp. 1-8.

Reisenauer, A.E., 1963, "Methods for Solving Problems of Multidimensional, Partially Saturated Steady Flow in Soils, *Journal of Geophysical Research*, Vol. 68, No. 20, pp. 5725-5733.

Shan, C., and Stephens, D. B., 1993, "Double Integration Method and Its Application to Determine Unsaturated Soil Properties," Submitted to *Water Resources Research* (in review process).

Stephens, D. B., 1985, "A Field Method to Determine Unsaturated Hydraulic Conductivity Using Flow Nets," *Water Resources Research*, Vol. 21, No. 1, pp. 45-50.

Stephens, D. B., and Rehfeldt, K. R., 1985, "Evaluation of Closed-Form Analytical Models to Calculate Conductivity in a Fine Sand," *Soil Science Society of America Journal*, Vol. 49, No. 1, pp. 12-19.

Toorman, A. F., Wierenga, P. J., and Hills, R. G., 1990, "Estimation of Soil-Water Parameters from Transient Laboratory Experiments," *Agronomy Abstracts*, p. 220.

van Genuchten, M. Th., 1980, "A Closed-Form Equation for Predicting the Hydraulic Conductivity of Unsaturated Soils," Soil Science Society of America Journal, Vol. 44, No. 5, pp. 892-898.

Watson, K. K., 1966, "An Instantaneous Profile Method for Determining the Hydraulic Conductivity of Unsaturated Porous Materials," Water Resources Research, Vol. 2, pp. 709-715.

Youngs, E. G., 1964, "An Infiltration Method of Measuring the Hydraulic Conductivity of Unsaturated Porous Materials," Soil Science, Vol. 109, pp. 307-311.

Stephen J. Trautwein[1] and Gordon P. Boutwell[2]

IN-SITU HYDRAULIC CONDUCTIVITY TESTS FOR COMPACTED SOIL LINERS AND CAPS

REFERENCE: Trautwein, S. J., and Boutwell, G. P., "In Situ Hydraulic Conductivity Tests for Compacted Soil Liners and Caps," Hydraulic Conductivity and Waste Contaminant Transport in Soil, ASTM STP 1142, David E. Daniel and Stephen J. Trautwein, Eds., American Society for Testing and Materials, Philadelphia, 1994.

ABSTRACT:

The most important geotechnical parameter for soil liners or caps for waste facilities is hydraulic conductivity. In typical practice, the vertical conductivity normally governs the barrier effect. Regulatory agencies are increasingly requiring in situ tests in addition to laboratory tests to verify hydraulic conductivity. Many different in situ tests have been proposed; seven have some usage. These are discussed with reference to their suitability for liner/cap evaluation. Of these, the Sealed, Double-Ring Infiltrometer and the Two-Stage Borehole test have received widest acceptance and use; each has already been used on 50 to 100 projects. Both tests are examined in detail, including theoretical basis, field procedures, and data reduction. Reasons for their preference over other methods are discussed.

The Sealed Double-Ring Infiltrometer measures one-dimensional infiltration from which vertical hydraulic conductivity is determined. Testing times are comparatively long but a larger volume of soil is tested, 1000 to 3000 times that typically tested in the laboratory. It is the better of the two methods for testing thin liners (<50 cm). One test is adequate for all but the most poorly constructed test media.

The Two-Stage Borehole method involves three-dimensional infiltration, from which both vertical and horizontal conductivities are determined. A typical study involves multiple tests, the total volume of soil affected by a 5-test group being 250 to 1000 times that typically tested in the laboratory. Testing times are comparatively short, and the method is the better of the two for evaluating variability in hydraulic conductivity with depth, on slopes, and where horizontal conductivity is needed. A five-test group is adequate for all but the most poorly constructed test media.

Comparative results are presented from 15 projects. The two preferred methods yield comparable values; both have been successful in verifying well-built barriers and identifying those with defects.

KEY WORDS: borehole test, case histories, clay liner, compacted clay, double-ring infiltrometer, field test, hydraulic conductivity, infiltrometer, permeability

INTRODUCTION

Over the past decade, two significant changes have occurred with respect to evaluating compliance of earthen liners and caps used in waste disposal facilities. First is the use of full scale test

[1]President, Trautwein Soil Testing Equipment Co., Houston, TX
[2]President, Soil Testing Engineers, Inc., Baton Rouge, LA

pads in the design phase for evaluating effectiveness of construction procedures. Second, and the focus of this paper, is the use of field tests for evaluating in situ hydraulic conductivity.

Traditionally, compliance evaluation of earthen liners and caps was based solely on results of laboratory tests on small diameter (7 to 10 cm) specimens. Field testing represents a significant change not only because it is more costly and subject to interpretation, but also because testing times are typically much longer and can adversely impact construction schedules. The change in emphasis from lab to field testing began in the mid-1980's after the reliability of laboratory tests on small diameter samples was questioned. Olson and Daniel (1981), Daniel (1984) and others reported on higher than expected leakage rates through earthen liners. Later investigations on prototype liners by Day and Daniel (1985), Elsbury et. al. (1988) demonstrated that in situ hydraulic conductivity, k_f, could be as much as 10,000 times greater than that determined by small scale laboratory tests, k_l. Disagreement arose in the technical community with regards to the significance of these findings with respect to full scale liners and caps [see discussions on Day and Daniel (1985)]. The debate continued as some investigators, such as Daniel and Trautwein (1986), Trautwein and Williams (1990) reported k_f/k_l greater than 10, while others including Lahti, et. al.(1987), Mundell and Boos (1990), and Johnson et. al. (1990), reported close agreement between k_f and k_l. Benson and Boutwell (1992) provided explanations reconciling these inconsistent results.

While the need for field testing was under debate within the engineering community, concern in the regulating community resulted in shifts towards emphasizing field testing. At the federal level, EPA(1984) suggested field testing in guidance documents. In 1986, California became the first state to require field testing as part of compliance evaluation. Other states, such as PA, TX, and OH soon followed. This was a bold step not only because it was unprecedented, but also because there were no widely accepted or standardized techniques for testing thin layers of low hydraulic conductivity soils.

In the rush to meet the new requirements, the engineering community responded with a variety of testing methods, some state of the art, others variations of tests in use for years. Many methods tried were originally developed for high conductivity soils and failed when used for measuring low flow rates. Since 1986, not only has considerable experience been gained in field testing methodology, but also in the understanding of factors that affect performance of earthen barriers.

The authors have been extensively involved with field testing and have had the unique experience of "hands on" involvement with a wide variety of testing methods on projects throughout the country. One of the problems noted is that due to the relative rapid implementation of field testing, the state of practice has not advanced uniformly throughout the country. Hence, lessons learned by some are being unnecessarily relearned by others. The purpose of this paper is to provide an updated report on the current state of practice. The paper is organized into three sections. First, a discussion on the debate over field versus lab testing is given. Next, a review of the more commonly used testing methods is presented. Comments on which tests work and which do not, along with testing details that can impact results are provided. Finally, a third section on testing strategy is presented. Issues such as when and where to test, frequency of testing, and the weight to be given a field test among the numerous factors used in evaluating compliance are discussed.

FIELD VERSUS LAB HYDRAULIC CONDUCTIVITIES

While much knowledge and experience with field testing and performance of earthen liners and caps has been gained in the last 10 years, a general consensus on the need for field testing has not been reached. The debate typically hinges on whether or not tests conducted on small samples in the laboratory are representative of the in situ hydraulic conductivity. Central to the debate are the following three issues: what is the effect of specimen size, is k_f really significantly greater than k_l, and what k should be measured. Each of these issues is discussed below.

Specimen Size

Olson and Daniel (1981) discuss factors that lead to errors when measuring k for both field and lab testing. The most important factor with respect to earthen liners and caps is the testing of non representative specimens. In fact, the authors argue that the issue is not lab versus field, but rather specimen size, i.e., testing large versus small sized specimens. Size is an issue because of compacted clay structure which consists of small and large pore spaces. To describe flow through the different size pore spaces, it is helpful to define the terms micropermeability and macropermeability.

Micropermeability refers to flow through micropores, the small void spaces between soil particles or aggregates of soil particles, most of which are in contact with adjacent soil particles. Macropermeability refers to flow through macropores, the larger void spaces corresponding to secondary structure, such as clod interfaces, lift interfaces, shear surfaces, and desiccation cracks. Auvinet and Espinosa (1981) noted that flow rate (and thus hydraulic conductivity) is proportional to the fourth power of effective pore diameter. Micropores are on the same scale as clay particles, e.g., effective diameters of 0.1 - 0.3 µm (Acar, et al, 1989). The latter authors show by pore-size distribution measurements that macrostructure virtually disappears when both the degree of saturation and of compaction are high. The size of macropores has a tremendous range, however; note that even small macropores, on the order of 10 - 50 µm (small enough that they can not be seen) can have a significant effect on the flow rate. Theoretically, if macropores are present and continuous, and represent only 1 per cent of the total pore volume, hydraulic conductivity will still be increased by a factor of 100 or more. If macropores are present the flow through micropores can be negligible in comparison.

Clearly, leakage through an earthen liner or cap will be controlled by macropermeability. If in situ k is to be measured, then the test specimen must be large enough to contain a representative density of secondary features. While no standard specimen size has been established, an educated guess can be made as to how large the specimen must be to contain a representative density of secondary features found in compacted liners and caps. Clods can range in size from .05 to 10 cm in diameter, and to allow for interconnection between interfaces, a specimen diameter of 20 cm should be allowed. Poor bonding at lift interfaces can result in pathways of higher flow along the interfaces than normal to them, thus connecting other secondary features and allowing for flow normal to the interface. To connect clod interfaces, a length of 20 to 30 cm would be needed. Desiccation cracking on lift surfaces creates vertical pathways, typically 2.5 to 5 cm in length, spaced 1 to 10 cm apart. These vertical pathways connect clod interfaces to lift interfaces. A specimen 10 to 15 cm in diameter would include desiccation cracking. Shear surfaces result from the kneading action of individual feet on a footed roller, bearing capacity failures due to the weight of the roller, and slippage due to quick starting or stopping of the roller. The individual feet may create shear surfaces 20 to 40 cm in length, while shear surfaces due to compactor weight and slippage may be 1 to 3 meters in length. Orientation of shear surfaces can be at any angle, and they are long enough to connect between clods, lift interfaces, and desiccation cracks. Specimens as large as 30 to 60 cm may be needed to account for effects of shear surfaces. Accounting for all the secondary features mentioned, an educated guess for a representative specimen size would be 20 to 60 cm in diameter; this correlates well with research cited below.

How large is large enough has not been fully established, although Williams (1988), Trautwein and Williams (1990), and Benson et al (1993) have investigated the influence of specimen size on k_f. Their results clearly indicate that if macropores are present, specimens at least 30 cm in diameter are needed to detect their influence. These investigators also found that in properly compacted liners and caps, where macropores were minimized or eliminated, specimen size had no influence on k_f.

Measurement of in situ hydraulic conductivity requires a test specimen large enough so that macropermeability (if present) is measured. This requires inclusion of a representative density of secondary features. Where the sample is tested, field or lab, is not a concern. However, most large scale tests are performed in the field and hence, the discussion seems to focus on field versus lab. Tests on large specimens can also be performed in the lab.

Is $k_f \gg k_l$?

Regulations requiring field testing were prompted by reports of large differences between in situ k_f and laboratory k_l. Clearly, some of these cases were the result of either thin liners, poor construction techniques, or inadequate protection from desiccation. Is k_f generally significantly greater than k_l? The answer is, it can be, but the probability of its being so can be greatly minimized. Almost 10 years have passed since those early findings, and significant improvements have been made in barrier construction methodology. More recent findings, Gordon et al (1990), Benson and Boutwell (1992), Trautwein and Williams (1990), Cartwright and Krapac (1990), demonstrate that by using construction procedures which minimize secondary features along with a thorough Construction Quality Assurance program, close agreement between laboratory and in situ hydraulic conductivity can be achieved. Again, clearly the key is minimizing or eliminating secondary features; direct verification requires testing of large specimens.

What k ?

Most regulations require earthen liners to have a hydraulic conductivity of either 1x10(-7) cm/s or 1x10(-6) cm/s. An important point often not appreciated by the regulating community is that hydraulic conductivity is not an intrinsic property which depends only on material type, but is dependent on a number of factors including: specimen preparation, degree of saturation, stress level, nature of the permeating fluid, and direction of flow. These factors should be accounted for when evaluating compliance.

Specimen preparation is a factor when comparing tests performed on laboratory compacted samples and those performed in the field. Williams (1988) lists a number of reasons why there are significant structural differences between laboratory compacted specimens and compacted earthen barriers. Pulverizing clods, removing of large clods, uniform mixing of moisture, and tempering all lead to specimens with no or minimal secondary structure. While tests on recompacted specimens are useful for qualifying materials, their k_l is often not representative of k_f.

Flow direction is also important when defining k. It is well established that hydraulic conductivity, k_h, parallel to the bedding (in natural soils) and compaction planes (in recompacted soils) is higher than that in the normal direction, k_v. Examples for compacted soils in the literature include Auvinet and Espinosa (1981), Day and Daniel (1985), and Boutwell and Rauser (1990). In general, flow normal to compaction planes controls the leakage rate. This is the case even for side liners (built with lifts parallel to the slope) where flow net studies (Boutwell and Rauser, 1990) confirm this point. Therefore, test methods in which k_h and k_v can be distinguished are preferred.

Effective stress level is important when evaluating k. Barriers tested with low overburden stress (which is the case for most field tests) may exhibit high flow rates. Permanence of this effect depends on the long term state of stress. For a waste retention liner, such as in a landfill, overburden typically increases with time. Fortunately, the macropores are the first to compress as the overburden pressure is increased. If that stress is high enough, k that corresponded to macropermeability can decrease to the point where it is controlled by micropermeability. Boynton (1983) presented a clear demonstration of this effect, where macropermeability controlled flow switched to micropermeability flow when the stress level attains 55 to 95 kN/m^2 (8 to 12 psi).

Some have argued that k corresponding to the long term stress conditions is the value on which compliance should be based. The counter argument is a significant amount of leakage may occur through the macropores before they are reduced to micropores. A comparison between short term and long term leakage can be made with a few simplifying assumptions. These assumptions include a linear increase in overburden with time, and that Log (k) decreases linearly with pressure from its initial macropore dominated value, k_i, to its final micropore dominated value, k_u ($k_i \sim k_f$, $k_u \sim k_l$).

Then:

$$Q_a/Q_u = [k_i/k_u - 1]/ \ln (k_i/k_u) \tag{1}$$

Where:

Q_a = actual flow volume per unit area during $0 < t < t_c$
Q_u = flow volume per unit area during $0 < t < t_c$ computed using k_u
t = time
t_c = time when macropores close

For $k_i/k_u = 10$, Q_a, the amount of leakage occurring up to t_c, is 4 times Q_u, increasing to 20 times Q_u for $k_i/k_u = 100$. Hence, short term leakage can be significant and it is reasonable to base design on k corresponding to low levels of stress, such as those obtained from field tests where stress levels are low. However, compliance evaluation should also account for minimization of long term leakage rates, which can be best evaluated in the laboratory where stress control is more practical. Presence of a synthetic liner overlying the clay can minimize Q_a/Q_u.

To summarize this section, the issue is not field testing versus laboratory testing, but how large a specimen is needed to account for secondary features. The test method should be capable of measuring the vertical hydraulic conductivity. It is reasonable to test at low overburden pressures; however the long term state of stress should be considered. While procedures have been established which increase the probability of obtaining close agreement between k_f and k_l, large scale testing should be performed as direct verification that the procedures worked.

FIELD TEST METHODS

In the rush to comply with regulatory requirements for field testing, the authors are aware of at least a dozen different methodologies used. The more significant were reviewed by Daniel (1989) and Sai and Anderson (1990). Most methods can be classified as either infiltration tests or borehole tests. Important distinguishing factors among the various methods include test geometry, i.e., one or three dimensional flow, volume of soil tested, and techniques for measuring low amounts of flow. The authors have selected the following test methods as worthy for discussion:

BAT Probe
Open Single Ring Infiltrometer (OSRI)
Sealed Single Ring (SSRI)
Air-Entry Permeameter (AEP)
Open Double Ring Infiltrometer (ODRI)
Sealed Double Ring Infiltrometer (SDRI)
Two-Stage Borehole Test (TSB)

The first five methods have found limited acceptance in certain parts of the country at one time or another. These methods are not recommended by the authors and discussion of them is brief, the intent being to caution users of their limitations. The last two methods, the SDRI and the TSB, have much broader acceptance and will be discussed in detail.

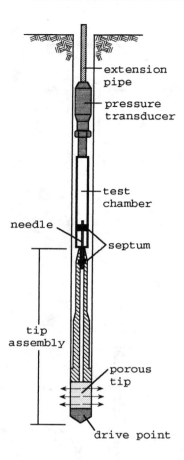

Figure 1. BAT Probe

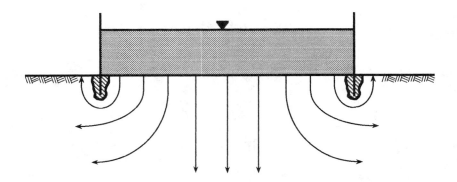

Figure 2. Open Single-Ring Infiltrometer

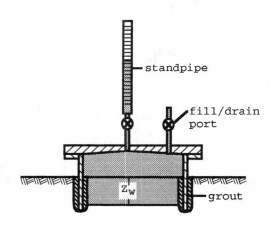

Figure 3. Sealed Single-Ring Infiltrometer

BAT Probe

The BAT probe, Fig. 1, was developed in Sweden and is named after its developer, B. A. Tortstensson (1984). Initially developed for sampling ground water, it provided a means for obtaining liquid samples in a hermetically sealed glass tube. The probe was modified and used in Sweden to measure hydraulic conductivity in saturated soft clays. In the mid 1980's, it was first used to measure the hydraulic conductivity of soil liners and caps.

The device works as follows. A tip assembly is inserted into the test zone by driving or pushing directly into the test zone or into an augered hole. The tip contains a porous element that is in direct contact with the soil. A drain line connected to the porous element runs up the center of the tip assembly and is sealed at the top with a rubber septum. Once the tip is installed, a test chamber, with a double ended needle positioned beneath it, is pushed down on the tip assembly. The test chamber contains a rubber septum seal and prior to being lowered into the hole, the chamber is partially filled with water and then pressurized. When pushed onto the tip assembly, the needle pierces both septums. At this point the tip and test chamber are hydraulically connected.

Pressure in the test chamber forces water out through the porous element into the test zone. This pressure decreases as water flows out and the change is measured with a transducer connected to the chamber top. Boyle's law is used to determine the amount of water flowing into the test zone. Hvorslev (1951) type equations are used to calculate k_h, the horizontal hydraulic conductivity. Unless the degree of anisotropy is known, the BAT value will be greater than k_v and less than k_h.

In some cases, soil suction in the test zone is measured and is used in calculating k. Suction is measured by lowering a pressure transducer assembly directly onto the tip. This is done prior to using the test chamber.

The primary advantage of the BAT probe is rapid measurements. Testing times can be as short as a few minutes and rarely are longer than several days.

While quick test times are appealing, this advantage is far outweighed by the disadvantages which are threefold. First, the test volume is small. The diameter of the probe is 1 inch (25mm), resulting in a test volume smaller than that typically tested in the laboratory. The size of the test zone is considered too small to be able to detect macro defects and hence too small to measure k_f. Second, the surface of the test zone is smeared during tip installation. This may result in an underestimate of k because the smearing may close macropores. Conversely, k may be overestimated due to opening of macropores by dragging of coarse grain particles across the test surface. The third disadvantage is that the measured k is somewhere between k_h and k_v. These disadvantages have limited widespread use of this device, and are the reasons for not recommending the BAT probe for liners and caps.

Open Single Ring Infiltrometers

Open single ring infiltrometers (OSRI), Fig. 2, are the simplest and probably most widely known type of infiltrometers, dating back to at least 1905 (Day and Daniel, 1985). The device consists of a ring which is grouted into the test zone and then filled with water. Testing consists of noting the drop in water level versus time as water infiltrates the test zone. The appeal of this method is that large volumes of soil can be tested, ponding of water on the surface models the actual field case, and a direct measurement of leakage is made. The OSRI method is used primarily in agricultural applications where infiltration rate is high and the test is performed to ensure that a soil has a high enough hydraulic conductivity to drain adequately.

Hydraulic conductivity, k_v is determined from infiltration I, which is a straight forward calculation:

$$I = Q/(A t) \qquad (2)$$

where:

I = infiltration, cm/s
Q = volume of flow, cm^3
A = area of flow, cm^2
t = time interval in which Q was determined, s

The calculation of k_v is also straight forward:

$$k_v = Q / (i A t) = I / i \qquad (3)$$

where:

i = gradient = $\Delta h / \Delta z$
Δh = head loss
Δz = length of flow path for which Δh is measured

The assumptions upon which eq. (3) are based include steady state, vertical one-dimensional flow, and saturated conditions.

While the equation for k_v is simple, its determination from OSRI data is complicated by two major problems. First, practical limitations make it difficult to measure I with acceptable accuracy in low-k soils. Water elevation changes for properly constructed liners and caps are extremely small. The drop in water level for water ponded on a soil with an infiltration rate of 1x10(-7) cm/s is 0.09 mm per day. Evaporation can be as much as 6 mm per day in arid regions. Attempts to account for evaporation have been made by using a control ring of the same diameter but with a sealed bottom. The drop in the water level in the control ring was subtracted from the drop in water level in the test ring. However, in many instances evaporation was significantly greater than infiltration. Subtracting two large numbers to obtain a small number generally results in significant error. Also, it is questionable whether or not the water level can be measured to the needed degree of accuracy because the water surface is not static in large rings.

The second major problem is three-dimensional flow; without knowing k_h, k_v cannot be determined. These two problems, accuracy and not being able to account for lateral flow, make OSRI testing unsuitable for evaluating k_v of liners and caps.

In addition to the two major problems just discussed, other problems include gradient determination and violations of eq. (3) assumptions. Problems associated with gradient determination along with various procedures used to determine it are discussed in a later section. Violations of eq. (3) include transient flow and unsaturated conditions. Minimizing effects of these violations can be accomplished by requiring testing to last until flow is steady and until the wetting front passes through the test zone. However, this can result in testing times of 1 to 6 months.

Described below are additional infiltration tests which have been used to overcome problems associated with the OSRI as well as to shorten testing times.

Sealed Single Ring Infiltrometer

Sealed single ring infiltrometers (SSRI), Fig. 3, have been used in an attempt to achieve better

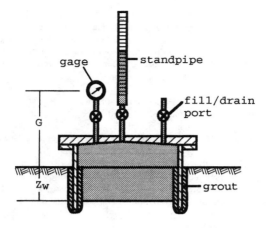

Fig. 4. Air-Entry Permeameter

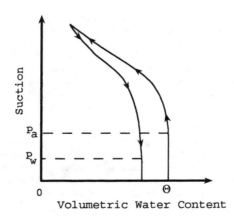

Figure 5. Suction versus Volumeteric Water Content

accuracy in flow measurements and to force flow to be one-dimensional. This device consists of a ring, a top plate, and a standpipe extending from the top plate. The ring is embedded in the test area, sealed with a bentonite grout, then filled with water. Flow is one-dimensional as long as the depth to the wetting front, Z_w, is less than the embedment depth. Flow measurements are made by noting the drop in water level in the standpipe versus time. Testing occurs until the wetting front reaches the bottom of the ring, at which time flow becomes three-dimensional.

A major problem with this device is that although high resolution can be obtained by selecting a small diameter standpipe, overall accuracy is not good. The poor accuracy is the result of several factors. First, standpipes for these devices are usually 60 to 90 cm high. It is difficult to seal the ring in low flow soils with this amount of head. Leaks are a common problem. Second, head in the ring causes an uplift pressure that tends to pop the ring out of the ground. This restricts ring size to about 60 cm in diameter. To prevent uplift, normally a counterweight is placed on the ring or alternatively, the ring is held down by a strap connected to stakes driven into the ground on either side of the ring. If the counterweight or strap causes the ring to settle into the test area, water will be forced back up the standpipe, resulting in an apparently low flow rate. Even small amounts of settlement can result in significant flow up into the standpipe. For example, 0.003 mm settlement of a 60 cm diameter ring results in over 7 ml of apparent negative flow.

Additional inaccuracies are caused by temperature fluctuation. Heating and cooling of the ring results in contraction and expansion of the test device as well as the water within it. When combined, errors resulting from leaks, ring movement, and temperature change outweigh improvements in resolution gained by using a standpipe. In addition, the lateral flow problem common to all single ring methods still applies. Overall, the accuracy of these devices is not sufficient for measuring infiltration rates less than $1 \times 10(-6)$ cm/s. They are therefore not recommended for liners and caps.

Other disadvantages include thin test zones, (embedment depths of 10 cm to 15 cm are common), and that steady state flow is not achieved before the wetting front reaches the bottom of the ring. Determining the position of the wetting front is also a problem. It is usually estimated during testing using weight-volume calculations. However, these calculations do not account for swell and are very sensitive to small inaccuracies in density, specific gravity, and degree of saturation. Precise location of the wetting front is important for determining when to terminate testing and for use in calculating the gradient. Additional discussion on gradient determination is contained in the following section.

Air-Entry Permeameter

The Air Entry Permeameter (AEP), Fig. 4, was developed by Bouwer (1966) to measure k_v in coarse-grained soils for which flow rates are high and the wetting front is sharp and easy to locate. The AEP also provides a means for determining suction at the wetting front, H_s, which is needed to evaluate the term Δh used in eq. (3). The AEP has been used to determine k_v in liners and caps and the advantages are claimed to be rapid measurements and that the suction measured at the wetting front can be used to determine the gradient.

The test is performed in two stages. The first stage is an infiltration test using a sealed single ring infiltrometer. Infiltration proceeds until the wetting front is estimated to be near the bottom of the ring, at which point the water supply is shut off. Suction builds up in the water contained in the permeameter and is registered on a gage mounted in the top plate. Suction will continue to increase until a maximum, P_{max}, is reached, at which point air bubbles should rise up out of the soil. P_{max} is noted, the test dismantled, and the test zone is sampled to locate the actual position of the wetting front.

An explanation of how H_s is determined is as follows. A typical relationship between suction and volumetric water content is shown in Fig. 5. The left leg of the curve corresponds to a soil

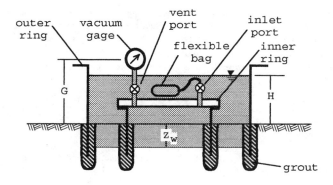

Figure. 6. Double Ring Air-Entry Permeameter

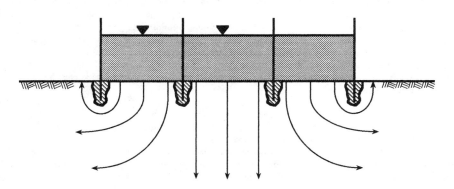

Figure 7. Open Double-Ring Infiltrometer

undergoing drying. Suction begins to increase, however θ, the volumetric moisture content, will remain constant until suction reaches a certain magnitude which causes larger pores to drain. The value of suction at this point is referred to as the air-entry suction, P_a, because air enters the soil as pores drain. As suction continues to increase, draining continues (soil drying) and θ decreases. The right leg of the curve corresponds to soil wetting, or infiltration. Water re-enters the soil, air drains and θ decreases. Not all air drains; some remains trapped in the soil, resulting in a different P - θ relationship for wetting. At some point, water ceases to enter the soil, and θ remains constant, even though suction continues to decrease. The suction corresponding to this point is known as the water-entry suction, P_w, and represents H_s. Bouwer (1966) presented limited data which suggests that for a coarse-grained soil, P_w is approximately $1/2\ P_a$.

During an air-entry test, it is assumed that the maximum suction read on the vacuum gage occurs at the point when P_a is reached at the wetting front. Correcting the gage reading for the height above the wetting front gives P_a. Dividing P_a by 2 then yields P_w. The gradient is calculated as follows:

$$i = (H+Z_w+P_w)/Z_w \tag{4}$$

$$\begin{aligned} P_w &= P_a/2 \\ P_a &= P_{max}-G-Z_w \\ P_{max} &= \text{maximum gage reading} \end{aligned} \tag{5}$$

Hydraulic conductivity is determined as follows:

$$k = I/i \tag{6}$$

where I is determined during the infiltration stage.

A number of uncertainties exist with this method. First, all the errors associated with performing a sealed single ring infiltration test apply to the determination of I. These errors include inaccuracy in flow measurements, non-steady-state flow, and difficulty in determining Z_w. Second, no data exist to verify the relationship between P_a and P_w for fine-grained soils. In fact, for clay liner material, bubbles do not emerge uniformly across the test surface as happens when testing coarse-grained soils, (the indication that the air-entry suction has been reached). In clay, a stream of bubbles usually emerges from one or two spots, probably where a macro defect exists or where air pockets have been trapped in the soil during compaction.

These uncertainties raise serious questions concerning the reliability and accuracy of this test method for low hydraulic conductivity soils. If this test method is required, the authors suggest that a double ring device, Fig. 6, be used to minimize inaccuracies associated with measuring flow using an SSRI (see later discussion on SDRI). However, note that difficulty in wetting front location and theoretical uncertainties with respect to determining suction at the wetting front are still significant problems, especially since suction is usually a large proportion of the gradient in this procedure.

Open Double Ring Infiltrometer

The Open Double Ring Infiltrometer (ODRI), Fig. 7, was developed to minimize the effect of lateral flow in the measurement of vertical infiltration. The device consists of two rings, the smaller ring centered within the outer ring so that it encompasses the area in which flow is virtually one-dimensional. Measurements of flow are made from the inner ring. Equal water levels must be maintained in both rings to eliminate errors caused by water flowing from one ring to the other.

An ASTM Standard Test Method (D3385) describes the use of this device for measurement of infiltration. Being a standardized test, this method gained limited acceptance early on. However, while the influence of the lateral component of flow can be minimized by using large diameter rings, the diameters of the rings recommended in ASTM D3385 (60 cm and 30 cm) are too small to ensure that flow is one-dimensional in liners and caps. Also the problem of measuring small flow rates is still present in this method. Recommended flow measuring devices in ASTM D3385 are a Hooke gauge or a Mariotte Bottle, neither of which have the accuracy needed for measuring low infiltration rates.

It is important to note that ASTM D3385 was written with the intention of measuring high flow rates, primarily those associated with irrigation studies. That fact that it was gaining acceptance for use on soils with low infiltration rates was brought to the attention of the subcommittee that oversees this standard. Because of this, a provision cautioning users that this method is unreliable for infiltration rates below $1 \times 10(-6)$ cm/s was added to the standard. This procedure is therefore recognized as not suitable for typical liners and caps. An example of why is Site 43 of Table 3. As measured by the SDRI and TSB, plus 47 cm, 30 cm and 7.6 cm diameter lab tests, k_v was about $1 \times 10(-8)$ cm/s. An ODRI test indicated $3 \times 10(-7)$ cm/s.

Sealed Double Ring Infiltrometer

Description

The Sealed Double-Ring Infiltrometer, Fig. 8, was developed specifically for soils with infiltration rates less than $1 \times 10(-6)$ cm/s, Daniel and Trautwein (1986). An ASTM Standard Test Method (D5093) was recently adopted for this device. A double ring design, incorporating the use of large, moderately deep rings (3.7 m and 1.5 m), is used to overcome the problem of lateral flow. Measurement of flow is made by connecting a flexible bag, filled with a known weight of water, to a port on the inner ring. As water infiltrates the ground from the inner ring, an equal amount of water is drawn into the inner ring from the flexible bag. After a known interval of time, the bag is removed and weighed. The weight loss, converted to a volume, is equal to the amount of water that has infiltrated the ground. An infiltration rate is then determined from this volume of water, the area of the inner ring, and the interval of time.

The design of the SDRI offers several other advantages over open ring systems. Evaporation effects are eliminated because the inner ring is sealed. The head at any elevation in the inner or outer ring is the same so there is no gradient to cause water to flow from one ring to the other. Also, since the head in the inner and outer rings is always the same, the pressure difference across the wall of the inner ring is constant; hence, the inner ring will not expand or contract even though the water level in the outer ring may change.

Testing Details

The most common size rings for testing soil liners and caps are a 3.7 m square outer ring and 1.5 m square inner ring. Square rings facilitate trench excavation; straight trenches are easier to dig than circular trenches. Metal panels that bolt together at the corners form the outer ring; the one piece inner ring is made of fiberglass. The outer ring is typically embedded 45 cm, the inner ring 10 cm. The deep embedment of the outer ring forces one dimensional flow to a depth of at least 45 cm, 50 to 75% of the total thickness of most liners and caps tested. Additional items used during testing include tensiometers, a swell gage, and a thermometer. All are used to collect additional information during testing which is then used to determine k_v from infiltration measurements.

Tensiometers are used to track the wetting front location during testing. Tensiometers consist of a sealed plastic tube with a porous tip on one end and a vacuum gage on the other. The tensiometer

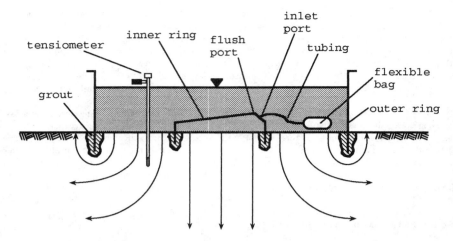

Figure 8. Sealed Double-Ring Infiltrometer

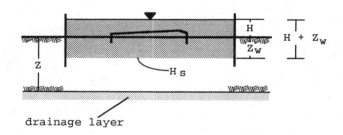

Figure 9. Parameters for Gradient Calculation

is saturated with water and inserted into the test zone so that the porous tip is in close contact with the unsaturated soil. Pore water is hydraulically connected to water in the tensiometer through the porous tip, thereby allowing for pore water pressure to be indicated on the gauge. Initially, soil around the tip is unsaturated and the gage will register a suction. When the wetting front passes the tip, the soil will become saturated and suction will decrease to zero. By installing tensiometers at several depths, wetting front position can be determined by noting when the gage reading goes to zero.

Tensiometers are widely used in the agricultural area and the typical method of installation involves driving a pipe into the ground to form a hole and then pushing the tensiometer in place. This installation procedure may cause cracking in a compacted soil and a preferred method of installation involves hand augering an oversized hole to within 50 mm of the tip location. A small pilot hole having the same diameter as the tip is then advanced to the tip depth. The tensiometer is then inserted into the hole. The annular space surrounding the tensiometer is then backfilled with soil that was removed during the augering. Backfilling is accomplished by placing loose soil in 50 mm lifts and compacting it thoroughly with a rod. A 25 mm collar of granular bentonite can be placed midway between the tip and ground surface as an extra measure to ensure that a good seal is obtained at the soil/tensiometer interface. Typically, nine tensiometers are used per test: three at each depth of 15 cm, 30 cm, and 45 cm.

Soil swell can have a significant effect on infiltration measurements and is therefore monitored during testing by noting changes in inner ring elevation. The inner ring is used as the reference because of difficulty in obtaining a stable reference on the soil surface. A pair of constant tension reference wires is strung across the test setup, positioned over the corners of the inner ring. The vertical distance between the wires and a reference point on each inner ring corner is then measured with a swell gage, which consists of a dial caliper mounted on a metal rod.

Infiltration measurements are also affected by changes in water temperature. Expansion and contraction of the inner ring water as well as of the inner ring itself occur as temperature changes. A thermometer is used to record temperature changes. The test set up is insulated to minimize temperature changes. Insulation consists of berming soil against the outer ring and topping the outer ring with a cover constructed from sheets of plywood and insulation. In cases of extreme temperature variations, shelters, such as a simple tent or plywood, are used. Where freezing may occur, heated shelters are required.

Data Reduction.

Infiltration and k_v are determined using eq. (2) and (3). As with other infiltration tests, calculation of gradient is complicated by the fact that the test zone is initially unsaturated. Three methods have been proposed for estimating the gradient. The parameters used for calculation using these approaches are shown graphically in Fig. 9 for a typical test pad. Expressed in the form of an equation, i is calculated as follows:

$$i = (H + Z_w + H_s) / Z_w \qquad (7)$$

The simplest procedure used for determining i is the Apparent Hydraulic Conductivity method; however, it yields the most conservative value of k_v. Gradient is calculated by assuming $Z_w = Z$ and $H_s = 0$. The advantage of this method is not having to know the wetting front location or suction at the wetting front. The disadvantage is that correct values of k_v are obtained only if the wetting front has passed through the test zone, otherwise i will be underestimated, resulting in a higher than actual k_v. Using this method to verify compliance is fine; however, it should not be used to judge non-compliance because of the unreasonably high values of k_v that can be obtained when Z_w/Z is small.

Figure 10a. Infil. Rate (I) and Hydraulic Con. (k) versus Time - Test A

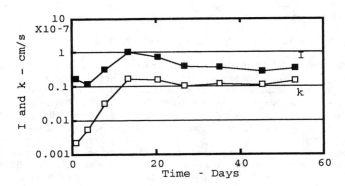

Figure 10b. Avg. Ten. Read. versus Time - 15, 30, 45 cm - Test A

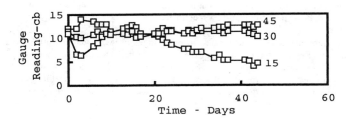

Figure 10c. Wetting Front Depth versus Time Test A

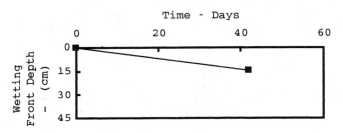

Figure 11a. Infil. Rate (I) And Hydraulic Con. (K) Versus Time - Test B

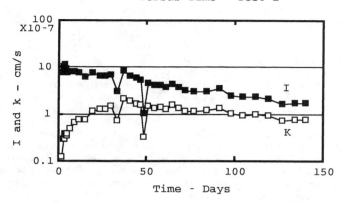

Figure 11b. Avg. Ten. Read. Versus Time - 15, 30, 45 Cm - Test B

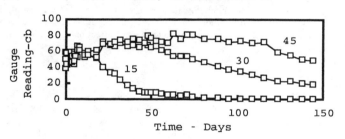

Figure 11c. Wetting Front Depth Versus Time - Test B

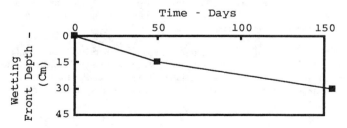

The second method, referred to as the Suction Head Method, requires that both wetting front location and ambient soil suction be known. The assumption made in this method is that Hs equals the ambient suction and that it can be added directly to the gravity terms. The appeal of this method is that high gradients can be calculated, yielding low hydraulic conductivities. It is not unusual to measure suction heads in excess of 6 meters, which correspond to gradients in excess of 100 for small wetting front depths. The disadvantage of this method is that suction at the wetting front is unknown, and is probably much less than the ambient suction, which results in calculation of a k_v less than actual. To properly account for the effect of suction on infiltration rate, a non-linear analysis is required in which the moisture characteristic curve as well as the relationship between the degree of saturation and hydraulic conductivity for the soil being tested is used. While the suction below the wetting front may have some effect on infiltration rate, the full impact of ambient suction on infiltration rate is not observed. The authors postulate that the effects of ambient suction are offset by low hydraulic conductivity in the unsaturated zone beneath the wetting front, which restricts downward flow. Also, the wetting front is diffuse, not sharp (as assumed in the Green-Ampt model). As water ahead of the "average wetting front depth" enters zones of high suction, the degree of saturation increases and suction decreases dramatically. This exact transition zone effect on permeability and suction was reported in capillarity tests by Lane and Washburn (1946). The Suction Head method is not recommended because of the unconservative values of k_v that can be obtained.

The third procedure is the Wetting Front Method. The assumption made using this method is that $H_s=0$. Data from tests in which the wetting front has passed completely through the test zone, thereby eliminating suction, supports the validity of this method. In these tests, the change in infiltration rate that occurs between the beginning and end of testing can be accounted for by the change in Zw. If the ambient suction were used in the gradient calculation, a much larger change would have been expected. Until suction effects are fully understood, the Wetting Front Method is the recommended procedure for gradient calculations. As suction may have a small effect on infiltration, the method is considered conservative.

Test Results

Results for two SDRI tests, Test A and Test B, performed on test pads are shown in Figs. 10 and 11. Both tests were performed to verify that k_v was less than $1 \times 10(-7)$ cm/s. Data reduction and analysis typically begins with plotting infiltration versus time as shown in Figs. 10a and 11a. Next, a plot of average tensiometer readings (for each depth) versus time is constructed, Figs. 10b and 11b. Nine tensiometers were used in each test, three at each depth of 15 cm, 30 cm, and 45 cm. Figures 10b and 11b are then used to construct wetting front depth versus time curves as shown in Figs. 10c and 11c. (Note: For simplicity, the wetting front depth was plotted linearly against time. Infiltration theory, however, suggests that infiltration varies linearly with the square root of time). Next, wetting front depths Zw are used to determine gradients at times when infiltration rate was determined. These gradient values are then used to calculate k_v which is also plotted in Figs. 10a and 11a. The "wetting front method" was used for gradient determination.

Note that for Test A, the determination that k_f was less than $1 \times 10(-7)$ cm/s could be made in a relatively short time, two to three weeks. For Test B however, that determination could not be made for four months. The reason for the high apparent infiltration rates recorded for Test B was soil swell. How soil swell affects infiltration measurements is discussed in the next section.

Swelling Soils

Swelling soils can exhibit high apparent infiltration rates which are not representative of water flow or leakage through the test zone. A significant portion of the water entering the ground may be held by the soil due to swell. A corrected infiltration rate, I_c, corresponding to water actually flowing

Figure 12. Vertical Movement of Inner Ring versus Time - Test B

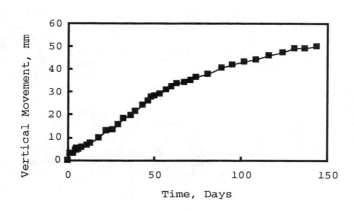

Figure 13. Infil. Rate (I) and Hydraulic Con. (k) versus Time
Ic - Infiltration corrected for swell
kc - hydraulic conductivity corrected for swell

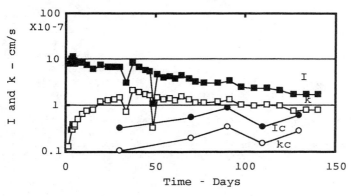

through the test zone, can be calculated by subtracting the swelling rate, I_s, from the measured infiltration rate. Swelling rate is equal to the rate of elevation change of the inner ring, which is measured as discussed previously. Hydraulic conductivity is then calculated using I_c.

The vertical movement of the inner ring during Test B is shown in Fig. 12. Over the course of 150 days of testing, approximately 5 cm of upward movement was observed. As the wetting front had only reached a depth of 30 cm, 5 cm is a significant amount and the soil being tested would be considered to be highly swelling. Swelling rate, I_s, was calculated by determining the slope of the curve (Fig. 12) at several points. For this case, the I_s was determined graphically by hand. Infiltration corrected for swell, I_c, was determined along with the corresponding hydraulic conductivity, k_c, and both are plotted in Fig. 13.

It should be noted that due to practical limitations, the accuracy of swell measurements is poor, particularly when comparing measurements made over a short time period. However, smoothing the data by constructing a best fit curve through the data points, either numerically or graphically, provides an adequate approximation of the swell rate. This rate can be used to determine if high infiltration rates are due to actual leakage through the test zone, or to soil swell. Hence, by accounting for swell, disqualification of suitable materials can be avoided.

Testing Times

Testing times for an SDRI test range from a few weeks to several months. The shorter testing times correspond to soils with low PI, low swell potential, and high degree of saturation at compaction. Long test times correspond to soils with high PI, high swell potential, and low degree of saturation at compaction. Testing time also depends on the information desired. If the test is being performed to demonstrate compliance, i. e., that k_v is below k_s, that decision can be made in a much shorter time than it takes to determine k_v corresponding to a long term equilibrium condition.

Two Stage Borehole Test

Description

As the name implies, the Two-Stage Borehole (TSB) test is a borehole test which is performed in two stages. The TSB combines procedures and techniques from two well established borehole test methods: U. S. Bureau of Reclamation E-18 and E-19 (USBR, 1984). As suggested by Olson and Daniel (1981), a varying geometry is used along with Hvorslev (1951) equations to estimate the (k_h /k_v) ratio.

Testing during the first stage is performed with the casing flush with the hole bottom as shown in Fig. 14. Flow during this stage is governed primarily by k_v. Second stage infiltration is measured with an extended borehole, also shown in Fig. 14. Flow during this stage is governed primarily by k_h. The two stages thus yield two equations by which the two unknowns, k_v and k_h, can be determined. The equations currently used are those of Hvorslev (1951), as modified for bounded media by Boutwell and Tsai (1992) using three-dimensional image potential theory following Carslaw and Jaeger (1959). Falling-head methodology has been used for simplicity in the field. A detailed discussion on data reduction is presented in a following section.

Testing Details

Installation and monitoring of TSB tests are covered briefly below; for details, see Boutwell (1992). The vertical geometry should meet at least the clearances shown on Fig. 14 and the test units should be at least 30 diameters apart and from any free edge. The standard test diameter (D) has been 10 cm; for that size, the tested unit should be at least 60 cm thick and the tests 3m apart.

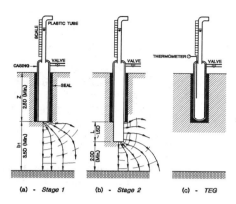

Figure 14 - *TSB Diagram*

First, a borehole with a diameter about (D+5) cm is augered to within 2 cm of the desired Stage 1 test level. The bottom is reamed flat, and the casing (normally monitoring well pipe) inserted. The top of the casing should be as close as possible to ground surface (usually about 2 cm). The casing is steadied while successive 2 - 3 cm layers of bentonite are placed in the casing-soil annulus, tamped and wetted. Powdered bentonite has not worked well; better sealing results with granulated or chipped bentonite, or 1/4" (0.6 cm) pellets. This seal extends 5 diameters above the casing bottom or to the surface, whichever is less. Any remaining annular space can be grouted. The seal is then allowed to hydrate at least 12 hours. Models of the annulus situation subjected to laboratory testing yielded hydraulic conductivity values of 1-3x10(-9) cm/s. The top of the casing is covered during the hydration period to prevent desiccation or entry of rainfall.

The cap assembly with a 1 - 2 cm ID clear standpipe is then added and the assembly filled with water. The water height above ground should not exceed 1 - 1.5 times the casing depth below ground to avoid hydraulic fracturing of the soil. Periodic readings of time and standpipe water level are taken and used to calculate limiting vertical conductivities *(K1)* -see later section. There will usually be a 0.5 - 1.0 order of magnitude drop in $K1$ from the first 30 - 60 minutes to a relatively steady arithmetic time-weighted average $K1'$ beginning some 48 - 96 hours later.

When K1' has become steady, Stage 1 is terminated. The cap assembly is removed, the casing bailed dry (if possible), and the (1.0D) hole advanced about (1.5D) as described for Stage 1. The exposed soil walls are roughened (usually with a wire brush), the cap assembly replaced, and Stage 2 conducted in the same manner as Stage 1. An often-noted problem with the TSB is smear on the sidewalls of the Stage 2 extension; care must be exercised in roughening the Stage 2 extension walls. Upon completion of Stage 2, the casing is usually removed and the remaining hole plugged with bentonite or grout.

Temperature affects the results of field tests for hydraulic conductivity through kinematic viscosity and by volumetric changes. These are handled through a separate TSB unit having the same size and depth of embedment as the test units, but sealed to prevent exfiltration. Standpipe level changes due solely to temperature and/or flowing barometric variations can be read directly from this unit. Water temperature measurements are also made in this "Temperature Effect Gauge", or TEG. It is monitored at the same times as the flowing units for direct correlation.

When testing must be performed in freezing conditions, use of antifreeze or heating the units is required. The primary consideration for antifreeze is that it not cause pollution in either the groundwater or leak detection system. Ethanol (as Vodka) has been accepted in several jurisdictions, but formic acid would be preferable since it is efficient at lowering the freezing point and has a small effect on kinematic viscosity. Of course, a temperature -kinematic viscosity relationship is needed for the actual mix employed. When five units were heated and insulated similarly on one report project, the temperature differential between them was on the order of 1 to 2°C.

Data Reduction

The field data for each reading of each stage is reduced to limiting hydraulic conductivity values *(K1, K2* - explained later). It is easily shown that any falling-head test can be described in an equation having the generic form:

$$k_v = R_T \, G \, Ln \, (H_1/H_2')/(t_2-t_1) \tag{8}$$

where: k_v = vertical hydraulic conductivity
R_T = viscosity factor (to water at 20°C)
H_1 = initial head (at time t_1)
H_2' = final head (at time t_2), corrected for volumetric effects
G = geometric constant for each stage

Equation (8), like eq. (3), is based on flow in a saturated medium. This is reasonably valid from the water entry point to the wetting front. Flow geometry is taken into account by the factor (G). While the flow in a falling-head test is not truly steady-state, the resulting transient effects are small in this test since the head change is relatively small compared with the total head. The effects of unsaturated flow conditions can be handled as described in Boutwell (1992).

The viscosity factor R_T is that given by ASTM D5084 (for water) for the average temperature during the time period t_1 to t_2. Head is measured from the water level in the standpipe to that in the ground or first pervious layer before testing, with a limitation that the head due to height of test above a boundary or the potentiometric level be limited to 20 test diameters. Any *increase*/decrease in TEG standpipe level is *subtracted from*/added to H_2 to obtain H_2'.

The geometric factor *(G)* depends on the test geometry, pad boundary conditions, and degree of anisotropy. For Stage 1,

$$G1(m) = (\pi d^2/11mD)[1 + a(D/4mb_1)] \tag{9}$$

where: d = standpipe ID
m^2 = k_h/k_v (not yet known)
a = -1 for permeable base at b_1
a = +1 for impermeable base at b_1
b_1 = thickness of tested soil below casing

Calculation of *(G)* for Stage 2 is somewhat more complex:

$$G2(m) = (d^2/16Lfm^2)[2Ln(U_1) + a \, Ln(U_2) + p \, Ln(U_3)] \tag{10}$$

where: $U_1 = [mL/(D+2T)]+\{1+[mL/(D+2T)]^2\}^{1/2}$ (11)

$$U_2 = \frac{4mb_2/D + mL/D + [1+(4mb_2/D + mL/D)^2]^{1/2}}{4mb_2/D - mL/D + [1+(4mb_2/D - mL/D)^2]^{1/2}}$$ (12)

$U_3 = \{mL/D + [1+(mL/D)^2]^{1/2}\}/U_1$ (13)

$f = 1.0 - 0.5623\ e^{-1.566L/D}$ (14)

L = length of Stage 2 extension

$b_2 = b_1 - L/2$

T = thickness of any smeared zone

p = ratio of (k_h) to conductivity of smeared zone

The factor (11) in eq. (9) can be traced back to electrical analog studies by Harza and flow net analyses by Taylor; both are reported in Taylor (1948). Equations (10) - (13) are based on approximating the cylindrical exfiltration zone of Stage 2 by an ellipsoid as discussed in Hvorslev (1951). The empirical correction of eq. (14) accounts for inconsistencies arising in Hvorslev (1951) as L/D approaches zero. The geometric factors given by eq.(15) and (16) are in good agreement with those for similar situations in the case of isotropic media with top and bottom boundaries as derived through electrical analog experiments by Youngs (1968). Equations (not presented herein) were developed (Boutwell 1992) to account for the primary effects of non-saturation (radius to "wetting front" and soil suction); these analyses are seldom necessary in practice, resulting in a somewhat conservative test value.

The limiting hydraulic conductivity values $K1, K2$ are calculated using eq's (8) and (9) or (10), as appropriate, plus assuming an isotropic medium $(m=1)$; they are thus the values which would represent $(k_h=k_v)$ if the tested medium were isotropic. This yields the following geometric constants for the limiting values by Stage 1 and Stage 2:

Stage 1: $G1(1) = (\pi d^2/11D)[1+a(D/4b_1)]$ (15)

Stage 2: $G2(1) = (d^2/16Lf)\ [2Ln(U_4) + a\ Ln(U_5)]$ (16)

$U_4 = L/D + [1+(L/D)^2]^{1/2}$ (17)

$$U_5 = \frac{4b_2/D + L/D + [1+(4b_2/D+L/D)^2]^{1/2}}{4b_2/D - L/D + [1+(4b_2/D-L/D)^2]^{1/2}}$$ (18)

In order to provide a consistent basis for final calculations and to avoid errors from minor fluctuations, time weighted averages over the relatively steady-state portions of each stage are calculated:

$$K'_j = \Sigma K_{j,i} T_i / \Sigma T_i$$ (19)

where: K'_j = arithmetic time-weighted average
j = 1 for Stage 1, 2 for Stage 2
i = time increment number
T_i = duration of time increment i

When the purpose of the test is solely to prove that k_v is less than some specified value k_s, no further data reduction is necessary if $K1'<k_s$. To obtain k_h, k_v values requires one more step. Assuming the soil to be homogeneous (although isotropic) the vertical hydraulic conductivity k_v must be the same w obtained using eq. (9) or eq. (10). It then follows that:

$$K1'[G1(m)/G1(1)] = k_v = K2'[G2(m)/G2(1)]$$

or

$$K2'/K1' = [G1(m)/G1(1)][G2(1)/G2(m)]$$

This is a complex but unique function of $K2'/K1'$ which is determined in the test, L/D which is known test geometry, p which is estimated (see Fig. 15), and the unknown (m). This function is solved by trial-and for m, which is then introduced into the left side of eq. (20) to obtain k_v. The k_h value then follows directly the definition of m. While all these calculations sound quite complex, they are easily programmed on a pe computer. Alternatively, m and $k_v/K1'$ can be read directly from a nomograph such as Fig. 15.

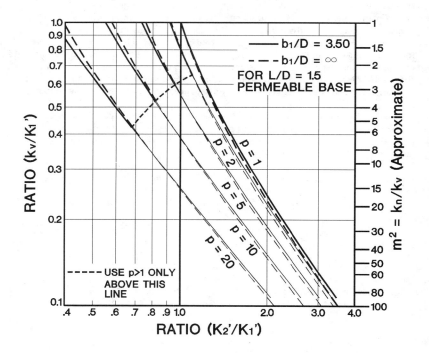

Figure 15 - *TSB Data Reduction Nomograph*

Test Results

As of January, 1993, the authors' files contained the results from TSB evaluations on some 70 liners/test pads totalling almost 400 individual tests. They are also aware of numerous other such projects. In addition, they have data from 7 projects comprising about 100 tests in natural deposits. Typical results from a test pad study are illustrated on Fig. 16.

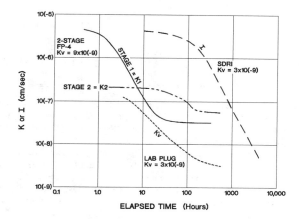

Figure 16 - *Time Comparision, Site 27*

A complete test, i.e. where both stages were performed and continued until equilibrium values of $K1$ and $K2$ were obtained, typically takes 10 days to 20 days to perform. However, an advantage of the TSB is that the decision time, i.e., determining when if $k_v < k_s$, usually occurs within 24 hours or less in cases where $k_v < 1/2\ k_s$. Note that $K1$ is an upper estimate of k_f hence the decision time occurs as soon as $K1$ is less that k_s. Then, if information on k_h is not needed, testing can be terminated after Stage 1.

SUMMARY OF TESTING METHODS

Size Comparison

As stated previously, the authors suggest that the important factor in determining k_f is not so much where the test is performed, but rather size of the test specimen. Differences in measured values of k_1 and k_f result primarily from scale effects. Listed below in Table 1 are various testing methods along with a comparison of the volume of soil tested, v_t, to that tested in a standard laboratory test, v_l, using a specimen 7.6 cm in diameter and 7.6 cm in length. An arbitrary rating system, ranging from very small to very large, is also presented.

TABLE I - *Relative volumes affected by hydraulic conductivity tests*

Method	v_t/v_l	Scale
3.5 cm x 7.0 cm (lab)	0.2	Very Small
BAT Probe	0.3 - 0.9	Very Small to Small
7.6 cm x 7.6 cm (lab)	1.0	Small
10.2 cm x 10.2 cm (lab)	2.4	Small
15.2 cm x 15.2 cm (lab)	8.0	Small
30.4 cm x 30.4 cm (lab)	64.	Intermediate
Air-Entry Permeameter	20 - 80	Intermediate
Two-Stage Borehole	50 - 200	Intermediate
SDRI	1000 - 4500	Large
Pan Lysimeters	12000 - 100000	Large to Very Large
Full-Scale Barriers	1,000,000+	Full

One of the biggest challenges in compliance evaluation of earthen liners and caps is determining an appropriate specimen size for testing. Sai and Anderson (1991) recommend a tested area of at least 60,000 cm^2, but presented neither statistical nor experimental results to support their assertion. For a 60 cm thick liner or cap, this specimen size represents a v_t / v_l ratio of 10,600. At the time of their recommendation, limited data was available to compare specimen size and its effect on measurement of k_f. More recent data, (Trautwein and Williams 1990, Benson et. al. 1993), has shown that scale effects can be detected with sample volumes in the intermediate range. Some of these results are plotted in Fig. 17. While additional research is needed in this area, it is clear that testing volumes less than "intermediate" are not useful for determining k_f when significant macrostructure is present.

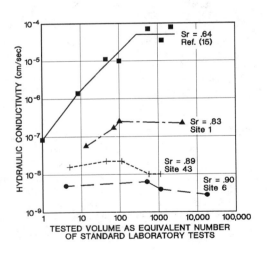

Figure 17 - *Scale Effects on Hydraulic Conductivity*

Applicability

While development of both the SDRI and the TSB was directed towards testing liners and caps, both methods have been used for other applications. The SDRI has been used to test natural soil deposits, and has been used to investigate variations in k_f with depth by performing tests at various levels within an excavation. Testing is restricted to relatively level areas; however, testing has been performed on slopes into which a level bench was excavated. While an SDRI can be used for testing soils with high k_f, its use is generally warranted where $k_f < 1 \times 10(-6)$ cm/s.

The TSB has been used for natural soils as well. It has the advantage that it can be used above or below the water table. It has been successfully used in soil layers as thin as 50 cm and at depths of 10 meters. The only major limitations to the TSB are obtaining a good seal in the annulus around the casing and having sufficient distance between the test zones and boundaries. The TSB can also be used on slopes with horizontal stratification or compaction planes; however, data reduction techniques for sloping anisotropy have not been developed as of yet.

Testing Time Comparison

Time for steady state conditions for both testing methods can be determined theoretically and provides a useful comparison. Both the SDRI and the TSB are infiltration tests and are therefore subject to transient flow effects. In a bounded medium, such as a test pad underlain by a pervious boundary, either will approach or attain steady-state flow after sufficient time. The transient flow condition can be important. It is usually analyzed using either the Heat Equation (or extended Laplacian Model - primarily for saturated soils) or the Green-Ampt Model for unsaturated soils. The SDRI has basically a one-dimensional flow field; the TSB can be approximated by three-dimensional flow from a spherical source (assumed isotropic for simplicity). The theoretical equations for the heat equation solutions are available in Carslaw and Jaeger (1959). That for the 1-D Green-Ampt Model is given e.g. in Johnson, et. al. (1990). The authors extended the 1-D solution to spherical flow for the 3-D solution. The equations each of these cases are presented below:

1-D Heat
$$I = k_v[1+H/Z][1+2\sum_{j=1}^{\infty} e^{-j^2\pi^2 T_{L1}}] \quad (22)$$
$$T_{L1} = (k_v t/s_s Z^2)$$

3-D Heat
$$q = 4\pi k_v r_o H[1+(\pi T_{L3})^{-\frac{1}{2}}] \quad (23)$$
$$T_{L3} = (k_v t/s_s r_o^2)$$

1-D Green
$$I = k_v[1+(H'/Z_w)] \quad (24)$$
$$T_{G1} = (k_v t/n_a H') = (Z_w/H') - Ln\,[1+(Z_w/H')]$$

3-D Green
$$q = 4\pi k_v r_o H'\,[(r_w/r_o)/(r_w/r_o-1)] \quad (25)$$
$$T_{G3} = (k_v t/n_a H')\,(H'/r_o)^2 = (1/3)(r_w/r_o)^3 - (1/2)(r_w/r_o)^2 + 1/6$$

where:
H = applied head
Z = thickness of test pad
r_o = radius of spherical source
H' = $H(1+H_s/H)$
H_s = soil suction
Z_w = depth of wetting front
r_w = radius to wetting front
n_a = effective air porosity of soil

t = time
S_s = soil specific storativity

These equations are illustrated in a dimensionless form on Fig. 18(a) in terms of ratios of the transient values (I,q) to those for the steady-state $(t = \infty)$ condition (I_u, q_u). The relationships are quite similar. A specific case using typical geometries for the two test types is shown on Fig. 18(b). Both theory and observations show that the TSB approached steady-state (and thus a test result) in about 4% of the time required for the SDRI.

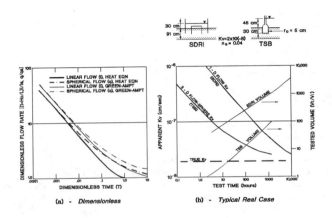

Figure 18 - *Theoretical Flow Behavior*

Testing times in practice are seldom as long as predicted by theory. As mentioned previously, length of testing is usually determined when the "decision time" occurs. For the TSB, compliance evaluation can usually be determined in 24 hours or less when k_f is less the half of the specified value. For the SDRI, compliance evaluation usually takes two weeks or more; longer times correspond to swelling soils. A comparison of testing times is presented in Table 2.

TABLE 2 - *Testing Times*

Site No.	Decision Time (days)		Total Test (days)	
	SDRI	TSB	SDRI	TSB
6	10	1	146	60*
20	<1	<1	92	14
21	12	<1	53	17
26	62	<1	134	30
27	30	4	160	31
31	<1	<1	56	4
43	7	<1	69	28*

* Extended for research purposes.

Advantages/Disadvantages

The advantages of the SDRI test are that large areas can be ponded, flow is one dimensional, and infiltration is a direct measurement of leakage rate, except when swelling is present. Disadvantages are the relatively higher costs and longer testing times. The lack of an overburden stress on the test zone may also be considered a disadvantage.

Advantages of the TSB are relatively low cost and short testing times, and the ability to measure both k_v and k_h. Also, the TSB can be used to investigate the variation in k_f with depth in a liner or cap and also to detect poor lift bonding. Disadvantages of the TSB can be that an intermediate rather than a large volume of soil is tested, and difficulty in testing thin soil layers because of interference from nearby boundaries.

CASE HISTORIES

Both the SDRI and the TSB have been used for test pad evaluation at sites all over the country. While a large number of tests have been performed [SDRI (300+); TSB (400+)], the number of projects where both have been used is relatively small. Data for comparative testing to which the authors have access are presented in Table 3. The AEP was used at several of these projects and that data is presented as well. A statistical analysis was performed using this data in order to draw general conclusions concerning the usefulness of these test methods.

TABLE 3 - *Data Base of Tests on Common Pads*

Site	Mean Classification Data					Mean Vertical Permeability (cm/s)				Ref
	LL (%)	PI (%)	Fines (%)	Comp. (%D698)	S_r (dec.)	SDRI $H_s = 0$	TSB	AEP	Lab (UD)	
1	24	10	65	98	0.83	2.3×10^{-7}	-	6.1×10^{-8}	6.0×10^{-8}	10
6	50	34	94	96	0.92	1.1×10^{-8}	5.0×10^{-9}	-	5.0×10^{-9}	10
20	39	17	86	98	0.94	1.3×10^{-8}	1.4×10^{-8}	-	1.3×10^{-8}	10
21	54	31	-	97	0.87	9.2×10^{-8}	1.2×10^{-8}	-	1.7×10^{-8}	10
22	33	13	-	95	0.86	4.7×10^{-8}	2.8×10^{-8}	-	5.7×10^{-8}	10
26	101	71	98	96	0.92	4.9×10^{-8}	1.6×10^{-8}	-	3.5×10^{-9}	*,D
27	85	58	99	96	0.94	3.3×10^{-9}	1.6×10^{-8}	-	4.4×10^{-9}	S
30	35	12	74	96M	-	1.2×10^{-7}	-	1.9×10^{-7}	1.5×10^{-7}	10
31	48	26	84	94	0.85	4.1×10^{-8}	1.6×10^{-7}	2.2×10^{-8}	1.1×10^{-7}	
32	28	17	31	97	0.79	6.7×10^{-7}	1.4×10^{-7}	4.8×10^{-8}	6.8×10^{-8}	*
33	28	17	30	102	0.83	3.1×10^{-7}	1.9×10^{-7}	1.2×10^{-7}	2.6×10^{-8}	10
34	37	17	90	100	0.83	3.2×10^{-7}	2.3×10^{-7}	9.3×10^{-7}	1.5×10^{-8}	10
40	36	20	-	103	0.87	1.2×10^{-8}	1.1×10^{-8}	-	1.1×10^{-8}	10
43	30	18	52	33	0.89	9.8×10^{-9}	9.1×10^{-9}	-	1.5×10^{-8}	*
45	-	-	-	-	-	2.6×10^{-8}	-	2.8×10^{-8}	-	10

M = Modified (ASTM D1557)   * Authors' Files
DS = Desiccation suspected H_s = Soil Suction   UD = Undisturbed Sample

The results on Table 3 were analyzed in two ways. First, the individual ratios of hydraulic conductivity as determined by the test method $(k_t = TSB, k_e = AEP, k_l)$ to that from the SDRI (k_s) were calculated. The geometric mean result ratio for each test method group $(k_t/k_s, k_e/k_s, k_l/k_s)$ is presented on Table 4. Also, least squares regressions of the form:

$$Log\ (k_s) = A + B\ Log\ (k)\qquad(26)$$

were performed for each test method group. The statistical parameters of each are given on Table 4; see Figures 19(a), (b), and (c) for a graphical presentation. Since the typical k_r of 1x10(-7) cm/s applied to many projects, the regressions were used to estimate the hydraulic conductivities (k_p) that each test type would yield, on the average for $[k_r = 1 \times 10(-7)\ cm/s]$. These results are given on Table 4, below:

TABLE 4 - *Mean Results by Method*

Item Measured	Test Method			
	SDRI	TSB	AEP	Lab
Result Ratio	1.00	0.72	0.40	0.41
k_p (cm/s)	1x10(-7)	8x10(-8)	4x10(-8)	5x10(-8)
Coeff. of Correlation	1.00	0.75	0.66	0.52
Std. Est. of Error*	0.00	0.48	0.48	0.64

* As logarithm

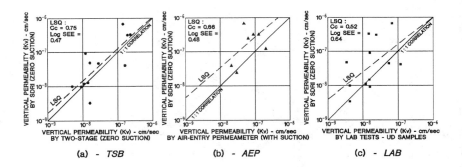

Figure 19 - *Hydraulic Conductivity by SDRI vs Other Methods*

Given the relatively small number of direct comparisons available and the uncertainty concerning the role of soil suction in the SDRI, only general evaluations can be made at the current time.

* The SDRI and TSB yield comparable results. The TSB is somewhat lower, as expected since it tests material not as subject to surface disturbance such as drying or swelling.

* Results from the AEP are notably lower than those from the SDRI. This may be partly attributable to the powerful effect of soil suction as used in the AEP equations.

* The laboratory test results compare closely with the relationships of Benson and Boutwell (1992). They predict a mean $(k_f/k_1 = 1.7)$ for this test group while the actual is $(k_f/k_1 = 2.4)$. This is not surprising, since 2/3 of the test group is from that reference.

In evaluating these numerical comparisons, the reader should keep in mind that the test zones for each of these testing methodologies differs. The AEP tests only the top portion of the test pad. The zone affected by the SDRI test typically is only the upper half of the test pad. The TSB tests the middle portion and the lower half of the test pad. Any differences in the test pad, such as material variation, compaction criteria, or environmental stress variation(wetting or drying) can have a strong influence on the results.

Since the SDRI tests the largest volume and hence should be a better predictor of scale effects, it was selected as the standard for comparing all the methodologies. The wetting front method (H_s = 0) used to analyze the SDRI data.

Site 6 of Table 3 was taken from Johnson, et.al. (1990). It involved the use of pan lysimeters, and provided an excellent opportunity for comparing results with the SDRI and the TSB. The lysimeter, SDRI, and TSB data correlated well. As pan lysimeters involve few procedural or theoretical uncertainties, their use is considered to be the best technique for obtaining k_f. However, because of the long testing times, years in many cases, they are not used for compliance evaluation.

TESTING STRATEGY

The introduction of field testing as part of compliance evaluation of earthen barriers has occurred relatively rapidly and has led to differing and sometimes confusing testing strategies. In developing a sensible testing strategy, the following questions should be addressed: why, what type, when, where, and how many tests should be performed. The authors offer the following observations based on their collective experience.

Why Perform Field Tests?

Qualifying a material for use in liner or caps typically involves determining specifications that include acceptable combinations of ranges in plasticity, grain size, and density - moisture content values that result in a k_1 less than k_s. Once qualified and placed according to specifications, no additional hydraulic conductivity testing should be needed for compliance evaluation. Compliance could be checked by performing index and moisture density tests which are simpler, quicker, and less costly. This would be true if the material could be placed in the field such that its structure was identical to that of the specimens tested when determining k_1. However, this is the big unknown; after placement are there macropores, and if so is k_f greater than k_1? This can only be determined by testing large specimens and hence is the reason for field testing.

What Methods to Use?

Having established that the purpose for testing after placement is to check for changes in

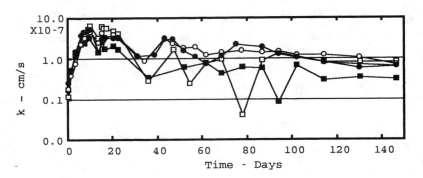

Figure. 20. Hydrualic Conductivity Determined with SDRI Four Tests on One Test Pad

structure that may result in $k_f > k_l$, a test method that accounts for macropores should be used. Laboratory tests on small diameter samples (7 cm in diameter) are not recommended. Referring to Table 1, the authors suggest methods that test a volume of material in the intermediate range or greater. Test volumes less than intermediate are not considered to be useful for determining k_f when soil structure is of concern.

When Should Testing Be Performed?

If compliance evaluation is part of a permitting processes, during what stage should testing be performed: before, during, or after construction? Ideally, testing during all stages would be preferred; however, economic and time limitations mandate a more judicious plan. Testing during construction has the advantage that areas of non-compliance can be detected and corrected while the contractor is on site. Unfortunately, field testing during construction is not practical because long test times cause delays in construction, equipment traffic may impact test results, and the tested area requires extensive repair upon removal of test equipment. Testing after construction has the advantage of evaluating the final product. However, disadvantages include excessive costs from requiring a contractor to wait on site for the results of long term testing or to remobilize to correct non-compliant areas. Also, field testing would prohibit other contractors from placing protective layers over the barrier layer until testing was complete.

Considering the limitations of field testing during or after construction, the current state of practice is to demonstrate compliance during the design phase by building and evaluating a test pad. The advantage of this approach is that field testing can be used as part of the design process to establish construction procedures that best minimize or eliminate secondary structure. These procedures are then used to construct the actual liner or cap. A well-conceived, strongly enforced Construction Quality Assurance (CQA) program is then relied on to ensure that k_f is less than k_s.

It should be noted that field testing should not be restricted to use in the design phase. If questions arise about how well CQA was enforced or if a substantial change in material properties or procedures occurs, additional field testing may be warranted to ensure compliance.

How Many Tests Should Be Performed?

An important question that has no widely accepted answer is how many field tests are needed to obtain a reasonable level of confidence that k_f is less than k_s. Sai and Anderson (1991) concluded that compliance should be evaluated by performing three SDRI tests on a test pad, but without statistical analysis or experimental data, which was limited at the time of that recommendation. In practice, when SDRI testing is used, typically only one test is performed. In some cases, two tests are used, one test serving as a back-up should the other be damaged. This eliminates losing valuable time by not having to start over again. The authors are aware of one testing program that consisted of four SDRI tests on one test pad. A plot of hydraulic conductivity for these tests is shown in Fig. 20. The test pad was constructed during hot, windy weather using highly plastic clay. The probability of secondary structure was high. The measured hydraulic conductivities for all the tests varied only by a factor of three. For three of the tests, the measured hydraulic conductivities were essentially the same. Data from three other test pads where two SDRI tests were performed yielded similar results. These test pads consisted of bentonite amended soils and variations between measured hydraulic conductivities ranged between 2 and 4.

It is commonly accepted in Construction Quality Assurance (CQA) work that the frequency and locations of sampling should have a sound statistical basis (EPA 1984). Statistical procedures have long been available to determine the required number of tests to achieve a given confidence level in the results (EPA 1984). The number of tests depends on the allowable error, the test group scatter, and the desired level of confidence that the real mean does not exceed the test group mean. Since, for all practical purposes, hydraulic conductivity may be taken as log-normally distributed (Boutwell and Hedges 1989), the standard equation in (EPA 1984) should be rewritten as:

$$N = z^2 \{Log\ Dev(k)/[Log(k_s/k_a)]\} \tag{27}$$

where:
- N = number of tests required
- z = standardized normal distribution factor
- $Log\ Dev(k)$ = standard deviation of test group (as log)
- k_s = specified value of hydraulic conductivity
- k_a = mean hydraulic conductivity of test group

Some value of k_a will be needed to determine the number of tests in the program. The probable value of k_a can be estimated from experience or by laboratory tests adjusted following Benson and Boutwell (1992). Mean standard deviations were computed from projects in the authors' files where multiple tests of a given type were performed on the same test pads. The mean Log Dev(k) values were 0.12 for the SDRI (2 sites), 0.26 for the TSB (45 sites), and 0.32 for the AEP (5 sites). Equation (27) then yields the number of tests required as shown in Table 5.

TABLE 5 - *Number of Tests by Method*

Confidence Level	Test Type	Number of Tests Required for Given (k_s/k_a)				
		10	5	2	1.5	1.33
90%	SDRI	1	1	1	1	2
	TSB	1	1	2	3	4
	AEP	1	1	2	3	5
95%	SDRI	1	1	1	2	3
	TSB	1	1	3	4	6
	AEP	1	2	3	5	7

The data bases for the SDRI and AEP are small, so that the above numbers are only a guide for these procedures. However, it appears that the standard practice of one SDRI or five TSB's per pad is adequate for all but the poorest pads. Clearly also, it is less expensive to prove that a good pad is acceptable than to evaluate a marginal pad.

Where To Test ?

When performing field tests on a test pad, the main criterion is to select an area representative of the liner or cap. Usually, testing is performed in the central area of the test pad, away from areas that may have edge effects. Preferred areas of testing are marginal zones, i.e., areas that just meet minimum criteria, as these will be areas of highest hydraulic conductivity. If compliance is verified in these zones, then compliance will be met in other areas which are constructed to standards exceeding those in the area actually tested.

When the TSB method is used for compliance evaluation, the normal pattern is 5 tests: 4 on each corner of the pad's central area and one in its center. This allows evaluation of variations in k_f.

When testing after construction, perhaps due to questions about CQA enforcement, indicators of high k_f / k_1 can be used to select test areas. Benson and Boutwell (1992) have been able to establish a correlation between k_l, (measured in the lab using undisturbed samples obtained with 7.6 cm Shelby tubes), and k_f, obtained using large scale field tests. This statistical correlation, based on 26 projects, is as follows:

$$Log\ (k_f/k_l) = 0.42 - 0.47\ Log\ (CI) + 109\ n_a^2 \qquad (28)$$

where: CI = plasticity index(%) x Fines content (%)
   n_a = air porosity = $n(1 - s_r)$
   n = total porosity
   s_r = degree of saturation at compaction (%)

This correlation can be made using parameters easily obtained from standard tests used in compaction control. This indirect method of determining k_f can be used when deciding on test methodologies and frequencies.

What to Use as Pass/Fail Criteria

Like all other testing programs, the field (large-scale) testing program should have a clear definition of Pass/Fail *before* the program beings. Pass/Fail determination would be clear if k_f were 1) perfectly accurate and 2) had no variability.

With respect to accuracy, it is generally accepted that k_f cannot be determined to a real accuracy of two significant figures, although individual test results are often so reported. Hence, if k_f equals 1.5x10(-7) cm/s, and k_s is 1x10(-7) cm/s, is it acceptable? Rounding should be acceptable. This may seem a trivial point; however, it does arise and significant costs and time delays rest on how this issued is determined. While most engineers and geohydrologists would not hesitate to round off, regulators unfamiliar with the accuracy associated with k_f determination may not consider anything above 1.00 k_s a passing test.

Things become more complex with multiple test results. Statistically, having no failing test simply means that no enough tests have been performed. With the SDRI as an example, if $k_s/k_a = 1.5$, 7% of all tests should fail. How should k_f be selected in this case? Should it be the highest test value, the arithmetic mean, or the geometric mean? Current practice for multiple SDRI testing is to use the highest value, based on the assumption that the test itself yields a field-determined mean. For TSB testing, the current recommendation is that no single test should be greater than k_s and the arithmetic mean should be less than k_s. The arithmetic mean is chosen over the geometric mean because it is more conservative. However, the effect of multiple lifts is to reduce the equivalent vertical hydraulic conductivity to less than the geometric mean (Boutwell and Rauser 1990). When the liner has more than four lifts, the arithmetic average may be too conservative.

A special case where $k_f > k_s$ may be acceptable is where the liner will be loaded quickly and testing has been performed to demonstrate a significant decrease in k_f with overburden, and a geomembrane is placed on top of the clay liner.

A situation which is often frustrating to geoscience practitioners is satisfying pass/fail criteria based on an arbitrary k_s which may be unreasonable. This situation arises often, as current practice is to design for k_s imposed by regulation, rather than one based on a rational study involving site specific parameters. It is interesting to note that the origin of the most commonly imposed k_s (1x10(-7) cm/s) is unknown. It was most likely selected because 1x10(-7) cm/s was generally accepted to be the dividing point between "permeable" and "impermeable" soil at the time regulations were first formulated. For some sites, 1x10(-7) cm/s is unduly low, for others it is not low enough. Use of a site specific k_s would make meeting strict pass/fail criteria more purposeful and more readily accepted by the technical community. It would result in liners that better protected the environment where k_s should be lower than

$1 \times 10(-7)$ cm/s, and save money and time where k_s could be higher.

SUMMARY AND CONCLUSIONS

The requirement of field hydraulic conductivity testing is a recent and significant change added to compliance evaluation of compacted soil liners and caps. Instituted because of questions concerning the reliability of laboratory tests on small specimens, field testing provides a means for better assessing the effects of secondary structure. The authors note that while debate may center on the need for field versus laboratory testing, the real issue is one of specimen size, small versus large.

The shift in emphasis towards field testing resulted from early reports of large differences between k_f and k_l. More recent findings reported herein and by others demonstrate that for well constructed liners and caps, acceptable ratios of k_f / k_l can be obtained. Because of the large numbers of factors that can result in secondary structure, field testing is needed to confirm acceptable k_f values.

Of the 20 plus field methods that have been tried, seven have been selected for discussion. Of these, two methods, the SDRI and the TSB, have gained the widest acceptance and use. The others: BAT Probe, open Infiltrometers both single and double, sealed single ring infiltrometers, and the air-entry permeameter are not recommended due to small test volume, limitation in accuracy, difficulty in data interpretation, or uncertainties in their theoretical assumptions.

The advantage of the SDRI is the ability to test large volumes. The TSB has the advantage of being able to determine both k_v and k_h. The SDRI is the better method for testing thin layers. The TSB is better for determining variation in k with depth. On projects for which both tests have been performed, comparable results have been obtained. Based on limited statistical data, the authors recommend that at least one SDRI test or at least 5 TSB's be performed to evaluate compliance.

The SDRI test is more costly and has longer testing times. Testing times for equilibrium values of k_f range from 1 to 6 months for the SDRI and 1 to 6 weeks for the TSB. However, if the purpose of testing is solely to demonstrate that $k_f < k_s$, testing times are reduced to 2 to 3 weeks for the SDRI and 24 to 48 hours for the TSB.

Calculations for the SDRI are simple; however, data interpretation is complicated by soil swell, determining the location of the wetting front, and accounting for suction. It is recommended that swell be monitored, tensiometers be used to track the wetting front positions, and that soil suction be ignored when calculating k_f. Calculations for the TSB are more complex. The primary concern in data interpretation is accounting for the effects of smear.

Because of practical limitations of testing during or after construction, current practice is to perform field testing during the design phase on a test pad. This has the advantage of being able to verify before construction that the selected soil criteria and construction procedures result in $k_f < k_s$. Compliance of the actual liner or cap then depends on strong enforcement of a thorough CQA program to ensure that the procedures and specifications actually used meet or exceed those used on the test pad.

The definition of Pass/Fail should be clear at the beginning of the test program. For the SDRI, the authors recommend that the test value (or highest value in the case of multiple tests) not exceed $1.5 k_s$. The recommended TSB program always involves multiple tests. The authors recommend for this method that the maximum k_v not exceed $1.5 k_s$ plus that the arithmetic average be no greater than k_s. The latter criterion may be too conservative for liners constructed in 4 or more lifts.

Field testing should not be limited to test pads. If concerns about proper enforcement of CQA arise during or after construction, field testing is recommended to ensure $k_f < k_s$.

References

Acar, Y. B., and Olivieri, I., 1989, "Pore Fluid Effects on the Fabric and Hydraulic Conductivity of Laboratory Compacted Clay," Transportation Research Record, No. 1219, Transportation Research Board, National Research Council, Washington D. C., pp. 144-159.

Auvinet, G., and Espinosa, J., 1981, "Impermeabilization of a 300-Hectare Cooling Pond," Permeability and Groundwater Contaminant Transport, ASTM STP 746, American Society for Testing and Materials, Philadelphia, pp. 151-167.

Benson, C. H., and Boutwell, G. P., 1992," Compaction Control Criteria and Scale-Dependent Hydraulic Conductivity of Clay Liners," Proceedings, 15th Annual Madison Waste Conference, Madison, WI.

Benson, C. H., Hardianto, F. S., and Motan, E. S., 1993, "Representative Sample Size for Hydraulic Conductivity Assessment of Compacted Soil Liners," Hydraulic Conductivity and Waste Contaminant Transport in Soils, ASTM STP 1142, ASTM, Philadelphia.

Boutwell, G. P., and Hedges, C. S., 1989, "Evaluation of Waste-Retention Liners by Multivariate Statistics," Proceedings, XII International Conference on Soil Mechanics and Foundation Engineering, Publications Committee of XII ICSMFE, A. A. Balkema, Pub., Vol. 2, 1990, pp. 815-818.

Boutwell, G. P., and Rauser, C. R., 1990, "Clay Liner Construction," Geotechnical Engineering in Today's Environment, American Society of Civil Engineers (Central PA Section), Hershey, PA.

Boutwell, G. P., 1992, "The STEI Two-Stage Borehole Field Permeability Test," Containment Liner Technology and Subtitle "D", Houston Section, ASCE, Houston, TX.

Boutwell, G. P., and Tsai, C. N., June 1992,"The Two-Stage Field Permeability Test for Clay Liners," Geotechnical News, Vol. 10, No. 2, pp. 32-34.

Bouwer, H., 1966, "Rapid Field Measurement of Air Entry Value and Hydraulic Conductivity of Soil as Significant Parameters in Flow System Analysis," Water Resour. Res., 2, 729-732.

Boynton, S. S., 1983, "An Investigation of Selected Factors Affecting the Hydraulic Conductivity of Compacted Clay, MS Thesis, University of Texas at Austin, p. 79.

Carslaw, H. S., and Jaeger, J. C., 1959, Conduction of Heat in Solids, 2nd Ed., Oxford University Press, London, UK.

Cartwright, K., and Krapac, I. G., 1990, "Construction and Performance of a Long-Term Earthen Liner Experiment," Waste Containment Systems, Geotechnical Special Publication No. 26, American Society of Civil Engineers, pp. 135-150.

Daniel, D. E., Feb. 1984, "Predicting Hydraulic Conductivity of Clay Liners," Journal of Geotechnical Engineering, Proceedings of the ASCE, Vol. 110, No. 2, pp. 285-300.

References (continued)

Daniel, D. E. and Trautwein, S. J., 1986, "Field Permeability Test for Earthen Liner," Proceedings, In-Situ '86, ASCE Specialty Conference on Use of In-Situ Tests in Geotechnical Engineering, Virginia Polytechnic Institute and State University, Blacksburg, ASCE New York, pp. 146-160.

Daniel, D. E., Sept. 1989, "In Situ Hydraulic Conductivity Tests for Compacted Clays," Journal of Geotechnical Engineering, American Society of Civil Engineers, Vol. 115, No. 9, pp. 1205-1226.

Day, S. R., and Daniel, D. E., Aug. 1985, "Hydraulic Conductivity of Two Prototype Clay Liners," Journal of Geotechnical Engineering, Proc. American Society of Civil Engineers, Vol. 111, No. 8, pp. 957-970.

Elsbury, B. R., Straders, G. A., Anderson, D. C., Rehage, J. A., Sai, J. O., and Daniel, D. E., 1988, "Field and Laboratory Testing of a Compacted Soil Liner," PB-89-125942, prepared for the Office of Research and Development, Hazardous Waste Engineering Research Laboratory, U. S. Environmental Protection Agency, Washington, D. C.

Gordon, M. E., Huebner, P. M., and Mitchell, G. R., 1990, "Regulation, Construction and Performance of Clay Lined Landfills in Wisconsin," Waste Containment Systems, Geotechnical Special Publication No. 26, American Society of Civil Engineers, pp. 14-27.

Hvorslev, M. J., 1951, Time Lag and Soil Permeability in Ground Water Observations, Bulletin No. 36, USA/COE WES Vicksburg, MS.

Johnson, G. J., Crumley, W. S., and Boutwell, G. P., 1990, "Field Verification of Clay Liner Hydraulic Conductivity," Waste Containment Systems, American Society of Civil Engineers, pp. 226-245.

Lahti, L. R., King, K. S., Reades, D. W., and Bacopoulos, A., 1987, "Quality Assurance Monitoring of a Large Clay Liner," Geotechnical Practice for Waste Disposal '87, ASCE, New York, pp. 640-654.

Lane, K. S., and Washburn, S. E., 1946, "Capillarity Tests by Capillarimeters and Soil Filled Tubes," Proceedings, Highway Research Board, Vol. 26, pp. 460-473.

Mundell, J. A., and Boos, T. A., "Interpretation of Field Permeability Test Results on Full-Scale Liner System," Environmental Aspects of Geotechnical Engineering, Proceedings of Ohio River Valley Soils Seminar XXI, American Society of Civil Engineers (Cincinnati Section), pp. 1-6.

Olson, R. E., and Daniel, D. E., 1981, "Measurement of the Hydraulic Conductivity of Fine-Grained Soils," Permeability and Groundwater Contaminant Transport, ASTM STP 746, American Society for Testing and Materials, Philadelphia, pp. 18-64.

Sai, J. O., and Anderson, D. C., Sept. 1990, "Field Hydraulic Conductivity Tests for Compacted Soil Liners," Geotechnical Testing Journal, Vol. 13, No. 3, pp. 215-225.

Sai, J. O., and Anderson, D. C., 1991, State-of-the-Art Field Hydraulic Conductivity Testing of Compacted Soils, EPA/600/S2-91/022, U. S. Environmental Protection Agency, Cincinnati, OH.

Taylor, D. W., 1948, Fundamentals of Soil Mechanics, John Wiley & Sons, Inc., New York, pp. 192-194.

References (continued)

Torstensson, B. A., 1984, "A New System for Ground Water Monitoring" in Ground Water Monitoring Rev., 4(4), pp. 131-138.

Trautwein, S. J., and Williams, C. E., 1990, "Performance Evaluation of Earthen Liners," Waste Containment Systems, Geotechnical Special Publication No. 26, American Society of Civil Engineers, pp. 30-49.

U. S. Bureau of Reclamation, 1974, The Earth Manual, Second Edition, pp. 573-578.

U. S. EPA, 1984, Geotechnical Quality Assurance of Construction of Disposal Facilities, EPA 600/2-80-040.

Williams, C. E., 1988, "Facts About the Design and Construction of Earthen Containment Structures," presented in the Fall meeting, Texas Section of the American Society of Civil Engineers, College Station, Tx.

Youngs, E. G., 1968, "Shape Factors for Kirkham's Piezometer Method for Determining the Hydraulic Conductivity of Soil In Situ for Soils Overlying an Impermeable Floor or Infinitely Permeable Stratum," Soil Science, Vol. 106, No. 3, pp. 235-237.

Other papers

Majdi A. Othman[1], Craig H. Benson[2], Edwin J. Chamberlain[3], and Thomas F. Zimmie[4]

LABORATORY TESTING TO EVALUATE CHANGES IN HYDRAULIC CONDUCTIVITY OF COMPACTED CLAYS CAUSED BY FREEZE-THAW: STATE-OF-THE-ART

REFERENCE: Othman, M. A., Benson, C. H., Chamberlain, E. J., and Zimmie, T. F., "**Laboratory Testing to Evaluate Changes in Hydraulic Conductivity of Compacted Clays Caused by Freeze-Thaw: State-of-the-Art**," Hydraulic Conductivity and Waste Contaminant Transport in Soil, ASTM STP 1142, David E. Daniel and Stephen J. Trautwein, Eds., American Society for Testing and Materials, Philadelphia, 1994.

ABSTRACT: Several laboratory studies have shown that the hydraulic conductivity of compacted clay may increase up to three orders of magnitude when subjected to freeze-thaw. In this paper, methods to freeze and thaw specimens of compacted clay are reviewed and compared. Methods to measure the hydraulic conductivity of the specimens are also reviewed. Only naturally formed clay soils are considered; soil-bentonite mixtures and other amended soils are not included.

A review of testing conditions present during freeze-thaw and their effect on hydraulic conductivity is also included. Testing conditions that are addressed include availability of an external supply of water (closed vs. open system), dimensionality of freezing (one-dimensional vs. three-dimensional), rate of freezing, ultimate temperature, number of freeze-thaw cycles, and state of stress. The rate of freezing, number of freeze-thaw cycles, and state of stress appear to have the largest effect on hydraulic conductivity.

The effect of sampling disturbance on the hydraulic conductivity of compacted clay subjected to freeze-thaw is also presented. Specimens removed in Shelby tubes may be disturbed during sampling and extrusion. As a result, the effects of freeze-thaw can be masked. Collecting block specimens of thawed clay or taking core specimens of frozen clay are suggested as alternative procedures. A method to collect block specimens is presented.

---

[1] Assistant Project Engineer, GeoSyntec Consultants, 5775 Peachtree Dunwoody Road, Suite 200F, Atlanta, GA 30342.

[2] Assistant Professor, Department of Civil and Environmental Engineering, University of Wisconsin, Madison, WI 53706.

[3] Research Civil Engineer, U. S. Army Cold Regions Research and Engineering Laboratory, Hanover, New Hampshire 03755-1290.

[4] Associate Professor, Department of Civil Engineering, Rensselaer Polytechnic Institute, Troy, New York 12180-3590.

KEYWORDS: freeze-thaw, compacted clay, hydraulic conductivity, consolidometer, permeameter

INTRODUCTION

Compacted clays are often used as hydraulic barriers. Examples of their use include liners and covers for landfills, liners for ponds and waste lagoons, and caps for remediation of contaminated sites. Because their primary purpose is to minimize flow, low hydraulic conductivity is of paramount importance. In cold regions, compacted clay barriers may be subjected to cycles of freeze-thaw during the winter months. Recent studies (Chamberlain et al. 1990; Zimmie and LaPlante 1990; Othman and Benson 1992,1993; Benson and Othman 1993a) have shown that freezing and thawing affects the structure and hydraulic conductivity of compacted clays. In particular, increases in hydraulic conductivity in excess of two orders of magnitude have been measured.

The objective of this paper is to review and compare laboratory procedures currently employed to evaluate changes in hydraulic conductivity caused by freeze-thaw. Methods to freeze and thaw specimens are reviewed and apparatus used to measure their hydraulic conductivity are compared. Freezing of specimens inside or outside the permeameter is considered. Advantages and limitations of each method are discussed. Conditions during freeze-thaw and their effect on hydraulic conductivity are also discussed. These conditions include: availability of an external supply of water (open vs. closed system), dimensionality of freezing (1-D vs. 3-D), number of freeze-thaw cycles, ultimate temperature, rate of freezing, and state of stress.

Although this paper focuses on testing specimens frozen and thawed in the laboratory, sampling and testing of specimens frozen and thawed in the field is also discussed. The effects of disturbance are described and methods to obtain undisturbed specimens are suggested.

BACKGROUND

Processes Occurring During Soil Freezing

When the temperature in moist soil drops below 0° C, the water supercools and ice crystals nucleate in larger pores. As the water changes phase to ice, its volume increases about 9% due to the development of a hexagonal crystalline structure. The crystals grow to form ice lenses as long as water is available and heat is being extracted. The thickness and spacing of the ice lenses depend on the relative magnitudes of the rate of freezing, temperature gradient, pressure, and availability of water (Penner 1960). The growing ice crystals interact with each other and surrounding soil particles (Andersland and Anderson 1978).

In the region adjacent to the growing ice lenses, large pore water suctions are generated, pulling water from the unfrozen soil into the freezing zone (Benson and Othman 1993a; Williams 1966; Konrad and Morgenstern 1980; Chamberlain and Gow 1979). Effective pore water suctions as large as 500 kPa have been observed (Chamberlain 1981). This often results in extremely high effective stresses in the zone adjacent to the growing ice lenses, and it results in drying and consolidation of the unfrozen soil in this region. Drying and consolidation cause changes in the soil structure; in particular, shrinkage cracks form perpendicular to the freezing zone (Benson and Othman 1993a; Chamberlain and Gow 1979). As the freezing zone advances into the unfrozen soil mass, these cracks become filled with ice and the soil develops an aggregated structure, with the aggregates bounded by the ice lenses and ice filled cracks (Fig. 1).

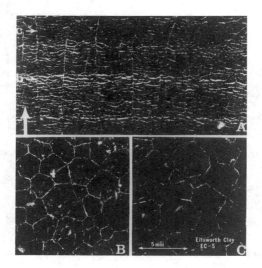

FIG. 1--Thin vertical (A) and horizontal (B and C) sections of a specimen of low plasticity clay frozen and thawed one-dimensionally in an open system (Chamberlain and Gow 1979). Scale is for all three sections.

The formation of ice lenses and shrinkage cracks is not the only evidence of structural changes occurring during freezing. In sandy or silty clayey soils where the skeletal structure is dominated by coarse particles, the structural changes may be related to the composition of the clay aggregates in voids formed by the coarser particles. During freezing and thawing, the space occupied by the small clay aggregates within these voids may decrease because of shrinkage of the aggregates. This too can result in an increase in void size and thus, an increase in the hydraulic conductivity.

Ice continues to grow in the frozen zone even when the flow of water from the unfrozen soil is cut off by the leading ice layer. This occurs because water progressively freezes as the temperature is lowered (Andersland and Anderson 1978). Some film water adjacent to the soil particles remains unfrozen even at very cold temperatures. The expansion of this unfrozen water in the frozen zone exerts pressure on the surrounding soil as it crystallizes and causes the soil aggregates to further deform (Andersland and Anderson 1978; Konrad 1989; Chamberlain 1981). As a result, the thickness of the ice lenses and the cracks increases.

While cracks and other structural changes may not be visible to the naked eye after thawing, they have been observed under SEM magnification (Hunsicker 1987). At low magnification, distinct cracks spaced at 0.5 mm were observed in Fort Edwards Clay specimens. At higher magnifications, voids as large as 0.005 mm were observed within the aggregates formed during the freezing process. Thus, freezing causes changes in the visible macro structure and the microstructure. The changes in the macrostructure, however, are most likely the cause of large increases in hydraulic conductivity that occur in clayey soils after freezing and thawing.

The cracks and voids that form in soil during freezing can change its effective porosity, which is defined as the volume of fluid-conducting pores divided by the total volume of soil. Kim and Daniel (1992) used tracer tests to calculate the effective porosity of compacted clay specimens before and after freeze-thaw. They found the effective porosity ranged between 0.18 and 0.22 for control specimens and between 0.23 and 0.33 for specimens subjected to 5 cycles of freeze-thaw. They concluded that more fluid-conducting pores are present in the soil after freeze-thaw, which causes the hydraulic conductivity to increase.

Effect of Freeze-Thaw on Hydraulic Conductivity

Several investigators have conducted laboratory studies to evaluate the effect of freeze-thaw on the hydraulic conductivity of compacted clays. Figure 2 shows these clays plotted on Casagrande's plasticity chart and Table 1 summarizes their properties. As shown in Table 1 and Fig. 2, the clays vary in classification, composition, plasticity, and compaction characteristics. The writers note that only naturally occurring clays have been considered; soil-bentonite and other admixtures have not been analyzed.

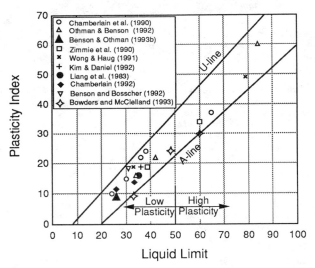

FIG. 2--Soils used in several laboratory studies on Casagrande's plasticity chart.

Figure 3 summarizes the change in hydraulic conductivity measured in each study in terms of hydraulic conductivity ratio, which is defined as the hydraulic conductivity after freeze-thaw divided by the hydraulic conductivity before freeze-thaw (Chamberlain et al. 1990). The graph shows hydraulic conductivity ratio as a function of initial hydraulic conductivity (i.e., hydraulic conductivity before freeze-thaw). As shown in Fig. 3, the hydraulic conductivity ratio ranges between 1 and 1400. It is generally largest when the initial hydraulic conductivity

TABLE 1--Characteristics of soils used in several laboratory studies.

Reference	Soil	USCS Class.	LL[1]	PI[1]	P$_{200}$[1]	Clay[1]	W$_o$[2] (%)	$\gamma_{d,m}$[2]
Chamberlain et al. (1990)	Green River	CL	24	10	89	40	13.0	18.6
	Durango	CL	38	24	65	30	15.9	17.6
	Slick Rock	CL	30	15	68	42	14.8	18.0
	Rifle	CL	36	22	61	34	14.0	17.8
	Fort Edwards	CH	65	37	93	62	25.0	15.7
Zimmie et al. (1990) and Zimmie (1992)	Niagara Clay	CL	39	19	90	50	16.5	17.6
	Brown Clay	CH	60	34	...	...	31.0	...
Othman & Benson (1992)	Wisconsin A	CL	34	16	85	58	16.0	18.0
	Wisconsin B	CL	42	19	99	77	18.5	16.8
	Wisconsin C	CH	84	60	71	58	26.0	14.7
Benson & Othman (1993b)	Portage	SC	26	9	47	...	13.0	18.3
Wong & Haug (1991)	Regina Clay	CH	79	49	97	46	30.6	13.7
	Battleford Till	CL	33	19	63	19	15.0	18.1
Kim & Daniel (1992)	Wisconsin A	CL	36	19	88	61	15.0	18.2
Liang et al. (1983)	Qinghe	CL	35	16	...	27	...	...
Chamberlain (1992)	Shakopee	CL	33	14	51	22	13.0	19.0
	Rosemont II	CL	26	12	63	28	8.9	20.7
Benson & Bosscher (1992)	Ridgeview Clay	CL	31	18	75	41	11.5	20.6
Bowders & McClelland (1993)	Kaolinite	CL-MH	58	24	90	...	31.0	13.5
	Wetzel County	CL-MH	33	9	50	...	11.0	19.1
	Monogalia Co.	CH	60	30	65	...	23.0	15.2

... = Not Available
[1] LL = Liquid limit, PI = Plasticity index, P$_{200}$ = % Finer than 0.075 mm, Clay = % Finer than 0.005 mm, W$_o$(%) = Optimum water content (%), $\gamma_{d,m}$ = Maximum dry unit weight (kN/m^3)[2].
[2] Optimum water content and maximum dry unit weight are based on the modified Proctor method (ASTM D1557) for Chamberlain et al. (1990), Chamberlain (1992), and Benson and Bosscher (1992). For all other studies, optimum water content and maximum dry unit weight are based on the standard Proctor method (ASTM D698).

is lowest and decreases as the initial hydraulic conductivity becomes larger.

Clayey soils with low hydraulic conductivity are generally compacted at water contents wet of optimum. In this condition, clods easily deform during compaction which results in a dense, relatively homogeneous mass with a very fine pore size. The fine (perhaps microscopic) pore size limits the conduction of fluid. Hence, large pores and cracks formed during freezing significantly increase the effective pore size. Consequently, the hydraulic conductivity increases dramatically.

Alternatively, specimens with relatively high initial hydraulic conductivity are generally compacted dry of optimum water content or

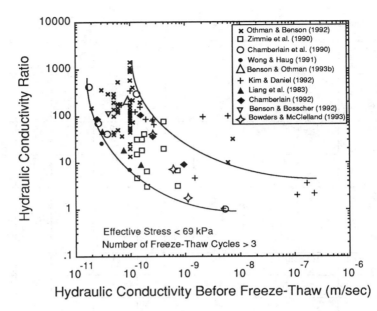

FIG. 3--Hydraulic conductivity ratio versus hydraulic conductivity before freeze-thaw from several laboratory studies.

with a low compactive effort. The drier aggregates of clay have higher strength and thus are more resistant to deformation during compaction. As a result, compaction dry of optimum results in a heterogeneous network of macroscopic pore and high hydraulic conductivity (Benson and Daniel 1990; Lambe 1962). In this condition, freeze-thaw results in greater aggregate definition and possibly even crack formation, but the size of the pores formed during freezing is not dramatically different than the size of the pores existing before freezing. Thus, for specimens with high initial hydraulic conductivity, the hydraulic conductivity remains unchanged or increases only slightly after freeze-thaw.

Effect of Soil Type

Changes in hydraulic conductivity caused by freeze-thaw are not unique for a given type of clayey soil. Figure 4 shows hydraulic conductivity ratio as a function of plasticity index for the soils summarized in Table 1. As shown in Fig. 4, increases in hydraulic conductivity of an order of magnitude or more have been measured for a variety of soils commonly used for compacted soil liners. Furthermore, hydraulic conductivity ratio appears to have no trend with plasticity index, suggesting that changes in hydraulic conductivity are controlled more by the freezing process than the characteristics of the soil. However, studies on soils of much higher plasticity such as bentonite and bentonite-soil mixtures indicate that freeze-thaw has little effect on their hydraulic conductivity (Shan and Daniel 1991; Wong and Haug 1991).

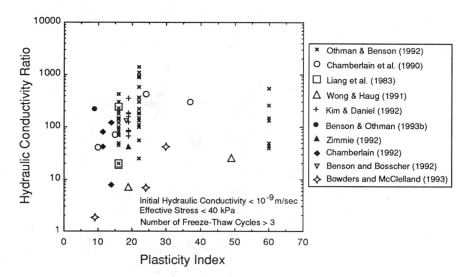

FIG. 4--Hydraulic conductivity ratio versus plasticity index from several laboratory studies.

TESTING APPARATUS

Different methods are currently employed to evaluate the effect of freeze-thaw on hydraulic conductivity. In this paper, the methods are segregated into two groups: (1) freeze-thaw in the permeameter and (2) freeze-thaw outside the permeameter.

Freeze-Thaw in the Permeameter

CRREL consolidometer

The consolidometer shown in Fig. 5 has been employed by the United States Army Cold Regions Research and Engineering Laboratory (CRREL). Chamberlain et al. (1990) used this apparatus to evaluate the effect of freeze-thaw on the hydraulic conductivity of compacted clays. A specimen is compacted in the consolidometer and then subjected to freeze-thaw cycling. After a selected number of cycles, the hydraulic conductivity of the specimen is measured.

The CRREL consolidometer consists of a Teflon-lined Plexiglas cylinder with inner and outer diameters of 63.5 mm and 152.4 mm. The base and the piston are made of aluminum, each containing a glycol cooling chamber, a thermoelectric cooling plate, and a porous stainless steel drainage plate. The piston, which is guided by a linear bearing, has a loading plate on which weights are placed for simulating overburden pressure. An air actuator can also be used for very high pressures. The drainage ports in the base and the piston are attached to a water supply. To measure hydraulic conductivity, the valves are opened to conduct falling or constant head tests. The valves can also

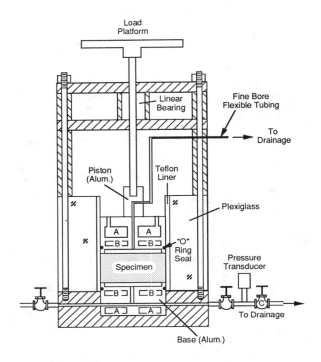

FIG. 5--CRREL consolidometer (Chamberlain et al. 1990).

be open or closed during freeze-thaw to simulate an open or closed system.

Thermoelectric cooling devices in the base and piston are used to provide precise control of the temperature during freezing and thawing. A solution of ethylene glycol and water is circulated at a temperature of about 1° C from a refrigerated bath through chambers in the piston and base and through a jacket surrounding the cell. This provides a steady ambient temperature near the freezing point of the soil water and a constant reference temperature for the thermoelectric plates. Two precision thermistors are positioned at the contact surfaces between the porous plates and the specimen to measure the top and bottom boundary temperatures. A computer controlled data acquisition/control system and a power supply are used in a closed loop system to monitor the temperatures and to make automatic adjustments to the current controlling the temperature of the thermoelectric devices. This system permits precise control of very small temperature gradients (< 0.02° C/mm) and very slow freezing rates (< 5 mm/day). More details regarding the CRREL consolidometer can be found in Chamberlain et al. (1990) and Chamberlain and Blouin (1976).

The CRREL consolidometer has several advantages (Table 2). The primary advantages of using the consolidometer is that compaction, consolidation, freeze-thaw, and measurement of hydraulic conductivity are all performed in the consolidometer and thus the specimen does not undergo any disturbance that may result during moving or handling.

TABLE 2--Advantages and disadvantages of testing methods.

Method	Advantages	Disadvantages
CRREL Consolid-ometer	• minimal specimen disturbance • precise control of temperature gradient • one-dimensional freeze-thaw • closed- or open-system of freezing • simulates overburden pressure	• side-wall leakage possible under low overburden pressure • expensive • complicated
Flexible-Wall Permeameter	• minimal specimen disturbance • simulates overburden pressure • no side-wall leakage • inexpensive • commercially available	• temperature gradient is difficult to control • three-dimensional freeze-thaw • closed-system of freezing
Fiberglass and Styrene Wraps	• one-dimensional freeze-thaw • simple • inexpensive	• time consuming (takes few days for a single freeze-thaw cycle) • disturbance of specimen possible • closed-system of freezing • temperature gradient is difficult to control • zero overburden pressure
Freestanding Specimen	• simple • inexpensive • fast	• three-dimensional freeze-thaw • disturbance of specimen possible • closed-system of freezing • temperature gradient is difficult to control • zero overburden pressure

Furthermore, because temperature readings and adjustments are computer-controlled, the temperature gradient and rate of freezing are precisely maintained. Also, the specimen is subjected to one-dimensional freezing (which simulates freezing in the field) and tests can be conducted in a closed- or open-system. Another advantage of this apparatus is the ability to simulate overburden pressure.

A disadvantage of this apparatus is that the specimen is compacted in a rigid Teflon cylinder and thus side wall leakage may occur during permeation. As a result, hydraulic conductivity measurements can be erroneous, especially if the surcharge pressure is not large enough to ensure the specimen conforms to the Teflon liner. Also, because temperature readings and adjustments are computer-controlled, the apparatus is more expensive and complicated than other apparatus currently being used. Because of concerns about side wall leakage, a test was conducted at CRREL with dye in the permeating fluid. A surcharge pressure of 14 kPa was applied to the piston. After about 1.5 pore volumes passed through the specimen, it was carefully examined. The examination revealed that most of the dye was sieved out at the initial contact surface. No dye passed through the test sample, nor did any dye appear at the cylindrical contact surface between the specimen and the Teflon liner. Nevertheless, the potential for short-circuiting remains a concern. Modifications are currently being made to replace the rigid Teflon liner with a rubber membrane.

Flexible-wall permeameter

Benson and Othman (1993a) used a flexible-wall permeameter to study the effect of freeze-thaw on the hydraulic conductivity of compacted clay specimens. The flexible-wall permeameter (Fig. 6) is

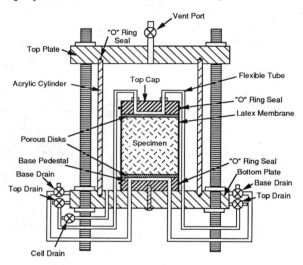

FIG. 6--Flexible-wall permeameter.

widely used for hydraulic conductivity testing, is commercially available, and standard procedures have been developed for its use (e.g., ASTM D5084). Daniel et al. (1985) describe flexible-wall permeameters in detail.

To permit freezing, the cell is filled with a 50-50 mixture of ethylene glycol and tap water. Pressure is applied to the water-glycol mixture and the specimen is allowed to consolidate. After consolidation is complete, the specimen is placed in a cold environment. As the temperature of the permeameter decreases, the clay specimen will freeze, while the ethylene glycol will remain unfrozen. Thus, the total stress on the specimen remains unchanged. After the specimen has frozen completely, it is removed from the cold environment and placed in a warm environment to thaw. When the specimen has warmed to the ambient temperature, one cycle of freeze-thaw is complete. Additional cycles are obtained by repeating this procedure.

The time required to freeze and thaw a specimen is determined by monitoring a control specimen. Thermocouple wires are inserted in the specimen and extended outside the cell through tubes in the base pedestal. Silicon caulk is used to seal the openings. Benson and Othman (1993a) found that the time required to freeze a specimen in a flexible-wall permeameter is approximately equal to the time required to freeze a specimen while freestanding. In general, one day was sufficient to reduce the temperature of a specimen from room temperature (~ 25° C) to - 18° C.

After the specimen has been subjected to a specified number of freeze-thaw cycles, a hydraulic conductivity test is conducted while the specimen remains in the permeameter. Benson and Othman (1993a) used the falling-head procedure without backpressure to conduct their tests. However, constant-head and falling-head/rising-head procedures can be used and backpressure can be applied.

Using a flexible-wall permeameter to conduct freeze-thaw tests has several advantages (Table 2). Disturbance of the specimen is minimal because one apparatus is used for consolidation, freezing and thawing, and permeation. Overburden pressure is also easily controlled and, because the specimen is surrounded by a flexible membrane, sidewall leakage is not a problem. This apparatus is also relatively inexpensive and is commercially available. As shown in Table 2, the disadvantages of this apparatus are: (1) freezing is not one-dimensional, (2) the rate of freezing is difficult to control, and (3) the specimen can only be frozen and thawed in a closed system (no external supply of water).

Three-dimensional freezing that occurs in the flexible-wall permeameter may yield a soil structure significantly different than occurs in one-dimensional conditions, such as in the field or the CRREL consolidometer. During one-dimensional freezing of clayey soils, large pore water suctions generated in the freezing zone cause shrinkage cracks to form in the unfrozen material. Positive pore water pressures probably do not develop until a zone is engulfed by frost. During three-dimensional freezing, the same type of shrinkage cracks may develop, but in a three-dimensional configuration. However, when a frozen shell has developed around the specimen, only positive pore pressures will develop in the center of the specimen because of expansion of the water during crystallization. Hence, the development of shrinkage cracks in the center core may cease to progress as the positive pressures develop. These pressures can be very high, particularly if the freezing progresses rapidly and the test specimen is saturated with water. Furthermore, high pore water pressures in the center of the specimen can also increase the temperature of the water (thus retarding freezing), or cause tension cracking of the outer frozen shell if the frozen soil cannot relieve the stresses that develop. Consequently, three-dimensional freezing in a permeameter may result in a structure much different than the structure occurring as a result of one-dimensional freezing.

Freeze-Thaw Outside the Permeameter

Fiberglass and polystyrene wraps

Othman and Benson (1992) and Zimmie and LaPlante (1990) have used similar apparatus to subject specimens to one-dimensional freeze-thaw outside the permeameter. Photographs of their apparatus are shown in

Fig. 7. Othman and Benson (1992) seal the specimen with several layers of plastic wrap to prevent desiccation. Then, the specimen is wrapped in a cylinder of fiberglass and placed on a heating pad sandwiched between two 50 mm-thick sheets of polystyrene. Othman and Benson (1992) have found that an insulation thickness of 70 mm and a heating pad temperature of 37° C is sufficient to yield one-dimensional freeze-thaw for a wide range of freezing temperatures. They verified that freezing occurred one-dimensionally by placing thermocouples at various locations in a specimen. The thermocouples showed a vertical gradient in temperature, but no variation in temperature in the horizontal plane.

The one-dimensional freeze-thaw apparatus used by Zimmie and LaPlante (1990) is composed of a thick Styrofoam sheet (thickness= 102 mm) that contains cylindrical holes (diameter= 83 mm) as shown in Fig. 7b. Soil specimens are placed in the holes while remaining in the compaction molds. A heating pad is placed underneath the Styrofoam sheet and an air gap 51 mm thick separates the heating pad and the bottom of the specimens. A Styrofoam sheet 51 mm thick is placed underneath the heating pad.

The apparatus is placed in a cold environment to freeze the specimen. The freezing front advances into the specimen from the top and moves downward due to the temperature gradient created by the difference in temperature at the two ends of the specimen. Thermocouples are installed in a control specimen to monitor temperature. Othman and Benson (1992) have found that two days are required for freezing and one day is required for thawing when the specimen is cooled from room temperature (~ 25° C) to -18° C. Thus, a total of three days is needed for one freeze-thaw cycle. Similar findings were obtained by Zimmie and LaPlante (1990). After a specimen has been subjected to a number of freeze-thaw cycles, the wrappings are removed. Then, the specimen is placed in a flexible-wall permeameter (Fig. 6) and its hydraulic conductivity is measured.

This type of apparatus is simple, inexpensive, and permits one-dimensional freeze-thaw. However, at least three days are required to complete one freeze-thaw cycle and after freeze-thaw, the hydraulic conductivity of the specimen has to be measured using another apparatus (e.g., flexible-wall permeameter). Thus, this procedure is time consuming and requires handling of the specimen which may result in disturbance. This apparatus has several other disadvantages. Specimens can only be frozen and thawed in a closed system, the rate of freezing is difficult to control, and the overburden pressure applied on the specimen during freeze-thaw is negligible.

To minimize disturbance, Zimmie and LaPlante (1990) remove the specimen while frozen and immediately place it in a flexible-wall permeameter. The specimen is allowed to thaw in the permeameter and then its hydraulic conductivity is measured.

Freeze-thaw of specimen while freestanding

The freestanding method has been used by Zimmie and LaPlante (1990), Othman and Benson (1992), and Kim and Daniel (1992). The specimen is sealed with several layers of plastic wrap and placed in a plastic bag. To freeze the specimen, it is placed in a cold environment and to thaw it is placed in a warm environment. For typical freezing temperatures (-1° C to -25° C), only a few hours are required for the entire specimen to cool from room temperature to the ambient temperature of the cold environment. To reduce the rate of freezing, the specimen can be wrapped with fiberglass insulation. Another method to control

(a)

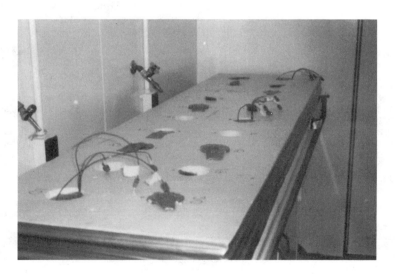

(b)

FIG. 7--Apparatus for one-dimensional freeze-thaw: (a) fiberglass wrap (Othman and Benson 1992), and (b) polystyrene wrap (Zimmie et al. 1991).

the rate of freezing is to slowly change the temperature of the freezing environment while monitoring temperatures within an instrumented control specimen.

This method is the easiest approach to freeze and thaw specimens. It is less time-consuming and least expensive. This method has similar disadvantages as the one-dimensional freeze-thaw procedures and, in addition, with this method one-dimensional freeze-thaw cannot be achieved.

CONDITIONS DURING TESTING

Conditions during freeze-thaw can affect the structure and hydraulic conductivity of compacted clay specimens. To estimate the effect of freeze-thaw on hydraulic conductivity, conditions expected in the field should be simulated. Factors to consider are: availability of an external supply of water (closed vs. open system), dimensionality of freezing (1-D vs. 3-D), number of freeze-thaw cycles, rate of freezing, ultimate temperature, state of stress, and depth of freezing.

Number of Freeze-Thaw Cycles

The number of cycles of freeze-thaw occurring in a compacted clay barrier depends on the local climate. In regions where the temperature varies significantly, many cycles of freeze-thaw can be expected. Alternatively, if the temperature remains below freezing for a significant portion of the winter season, soil will be subjected to fewer freeze-thaw cycles. The number of cycles also depends on the depth of the barrier. For barriers located near the surface, the soil temperature is more sensitive to changes in air temperature and thus more cycles of freeze-thaw occur. Deeper in the soil, fluctuations in temperature that exist at the surface are tempered and hence less cycling occurs.

Experiments have shown that an increase in hydraulic conductivity of one order of magnitude or more may occur in the first cycle of freeze-thaw (Fig. 8) and most changes in hydraulic conductivity occur during the first few cycles. Increases in hydraulic conductivity after 3 to 10 freeze-thaw cycles are usually not significant (Chamberlain et al. 1990; Zimmie and LaPlante 1990; Othman and Benson 1992; Wong and Haug 1991).

The most significant increase in hydraulic conductivity occurs in the first cycle because of the radical change in soil structure that occurs during the initial freeze. Apparently, the network of cracks formed during the first cycle of freeze-thaw causes the greatest change in a soil's fabric. During subsequent cycles, more cracks form, but these additional cracks have much less impact on hydraulic conductivity. Furthermore, at some point, the soil becomes so aggregated that no more cracking occurs, and thus the hydraulic conductivity ceases to change.

Availability of an External Supply of Water

The presence of water during freezing affects changes in structure and hydraulic conductivity of compacted clays because the thickness and spacing of ice lenses depends on the rate of freezing and the availability of water (Penner 1960). In the field, compacted clay barriers have some access to water because of infiltration caused by rainfall, snowmelt, or the existence of a shallow water table. In most cases, however, the clay will not have continuous and unrestricted

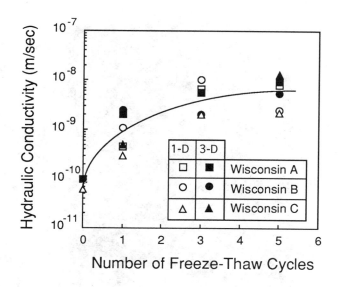

FIG. 8--Hydraulic conductivity versus number of freeze-thaw cycles for specimens of three Wisconsin clays frozen and thawed one- and three-dimensionally (Othman and Benson 1992).

access to water. Thus, neither the closed or open systems simulate the condition of a compacted clay in the field.

Chamberlain et al. (1990) conducted experiments in an open system using the CRREL consolidometer (Fig. 5). In all other laboratory studies the writers have reviewed (Table 1), a closed system has been used. Surprisingly, however, increases in hydraulic conductivity are similar for specimens frozen and thawed in both systems provided other testing conditions are held constant (Figs. 3 & 4). Figure 9a is a photograph (from Othman and Benson 1993) of a thin vertical section of a specimen of compacted clay (Wisconsin C, Table 1) after one cycle of freeze-thaw. The specimen was frozen and thawed one-dimensionally in a closed system using the fiberglass-wrap technique. The thin section was prepared using a procedure introduced by Chamberlain et al. (1990). As shown in Fig. 9a, ice lenses have formed in the specimen although it was frozen without access to an external supply of water.

Apparently, at the high degree of saturation at which clays are ordinarily compacted (typically above 85%, Benson, et al. 1992), sufficient water is available in the pores to form ice lenses. Further increases in saturation or water content occurring in an open system appear to have little effect on hydraulic conductivity. Also, because the hydraulic conductivity of compacted clays is typically low, the migration of water in response to freezing suction is highly localized in the region near the freezing zone.

Further evidence suggesting that availability of water is not very significant was provided by Zimmie et al. (1992). They compacted specimens of Niagara Clay at a molding water content exceeding optimum water content and then allowed them to sit under a column of water 1.5 m

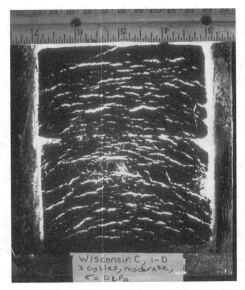

(a)

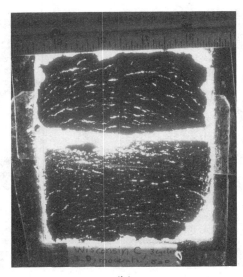

(b)

FIG. 9–Thin vertical sections of specimens of compacted clay (Wisconsin C, see Table 1 for properties) frozen and thawed (a) one-dimensionally and (b) three-dimensionally at a freezing rate of 4×10^{-7} m/s in a closed system (Othman and Benson 1993).

high for two years. The specimens were then subjected to freeze-thaw and their hydraulic conductivity was measured. Zimmie et al. (1992) measured increases in hydraulic conductivity of the saturated specimens that were essentially the same as increases in hydraulic conductivity of the specimens frozen and thawed in a closed system at their molding water content.

Dimensionality of Freezing

In the field, soil freezes and thaws one-dimensionally. To simulate one-dimensional freezing in the laboratory, special care is taken to ensure the specimen is subjected to a one-dimensional temperature gradient. However, this process is time consuming and requires a special testing apparatus. Hence, from a practical perspective, it is useful to determine if three-dimensional freezing results in similar changes in hydraulic conductivity as observed in specimens frozen and thawed one-dimensionally.

The effect of dimensionality of freezing has been studied by Zimmie and LaPlante (1990) and Othman and Benson (1992). They conducted tests to determine if specimens frozen three-dimensionally have hydraulic conductivities similar to specimens frozen under more realistic one-dimensional conditions. Results of one- and three-dimensional tests conducted by Othman and Benson (1992) are shown in Fig. 8. Similar findings have been presented by Zimmie and LaPlante (1990). They concluded that similar changes in hydraulic conductivity occur if specimens are frozen and thawed using a one- or three-dimensional temperature gradient.

Figure 9b shows a thin vertical section of a specimen of compacted clay (Wisconsin C, Table 1) frozen and thawed three-dimensionally while freestanding. Testing conditions for this specimen, with the exception of dimensionality of freezing, were the same as the conditions to which the specimen in Fig. 9a was subjected. To achieve the same rate of freezing, the 3-D specimen was wrapped with fiberglass insulation.

The photographs in Fig. 9 show that ice lenses (and subsequently the cracks controlling hydraulic conductivity) have slightly different geometry when subjected to one- or three-dimensional freeze-thaw. For the specimen frozen one-dimensionally from top to bottom, many thin ice lenses are found near the top, but with increasing depth, the frequency of the ice lenses decreases and their width increases (Fig. 9a). Although this specimen was frozen one-dimensionally, the ice lenses are not completely horizontal but are curved near the outer edges.

For the specimen frozen three-dimensionally, however, the freezing front advances into the specimen from all directions and thus ice lenses form parallel to boundaries of the specimen. As a result, the ice lenses show more curvature (Fig. 9b) and have approximately equal width and spacing. Apparently, however, the differences in geometry of the cracks between one- and three-dimensional freezing does not have a great effect on the hydraulic conductivity of the specimen (Fig. 8).

Rate of Freezing

Research has shown that rate of freezing affects the magnitude and rate of frost heave. At greater rates of freezing, larger pore water suctions develop and a greater number of ice lenses form provided a sufficient quantity of water is available (Dirksen and Miller 1966). Othman and Benson (1992) studied the effect of rate of freezing on the hydraulic conductivity of compacted clays subjected to three-dimensional freeze-thaw. They used the free-standing method and controlled the rate of freezing by varying the thickness of fiberglass insulation surrounding the specimen. Specimens were subjected to rates of freezing of 2×10^{-6} m/s (fast), 4×10^{-7} m/s (moderate), and 2.6×10^{-7} m/s (slow). The rate of freezing was defined as the distance to farthest point to which the freezing front travels divided by the time it takes to change the temperature of this point from 0° C to the ambient temperature of the cold environment.

Figure 10 shows the hydraulic conductivity of these specimens as a function of number of freeze-thaw cycles. As shown in Fig. 10, the hydraulic conductivity was largest in the case of fast freezing. Examination of the specimens showed that fast freezing resulted in a greater number of ice lenses than occurred during slow freezing. The writers note, however, that it is not known if the hydraulic conductivity will continue to increase or if it will decrease if the freezing rate is increased beyond the maximum value used by Othman and Benson (1992).

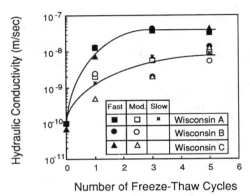

FIG. 10--Effect of rate of freezing on the hydraulic conductivity of three clays from Wisconsin (Othman and Benson 1992).

Ultimate Temperature

Although ice crystals nucleate in large pores when the soil temperature drops below 0° C, an unfrozen water phase remains that separates ice from solid particles. As the temperature of the soil decreases, more of the unfrozen water changes to ice. Hence, the quantity of ice and size of the ice structures depend on the temperature of the specimen. In some cases, unfrozen water has existed at temperatures as low as -70° C (Lundarini 1981). Because an increasing amount of ice is formed as the temperature is lowered, it is possible that the ultimate temperature of the specimen can affect its hydraulic conductivity.

Othman and Benson (1992) investigated the effect of ultimate temperature on the hydraulic conductivity of three compacted clays (Wisconsin A, B, and C). Compacted specimens of the three clays, with an initial hydraulic conductivity of approximately 1×10^{-10} m/s, were subjected to three cycles of three-dimensional freeze-thaw in a freezer that had an ambient temperature of -23°C. The specimens were removed from the freezer at temperatures of -1, -9, -18, and -23°C and were allowed to thaw at room temperature.

Figure 11 shows the hydraulic conductivity of these specimens as a function of ultimate temperature. The hydraulic conductivity increased about two orders of magnitude for each ultimate temperature and the increase in hydraulic conductivity was slightly larger for lower ultimate temperatures. Thus, it appears that most of the change in hydraulic conductivity is caused by the initial freezing of free water and subsequent formation of cracks and only a small increase in hydraulic conductivity occurs as more unfrozen water becomes ice.

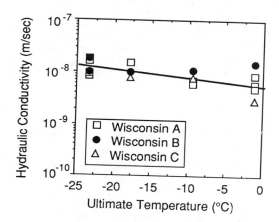

FIG. 11--Effect of ultimate temperature on the hydraulic conductivity of three clays from Wisconsin (Othman and Benson 1992).

State of Stress

Overburden pressure after freeze-thaw

Increases in hydraulic conductivity that occur because of freeze-thaw can be reversed if a large overburden pressure is applied to the soil. Overburden pressure reduces the size of voids and cracks and thus the hydraulic conductivity decreases. Figure 12 shows the influence of effective stress applied after freeze-thaw on the hydraulic conductivity of twenty-one clays. As the effective stress increases, the hydraulic conductivity ratio decreases significantly. However, very large stresses (> 200 kPa) are required to reach hydraulic conductivity ratios near unity.

The sensitivity of hydraulic conductivity to effective stress has important ramifications. For compacted soil liners used in municipal and hazardous waste landfills, exposure to freeze-thaw will result in only temporary increases in hydraulic conductivity, provided a sufficient depth of waste is placed on the liner. However, in cases where overburden pressure is low, such as a final covers and surface impoundments, changes in hydraulic conductivity are likely to be permanent.

Overburden pressure during freeze-thaw

The effect of overburden pressure applied during freeze-thaw is three-fold. First, it inhibits the formation and growth of ice lenses

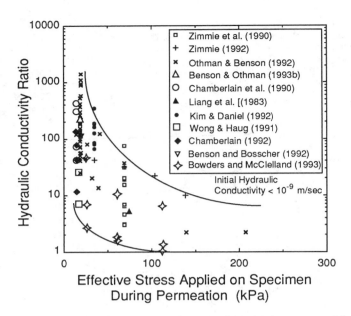

FIG. 12--Influence of effective stress applied during permeation on the hydraulic conductivity ratio of twenty-one clays.

because it reduces suctions below the growing ice lens (Konrad and Morgenstern 1982). Second, overburden pressure reduces the hydraulic conductivity of the frozen fringe, a partially frozen zone separating frozen soil from unfrozen soil (Konrad and Morgenstern 1982; Penner and Ueda 1977). Both of these factors impede the migration of water to the frozen soil. Third, large pores and cracks formed during freeze-thaw compress during thawing when overburden pressure is applied. Hence, movement of water to the freezing front during the next freezing cycle is restricted.

Othman and Benson (1993) evaluated the effect of overburden pressure applied during freeze-thaw on the hydraulic conductivity of a compacted clay (Wisconsin A, Table 1). They froze specimens three-dimensionally in a flexible-wall permeameter (Fig. 6) under different overburden pressures. They found that the hydraulic conductivity after freeze-thaw was largest at zero overburden pressure and decreased significantly with increasing overburden pressure (Fig. 13). Furthermore, the decrease in hydraulic conductivity caused by pressure applied during freezing is greater than the reduction in hydraulic conductivity which occurs when pressure is applied after freezing (Fig. 13). Similar results were obtained by Chamberlain et al. (1990).

SAMPLING FOR FREEZE-THAW EVALUATION

Laboratory testing is often conducted to evaluate if freeze-thaw has affected the hydraulic conductivity of liners and covers which have been exposed to winter conditions. For example, in Wisconsin, specimens are removed from an exposed liner using Shelby tubes and hydraulic conductivity tests are conducted on the specimens using standard

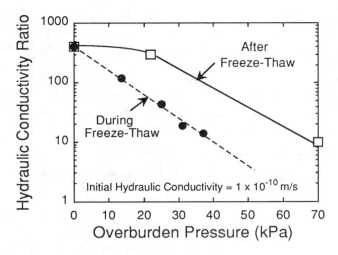

FIG. 13--Effect of overburden pressure during and after freeze-thaw on the hydraulic conductivity of Wisconsin A clay (adapted from Othman and Benson 1992 and Benson and Othman 1993a).

procedures (e.g., ASTM D 5084). Because the hydraulic conductivity of specimens frozen and thawed is governed by cracks in the soil fabric, methods to collect and prepare specimens for testing need to be carefully selected to ensure sampling disturbance does not affect the fabric and the test results.

Thin-Wall Sampling Tubes

Starke (1989) conducted hydraulic conductivity tests on specimens removed from a soil liner in Wisconsin that was exposed for one winter. Specimens were removed in sampling tubes (71 mm diameter) before and after winter, extruded in the laboratory, and their hydraulic conductivity was measured in a flexible-wall permeameter. Specimens removed before and after winter had average hydraulic conductivities of 2.5×10^{-10} m/s and 1.5×10^{-10} m/s, respectively. From this data it may be inferred that: (1) hydraulic conductivity decreases slightly, or remains unchanged, as a result of freeze-thaw, a result in direct contrast to all other studies, (2) the specimens were disturbed and hence the changes in hydraulic conductivity were masked, or (3) the specimens were too small to adequately capture the change in soil fabric controlling hydraulic conductivity.

To investigate the effect of sampling disturbance, 4 specimens of clay (Wisconsin A, Table 1) were compacted in steel molds (diameter = 102 mm) at a water content 4% wet of standard Proctor optimum. Two specimens were extruded from the molds and one of these was subjected to 3 cycles of freeze-thaw. Hydraulic conductivities of the two specimens were then measured in flexible-wall permeameters. The other two specimens were kept in the molds and one of them was subjected to 3 cycles of freeze-thaw. Then, thin-wall sampling tubes were pushed into the specimens. Afterwards, the specimens were extruded from the Shelby tubes and their hydraulic conductivity was measured in a flexible-wall permeameter.

A summary of the hydraulic conductivities is presented in Table 3.

TABLE 3--Effect of sampling disturbance on hydraulic conductivity.

Number of Freeze-Thaw Cycles	Specimen Diameter (mm)	Sampling Technique	Hydraulic Conductivity (m/s)
0	102	As Compacted	1×10^{-10}
3	102	As Compacted	2×10^{-8}
0	71	Shelby Tube	1×10^{-10}
3	71	Shelby Tube	6×10^{-11}

In each case, the average effective stress was 21 kPa and the hydraulic gradient was maintained between 21 to 27. No backpressure was used. For the specimens that were never frozen, the same hydraulic conductivity was measured (1×10^{-10} m/s). However, for specimens subjected to 3 cycles of freeze-thaw, the specimen extracted with a Shelby tube showed much lower hydraulic conductivity (6×10^{-11} m/s) than the specimen tested directly after freeze-thaw (2×10^{-8} m/s) and was even lower in hydraulic conductivity than the specimens that were never frozen and thawed.

The decrease in hydraulic conductivity observed for the specimen frozen, thawed, and then sampled with a Shelby tube is in direct contrast with the results of the previously described laboratory tests. During the sampling procedure, it was observed that penetration of the tube caused significant deformation of the soil. Furthermore, the specimen extruded from the Shelby tube showed no signs of cracking and was much stiffer than the original specimen.

Although these limited results do not discredit the findings of Starke (1989), they do suggest that hydraulic conductivity measurements conducted on specimens obtained with Shelby tubes may be misleading. Disturbance during sampling may dramatically change the soil structure and hydraulic conductivity. Hence, when testing is conducted to evaluate the effect of freeze-thaw on actual soil liners and covers, sampling techniques that result in minimal disturbance should be employed.

Block Sampling and Frozen Cores

Specimens that are essentially undisturbed and are large enough to yield field-scale hydraulic conductivity can be obtained as block specimens. Benson et al. (1993) have found that block specimens with a diameter of 0.30 m and a thickness of 0.15 to 0.20 m are large enough to represent field-scale conditions for a variety of construction conditions. Benson et al. (1993) use the following procedure to obtain large block specimens:

1. The location of the block specimen is selected and marked and the ground surface is cleaned and leveled. A trimming ring, similar to a consolidation ring, is placed on the surface. The ring is manufactured from PVC pipe and has a tapered cutting edge.

Typically a ring with an inside diameter of 0.36m and height of 0.30 m is used.

2. Trenches are excavated around the marked area using a backhoe or hand tools (Fig. 14a).

3. Soil is carefully trimmed away near the cutting edge until the ring can be pushed over the soil with light effort (Fig. 14b).

4. When the trimming ring is at full depth, the block specimen is separated from the underlying soil. If the soil is very hard, a sharpened steel plate is tapped into the soil with a hammer. For softer soils, a spade or carbide-tipped saw is used.

5. The soil specimen is sealed with plastic or wax to prevent desiccation and is moved to the laboratory for hydraulic conductivity testing.

6. In the laboratory, the specimen is removed from the ring and further trimmed to remove disturbed soil on the outer edges. The specimen is trimmed to a diameter of 0.30m and a height of 0.15m.

7. The hydraulic conductivity of the specimen is measured in a large-scale flexible-wall permeameter.

At CRREL, the current standard practice to obtain specimens from the field in the frozen condition. The frozen specimens are extracted with a special power auger. Carbide bits cut the soil cleanly and spiral flights on the outside of the core barrel carry the cuttings to the ground surface. The samples are kept frozen during shipping and during the trimming process. This maintains the frozen structure with little disturbance. The specimen is thawed in a flexible-wall permeameter with a small positive deviator stress to insure one-dimensional consolidation.

SUMMARY AND CONCLUSIONS

Several laboratory studies have shown that freeze-thaw may cause increases in hydraulic conductivity of up to three orders of magnitude. Freeze-thaw creates larger pores and a network of cracks in the soil. As a result, the hydraulic conductivity increases. The largest increases in hydraulic conductivity occur for soils with low initial hydraulic conductivity and the increase in hydraulic conductivity diminishes with increasing initial hydraulic conductivity. Changes in hydraulic conductivity are similar for a wide variety of soils ordinarily used to construct hydraulic barriers, but depend on the conditions (e.g., rate of freezing, number of freeze-thaw cycles, and overburden pressure) imposed on the specimen during freeze-thaw.

Laboratory studies indicate that the number of freeze-thaw cycles, rate of freezing, and state of stress have the largest effect on the change in hydraulic conductivity. The hydraulic conductivity increases as the rate of freezing and number of freeze-thaw cycles are increased and as the overburden pressure is decreased. Other factors such as the ultimate temperature, dimensionality of freezing, and availability of an external supply of water do not appear to have a significant effect on the change in hydraulic conductivity.

(a)

(b)

FIG. 14--Excavating trench around block specimen (a) and trimming ring near full depth (b).

Four methods currently employed to evaluate the effect of freeze-thaw on hydraulic conductivity were reviewed in this paper. The advantages and limitations of each method have been discussed. Apparatus that are currently employed include the CRREL consolidometer, the flexible-wall permeameter, fiberglass and polystyrene wraps, or freeze-thaw while freestanding. Data collected to date suggests that similar changes in hydraulic conductivity are obtained with these methods provided the primary factors affecting the specimens are held constant.

The devices currently being used vary in complexity and in ability to control conditions during freeze-thaw. The CRREL consolidometer is the most complex apparatus and permits greatest control of the freeze-thaw process. Freezing and thawing of a specimen while freestanding is the simplest procedure, but conditions during freezing are difficult to control. The type of apparatus to select for testing depends on the level of control that is needed during testing. Worst case changes in hydraulic conductivity can be evaluated quickly and easily using three-dimensional freezing of a freestanding specimen.

Sampling disturbance can mask changes in hydraulic conductivity of soil that has been frozen and thawed. Experiments have shown that specimens sampled with a thin-walled sampling tube may show a decrease in hydraulic conductivity after freeze-thaw even when the undisturbed frozen and thawed soil shows an increase in hydraulic conductivity. Hence, when evaluating the effects of freeze-thaw, sampling techniques resulting in minimal disturbance should be used. A procedure to collect undisturbed block specimens was presented in this paper. Good quality specimens can also be collected with a coring auger when the clay is in the frozen condition.

ACKNOWLEDGMENTS

Various sponsors have provided financial support for the studies described in this paper. Financial support for the work by Othman and Benson has been provided by the Graduate School at the University of Wisconsin-Madison, Waste Management of North America, Inc., and Waste Management of Wisconsin, Inc. Many of the experiments conducted by Othman and Benson were performed in the University of Wisconsin's BIOTRON Facility. Support for the use of BIOTRON was provided by the College of Engineering at the University of Wisconsin-Madison.

Financial support for the studies by Chamberlain have been provided by the US Army Corps of Engineers under its CPAR Program, by the U.S. Department of Energy under its UMTRA program, and by Ardaman and Associates, Inc. All studies were performed at the US Army Cold Regions Research and Engineering Laboratory in Hanover, New Hampshire.

Financial support for the work by Zimmie was provided by the New York Center for Hazardous Waste Management and the National Science Foundation.

The writers gratefully acknowledge this support. However, the opinions expressed in this article are those solely of the authors and do not necessarily reflect the policies and opinions of the sponsors.

REFERENCES

Alkire, B. D., and Morrison, J. M., 1982, "Changes in Soil Structure Due to Freeze-Thaw and Repeated Loading," Transportation Research Record, No. 918, pp. 15-22.

Andersland, O. B., and Anderson, D. M., 1978, *Geotechnical Engineering for Cold Regions*, McGraw Hill, New York, p. 75.

Benson, C. H., Zhai, H. and S. Rashad 1992, "Assessment of Construction Quality Control Measurements and Sampling Frequencies for Compacted Soil Liners," *Environmental Geotechnics Report*, No. 92-6, Department of Civil and Environmental Engineering, University of Wisconsin, Madison, Wisconsin.

Benson, C. H., and Daniel, D. E., 1990, "Influence of Clods on Hydraulic Conductivity of Compacted Clay," *Journal of the Geotechnical Engineering Division*, American Society of Civil Engineers, Vol. 116, No. 8, pp. 1231-1248.

Benson, C. H., and Othman, M. A., 1993a, "Hydraulic Conductivity of Compacted Clay Frozen and Thawed In Situ," *Journal of Geotechnical Engineering*, American Society of Civil Engineers, Vol. 119, No. 2, pp. 276-294.

Benson, C. H., and Othman, M. A., 1993b, "Hydraulic and Mechanical Characteristics of a Compacted Municipal Solid Waste Compost," *Waste Management and Research*, Vol. 11, pp. 127-142.

Benson, C. H., Hardianto, F. S. and E. S. Motan, 1993, "Representative Sample Size for Hydraulic Conductivity of Compacted Clay," *Hydraulic Conductivity and Waste Contaminant Transport in Soils, ASTM STP 1142*, David E. Daniel and Stephen J. Trautwein, eds., American Society for Testing and Materials, Philadelphia, in press.

Benson, C. H., and Bosscher, P. J., 1992, "Effect of Winter Exposure on the Hydraulic Conductivity of a Test Pad," *Environmental Geotechnics Report*, No. 92-8, Department of Civil and Environmental Engineering, University of Wisconsin, Madison, Wisconsin, 33 p.

Bowders, J. J. Jr., and McClelland, S., 1993, "The Effects of Freeze/Thaw Cycles on the Permeability of Three Compacted Soils," *Hydraulic Conductivity and Waste Contaminant Transport in Soils, ASTM STP 1142*, David E. Daniel and Stephen J. Trautwein, Eds., American Society for Testing and Materials, Philadelphia, in press.

Chamberlain, E. J., 1981, "Overconsolidation Effects of Ground Freezing," *Engineering Geology*, Vol. 18, pp. 97-110.

Chamberlain, E. J., 1992, "Freeze-Thaw Effects on the Permeability of Shakopee and Rosemont II Soils," *CRREL Final Report*, US Army Cold Regions Research and Engineering Laboratory, Hanover, NH, 9 p.

Chamberlain, E. J., and Blouin, S. E., 1976, "Freeze-Thaw Enhancement of the Drainage and Consolidation of Fine-Grained Dredged Material in Confined Disposal Areas," *CRREL Final Report*, U.S. Army Cold Regions Research and Engineering Laboratory, Hanover, NH.

Chamberlain, E. J., and Gow, A.J., 1979, "Effect of Freezing and Thawing on the Permeability and Structure of Soils," *Engineering Geology*, Vol. 13, pp. 73-92.

Chamberlain, E. J., Iskander, I., and Hunsiker, S. E., 1990, "Effect of Freeze-Thaw on the Permeability and Macrostructure of Soils," *Proceedings, International Symposium on Frozen Soil Impacts on Agricultural, Range, and Forest Lands*, March 21-22, Spokane, WA, pp. 145-155.

Daniel, D. E., Anderson, D. C., and Boynton, S. S.,1985, "Fixed-Wall versus Flexible-Wall Permeameters," *ASTM STP 867*, pp. 107-126.

Dirksen, C., and Miller, R. D., 1966, "Closed-System Freezing of Unsaturated Soil," *Proceedings, Soil Science Society of America*, Vol. 30, pp. 168-178.

Hunsicker, S. E., 1987, "The Effect of Freeze-Thaw Cycles on the Permeability and Macro Structure of Fort Edwards Clay," *Master of Engineering Thesis*, Thayer School of Engineering, Dartmouth College, Hanover, NH, 168p.

Kim, W-H, and Daniel, D. E., 1992, "Effects of Freezing on the Hydraulic Conductivity of a Compacted Clay," *Journal of Geotechnical Engineering*, American Society of Civil Engineers, Vol. 118, No. 7, pp. 1083-1097.

Konrad, J.-M., and Morgenstern, N. R., 1980, "Mechanistic Theory of Ice Lens Formation in Fine-Grained Soils," *Canadian Geotechnical Journal*, Vol. 17, pp. 473-486.

Konrad, J.-M., and Morgenstern, N. R., 1982, "Effects of Applied Pressure on Freezing Soils," *Canadian Geotechnical Journal*, Vol. 19, pp. 494-505.

Lambe, T. W., 1962, *Foundation Engineering*, Leonards, G. A., Editor, Chapter 4, Soil Stabilization, pp. 351-437.

Liang, W., Bomeng, X., and Zhijin, W., 1983, "Properties of Frozen and Thawed Soil and Earth Dam Construction in Winter," *Proceedings, 4th International Conference on Permafrost*, pp. 1366-1371.

Lundarini, N. A., 1981, *Heat Transfer in Cold Climates*, Litton Educational Publishing, New York.

Othman, M. A., and Benson, C. H., 1992, "Effect of Freeze-Thaw on the Hydraulic Conductivity of Three Compacted Clays from Wisconsin," *Transportation Research Record*, No. 1369, pp. 118-129.

Othman, M. A., and Benson, C. H., 1993, "Effect of Freeze-Thaw on the Hydraulic Conductivity and Morphology of Compacted Clay," *Canadian Geotechnical Journal*, in press.

Penner, E., 1960, "The Importance of Freezing Rate in Frost Action in Soils," *Proceedings of ASTM*, Vol. 60, pp. 1151-1165.

Penner, E., and Ueda, T., 1977, Proceedings, Symposium on Frost Action in Soils, University of Lulea, Lulea, Sweden, Vol. I, pp. 91-100.

Shan, H. Y., and Daniel, D. E., 1991, "Results of Laboratory Tests on a Geotextile/Bentonite Liner Material," Proceedings, Geosynthetics 91, Industrial Fabrics Association International, St. Paul, MN, Vol. 2, pp. 517-535.

Starke, J. O., 1989, "Effect of Freeze-Thaw Weather Conditions on Compacted Clay Liners," Proceedings, 12th Annual Madison Waste Conference, University of Wisconsin-Madison, WI, pp. 412-420.

Williams, P. J., 1966, "Pore Pressures at a Penetrating Frost Line and their Prediction," Geotechnique, Vol. XVI, No. 3, pp. 187-208.

Wong, L. C., and Haug, M. D., 1991, "Cyclical Closed-System Freeze-Thaw Permeability Testing of Soil Liner and Cover Materials," Canadian Geotechnical Journal, Vol. 28, pp. 784-793.

Zimmie, T. F., 1992, "Freeze-Thaw Effects on the Permeability of Compacted Clay Liners and Covers," Geotechnical News, pp. 28-30.

Zimmie, T. F., and LaPlante, C. M., 1990, "The Effect of Freeze-Thaw Cycles on the Permeability of a Fine-Grained Soil," Proceedings, 22nd Mid-Atlantic Industrial Waste Conference, Philadelphia, PA, pp. 580-593.

Zimmie, T. F., LaPlante, C. M., and Bronson, D. L., 1991, "The Effects of Freezing and Thawing on Landfill Covers and Liners," Proceedings, 3rd International Symposium on Cold Regions Heat Transfer, University of Alaska Fairbanks, Fairbanks, Alaska, pp. 363-371.

Zimmie, T. F., LaPlante, C. M., and Bronson, D. L., 1992, "The Effects of Freezing and Thawing on the Permeability of Compacted Clay Landfill Covers and Liners," Proceedings, Mediterranean Conference on Environmental Geotechnology, CESME, Turkey, pp. 213-217.

D. Roger Bruner[1] and Alan J. Lutenegger[2]

MEASUREMENT OF SATURATED HYDRAULIC CONDUCTIVITY IN FINE-GRAINED GLACIAL TILLS IN IOWA: COMPARISON OF IN SITU AND LABORATORY METHODS.

REFERENCE: Bruner, D. R., and Lutenegger, A. J., "**Measurement of Saturated Hydraulic Conductivity in Fine-Grained Glacial Tills in Iowa: Comparison of In Situ and Laboratory Methods,**" Hydraulic Conductivity and Waste Contaminant Transport in Soil, ASTM STP 1142, David E. Daniel and Stephen J. Trautwein, Eds., American Society for Testing, Philadelphia, 1994.

ABSTRACT: Nested-standpipe and vibrating-wire piezometers were installed in Pre-Illinoian Wolf Creek and Albernett formations at the Eastern Iowa Till Hydrology Site located in Linn County, Iowa. These surficial deposits are composed of fine-grained glacial diamicton (till) with occasional discontinuous lenses of sand and silt. They overlie the Silurian (dolomite) aquifer which provides private, public, and municipal drinking water supplies in the region. The saturated hydraulic conductivity of the Wolf Creek Formation was investigated in a sub-area of the Eastern Iowa Till Hydrology Site. Calculations of saturated hydraulic conductivity were based on laboratory flexible-wall permeameter tests, bailer tests, and pumping test data. Results show that bulk hydraulic conductivity increases by several orders of magnitude as the tested volume of till increases. Increasing values of saturated hydraulic conductivity at larger spatial scales conceptually support a double-porosity flow model for this till.

KEYWORDS: hydraulic conductivity, fine-grained till, bailer test, pumping test, flexible-wall permeameter test

Many areas of North America glaciated during the Pleistocene have fine-grained deposits (tills) that are considered to act as aquitards and restrict the flow of ground water. Numerous underground storage tanks, landfills, and lagoons are constructed in these tills. Results of recent investigations in Iowa and elsewhere in the mid-continent however show contamination of ground water from routine use of agricultural chemicals and from landfill leachate in areas underlain by fine-grained glacial tills (Kross et al. 1990, Hallberg 1989, Herzog et al. 1989). Past studies have shown that the bulk permeability of most sediments is generally much greater when measurements are integrated over larger

---

[1]Hydrogeologist, Geological Survey Bureau, Iowa City, IA
[2]Associate Prof. of Civil Engineering, U. Mass., Amherst, MA

areas (Bradbury and Muldoon 1990, Keller et. al 1988, 1989, Olson and Daniel 1981). The purpose of this study is to investigate the effect of spatial scales on the measurement of saturated hydraulic conductivity in a Pre-Illinoian till in Iowa. A site in eastern Iowa was identified and instrumented. Results of laboratory permeameter tests were compared to results from field bailer tests and pumping tests. For the purposes of this paper, a subset of results from piezometers that were finished within a narrow range of depths, and had data collected from multiple scales are presented. This study was supported by the Iowa Department of Natural Resources-Aquitard Hydrology Project which has as its primary objective to investigate the hydrogeology of fine-grained tills in Iowa. A secondary objective of the Aquitard Hydrology Project addressed by this study was the assessment of new and innovative methods and technologies that could be used for hydrogeologic and geotechnical site investigations in fine-grained tills. This project was developed in response to the 1987 Iowa Groundwater Protection Act.

SITE DESCRIPTION

Location and Geology

Two sites were chosen to bracket the depositional ages of glacial tills in Iowa. One site is in central Iowa, located on the Wisconsinan-age Des Moines Lobe (14 000 to 12 000 years before present), and the other on older (>500 000 YBP)Pre-Illinoian till in eastern Iowa. Results from the eastern Iowa till site are presented in this paper. This site is located in southern Linn County, Iowa (Figure 1).

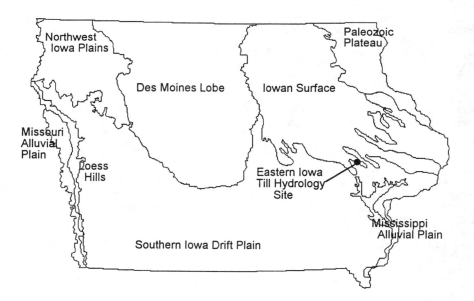

FIG. 1--Location of Eastern Iowa Till Hydrology Site.

on the southern margin of the Iowan surface landform region (Prior 1991). The Pleistocene stratigraphy at the eastern Iowa till site consists of approximately 29 m of Pre-Illinoian till. The top 17 m is Wolf Creek Formation till. The upper 7 m is oxidized, jointed (fractured), and leached (weathered), followed by 5 m of a reduced and unleached transition zone. The lower 5 m of the Wolf Creek Formation is unoxidized and unleached (unweathered). This unit overlies 12 m of unoxidized and unleached Albernett Formation till. The Pre-Illinoian tills at this site have similar grain-size distribution and are matrix-dominated with less than 8% gravel. The grain-size distribution is shown in Figure 2 by volume of clay, fine and coarse silts and sand fractions (100% total). The porosity measured from undisturbed samples collected for this study range from 26.1% to 33.3% which is typical for fine-grained till formations in the Midwest. The two till formations at this site are differentiated by their clay mineralogy (Hallberg 1980). The entire till package overlies about 50 m of undifferentiated Silurian dolomite which is a regional aquifer in eastern Iowa. Below the Silurian dolomite is the Maquokata shale, which acts as a regional confining unit in this part of Iowa. The water table is normally located approximately 1.5 m below ground surface but varies throughout the year in response to precipitation.

Site Layout

Twenty-two standpipe piezometers in six nested locations and 10 vibrating wire piezometers in two nested locations were constructed at this 0.5 ha site. Forty-eight additional 2.5 cm-diameter threaded PVC piezometers with 30.5 cm-long screens were constructed in a grid pattern across a 12 m by 8 m section of the site. Total depths for these piezometers ranged from 4.5 m to 10.7 m. One piezometer was constructed of 5.0 cm by 6.0 m threaded PVC with a 3.1 m screen. Piezometers were nested in a grid pattern about the 5.0 cm well. The mesh size of the grid varies from one to two meters. This sub-area was designated as the "tracer site," and was designed so both pumping tests and solute transport studies could be conducted. Data for this paper were collected from this sub-area.

Monitoring Well Construction and Sample Collection

A 19.3 cm outside diameter hollow stem auger was used to drill the hole for the 5.0 cm monitoring well. An 8.9 cm continuous flight auger was used to drill the holes for the rest of the piezometers at the tracer site. The auger was advanced to just above the interval to be screened, where a 7.6 cm thin-walled Shelby tube sampler was hydraulically pushed to collect an undisturbed sample. The tube was removed and plastic end caps placed over both openings. Duct tape was used to seal the end caps to the tubes. The auger was then advanced to the design total depth to finish the hole. The piezometers were finished from 3 m to 9 m deep and were equipped with screens approximately 30.5 cm long, with 8 mm (30 slot) wide factory-cut slots. Clean sand was used as a sand pack around the screens and pelletized bentonite was placed above the sand pack to form a seal. A combination of drill cuttings and bentonite was used to fill most of the remaining annular space, and powdered bentonite was place last as a surface seal.

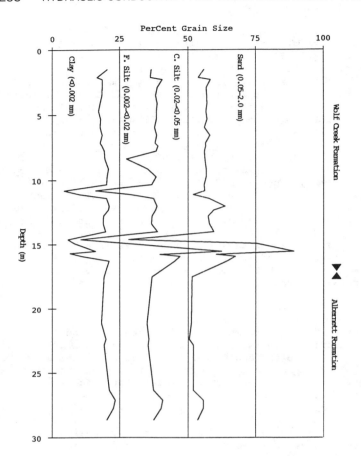

FIG. 2--Per cent particle-size distribution with depth.

TEST PROCEDURES

Laboratory Tests

Hydraulic conductivity tests were performed on the undisturbed Shelby tube samples using a triaxial cell as a flexible-wall permeameter. Tests were conducted following the procedure described by Daniel et al. (1984). Samples were removed from the thin-walled Shelby tubes with a hydraulic piston-jacking system. Samples were hand-trimmed to fit the permeameter and had a nominal diameter of 7 cm with an arbitrary height-to-diameter ratio of 0.7. The samples were placed into the permeameter and an effective stress equal to the in situ overburden stress was incrementally applied to obtain an average final back pressure of 410 kPa applied to both inflow and outflow ports to ensure sample saturation. After saturation, a differential pressure of about 10 kPa was applied between the pedestal and top cap. This induced a gradient on the sample of between 20 and 30. A standard laboratory permeant consisting of a solution of 0.005 N $CaSO_4$ and distilled, deaired water was used for all tests. In most cases, sample flow was allowed to continue for a period of 2 to 3 weeks to insure steady-state flow conditions. Adjacent samples were trimmed from individual Shelby tubes in order to perform flow tests in both the vertical and horizontal directions to evaluate anisotropy. Data were analyzed by the well known constant head equation and corrected to a constant temperature of 20^0 C.

Subsamples of till from each Shelby tube were analyzed for basic geotechnical properties. Grain-size analyses were performed by the hydrometer method using the procedures described in ASTM Standard Test Method for Partical-Size Analysis of Soils (D 422). Atterberg limits were obtained using ASTM Standard Test Method for Liquid Limit, Plastic Limit, and Plasticity Index of Soils (D 4318). Soil pH was determined using a digital pH meter on a 1:1 soil:distilled water (pH=7.0) suspension. Carbonate analyses were conducted using the gasometric Chittick apparatus described by Dreimanis (1962).

Field Tests

Bailer test -- Bailer tests are one form of single well hydraulic tests. Single well tests provide a relatively easy and cost-effective method for local scale testing of in situ hydraulic conductivity in a variety of geologic materials (Kraemer et al. 1990). A 2.3 cm diameter by 30.5 cm long bailer was slowly lowered into each of the 2.5 cm diameter piezometers and a full bailer of water was quickly removed. This lowered the water level approximately 24.4 cm and removed an average of 103 ml of water. Water-level recovery measurements were made for several hours in each piezometer. Data were analyzed by both Hvorslov (Hvorslov 1951) and Bouwer and Rice methods (Bouwer and Rice 1976, Bouwer 1989).

Pumping test -- Pumping tests estimate the bulk hydraulic conductivity by integrating relatively large volumes of aquifers. However, this method has not been extensively used on low permeability materials because of the actual and perceived problems in producing water from low permeability materials. In this study a constant drawdown test was conducted, where the head in the 5 cm well was lowered 3.80 m and held constant for 22 hours. The static water level was initially lowered rapidly with a bailer in order to negate borehole storage effects. A battery operated peristaltic pump was then used to maintain the constant drawdown. A low but steady pumping rate of 11 to 12 ml/min. was maintained throughout the test. The volume of water extracted was accounted for throughout the test. Water levels were measured through time at the pumping well and the nearest 33 PVC piezometers.

Initial analysis of the time-drawdown data used the Theis curve-matching method (Theis 1935, as presented in Freeze and Cherry 1979). The appropriateness of all the assumptions built into this model may be questionable. However, work by Bourdet and Gringarten (in Kruseman

1990) indicates that the double-porosity type drawdown response in piezometers monitoring a pumping well reduces to the Theis equation for early time-drawdown data in fractured aquifers.

RESULTS

Results of the basic geotechnical analyses for the undisturbed samples are presented in Tables 1 and 2. The stratigraphic and geotechnical description of the tills at this site show that they are typical of matrix dominated Pre-Illinoian tills in Eastern Iowa (Hallberg 1980, Hallberg et al. 1980). The hydraulic conductivity calculated from the till samples collected with shelby tubes during construction of the various piezometers vary by less than one order of magnitude, both vertically and horizontally. There was very little difference between the hydraulic conductivity measured in either the vertical or horizontal direction. The geometric mean hydraulic conductivity when the samples were oriented vertically was 1.21E-08 cm/s. The geometric mean hydraulic conductivity when the samples were oriented horizontally was 2.84E-08 cm/s. These values are consistent with values reported to the Iowa Department of Natural Resources (IDNR) as part of landfill site evaluation reports.

TABLE 1--Basic geotechnical data of till samples.

Sample I.D.	Grain-Size (%)			Atterberg Limits (%)		Plasticity Index
	Clay <0.002 mm	Silt 0.002-0.074 mm	Sand >0.074 mm	Liquid Limit	Plastic Limit	
B1-25	28.9	35.3	35.8	30.0	14.6	15.4
B2-25	28.0	35.5	36.5	31.0	13.8	17.2
B3-25	27.3	29.8	42.9	30.8	18.9	11.9
C2-25	26.8	34.1	39.1	30.6	15.1	15.5
C4-25	29.0	32.2	38.8	30.1	14.9	15.2
D1-25	18.1	39.2	42.7	23.8	19.8	4.0
D2-25	23.4	21.6	55.0	27.6	15.3	12.3
D3-25	29.0	35.7	35.3	31.5	15.7	15.8

TABLE 2--Carbonate content and pH of till samples.

Sample I.D.	pH	Carbonate Content (%)		
		Total	Calcite	Dolomite
B1-25	7.3	14.4	4.6	7.8
B2-25	7.1	14.0	4.8	9.2
B3-25	7.2	12.5	5.0	7.5
C2-25	6.5	13.6	3.9	9.7
C4-25	6.9	12.6	4.9	7.7
D1-25	7.1	12.1	3.7	8.4
D2-25	7.1	12.9	4.2	8.7
D3-25	7.1	14.0	4.7	9.3

Bailer tests were conducted for three hours, during which the water levels generally recovered 50% to 90% of the original head. Data from the piezometers produced linear segments when plotted on semi-log graphs (Fig. 3). The first multiple-point linear segment was used to calculate the hydraulic conductivity (generally between 10 and 60 minutes). The geometric-mean hydraulic conductivity from the bailer test results was 3.22E-06 cm/s.

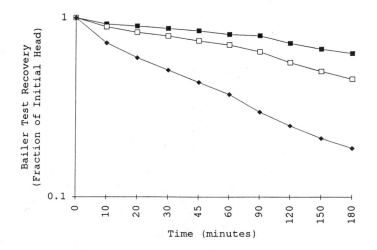

FIG. 3--Typical bailer test time-drawdown data.

Measurable drawdowns from the constant head pumping test occurred within 60 minutes in a few piezometers 1 to 3 meters from the pumped well. Time-drawdown data plotted on log-log scale approximated the characteristic well function curve of Theis, as shown in Figure 4. Although there was good fit of the data to the Theis well function equation, this does not necessarily make it the most appropriate model, as previously discussed.

The three methods used in this study yielded values of hydraulic conductivity that varied by more than three orders of magnitude. Values ranged from 9.3 E-09 cm/s for laboratory tests (K_V) to 4.9 E-05 cm/s from pumping test analysis. The variability within a particular test method was less than one order of magnitude. The geometric means, minimum values, and maximum values of hydraulic conductivity (K) for the eight closest piezometers are summarized in Table 3.

TABLE 3--Descriptive statistics of hydraulic conductivity, by method.

	Flexible-wall Permeameter (K_H)	Bailer Tests	Pumping Test
Geometric Mean K (cm/s)	2.84E-08	3.22E-06	1.72E-05
Minimum K (cm/s)	1.30E-08	1.70E-06	1.98E-06
Maximum K (cm/s)	2.10E-07	1.20E-05	4.92E-05

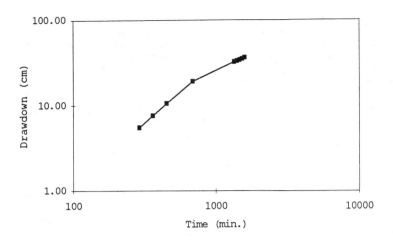

FIG. 4--Typical drawdown response in nearby piezometers.

An increase of two orders-of-magnitude was observed in the geometric mean hydraulic conductivity between laboratory permeameter and field bailer test methods. An additional order-of-magnitude increase was observed between the geometric mean hydraulic conductivity of the bailer and pumping tests. These field results are consistent with data from the companion till hydrology site located in central Iowa and conducted in much younger Late-Wisconsinan weathered till (Jones et al. 1992) which also exhibit fractures. Each of the three methods evaluated in this study test differing volumes of saturated sediment, with the laboratory permeameter testing volumes on the order of 10^{-4} m^3, bailer tests 10^0 m^3, and pumping tests 10^2 m^3 (Bradbury and Muldoon 1990). Results show that hydraulic conductivity increases with increasing volume of sediment tested as shown in Figure 5.

CONCLUSIONS

Typical values of saturated hydraulic conductivity in Pre-Illinoian tills in Iowa have been reported as 10^{-8} cm/s in many site evaluation reports (Iowa Department of Natural Resources) for landfills, lagoon, and underground storage tanks. These values are based on either laboratory tests performed on undisturbed samples from the site or other investigations on sites in other areas. Typically laboratory values reflect vertical permeability, while field tests primarily reflect horizontal permeability. This study indicates that the matrix anisotropy ratio averages 2.5 which is very low. These values adequately reflect the matrix permeability. However, in fractured tills which are common in Iowa and many other areas, fracture flow dominates the advection system. Therefore as different methods test larger volumes of sediment, there is more likelihood of encountering fractures, sand laminae, or other higher permeability features. This

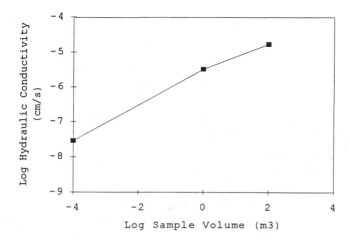

FIG. 5--Change in Hydraulic Conductivity With Sample Volume.

study shows that pumping test methods can be used to assess the bulk permeability of relatively large volumes of fractured tills. For contaminant migration investigations, site remediation design, and other site evaluation studies, matrix values of hydraulic conductivity severely underestimate the bulk hydraulic conductivity of Pre-Illinoian tills. This study indicates that a value three orders of magnitude larger than laboratory values should be used as a starting place for initial design purposes. Calculations of travel time based on the hydraulic conductivity estimated by the three methods used in this study would result in significantly different distances that conservative solutes could advect. Further work is needed to better understand the response of tills in Iowa to pumping. Longer term tests are scheduled which will include varying the pumping well drawdown to observe what effect changing these variables might have on the calculation of hydraulic conductivity. Natural gradient tracer tests are also planned in the near future to determine effective porosity of fractured tills at this site.

ACKNOWLEDGMENTS

The authors express their appreciation to Dr. George Hallberg IDNR for overall project support and CSRS for partial funding. We thank the Iowa Department of Transportation for allowing access to their property for this study. We would also like to thank Dr. Tim Kemmis and E. Arthur Bettis III IDNR for describing the glacial stratigraphy of both till hydrology sites. Lastly, appreciation is expressed to the USGS district office for installation of a majority of the standpipe piezometers at the Eastern Iowa Till Hydrology Site.

REFERENCES

Bouwer, H, 1989, "The Bouwer and Rice Slug Test - An Update," Ground Water, Vol. 27, no. 3, pp. 304-309.

Bouwer, H, and R. C. Rice, 1976, "A Slug Test for Determining Hydraulic Conductivity of Unconfined Aquifers with Completely or Partially Penetrating Wells," Water Resources Research, Vol 12, no.3, pp. 423-428.

Bradbury, K. R., and M. A. Muldoon, 1990, "Hydraulic Conductivity Determinations in Unlithified Glacial and Fluvial Materials," Ground Water and Vadose Zone Monitoring ASTM STP 1053, D. M. Nielsen and A. I. Johnson, Eds., American Society for Testing and Materials, Philadelphia, pp. 138-151.

Daniel, D. E., S. J. Trautwein, S. S. Boynton, D. E. Foreman, 1984, "Permeability Testing with Flexible-Wall Permeameters," Geotechnical Testing Journal, ASTM, Vol. 7, pp. 113-122.

Dreimanis, A, 1962, "Quantitative Gasometric Determination of Calcite and Dolomite using Chittick Apparatus," Journal of Sedimentary Petrology, Vol. 32, pp. 520-529.

Freeze, R. A. and J. A. Cherry, 1979 Ground water, 604 pp., Prentice-Hall, Englewood Cliffs, N.J.

Hallberg, G. R., 1980, "Pleistocene Stratigraphy in East-Central Iowa," Iowa Geological Survey, Technical Information Series no. 10, 168 pp.

Hallberg, G. R., Wollenhaupt, N. C., and Wickham, J. T., 1980, "Pre-Wisconsinan Stratigraphy in Southeast Iowa," in Illinoian and Pre-Illinoian Stratigraphy of Southeast Iowa, Iowa Geological Survey, Technical Information Series no. 11, pp. 1-110.

Hallberg, G. R., 1989, "Pesticide Pollution of Groundwater in the Humid United States," Agriculture, Ecosystems and Environment, Vol. 26, pp. 299-367.

Herzog, B. L., Griffen, R. A., Stohr, C. J., Follmer, L. R.,Morse, W. J., and Su, W. J.,1989, "Investigation of Failure Mechanisms and Migration of Organic Chemicals at Wilsonville, Illinois," Ground Water Monitoring Review, Vol 9, no. 2, pp. 82-89.

Hvorslov, J. M., 1951, "Time Lag and Soil Permeability in Ground-Water Observations," Bulletin No. 36, Waterways Experiment Station, Corps of Engineers, U. S. Army, 50 pp.

Iowa Department of Natural Resouces, Landfill Siting and Expansion Reports, Des Moines, IA. (unpublished).

Jones, L., T. Lemar, C. Tsai, 1992, "Results of Two Pumping Tests in Wisconsin Age Weathered Till in Iowa," Ground Water, Vol. 30, no.4, pp. 529-538.

Keller, C. K., G. Van Der Kamp, J. A. Cherry, 1988, "Hydrogeology of Two Saskatchewan Tills, I. Fractures, Bulk Permeability, and Spatial Variability of Downward Flow," Journal of Hydrology, Vol. 101, no.1, pp. 97-121.

Keller, C. K., G. Van Der Kamp, J. A. Cherry, 1989, "A Multiscale Study of the Permeability of a Thick Clayey Till," Water Resources Research, Vol. 25, no. 11, pp. 2299-2317.

Kraemer, C. A., Hankins, J. B., and Mohrbacher, C. J., 1990, "Selection of Single-Well Hydraulic Test Methods for Monitoring Wells," *Ground Water and Vadose Zone Monitoring, ASTM STP 1053*, D. M. Nielsen and A. I. Johnson, Eds., American Society for Testing and Materials, Philadelphia, pp. 125-137.

Kross, B. C., G. R. Hallberg, D. R. Bruner, R. D. Libra, K. D. Rex, L. M. B. Weih, M. E. Vermace, L. F. Burmeister, N. H. Hall, K. L. Cherryholmes, J. K. Johnson, M. I. Selim, B. K. Nations, L. S. Seigley, D. J. Quade, A. G. Dudler, K. D. Sesker, M. A. Culp, C. F. Lynch, H. F. Nicholson, and J. P. Hughes, 1990, "The Iowa State-Wide Rural Well-Water Survey Water-Quality Data: Initial Analysis," Iowa Department of Natural Resources, Geological Survey Bureau, Technical Information Series no. 19, 142 pp.

Kruseman, G. P., 1990, *Analysis and Evaluation of Pumping Test Data*, 2nd Edition, International Institute for Land Reclamation and Improvement.

Olson, R.E. and Daniel, D.E., 1981, "Measurement of the Hydraulic Conductivity of Fine-Grained Soils," *Permeability and Groundwater Contaminant Transport, ASTM STP 746*, T. F. Zimme and C. O. Riggs, Eds., American Society for Testing and Materials, Philadelphia, pp.18-64.

Prior, J. C., 1991, *Landforms of Iowa*, University of Iowa Press, pp. 153.

Theis, C. V., 1935, "The Relationship Between the Lowering of the Piezometric Surface and the Rate of Discharge of a Well Using Groundwater Storage," *American Geophysical Union, Transactions*, Vol. 2, pp.519-524.

Shi-Chieh Cheng[1], Jesus L. Larralde[1], and Joseph P. Martin[2]

HYDRAULIC CONDUCTIVITY OF COMPACTED CLAYEY SOILS UNDER DISTORTION OR ELONGATION CONDITIONS

REFERENCE: Cheng, S. C., Larralde, J. L., and Martin, J. P., "Hydraulic Conductivity of Compacted Clayey Soils Under Distortion or Elongation Conditions," Hydraulic Conductivity and Waste Contaminant Transport in Soil, ASTM STP 1142, David E. Daniel and Stephen J. Trautwein, Eds., American Society for Testing and Materials, Philadelphia, 1994.

Abstract: Models of fine-grained landfill cap response to differential subsidence are discussed, and experimental results are presented. Soils used include a clayey silt, a silty fine sand, and the latter with admixed kaolinite. Tests were conducted on statically compacted soils in a permeameter that induced axisymmetric distortion. Hydraulic conductivity increased with the increase in flexural stresses. Sensitivity to plasticity and compaction moisture content was observed. Tests were also conducted to obtain mechanical properties, including tensile strength (split-cylinder), unconfined compressive strength, modulus of rupture (beam flexural strength), and several deformation moduli. The results are normalized in terms of compaction moisture relative to optimum moisture content.

Keywords: Hydraulic Conductivity, Landfill Cap, Subsidence, Distortion, Flexure, Tension, Plasticity, Compaction

INTRODUCTION

The hydraulic conductivity of fine-grained soils is customarily measured in an undistorted condition, either undisturbed or as-compacted, or under increasing normal

---

1 Asst. Professor, 2 Assoc. Professor, Dept.of Civil & Arch. Engineering., Drexel Univ., Phila. PA 19104

stresses. However, there are situations where hydraulic barriers are subject to tensile, shear, or flexural stresses. This is of special concern at closed, unlined landfills where the cap or final cover is the sole restriction on surface infiltration and leachate impact on groundwater. Not only does waste decomposition result in global subsidence (Murphy and Gilbert 1984), but differential settlement also distorts the surface alignment (Lutton et al. 1979). This retards runoff and induces tensile elongation or flexural distortion as illustrated in Figures 1 and 2.

Compacted fine-grained soils generally have low resistance to tensile or flexural stresses (Leonards and Narin 1963; Fry et al. 1992). Consequently, it is expected that the hydraulic conductivity of unreinforced landfill caps is increased by elongation or distortion, first by loosening, and ultimately, by cracking (Murphy and Gilbert 1987; Cressman et al. 1992).

In acting as a landfill roof, cap performance depends upon initial hydraulic conductivity and also on durability, structural support, and continuity or integrity. All of these factors are at risk in older landfills. Composition, compaction and cross-section standards have been improved, implying that caps built earlier were less effective. Often, the nearest available fine-grained material was used and compacted as well as possible on a yielding waste subgrade (Lutton et al. 1979). The discontinuities between "peds" in fine-grained soils define a macrostructure that dominates hydraulic conductivity, rather than the properties of intact peds themselves (Elsbury et al. 1990). The high-stress compaction at high moisture contents required to remold the macrostructure was not always specified or even possible. Furthermore, unprotected fine-grained barriers can deteriorate by desiccation (Kleppe 1985) and freeze-thaw stresses (Zimmie 1990), concerns which are now addressed by multilayer caps (U.S. EPA 1989,1991).

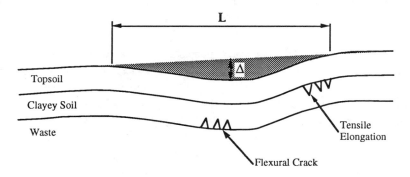

Figure 1--Flexural Distress in Landfill Cap

This paper describes the results of a laboratory investigation of distortion effects on fine-grained soil permeability. It includes the results of experiments on the sensitivity of hydraulic conductivity to axisymmetric flexure. It also includes a study of the mechanical properties that have been identified by researchers (Leonards and Narin 1963; Murphy and Gilbert 1987; Ajaz and Parry 1975; Walton 1987) as influencing the response to differential subsidence: compressive, tensile and flexural strengths, stiffness, and ductility (strain at cracking). The focus is on low plasticity soils, and on the effects of compaction moisture content on both hydraulic and mechanical properties (Lambe and Whitman 1969; Das 1990).

CONCEPTUAL RESPONSE MODELS

Deep within aged, highly stressed zones of a landfill, cells compress to different degrees. The effects of anomalies propagate, ultimately reflected at the surface as local differential subsidence. While many models of global settlement of landfills have been derived, the origin or magnitude of local subsidence cannot be predicted in detail. The problem arises randomly (Bredariol et al. 1990) and can only be practically studied in its effects at the surface.

Cap depressions occur in various shapes, but often, they can be approximated as circular or trough-like in shape. Surface distortion is often indexed by the deflection from the predominate surface (Δ) and the depression width (L) as shown on Figure 1, yielding a distortion index Δ/L.

The effect of a surface depression may only be to reduce runoff; so grading to restore the alignment might suffice. However, if the underlying hydraulic barrier is cracked, (Figure 1), an impermeable overlay is required. It must be assumed that subgrade instability will continue, perhaps worsened by the weight of the overlay. To assure survivability, tensile reinforcing is probably needed (Giroud 1990; Gaind and Char 1983).

The barrier material and the installation specifications should also be carefully considered. Plasticity provides finite tensile strength and increases ductility (Leonards and Narin 1963; Walton 1987; Augenbaugh 1989), but high plasticity presents risks of damage due to climatic factors. Compaction of a fine-grained soil wet of optimum moisture content (OMC) provides lower hydraulic conductivity, but also produces low stiffness (Lambe and Whitman 1969; Das 1990). Whether low or high stiffness is desired depends upon how the cap is seen to respond to local subsidence.

There are two general ways to describe cap response: as compliant or flexible, like a foil or membrane, carrying either tension or compression only in its plane, or as a beam with some degree of flexural rigidity. In the former view, the cap elongates (or shortens) to follow the subsidence and warping of the waste surface. The observed Δ/L can be converted to a tensile strain in the plane of the cap as illustrated in Figure 2.

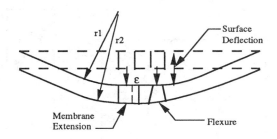

Figure 2--Possible Responses: Flexure or Elongation

In this model, low stiffness and high ductility would be desired to allow the cap to deform without rupturing. In modern caps, geomembranes of low modulus materials such as PVC or VLDPE also tend to be more suitable in this regard than stiffer HDPE. However, with a soil cap, the hydraulic conductivity must progressively increase due to loosening and particle or ped separation during elongation. Ultimately, of course, the cap can fail by tensile cracking through the barrier.

One inexpensive method to index the ability of the cap to resist rupture would appear to be to measure the tensile strength (σ_{tb}) with the split-cylinder (Brazilian) test. However, the tensile strength itself is only of secondary importance. The true measure of the limit of mechanical integrity is the tensile strain at failure, which can be directly compared to the tensile strain computed from Δ/L.

To estimate that maximum tolerable tensile strain, the simplest method is to assume that the deformation modulus in tension (E_t) is similar to the modulus in compression (E_c). The unconfined compressive strength, q_u, is typically much higher than the tensile strength. The stress-strain curve obtained in unconfined compression could thus be projected into the tension range. Using data obtained with soil used in this study, Figure 3 shows the extended secant modulus terminating at a stress equal to the separately evaluated σ_{tb}. Extending the initial tangent modulus would probably be too conservative. The effect of tensile strain on the hydraulic conductivity could be estimated by its measurement at different porosities, although this might result in large errors. It would appear that evaluating the sensitivity of hydraulic conductivity to pure tension would be best done in extension triaxial tests.

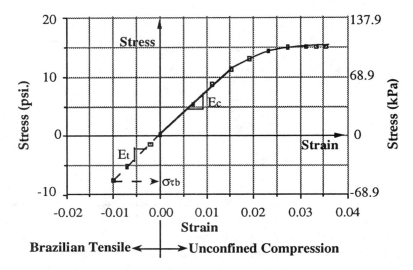

Figure 3 Approximating Maximum Tensile Strain

A thick cap under distortion could also be studied as if it were in flexure. Several models treating a cap as a beam have been derived (Murphy and Gilbert 1987), including cantilever models and adaptations of the classical beam on elastic foundation analysis, which appears to more accurately represent the condition of deep-seated differential compression. This implies a cap-subgrade interaction, such that a higher flexural stiffness (E_f) would modify the observed vertical deflection (Δ) from that attempted by unimpeded waste subsidence alone. This also tends to extend the width of the impact surface (L), causing a net reduction in Δ/L and the beam curvature as shown in Fig. 2. This also implies lower tensile stresses in the "outer fiber" and also mitigates formation of deep rainfall impoundments on the surface.

Under this model, cracking is initiated in the tensile zone of the "beam" when the flexural strength is exceeded. With the cap regarded as a continuous beam, tension also occurs on the top of the hydraulic barrier at the points of inflection at the edge of the depression, as shown in Fig. 1. The beam will ultimately fail along these cracked surfaces when the remaining shear interlock forces are exceeded. However, a complete rupture, causing total loss of infiltration resistance, does not necessarily occur. An increase in Δ/L and the radius of curvature will increase the normal stresses on the compression side, which may locally decrease hydraulic conductivity.

To perform the analysis with this model, up to the point of initial cracking, several parameters are required, including elastic moduli and compressive and tensile strengths. The material property values can be generated in separate tests.

An alternate method is to conduct a complete flexural or Modulus of Rupture (MOR) test, obtaining the flexural strength (MOR), and flexural modulus (E_f), similar to the test used for concrete and soil-cement (Walton 1987). This test is a reasonable approximation of the field situation of a cap responding to subsidence in flexure. Parametric description of flexural stiffness can be further discretized into a secant modulus (E_{fs}), and a ductility modulus (E_{fu}), measured up to the point of rupture.

SOILS USED

Plastic soils are less likely to crack under subsidence than low plasticity or nonplastic finer materials. The concentration in this study is on the latter, presumably more brittle soils. It is assumed that if they were shown to be of low hydraulic conductivity, such materials might have been considered acceptable for cap hydraulic barriers in the past, and they would be in landfills that are now closed.

The first soil, called "Tacony" herein, is an alluvial, nonplastic silty fine sand derived from shale weathering. It has a gritty texture, but a finite tensile strength and a fairly low hydraulic conductivity (10^{-6} cm/sec) can be obtained. The second soil, ("Coatesville") is a residual product of dolomite weathering and has a clayey texture.

The index properties are as follows:

Soil	Fines (-0.075 mm)	Clay Size (-0.002 mm)	Liquid Limit	Plastic Index	Max. Dry Unit Wt.	Optimum Moisture
Tacony	40%	3%	21	2	118 pcf (18.9 kN/m^3)	14%
Coatsville	77%	30%	32	9	108 pcf (17.3 kN/m^3)	18%

The Coatesville soil classifies on the ML-CL borderline, while the Tacony is ML. To vary plasticity without radically changing the gradation or fabric, the Tacony soil was also admixed with specific percentages of kaolinite. An almost linear relationship between the plasticity index and kaolinite content was found, with the PI increasing to 10 with 25% kaolinite. Increasing clay content also increased tensile strength as measured with the split cylinder test. Samples compacted at optimum moisture content (OMC) and about 92% of standard proctor unit weight ($\gamma_{dmax.}$). The tensile strength (σ_{tb}) varied from 0.64 psi (4.45 kPa) for the basic Tacony soil itself up to 1.08 psi (7.15 kPa) with 15% Kaolinite. The unconfined compressive strength for the similar compaction conditions was not as sensitive to plasticity, following a peaked proctor-like curve ranging from 8.5 psi (58.61 kPa) to 9.5 psi (65.51 kPa), but the secant modulus steadily decreased with plasticity. This confirms the assertion that slight changes in plasticity could have major effect on the distortion response (Augenbaugh 1987).

Figure 4 shows the effect of compaction moisture content on both unit weight and hydraulic conductivity (compacted with 100% proctor effort) on the Coatesville soil. The hydraulic conductivities varied only within an order of magnitude. Samples were compacted statically with a universal testing machine. More sensitivity to compaction moisture content might be possible with dynamic (proctor hammer) compaction.

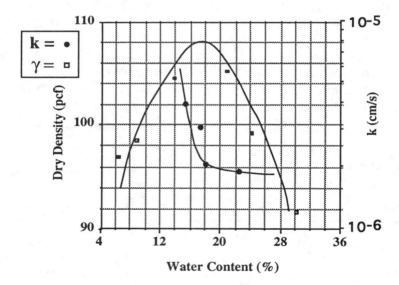

Figure 4. Standard Proctor and Hydraulic Conductivity vs. Water Content (Coatesville)

Samples of the Tacony soil, prepared by the same methods, did not display high sensitivity of hydraulic conductivity to compaction moisture content. Values in the range of 10^{-6} cm/sec were found at and above the OMC. This low hydraulic conductivity of a nonplastic soil is probably due to a very dense compacted structure. Adding 20% kaolinite also increased the OMC to 15%. Samples compacted at 1% and 5% above OMC had hydraulic conductivities of 6×10^{-7} cm/sec and 4×10^{-7} cm/sec, respectively.

It can be seen that either of these two soils would have been judged suitable for use as a cap on the basis of permeability, despite low plasticity.

DISTORTION EFFECTS ON HYDRAULIC CONDUCTIVITY

An apparatus as shown in Figure 5 was used to determine the effects of distortion on hydraulic conductivity, and also to show the relative susceptibility of soils of different plasticity, compaction moisture content, etc. Samples 1" (25.4 mm) thick were statically compacted to γ_{dmax} in 6" (152.4 mm) diameter plexiglas permeameters, and placed on porous stones with centerline holes drilled to diameters from 1.0" to 3.5" (25.4 to 88.7mm). This provided an unsupported section loaded by self-weight (gravity) of the soil itself plus the seepage force imparted by a gradient averaging 20:1.

Results are indexed by the ratio of the hole diameter (d) to the sample thickness (T), yielding an axisymmetric distortion index (d/T). To assure that the boundary (edge) conditions did not interfere with sagging over the opening, the ratio of the sample thickness to the diameter was checked with a finite-element program developed for concrete modeling. The intent was to assure that the the sides would not pull away, i.e., be in tension at the maximum d/T ratio.

The mechanical conditions are not fully understood at this time. It could be postulated that, as d/T increased, the mechanical condition varied from that of a deep beam "arching" over a void (Handy 1985), to a thin beam in flexure. As described below, the sample did not become a plate or membrane in pure tension at the center of the hole even at very high d/T ratios. The deformation response as measured by linear variable displacement transducers (LVDT's) has not been fully operational yet, so conversion to a Δ/L index is not yet possible. The field analogy is also complicated by the apparently rigid boundary condition at the edge of the holes. It was necessary to use a geotextile filter over the hole, which, despite a lack of anchorage, light weight and stiffness, would provide some subgrade support after initial cracking.

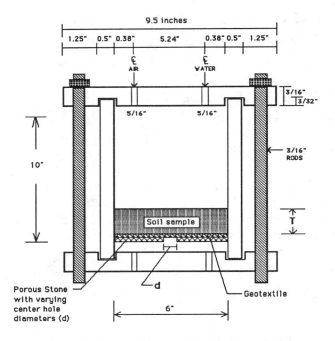

Figure 5. Distorted Permeability Test Apparatus

Figure 6 illustrates the results with the Tacony soil and the more plastic Tacony with 20% added kaolinite (Tacony +20), each tested at two moisture contents. The gross hydraulic conductivity of the entire 6" (152.4 mm) sample is shown. If it were assumed that the soil over the opening reacted uniformly to the loss of support, then parallel flow equations applied to numerically isolate this section would show more dramatic results.

Each soil maintained its undistorted hydraulic conductivity at low d/T ratios. The nonplastic Tacony soil then responded sharply at a distinct d/T value, with the hydraulic conductivity increasing by almost an order of magnitude. This brittle effect is presumably the result of microfissuring. The more plastic Tacony +20 showed earlier but more gradual response to the loss of support, and less increase in hydraulic conductivity. This gradually varying behavior could be a result of being a softer material (lower E_c), so loosening commences at low stresses. The lower ultimate change in hydraulic conductivity could be the result of densification of the plastic material on the compression side, and also to smearing along cracks as internal shear strains increased.

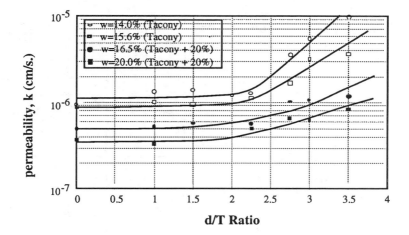

Figure 6. Distorted Permeability Results, Tacony Soils

During the tests, a depression with gently inflected sides was visible on the top at high distortion ratios, much like the conceptual representation of Figure 1. Upon demounting the samples, it was observed that the soil over the holes had thickened about 35%, i.e. soil creeped into the distressed area, but thinned about 10% at the hole boundary. This showed that tension did not extend across the whole section thickness at the center. The thickening in itself would mitigate the overall tendency toward increased hydraulic conductivity. This sagging also indicates that the geotextile filter did not have as much mechanical effect in providing an artificial support as initially feared. The boundaries of the distorted area around the holes are not perfectly fixed.

Figure 7 shows the results of the same type of tests with the Coatesville soil. There was little response up to a d/T ratio of 2 with the Tacony soils. Therefore, except for an undistorted reference sample on a complete porous stone base, the distortion test series tests started at d/T =2.0. The Coatesville samples bracketed the OMC (18%). The two samples compacted below OMC showed the a high resistance to low distortions, but ultimately displayed a brittle response. The sample compacted above the OMC (ω=22.5%) showed more ductile response. The high experimental scatter in the sample compacted at OMC possibly indicates a transitional (nonplastic-plastic) soil fabric effect.

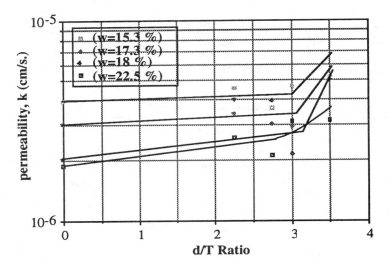

Figure 7. Distorted Permeability Results, Coatesville Soils

MECHANICAL PROPERTIES

The Coatesville soil was more plastic in texture than the index properties imply. Two effects of the compaction moisture content on the stability of hydraulic conductivity were observed: Compaction dry of OMC delayed the onset of distortion effects, but wetter compaction produced less total deterioration. Where strength is clearly important, it was noted that the peak values of both q_u and σ_{tb} for the Tacony soil were reached below the OMC (Cressman et al. 1992)). Consequently, the "best" compaction moisture content is an open question.

Figures 8 and 9 show that the Coatesville soil is much stronger than the silty Tacony, with maximum values of q_u of over 30 psi (207kPa), and a maximum σ_{tb} of about 7.5 psi (51.7 kPa).

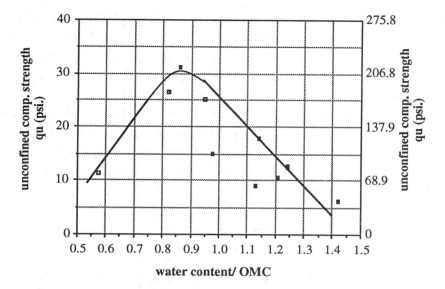

Figure 8. Unconfined Compressive Strength vs. Normalized Water Content, Coatesville

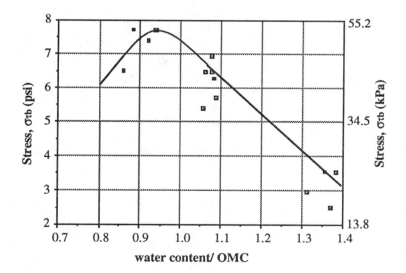

Figure 9. Split-Cylinder Tensile Strength vs. Normalized Water Content, Coatesville

Despite applying 100% proctor effort (on an applied energy per unit volume basis) with a Harvard miniature hammer to samples in 2" (50.8 mm) diameter teflon molds, the unit weights of the Tacony samples averaged about 92% of γ_{dmax}, while the same procedure with the Coatesville soil produced samples about 96% of γ_{dmax}. Results are normalized with respect to the OMC.

A high ratio of tensile to compressive strengths was also observed with a CL-ML residual soil of different geologic origin in the same area (Bredariol et al. 1992). It can be seen that q_u peakes well below OMC, while σ_{tb} peaks just below it.

The secant modulus (E_s) of the Coatesville material decreased almost linearly with compaction moisture content, as shown on Figure 10.

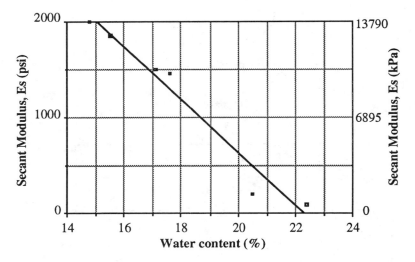

Figure 10. Stiffness in Unconfined Compression vs. Moisture Content, Coatesville

To more directly model soil response to distortion, the soil-cement beam or Modulus of Rupture (MOR) test was used to investigate flexural response as was done by other researchers (Walton 1987; Fry 1992). Figure 11 shows the apparatus. Samples were prepared by static densification to 97%-100% γ_{dmax} with a universal test machine, as dynamic compaction did not yield stable samples.

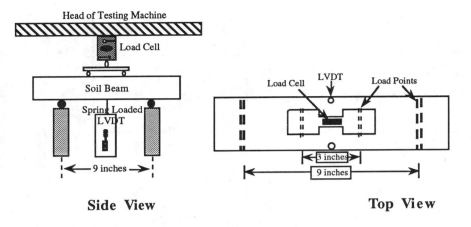

Figure 11. MOR Test Apparatus.

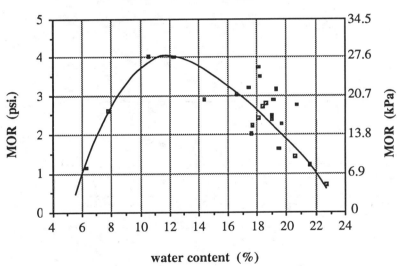

Figure 12. Modulus of Rupture (Flexural Strength) vs. Water Content, Coatesville

Figure 12 shows flexural strength or MOR results. The MOR values are less than the values of tensile strength obtained with dynamically compaction, but the values of the flexural secant modulus E_{fs} derived from this test were three times the E_c values.

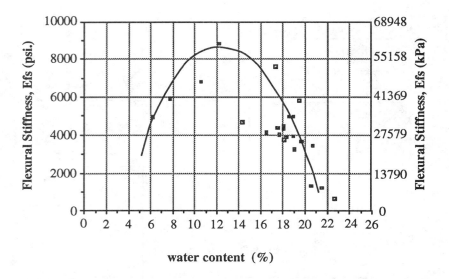

Figure 13. Flexural Stiffness (Secant Modulus) vs. Water Content, Coatesville

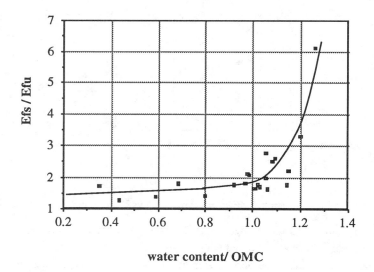

Figure 14. Flexural Ductility vs. Normalized Water Content, Coatesville

Compaction at high moisture contents provided higher ductility. An index E_{fu} called the ductility modulus can defined as the ratio of the MOR to the strain at failure Figure 14 is a plot of E_{fs}/E_{fu} versus normalized compaction moisture content and indicates that samples compacted wet of OMC displayed considerable nonlinearity and implicitly, reserve integrity, after initial "yielding" in flexure. This is compatible with the limited increase in hydraulic conductivity with high distortion for the wet-compacted samples shown on Figure 8.

CONCLUSIONS

The problem of assessing the impact of closed disposal sites on groundwater resources includes evaluating the integrity of the unreinforced caps placed on unstable waste subgrades. Consequently, there is concern about deterioration of the hydraulic barrier due to distortion and accompanying mechanical distress.

The focus of this work was on low plasticity fine-grained materials, and on the sensitivity of hydraulic and mechanical behavior to compaction moisture content. It is axiomatic that compaction wet of optimum moisture content will greatly decrease hydraulic conductivity. This a desirable result, but effects on mechanical properties may be adverse. Compaction dry of optimum tends to produce higher stiffness and both compressive and tensile strengths. A question then arises of focusing on initial low permeability or on survivability, at least in terms of mechanical distress caused by uneven subsidence. Various models tend to emphasize different properties in the latter regard, including stiffness, ductility, and tensile or flexural strength.

An apparatus was developed to investigate the effects of axisymmetric flexural distortion under gravity loads. It was seen that all sample variations showed resistance to low levels of distortional stress by displaying stable hydraulic conductivity up to the stress levels induced in a 1"(25.4 mm) thick sample bridging a 2" (50.8 mm) diameter opening. Thereafter, samples of higher plasticity or wet compaction tended to loosen and gain hydraulic conductivity gradually with increased stress. Samples of lower plasticity or dry compaction tended to sustain more flexural stress, then displayed more brittle internal cracking. Only minimal difference in the response was observed between the nonplastic silty sand and the clayey silt, although the latter had a tensile strength about an order of magnitude higher.

In investigating the mechanical properties, unconfined compressive, split cylinder tensile, and beam strength (MOR) tests were conducted. All three strength parameters showed high sensitivity to compaction moisture content, with peak values below the optimum moisture content. The stiffness, as indicated by either the secant modulus in unconfined compression or the flexure test, steadily decreased with increasing moisture content, and ductility increased rapidly with the implied remolding.

It was demonstrated that flexure can cause deterioration in hydraulic conductivity, but not complete loss of resistance to permeation. Survivability in terms of maintaining initial effectiveness as a hydraulic barrier is influenced by the factors that influence relevant mechanical parameters, but not overly sensitive to them. A condition not directly addressed in this paper was the effect of superimposed loads, where strength and stiffness would appear to be more important. Therefore, it would appear that compaction at about optimum moisture content, which is often assumed unless other construction condition information is available, represents an optimization of the concerns for initial low hydraulic conductivity and ability to withstand both loss of support by subsidence and surface loads.

ACKNOWLEDGEMENTS

The study performed in the paper is under the support of Grant N0. MSS-9245546 of the National Science Foundation. The authors express their appreciation of this support. The opinions expressed in this paper do not represent the opinions of the NSF and the authors take full responsibility for the content of this paper. The authors also like to acknowledge the laboratory work done by H. M. Arkan, and the graphical work done by G. T. Corcoran.

REFERENCES

Ajaz, A., and Parry, R.H.G., 1975, "Stress-Strain Behavior of Two Compacted Clays in Tension and Compression" Geotechnique. V.25, pp 495-512 and 586-591

Augenbaugh, N.B., 1989, "Flexural Integrity of Compacted Landfill Covers" Fifth Int'l Conf.Solid Waste Management, Phila, PA

Bredariol, A.S., Larralde, J.L., Martin,J.P. and Fiori C., 1992, "Predicting Effects of Subsidence on Cracking of Landfill Caps" Proc. ASCE Specialty. Conf.on Probabilistic Mechanics, Denver CO, pp. 360-364

Cressman,G.C., Cheng, S.C., Martin J.P., and Bredariol, A.S., 1992 "Effects of Subsidence Distortion on Integrity of Landfill Caps" Mediterranean Conf. on Environmental Geotechnology, Cesme Turkey, pp. 229-235

Das, B.M., 1990, Principles of Geotechnical Engineering, 2nd Edition, PWS-Kent

Elsbury, B.R., Daniel, D.E., Sraders, G.A., and Anderson, D.C., 1990, "Lessons Learned From Compacted Clay Liners" J. Geotech Engineering, ASCE, V 116, GT11, pp 1641-1677

Fry, P.A., Martin J.P., Bredariol, A.S., and Arkan, H.M., 1992 ,"Experimental Studies & Numerical Analysis of Landfill Cap Distortion" Proc. Air & Waste Management Assn., 38th Reg.Conf. Atlantic City NJ

Gaind K.J. and Char, A.N.R., 1983, "Reinforced Soil Beams" J. Geotech Engineering, ASCE, V 109, GT7, pp. 977-982

Giroud,J.P., Bonaparte, R., Beech, J.F., and Gross , B.A., 1990, "Design of Soil Layer-Geosynthetic Systems Overlying Voids" Geotextiles and Geomembranes, Elsevier Science pp. 11-50

Handy, R.L., 1985, "The Arch in Soil Arching" Jour. of Geotechnical Engineering, ASCE, V 111. pp. 302-318

Kleppe, J.H., and Olson, R.E., 1985, "Desiccation Cracking of Soil Barriers" Hydraulic Barriers in Soil and Rock, ASTM STP 874, pp. 263-275

Lambe, T.W., and Whitman, R.F. 1969, Soil Mechanics John Wiley, New York

Leonards, G.A., and Narin, J.,1963, "Flexibility of Clay and Cracking of Dams," Jour.. Soil Mechanics and Foundations Div., ASCE, Vol. 89, No. SM2, pp. 47-89

Lutton, R.J., Regan ,G.L., and Jones L.W., 1979, "Design and Construction of Covers for Solid Waste Landfills" U.S Envir. Protection Agency, EPA-600/2-79-165

Murphy, W.L., and Gilbert, P.A,1984, "Estimation of Maximum Cover Subsidence Expected in Hazardous Waste Landfills" U.S. EPA 600/9- 84-007

Murphy,W.L., and Gilbert, P.A,1987, "Prediction/ Mitigation of Subsidence Damage to Hazardous Waste Landfill Covers" U.S. EPA 600/2-87-025

U.S. EPA (1989) "Design and Construction of RCRA/CERCLA Final Covers" Seminar Publication EPA 625-4-91-025

U.S. EPA (1991) "Final Covers on Hazardous Waste Landfills and Surface Impoundments" Technical Guidance Document EPA 530-SW-047

Walton, K.H., 1987, "Laboratory Test for the Flexibility of Compacted Clays" M.S. Thesis, Univ. of Mississippi

Zimmie,T., and LaPlante,C., 1990, "Effects of Freeze-Thaw Cycles on Permeability of Fine-Grained Soil" 22nd Mid-Atlantic Indus. Waste Conf. Phila, pp. 580-594

Steven R. Day[1]

THE COMPATIBILITY OF SLURRY CUTOFF WALL MATERIALS WITH CONTAMINATED GROUNDWATER

REFERENCE: Day, S. R., "The Compatibility of Slurry Cutoff Wall Materials with Contaminated Groundwater," Hydraulic Conductivity and Waste Contaminant Transport in Soils, ASTM STP 1142, David E. Daniel and Stephen J. Trautwein, Eds., American Society for Testing and Materials, Philadelphia, 1994.

ABSTRACT: Slurry cutoff walls are frequently relied upon to block groundwater flows from toxic waste sites and landfills. The long-term effectiveness of slurry cutoff wall materials is critical to the successful containment of these facilities and the protection of groundwater resources. A variety of laboratory indicator tests have been attempted by engineers and academia to make compatibility determinations but at present there has been little published experience to show which tests produce meaningful results and how these tests can be used to demonstrate compatibility.

Hydraulic conductivity is a useful measure of chemical/soil compatibility but permeability tests alone cannot assure the long-term stability of a slurry cutoff wall. A suite of indicator tests are used where the leachate and the proposed materials are combined and tested in immersion, desiccation, sedimentation, and other modes. Each indicator test attempts to model a different scenario of the slurry cutoff wall installation and operation.

This paper presents the experience of a specialty contractor from a number of projects, where an incompatibility was discovered and alternate materials were used to find a successful solution. Monitoring results from these sites has proven the effectiveness of the chosen solution. The laboratory test methods described are relatively simple and rely on worst-case scenarios, performed in a step-by-step process, that culminates with flexible wall permeability tests. Based on the methods described and the results from successful projects where these methods were used, engineers, owners and the public may better rely on long-term slurry cutoff wall performance with an increased level of confidence.

KEYWORDS: attapulgite, bentonite, compatibility, containment, deep soil mixing, hydraulic conductivity, slurry cutoff wall

---
[1]District Manager, Pittsburgh Office, Geo-Con, Inc., 4075 Monroeville Boulevard, Pittsburgh, PA 15146

INTRODUCTION

Slurry cutoff walls are permanent subsurface structures used to direct and control groundwater flow. Since the inception of this technique in the 1940's, slurry cutoff walls have been used where relatively unpolluted groundwater was diverted for civil works such as dams, dikes and dewatering structural excavations (Ressi di Cervia 1992). With the beginnings of CERCLA legislation and the environmental movement in the 1970's, more and more slurry cutoff walls are built to contain contaminated groundwater at landfills, hazardous waste and industrial facilities (Ryan 1987). The hydraulic conductivity or permeability of slurry cutoff walls is usually the performance criterion relied upon in the design, construction and contracting of these structures. For projects with an environmental function, the lowest practical hydraulic conductivity is typically specified for the maximum protection of the public and groundwater resources.

Hydraulic conductivity (permeability) testing has significantly improved over the last decade but is of limited use in determining incompatibility. The time and expense required for hydraulic conductivity tests limit the user in formulating compatible mixtures and complicates feasibility estimates. Furthermore, the flexible wall permeability test, the industry standard, requires the imposition of a confining stress, which can mask certain incompatibilities (Evans 1993).

In this paper, compatibility is defined as when two materials, i.e., contaminated groundwater (or leachate) and soil-bentonite, can be mixed together or coexist without reacting chemically or interfering with the performance of the soil-bentonite. An incompatible result is an increase in permeability in the soil-bentonite or chemical reaction which produces a degradation in the physical properties of the soil-bentonite.

Predetermining the compatibility of slurry wall materials with contaminated groundwater is generally recognized as good engineering practice (Ryan 1987; D'Appolonia 1980; Grube 1992; Millet and Perez 1981; Tallard 1984). Some methods, other than hydraulic conductivity testing, have been proposed to determine compatibility (McCandless and Bodocsi 1988; Khera and Thilliyar 1990; Wu and Khera 1990) but these have had limited experience and the results of some test are poorly understood. This paper presents a suite of relatively simple and quick indicator-type tests which can be used in concert with hydraulic conductivity tests to more quickly and better determine the most applicable materials for the containment of contaminated groundwater with slurry cutoff walls.

PURPOSE OF COMPATIBILITY TESTING

Compatibility tests should simulate the long-term, worst-case performance of slurry walls in a contaminated groundwater environment. As yet, no standards exist which can guide the user to determine compatibility.

The primary reason for performing compatibility tests is to ensure that the slurry cutoff wall performs as intended. Compatibility testing also makes the planning and construction effort more efficient and results in a higher quality installation. The most important reasons for completing compatibility tests are as follows:

1. ensure permanence of the materials,
2. estimate long-term performance,
3. estimate material and additive types and amounts,
4. ensure success of construction,
5. accelerate feasibility studies, and
6. address regulatory concerns.

In general, incompatibilities result from chemical reactions. It may be assumed that superior knowledge of the chemicals involved will preclude compatibility testing but practical experience has shown the current state of knowledge to be limited (Ryan 1987). In some cases (e.g. landfills) the types and concentrations of chemicals varies widely. On other sites with more definable chemistry, the subspecies which result from mixing with groundwater cause similar uncertainty. Therefore, while a thorough understanding of soil/waste chemistry is important, studies to detect incompatibilities must rely on experimental methods.

It is, therefore, the purpose of this paper to explain and illustrate, by example, tests which can be used to determine the gross compatibility or incompatibility of slurry cutoff wall materials when used in contaminated groundwaters.

POTENTIAL FAILURE MECHANISMS

Slurry cutoff walls are susceptible to failure during construction and operation as a result of groundwater contamination. Because of the specialized nature of the construction process, the materials selected for the installation must meet workability restraints. In practice, this means that the materials must be suitable for the specialty contractors' requirements as well as the designers objectives for the installation to be effective.

The first and most important ingredient in slurry cutoff wall construction is the bentonite slurry. Ineffective slurry results in excessive material usage, the necessity for additives and/or the loss of slurry workability. Fresh water for mixing and premium grade bentonite are the primary slurry ingredients. Poor quality water (e.g. hard or polluted water) and/or poor quality bentonite can usually be identified by testing trial mixtures.

Excavating through refuse or concentrated wastes can have a detrimental effect on slurry performance. Unusual or excess material usage can result. Flocculation of bentonite in a slurry trench will often result in a trench collapse and/or massive settlement of solids on the bottom of the trench which limits backfilling. Contaminated groundwater has been a cause of bentonite flocculation and, therefore, tests to predetermine the potential for construction failures, material usage estimates and the need for additives is critically important.

Contaminants may react with the key ingredient, bentonite clay, more slowly, in a manner where the effect may be more gradual and not readily apparent during construction. The impermeability of slurry walls relies to a considerable degree on the swelling properties of bentonite. Contaminants which reduce or restrict bentonite swelling may increase permeability but also can damage the self-healing properties of bentonite.

Finally, contaminants can effect not only construction practice and bentonite behavior, but also the properties of the backfill. The slurry cutoff wall backfill may lose plasticity, shrink, experience weight changes, dissolve, or petrify in response to leachates all of which can affect the slurry cutoff walls' performance. Mixtures which use cementacious ingredients (i.e. cement and fly ash) require additional considerations. The more complex the blend of materials in the slurry wall (e.g. plastic concrete > cement-bentonite > soil-bentonite) the more critical the need for examining properties of the backfill other than hydraulic conductivity as they relate to compatibility.

The system used to enact and direct the testing program is critical to successful implementation as well as the timely completion of the project. By testing the materials systematically, under worst-case scenarios, the program quickly becomes focused on workable solutions. Relatively large numbers of simple and rapid tests can be performed to eliminate borderline materials.

INDICATOR TESTS FOR COMPATIBILITY

Various indicator tests have been proposed to investigate the effect of contaminants on slurry cutoff wall materials; but to date, there is limited understanding of their applicability and even less experience to document the success of one method over another. The basis for these tests was previously developed by the petroleum, well drilling, and geotechnical disciplines. These are relatively simple tests which rely on observations and comparative results. In general, comparisons are made between performance or observations with tap water as a control (or 0.005 N $CaSO_4$) compared to a leachate. These tests are by intent worst-case models of assumed field conditions; therefore, the user must be knowledgeable to interpret and apply the results. The tests described below are those most often used by the author to evaluate compatibility.

Construction

Construction compatibility can be modeled by comparing the performance of a standard bentonite slurry in dilution with water and leachate using conventional bentonite slurry test procedures (API, RP13B-1 1990). Generally, a slurry with B/W = 5% (Bentonite/Water ratio by weight) is used and diluted 1:1 with tap water and with leachate. Depending on the application, variations in the B/W and dilution ratios may be appropriate. Because of the uncertainty in interpreting test results, it is often best to run a suite of tests. The usual tests include:

- relative filtrate loss (D'Appolonia, 1980),
- viscosity by rotational viscometer (McCandless and Bodocsi 1988), and
- sedimentation (Ryan 1987; Bowders 1985).

These tests generally give a gross indication of the expected performance of the bentonite slurry during construction and generally require only a few hours or days to perform.

The filtrate lost test is performed by pressurizing a chamber filled with slurry until a cake of pure bentonite (filter cake) is formed. The volume of water which flows out of the cake during the 30 minute long test is called the filtrate. Trench stability is dependent on a low filtrate. A second and longer test of two identical filter cakes can be performed by permeating the filter cakes with leachate and water. A ratio of flow rate with water and leachate is calculated. Generally, a rate which exceeds two indicates an incompatibility. See Fig. 1.

Similarly, a change in viscosity as measured by a rotational viscometer, may indicate the potential for construction difficulties. Identical slurries are made and then diluted with water and leachate. The viscosity of each diluted slurry is tested and compared. Changes in viscosity can be subject to various interpretations. A decrease in viscosity may result from flocculation or from a beneficial thinning of the slurry. Increases in viscosity can be the result of a viscose contaminant (e.g. petroleum) which may have no real effect on compatibility.

The sedimentation test has been used to model the construction process when the slurry is used to support the trench walls. Two identical bentonite slurries are diluted with leachate and water and observed. In this test, it is often informative to use a variety of B/W ratios for the slurry prior to dilution with the leachate because sedimentation or flocculation may be controlled to some extent by using a thicker (higher B/W) slurry or additives. Evidence of flocculation is by observation of the slurry in glass cylinders usually over a period of days.

In all of the above tests, the user must balance workability constraints (primarily viscosity and filtrate loss) with the need to address compatibility. These needs may conflict and require new materials or slurry additives to achieve the desired result.

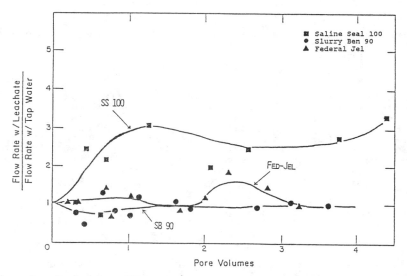

Fig. 1: Relative filtrate loss test using three bentonite clays with a landfill leachate.

Commercial Clay

Direct observations of the commercial clay product (bentonite, attapulgite, etc.) in contact with the leachate may also be used to indicate compatibility. These tests generally require a few days to complete. Again, multiple tests are used and includes:

- chemical desiccation (Alther et al. 1985), and
- free swell (McCandless and Bodocsi 1988).

These tests tend to model the most severe exposure and must be considered with some caveats. The chemical desiccation test is the drying of the bentonite slurry in contact with the leachate on a glass plate. The same standard slurry and dilution described above are used. Often severe cracking, chemical reactions, or dissolution of the clay particles can be observed. See Fig. 2. The clay is prehydrated in this test and then air dried which may be analogous to the field situation near the water table. The desiccation pattern of all clays are not identical. Some clays (e.g. sepiolite) appear unsuitable even when tested with tap water.

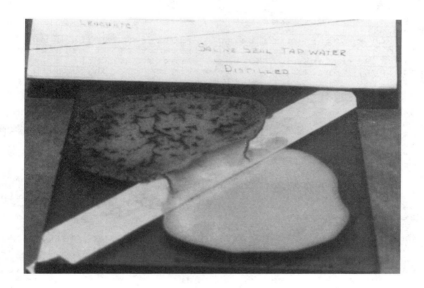

Fig. 2: Chemical desiccation test. Sample on left with leachate. Sample on right with water.

The free swell test has been used to investigate compatibility but is limited in its application since the bentonite is not prehydrated. In this test, dry bentonite particles are sprinkled into a graduated cylinder filled with water or leachate. If the bentonite does not swell, an incompatible result is indicated. In general, there is no field situation analogous to this test.

These two tests can often be used to confirm results obtained from the construction compatibility testing. The appearance of the bentonite filter cake from the filtrate loss test can be compared to the appearance of the desiccation test. Prehydrated bentonite in the sedimentation test can be compared to results from the free swell test. It is not uncommon to have apparently contradictory results.

Backfill

The slurry wall backfill material can be tested for compatibility using procedures which test the stability of the material when in contact with the leachate. Modified versions of ASTM standard tests can be used as follows:

- immersion test (ASTM Annual Book of Standards, C-267, 1991),
- fixed-wall test (ASTM D-2434, 1991), and
- plasticity (ASTM D-4318, 1991; Bowders 1985).

These tests usually require a week to a few months to complete, although typically much less time than the flexible wall test. Experience has shown that indications of incompatibility with these tests usually occurs quite early in the procedure, thereby reducing the overall testing schedule.

With cement-bentonite (CB), soil-cement (SC), and plastic concrete mixtures, a modified version of ASTM C-267, Chemical Resistance of Mortars, Grouts, and Monolithic Surfacings, can be used to investigate the physical stability of the slurry wall material. This is an immersion test where the weight and strength of the sample is measured over time in response to immersion in a leachate, as compared to immersion in water. Observations of the samples may give dramatic evidence of incompatibility. See Fig. 3. While immersion may model some conditions below the water table, only materials with a minimum unconfined strength (approximately 200 kPa) are applicable since slaking with water can produce similar weight changes in softer materials.

Soil-bentonite and other soft slurry wall materials may be tested in the fixed wall permeability cell to determine compatibility. The hydraulic conductivity developed in these tests is often of secondary importance, what is gained are observations of the potential of the material to swell, shrink, or chemically react with the leachate (Anderson et al. 1985). Since limited (or uncontrolled) effective stress is imposed, gross changes in the sample are possible which may not be possible with flexible wall permeability tests. The author has observed cases where the reaction to the leachate was so severe the sample foamed and then petrified (turned to stone), whereas no similar effect was observed in a flexible wall test. Other important physical characteristics such as resistance to high hydraulic gradients may be observed.

Replacement of pore water with leachate can change the plasticity of the backfill and therefore, hydraulic conductivity. This test works best with soil-bentonite in accordance with a modified ASTM D-4318, Liquid Limit, Plastic Limit and Plasticity Index of Soils. The user must take care to avoid imposing artificially induced effects as a result of drying. In general, the materials are slowly air dried and rewetted with tap water and contaminated

groundwater and the results compared. Some mixtures can
lose considerable plasticity yet retain a low permeability.

Fig. 3:   Immersion test with soil-cement sample soaked in
          corrosive groundwater.

It has been the author's tactic to use these tests in
approximately the sequence described above, using
incompatible results from earlier tests, to guide in the
elimination of materials with a low probability of success.
The testing program usually culminates with a limited number
of flexible wall permeability tests to document long-term
hydraulic conductivity in the presence of the leachate.
With a knowledgeable selection of tests, materials and
additives based on the indicator tests, the final flexible
wall tests are nearly always successful.

CASE STUDIES

The projects described below have been selected from
the author's files of over a hundred successful projects.
These case studies have been selected because they represent
projects where an incompatibility was discovered and/or
alternate materials were used to provide a suitable
solution. The author has, by intent, limited the discussion
to the facts of the case related to the determination of
incompatibility and the finding of an alternate solution.

Case Study No. 1:   Southern Wisconsin Landfill

An operating sanitary landfill was closing a formerly
uncontrolled landfill cell which had received hazardous
wastes. Physical and hydraulic isolation of the cell was
necessary to comply with regulatory directives to protect
the environment. Closure of the cell included a RCRA cap,

groundwater collection trench and soil-bentonite slurry cutoff wall.

Leachate from the landfill was generally characterized by a black color and pungent odor with high chloride (about 500 mg/l) and sulfate (about 10 mg/l) contents. The groundwater plume emanating from the site was found to contain toxic levels of organic chemicals including vinyl chloride. Contaminant levels were high enough that reuse of trench spoil in the soil-bentonite backfill was not permitted. Compatibility testing of the soil-bentonite backfill began with the development of a bentonite slurry for trenching. Three products were tested; two premium grade, sodium (API 13A) bentonites and one "contaminant-resistant," SS100 bentonite. A stable slurry with a B/W = 5% was produced from all three bentonites with a viscosity (Marsh Funnel) of 40 to 50 seconds without the use of additives.

Relative filtrate loss tests using the leachate and tap water are shown in Fig. 1. It was observed that the SS100 bentonite permeated with the leachate produced a relative filtrate loss three times greater than with tap water and much higher than either of the premium bentonites. In the desiccation test, a pattern of small cracks was observed with the SS100 which was not present in tests of the other bentonites. Finally, a sedimentation test of the bentonites was performed as shown in Fig. 4. In this test, all three bentonites performed similarly.

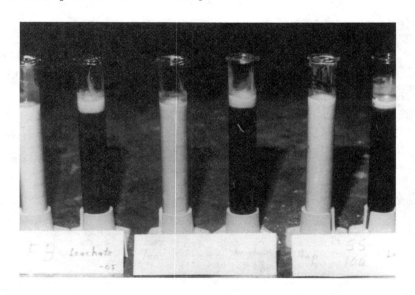

Fig. 4: Sedimentation test with three bentonite clay mixed with landfill leachate and with water.

Based on these results, SS100 bentonite was excluded from further consideration. The remainder of the test program, including hydraulic conductivity testing, proceeded successfully.

A 1200 meter (4000 ft) long by 10 meters (35 ft) deep slurry cutoff wall was installed which has, since 1987, prevented the further contamination of the area. Tests show that the slurry cutoff wall was effective and the vinyl chloride plume dissipated.

Case Study No. 2: Eastern Michigan Chemical Facility

A chemical plant was operating a system of treatment lagoons which abutted a former brine production area separated by a relatively narrow earthen dike. Closure of the brine ponds without disturbance to the treatment lagoons, using a slurry cutoff wall, was the aim of the project. The brine contained high levels of metals including calcium (8.3%), magnesium (0.60%), and sodium (1.61%). Total dissolved solids in the leachate was 25 to 30% and the density of the brine was 1.04, gm/cc.

Implementation of the project was complicated by at least three compatibility concerns:

1. brine is known to flocculate bentonite slurry,
2. chemicals in the treatment lagoons could have an unknown effect on the slurry wall, and
3. the dike was unstable (safety factor < 1.0) and required reinforcing.

The compatibility testing for this project began with the selection of an alternate clay to replace bentonite. Testing of premium bentonite, "saline-resistant" bentonite and attapulgite was conducted as shown in Fig. 5. In this case, attapulgite, a nonswelling montmorilite clay (Tobin and Wild 1986) was found to be most effective. In addition, attapulgite could be mixed with brine water for the trenching slurry. Using attapulgite with the brine water and wastewater also produced successful results in the desiccation and sedimentation tests.

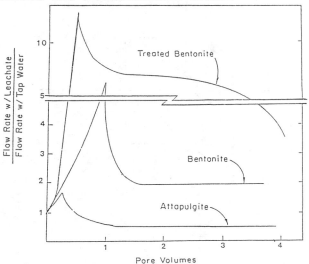

Fig. 5: Relative filtrate loss test using three commercial clays with brine water leachate.

Stabilization of the dike required a cementacious backfill which would reinforce the dike and increase the factor of safety against sliding. Cement-attapulgite (a variation of cement-bentonite self-hardening slurry) and plastic concrete mixtures were tested with permeabilities less than 1×10^{-6} cm/sec. Results of the unconfined compressive strength tests are shown in Fig. 6. Immersion tests and long-term permeability tests with the leachate were performed which demonstrated the compatibility of the cement-attapulgite with the brine water.

Based on the results described above, a 700 m (2,000 ft) long cement-attapulgite slurry trench about 10 m (30 ft) deep was constructed through the center of the dike. Brine water was used as the mix water for the slurry. Since 1988, the project has served to separate the wastewater pond and the brine pond. The stability of the dike has been ensured by the use of the cement-attapulgite.

Case Study No. 3: Upstate New York Lagoon Closure

A former mine and processing plant produced two byproducts which were co-mingled in a single earthen-lined lagoon. One byproduct, semet, has a pH < 0.5 and the other byproduct has a pH > 13. Storage of the two byproducts in a single lagoon did not produce neutralization and the leachates were found to be existing separately and seeping out of the lagoon into the groundwater.

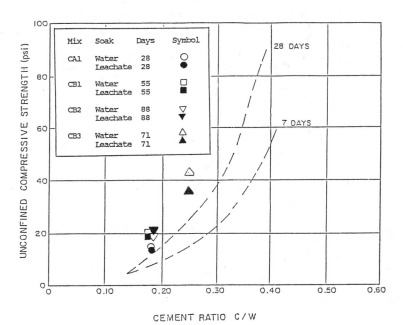

Fig. 6: Unconfounded compressive strength of cement-attapulgite immersed in water and low pH leachate. Comparative trends for Millet and Perez (Millet and Perez 1981)

At this time, one of the potential remedies to the sites is containment with a cutoff wall. The wall will be more than 30 m (100 ft) deep so deep soil mixing (DSM) and plastic concrete are considered as prime candidates for the cutoff wall. Compatibility testing for this site provides an opportunity to test the limits of the testing methods.

Testing began with separate tests of the high and low pH leachates with a variety of commercial clay products. As previously described, a step-by-step process was enacted which focused the program on the most critical compatibility challenge. The high pH leachate was compatible with all clays in the filtrate, sedimentation, and desiccation tests. Therefore, the majority of the program was focused on compatibility of materials with the low pH semet leachate. Filtrate, sedimentation, and desiccation testing proved that attapulgite was the best commercial clay to resist the semet. What remained, therefore, was to find a combination of soil and/or cement to complement the attapulgite. Initial tests with soil-attapulgite were carried out with fixed wall permeameters. The results were dramatic and unsuccessful. The leachate reacted violently with soil-attapulgite producing a gas and turning the sample into a petrified mass. Immersion tests with soil-cement-attapulgite (at relatively low total cement contents) were equally unsuccessful. As shown in Fig. 3, many of the samples dissolved in the immersion tests. Finally, cement-attapulgite blends (with relatively high total cement contents) were found which survived the immersion tests. The strength of immersed cement-attapulgite was similar to cement-bentonite mixture as shown in Fig. 6. Long-term flexible wall permeability tests confirmed the compatibility of the cement-attapulgite by the display of a stable hydraulic conductivity over three pore volumes of flow. Dissection of a cement-attapulgite sample after permeation is shown in Fig. 7 to illustrate the success of the compatible mixture.

Fig. 7: Dissected cement-attapulgite sample after permeation by low pH leachate for three pore volumes.

Case Study No. 4:   Former Industrial Site in Vancouver, B.C.

A site which borders the bay in the center of Vancouver had been used since the city's founding for a variety of industrial purposes including coal gasification, wood treatment, and fuel storage. A variety of toxins were found in the soils and groundwater including cyanide (10 ppm), hydrocarbons (100 ppm), pentachlorophenol (20 ppm), arsenic (1 ppm), lead (4 ppm), and zinc (6 ppm). In order to reclaim and develop the site, A DSM and jet grout wall was constructed to contain the contaminants. Development of the site requires excavation of an area of significant contamination and eventually building foundations; therefore, the cutoff walls were specified to have an unconfined compressive strength of up to 1.4 MPa (200 psi) as well as a hydraulic conductivity less than 10^{-6} cm/sec.

Due to the structural requirements and the availability of resources, the testing program focused on soil-cement blends which used a grout composed of Canadian calcium bentonite, Wyoming sodium bentonite, gypsum, fly ash, and cement. The use of gypsum was selected to provide improved strength with reduced permeability. Calcium bentonite is a low swelling bentonite clay which provides stability to the grout and reduces permeability. Concerns about the use of these innovative materials, as well as requirements for compatibility, resulted in an extensive testing program.

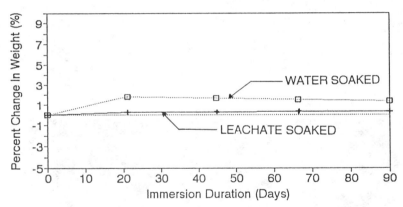

Fig. 8:   Immersion test result of DSM sample in water and hazardous leachate.

Testing of the bentonite resulted in the finding that at least three times as much Canadian calcium bentonite (B/W = 15%) as Wyoming sodium bentonite was necessary to produce a workable slurry. The addition of cement and fly ash to this slurry required thinners including both phosphate and lignosulfate based products.

The addition of gypsum provided beneficial thinning of the grout; and, therefore, the use of a relatively dense grout with no loss in workability. Once blended into the mix, the gypsum becomes a part of the cement matrix. No dissolution or other detrimental effects were noted with the use of gypsum.

Compatibility testing focused on immersion testing and flexible wall permeability testing of the soil-cement. Immersion tests were conducted for up to 90 days in the leachate. The immersed samples appeared identical in water and leachate with an average weight change of less than 1%. The majority of any weight change was usually discovered within the first 28 days of immersion. See Fig. 8. Hydraulic conductivity tests on the hardened soil-cement confirmed the long term stability of the materials.

The cutoff wall was constructed in the summer of 1992. Each type of cutoff wall and grout mixture was subjected to extensive field testing including test sections which were excavated and examined. In total, over 600 m (2,000 ft) of cutoff wall were installed up to 16 m (50 ft) deep. Insitu testing and monitoring to date has shown the cutoff wall to be highly effective.

CONCLUSIONS

A systematic approach to compatibility testing includes indicator tests along with permeability tests. Compatibility testing using indicator tests provides a relatively rapid and rational method for predetermining the compatibility of slurry cutoff wall materials with contaminated groundwater. Not all indicator tests are applicable on every project. Furthermore, some tests model situations which are impossible on some sites. The tests are relatively simple and rapid, but the application of the results to real remediation projects requires the expertise of a knowledgeable engineer and specialty contractor with experience in the materials selected for installation.

References

Alther, G., J. C. Evans, Fang, H. Y. and K. Witmer, 1985, "Influence of Inorganic Permeants upon the Permeability of Bentonite," _Hydraulic Barriers in Soil and Rock, ASTM STP 874_, pp. 64-73.

Anderson, D. C., W. Crawley, and J. D. Zabcik, 1985, "Effects of Various Liquids on Clay Soil: Bentonite Slurry Mixtures," _Hydraulic Barriers in Soil and Rock, ASTM STP 874_, pp. 93-103.

API, 13B-1, June 1990, "Recommended Practice Standard Procedure for Field Testing Water-Based Drilling Fluids."

ASTM, 1991, Annual Book of Standards, ASTM, Philadelphia, Pa.

Bowders, J., 1985, "The Influence of Various Concentrations of Organic Liquids on the Hydraulic Conductivity of Compacted Clay," PhD Dissertation, University of Texas, Austin, Texas.

D'Appolonia, D. J., 1980, "Soil-Bentonite Slurry Trench Cutoffs," _Journal of the Geotechnical Engineering Division_, Proceedings of ASCE, Vol. 106, No. GT4.

Evans, J., 1993, "Hydraulic Conductivity of Vertical Cutoff Walls," *Hydraulic Conductivity and Waste Contaminant Transport in Soils ASTM STP 1142,* David E. Daniel and Stephen J. Trautwein, Eds., American Society for Testing and Materials, Philadelphia.

Grube, W. E., Jr., 1992, "Slurry Trench Cut-Off Walls for Environmental Pollution Control," *Slurry Walls: Design, Construction, and Quality Control, ASTM STP 1129,* David B. Paul, Richard R. Davidson, and Nicholas J. Cavalli, Eds., American Society for Testing and Materials, Philadelphia.

Khera, R. P. and Thilliyar, M., 1990, "Slurry Wall Backfill Integrity and Desiccation," *Physico-Chemical Aspects of Soil and Related Materials, ASTM STP 1095,* K. B. Hoddinott and R. O. Lamb, Eds., American Society for Testing and Materials, Philadelphia, pp.81-90.

McCandless, R. M. and Bodocsi, A., August 1988, "Quick Indicator Tests to Characterize Bentonite Type," *Office of Research and Development, U.S. EPA.*

Millet, R. A. and J. Y. Perez, 1981, "Current USA Practice: Slurry Wall Specifications," *Journal of the Geotechnical Engineering Division,* ASCE, Vol. 107, No. GT8.

Ressi di Cervia, A. L., 1992, "History of Slurry Wall Construction," *Slurry Walls: Design, Construction, and Quality Control, ASTM STP 1129,* David B. Paul, Richard R. Davidson, and Nicholas J. Cavalli, Eds., American Society for Testing and Materials, Philadelphia.

Ryan, C. R., 1987, "Vertical Barriers in Soil for Pollution Containment," *Hydraulic Barriers in Soil and Rock, ASTM STP 874.*

Tallard, G., February 1984, "Slurry Trenches for Containing Hazardous Wastes," *Civil Engineering,* pp. 41-45.

Tobin, W. R. and P. R. Wild, February 1986, "Attapulgite: A Clay Liner Solution?," *Civil Engineering,* pp. 56-58.

Wu, J. Y. and Khera, R. P., 1990, "Properties of a Treated-Bentonite/Sand Mix in Contaminated Environment," *Physico-Chemical Aspects of Soil and Related Materials, ASTM STP 1095,* K. B. Hoddinott and R. O. Lamb, Eds., American Society for Testing and Materials, pp. 47-59.

THE COMPATIBILITY OF SLURRY CUTOFF WALL MATERIALS WITH CONTAMINATED GROUNDWATER

Steven R. Day, Geo-Con, Inc., Pittsburgh, Pennsylvania, U.S.A.

REFERENCE: Slurry cutoff walls are frequently relied upon to block groundwater flows from toxic waste sites and landfills. The long-term effectiveness of slurry cutoff wall materials is critical to the successful containment of these facilities and the protection of groundwater resources. A variety of laboratory indicator tests have been attempted by engineers and academia to make compatibility determinations but at present there has been little published experience to show which tests produce meaningful results and how these tests can be used to demonstrate compatibility.

Hydraulic conductivity is a useful measure of chemical/soil compatibility but permeability tests alone cannot assure the long-term stability of a slurry cutoff wall. A suite of indicator tests are used where the leachate and the proposed materials are combined and tested in immersion, desiccation, sedimentation, and other modes. Each indicator test attempts to model a different scenario of the slurry cutoff wall installation and operation.

This paper presents the experience of a specialty contractor from a number of projects, where an incompatibility was discovered and alternate materials were used to find a successful solution. Monitoring results from these sites has proven the effectiveness of the chosen solution. The laboratory test methods described are relatively simple and rely on worst case scenarios, performed in a step-by-step process, which culminates with flexible wall permeability tests. Based on the methods described and the results from successful projects where these methods were used, engineers, owners and the public may better rely on long-term slurry cutoff wall performance with an increased level of confidence.

Key Words: attapulgite, bentonite, compatibility, containment, deep soil mixing, hydraulic conductivity, jet grouting, slurry cutoff wall

Don J. DeGroot[1] and Alan J. Lutenegger[2]

A COMPARISON BETWEEN FIELD AND LABORATORY MEASUREMENTS OF HYDRAULIC CONDUCTIVITY IN A VARVED CLAY

REFERENCE: DeGroot, D. J., and Lutenegger, A. J., "A Comparison Between Field and Laboratory Measurements of Hydraulic Conductivity in a Varved Clay," Hydraulic Conductivity and Waste Contaminant Transport in Soil, ASTM STP 1142, David E. Daniel and Stephen J. Trautwein, Eds., American Society for Testing and Materials, Philadelphia, 1994.

ABSTRACT: The measurement of hydraulic conductivity using a variety of field and laboratory techniques was evaluated at a site consisting of medium stiff and soft lacustrine varved clay in western Massachusetts. Field measurements were obtained by conducting "slug" tests in both predrilled and push-in piezometers and also from pore pressure dissipation tests using the piezocone and flat dilatometer. Laboratory hydraulic conductivity values were obtained for both vertical and horizontal flow conditions using a flexible wall permeameter and by indirect estimation from 1-dimensional consolidation tests. Based on a comparison of tests conducted throughout the profile, laboratory flexible wall tests with vertical flow gave the lowest values of hydraulic conductivity whereas the predrilled piezometers yielded the highest values. Of all the field techniques, the push-in piezometers gave the lowest values. Hydraulic conductivity values interpreted from piezocone and flat dilatometer dissipation tests tended to be between those obtained from the predrilled and push-in piezometers. Results from tests in predrilled piezometers show that the hydraulic conductivity increases with increasing screen length, showing the influence of scale effects. The results of this study clearly show that estimation of hydraulic conductivity for this soil is highly dependent on scale effects, the test technique used and on the direction of flow (i.e., parallel versus perpendicular to the orientation of the varves).

KEYWORDS: hydraulic conductivity, piezometers, piezocone, dilatometer, varved clay.

Accurate determination of the hydraulic conductivity of saturated cohesive soils presents many challenges. This is particularly a problem if the soil is expected to exhibit hydraulic anisotropy as for example in a varved clay. Deciding whether to conduct in situ tests and/or laboratory tests involves many considerations and will directly impact the anticipated results. For example, Olson and Daniel (1981) summarized results given in the literature comparing field and laboratory measurements of hydraulic conductivity, and found that the range in the

---

[1]Assistant Professor, Department of Civil Engineering, University of Massachusetts at Amherst, Amherst, MA 01003.

[2]Associate Professor, Department of Civil Engineering, University of Massachusetts at Amherst, Amherst, MA 01003.

ratio of field hydraulic conductivity to laboratory hydraulic conductivity was between 0.3 and 46,000, but nearly 90 percent of the observations lie in the range from 0.38 to 64 (i.e., approximately two orders of magnitude difference).

Field techniques generally measure the horizontal hydraulic conductivity and can involve "permanent" installations such as predrilled or push-in piezometers for conducting slug tests. However, predrilled piezometers are time consuming and costly to install and involve other problems including disposal of drill cuttings, positioning and alignment of the screen and proper construction of the sand pack and isolation seal. Use of push-in piezometers eliminates many of these problems and are much quicker to install but have the disadvantage of disturbing the soil during installation and hence altering the natural hydraulic response of the soil (Lutenegger and Degroot 1992). Other in situ techniques such as the piezocone, the piezoblade and the flat dilatometer offer an indirect measurement of hydraulic conductivity through interpretation of the time rate of dissipation of excess pore pressures that are generated during penetration. The main advantage of these techniques is that they are relatively quick to perform and hence allow detailed profiling of hydraulic conductivity in a short period of time, e.g., during a site investigation. However, like push-in piezometers, they are full displacement techniques and hence the results may be influenced by disturbance of the soil during penetration. Furthermore, interpretation of the results requires assumptions to be made relative to soil compressibility and consolidation in order to compute the hydraulic conductivity from the pore pressure dissipation data.

Laboratory tests such as those conducted using a flexible wall permeameter offer the potential to test specimens under controlled boundary conditions (i.e., saturation, effective stress level, hydraulic head, flow direction, etc.). However, laboratory testing has many inherent disadvantages the most notable of which are scale effects when testing small laboratory size specimens (Olson and Daniel 1981).

Given all the relative advantages and disadvantages of the many field and lab techniques available to estimate hydraulic conductivity, engineers are often faced with the difficult question of which technique(s) should be used to obtain appropriate values for design. Continued documentation of case studies evaluating field and laboratory techniques in a variety of geologic materials will provide guidance to engineers in making appropriate choices. The purpose of the study presented herein is to provide a comparison of field and laboratory measurements of hydraulic conductivity of a cohesive soil. The soil tested was a varved clay which was specifically selected because varved clays generally have high hydraulic anisotropy ratios thus making selection of appropriate techniques to evaluate the hydraulic conductivity and the interpretation of results all the more challenging.

This paper presents a description of the test site soil conditions followed by a description of the different laboratory and in situ techniques used to evaluate the hydraulic conductivity and a discussion of test results obtained throughout an 20 m profile.

TEST SITE

The investigation described herein was conducted at the University of Massachusetts at Amherst permanent test site. The site is located on the Amherst Campus in the Connecticut River Valley and has recently been designated a U.S. National Geotechnical Experimental Test Site. The subsurface stratigraphy generally consists of about 1 m of mixed cohesive and cohesionless fill overlying a thick deposit of late Pleistocene varved silt and clay. The varved clay is locally referred to as Connecticut Valley Varved Clay (CVVC) and is a result of lacustrine deposition into glacial Lake Hitchcock during the last glaciation. The thickness of the individual silt or clay varves is typically on the order of 2 to 8 mm and the varves generally lie in a horizontal orientation. The upper 5 to 6 m of the deposit is overconsolidated as a result of surface erosion, desiccation, and seasonal fluctuations in the groundwater table. Below this weathered crust the soils become very soft and near normally consolidated with increasing depth. The groundwater table at the site typically occurs in the upper 3 m and varies by approximately ± 1.2 m throughout the year coinciding with changes in seasonal precipitation. Figure 1 gives a general soil profile of the site including grain-size distribution, Atterberg limits, and water contents of the bulk soil.

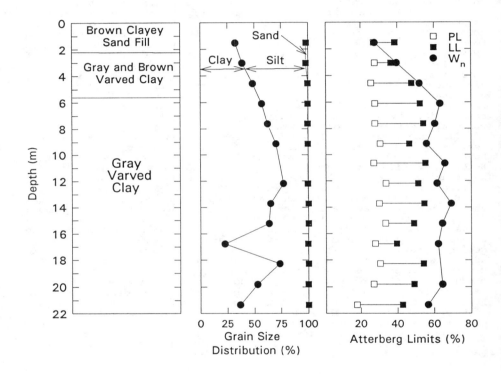

FIG. 1--Soil Profile and Index Properties at Test Site.

TEST PROCEDURES AND INTERPRETATION

Laboratory Tests

All laboratory tests were conducted on undisturbed samples that were obtained from the site using a 76 mm (3 in.) diameter fixed piston sampler. The tubes were sealed with wax, capped and stored in a humid room prior to extrusion for laboratory testing. All laboratory reported values of hydraulic conductivity are corrected to 20°C.

Flexible Wall Permeameter—Constant head flexible wall tests were performed on undisturbed samples using the procedures outlined by Daniel et al. (1984). Specimens were trimmed to a nominal diameter of 71 mm and height of 22 mm for an H/D ratio of 0.31 and a total specimen volume of 87,100 mm^3. They were isotropically consolidated to the in situ vertical effective stress and were back pressure saturated using a back pressure on the order of 400 kPa. Saturation was usually completed in one to two days and was verified by measuring the B value, which typically equaled 0.95 or greater. In order to determine anisotropic flow characteristics, companion specimens from the same tube were trimmed with vertical and horizontal orientations to allow water flow perpendicular and parallel to the varves. Tests were conducted using a hydraulic gradient ranging from 25 to 30. All tests were conducted until steady state conditions were reached with equal inflow and outflow and until the cumulative volume of flow exceeded about twice the pore volume of the specimens.

Oedometer Consolidation—Indirect measurements of horizontal hydraulic conductivity were obtained from one dimensional incremental loading oedometer tests which were performed in general accordance to ASTM D2435 using a load increment ratio of one. Specimens were trimmed in the horizontal direction to allow water flow parallel to the varves for each loading increment (typical diameter and height equalled 64 mm (2.5 in.) and 19 mm (0.75 in.) respectively). The hydraulic conductivity was calculated using Taylor's square root time fitting method to estimate the coefficient of consolidation. In this test, hydraulic conductivity is typically calculated for the increment of loading which brackets the in situ effective stress, which in this case would be the horizontal direction. Estimates of K_o were used to compute the in situ horizontal effective stress and the hydraulic conductivity was calculated from the load increment that corresponded to this stress level. While this procedure does not exactly reproduce the true in situ state of stress, since the stress in the horizontal direction in the oedometer is not equal to the in situ vertical effective stress, it was selected as the most appropriate stress level to use. Using higher stress levels would have exceeded the horizontal yield stress of the specimens and caused significant changes in void ratio and hence hydraulic conductivity. Preconsolidating to the stress levels used caused no significant volume changes resulting in specimens which were tested at a void ratio approximately equal to the in situ void ratio.

In Situ Tests

Four different field testing techniques were used to evaluate the in situ horizontal hydraulic conductivity of the soil. The techniques consisted of two permanent installation techniques which included predrilled and push-in piezometers and two "rapid" profiling techniques which included pore pressure dissipation tests using the piezocone and dilatometer. A schematic of the equipment used for each of the in situ testing techniques is given in Figure 2. All field reported values are corrected to 20°C from a nominal in situ temperature of 13°C to allow direct comparison with laboratory results.

Predrilled and Push-in Piezometers—Predrilled piezometers were constructed of a 305 mm (12 in.) long section of nominal 40 mm inside diameter slotted PVC pipe that was attached to a PVC riser pipe of the same diameter in boreholes. The filter element of shallow piezometers was packed using a uniform clean Ottawa sand with a median grain size of 0.6 mm. Deeper piezometers were packed using a clean pea gravel with a maximum diameter of 9.5 mm. Shallow piezometers (up to a depth of 6.1 m) were installed in boreholes drilled with a 76 mm (3 in.) bucket-type hand auger. Boreholes for deeper piezometers were drilled using a 102 mm (4 in.) dia. wash boring. Prior to installation of the slotted screens, the boreholes were flushed repeatedly with clean water to remove dirty water from drilling. In all cases, an isolation seal with a minimum length of

1.5 m was placed above the sandpack using bentonite pellets.

Push-in piezometers were constructed using a standard Geonor M-206 piezometer fitted with a sintered bronze porous element (314 mm length by 32 mm diameter) with an effective opening size of 50 μm flush mounted onto the body. The piezometers were retrofitted to act as simple standpipes by attaching 25 mm (1 in.) diameter waterpipes. During installation the extension pipes were filled with water and capped to minimize clogging of the filter element. A penetration rate of approximately 20 mm/s was used and a total time of about 10 to 15 minutes was needed to install each piezometer.

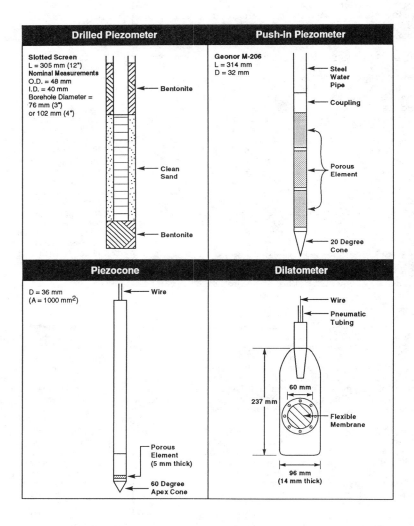

FIG. 2--Schematic of In Situ Hydraulic Conductivity Measurement Techniques.

Falling head hydraulic conductivity tests were performed in each of the piezometers by adding a "slug" of water to the standpipe. The initial excess heads were kept low enough to avoid hydraulic fracture of the soil. Changes in total head with time were obtained using an electric water level indicator with each test taking approximately 1 to 2 days depending on the initial excess head applied and the hydraulic response. Test results were plotted in the form of the change in head (log scale) as a function of time. The hydraulic conductivity was calculated using the following equation (Hvorslev 1951):

$$k = \frac{-2.3 \, a \, m}{F} \qquad (1)$$

where
a = cross sectional area of standpipe, cm^2,
m = slope of log head loss or gain versus time curve,
F = shape factor, cm.

Chapius (1989) presented a detailed summary of the available shape factors in determining hydraulic conductivity in piezometers. For the specific case of horizontal flow (i.e., impervious bottom) out or into a piezometer (i.e., slug or bail tests) Chapius recommends the following equation:

$$F = \frac{2 \pi L}{\ln\left[\frac{L}{D} + \left[1 + \left(\frac{L}{D}\right)^2\right]^{\frac{1}{2}}\right]} - 2.75 \, D \qquad (2)$$

where
F = shape factor, cm,
L = length of "injection" zone,
D = diameter of "injection" zone.

The L and D values in Eq. 2 correspond to the length and diameter of the sand or pea gravel pack or the porous element in the predrilled and push-in piezometers respectively (Fig. 2). Eq. 2 is particularly relevant to the tests conducted in this study since the varved nature of the soil results in predominantly horizontal flow and also because both the push-in piezometers and the predrilled piezometers have an impervious bottom as a result of the stainless steel tip on the push-in piezometers and the bentonite plug placed at the bottom of the predrilled piezometer hole (Fig. 2). Multiple tests were performed in each piezometer and indicated that the results displayed very good repeatability and were independent of the initial excess head and test sequence. Detailed results for individual tests showing this repeatability are given in Lutenegger and DeGroot (1992).

In order to evaluate the effect of "scale" on the results of field tests, a series of five slotted screens of varying length were installed with the center of the screen at a depth of 3 m (10 ft). Screen lengths varied from 76 mm (3 in.) to 1524 mm (60 in.) (i.e., L/D ranging from 1 to 20). Multiple slug and bail tests were conducted in each of these piezometers.

Piezocone--A 36 mm (1.4 in.) dia. piezocone (CPTU) with porous element located 5 mm behind the cone base was used in this investigation (Fig. 2) and was prepared for use by placing it in a glycerol filled de-airing and calibration chamber. The glycerol was used to keep the element saturated between the laboratory and installation in the field. A vacuum pressure of 207 kPa was applied to the chamber for 24 hours prior to use in the field. To insure saturation and to avoid possible cavitation which might occur if the probe were pushed through unsaturated soil, a borehole was prepared by hand angering to a depth below the groundwater table (approximately 1.5 m) and then filled with water to the ground surface. The CPTU was then removed from the chamber and lowered to the bottom of the hole without removing the porous element from fluid. The CPTU was advanced using a hydraulic jacking system at a nominal rate of 20 mm/s. Dissipation tests were conducted by pushing the CPTU to a preselected depth; stopping the advancement of the piston; and simultaneously starting a stop watch. Readings of pore pressure versus time were taken using a digital read out box until the excess pore pressure approximately returned to the hydrostatic value which typically occurred within 3 to 5 hours. After each dissipation test the CPTU was advanced to the next testing depth.

The hydraulic conductivity may be estimated from CPTU dissipation tests using the following equation:

$$k_h = c_h \, \gamma_w \, m_h \tag{3}$$

where

k_h = horizontal hydraulic conductivity, m/s,
c_h = horizontal coefficient of consolidation, m²/s,
γ_w = unit weight of water, kN/m³,
m_h = horizontal coefficient of volume change, m²/kN.

For dissipation tests conducted using full displacement instruments the horizontal coefficient of consolidation is determined as (Torstensson 1977):

$$c_h = \frac{T \, r^2}{t} \tag{4}$$

where

T = theoretical time factor,
r = radius of instrument, m²,
t = actual clock time required to dissipate a given percentage of excess pore pressure corresponding to a particular value of T, s.

While the CPTU provides estimates of the vertical coefficient of volume change, m_v, Mitchell and Gardner (1975) suggested that there is little anisotropy in the coefficients of volume changes (i.e., $m_h/m_v \approx 1$). Hence, estimates of m_h were obtained by equating it to values of m_v which in turn were estimated from the relationship suggested by Mitchell and Gardner (1975) that relates the constrained modulus M (= $1/m_v$) to the cone tip resistance by a factor α which is a function of tip resistance and soil type. The theoretical time factor, T_{50} corresponding to fifty percent of excess pore pressure dissipation, was used in Eq. 4 and taken equal to 1.2 as suggested by Gupta and Davidson (1986) for a cone with a sixty degree apex angle and the porous element located at the base of the tip. The value of t_{50} corresponding to T_{50} in Eq. 4 was determined from plots of the normalized pore pressure, U_n, versus time plots where U_n is defined as:

$$U_n = \frac{U_t - U_0}{U_{max} - U_0} \tag{5}$$

where
U_n = normalized pore pressure,
U_t = field pore pressure at any time t,
U_0 = in situ pore pressure at the test depth,
U_{max} = maximum observed pore water pressure during the dissipation test.

Piezocone dissipation tests were conducted at the site in 1.52 m (5 ft) intervals beginning at a depth of 3.0 m (10 ft).

Dilatometer--Figure 2 shows a schematic of the flat dilatometer (DMT) used in this study. The probe consists of a flat plate that houses a stainless steel flexible membrane on one face of the probe. After calibration of the circular membrane resistance the DMT was advanced from the ground using a hydraulic pushing frame at a nominal rate of 20 mm/s. At each desired test depth, a sequence of readings was taken by pressurizing the cavity behind the membrane with nitrogen gas that was regulated from the surface. Standard DMT test practice involves taking three readings that correspond to:
(1) the pressure required to just lift the membrane off the face of the probe (A-reading),
(2) the pressure to expand the membrane outward 1 mm (B-reading), and
(3) the pressure which reestablishes contact with the face of the probe (C-reading) to the original A position.
The C-reading is obtained by controlled deflation of the membrane after obtaining both the A- and B-readings. Evaluating the hydraulic conductivity consists of performing a "dissipation test" by taking a series of C-readings with time at each test depth. These tests were conducted at the site in 1.52 m (5 ft) intervals beginning at a depth of 3.0 m (10 ft).

Robertson et al. (1988) and Lutenegger and Kabir (1988) have shown that the C-reading pressures measured with time using a DMT are very similar to the pore pressures measured on the face of a piezoblade (or a DMT blade equipped with a pore pressure transducer); i.e., the C-reading or closing pressure is approximately equal to the pore water pressure on the face of the blade. Hence, results from C-reading dissipation tests can be used in a similar fashion to CPTU dissipation tests to estimate the horizontal hydraulic conductivity using Eq. 3. The value of the constrained modulus M (= $1/m_v$) was estimated from empirical relationships developed between the DMT elastic modulus E_D, material index I_D, and horizontal stress index K_D which are all measured from DMT A- and B-readings (Marchetti 1980). Although the DMT measures soil response in the horizontal direction, these correlations were developed for estimating m_v which in this investigation was assumed to be equal to m_h for interpreting the DMT results (i.e., similar to the CPTU interpretation procedure discussed in the previous section). Values of c_h were estimated using Eq. 4 assuming the DMT to have an equivalent radius equal to one-half its thickness and a time factor T_{50} = 25 which corresponds to a CPTU with an eighteen degree apex angle and the porous element located along the shaft (Baligh and Levadoux 1986).

Values of t_{50} were obtained from plots of the normalized pore pressure versus log time. Values of P_2 (defined as C-readings corrected for membrane stiffness) were normalized similar to CPTU results using the following equation:

$$P_{2(n)} = \frac{P_{2(t)} - U_0}{P_{2\,max} - U_0} \qquad (6)$$

where
  $P_{2(n)}$ = normalized P_2 values,
  $P_{2(t)}$ = P_2 reading at time t,
  $P_{2\,max}$ = interpreted maximum value of P_2
  U_0 = in situ pore pressure.

In the DMT test it is not possible to measure $P_{2\,max}$ (i.e., initial corrected C-reading) because of the time it takes to inflate the membrane out to the one millimeter expansion point and then deflate the membrane to obtain the C-reading. Therefore, $P_{2\,max}$ was estimated by plotting the P_2 readings versus square root of time and extrapolating linearly back to t=0 and setting this pressure equal to $P_{2\,max}$. This procedure is similar to Taylor's square root of time method for evaluating the coefficient of consolidation from laboratory consolidation tests.

PRESENTATION OF RESULTS

Laboratory Tests

Figure 3 presents the results obtained from the flexible wall and oedometer laboratory tests. As expected for this varved clay the hydraulic conductivity measured with flow perpendicular to the varves (k_v) gave lower values than for flow parallel to the varves (k_h) based on the flexible wall tests. The anisotropy ratio $r_k = k_h/k_v$ ranges from approximately 2 to 14 with an average value of 6. These r_k values are very similar to those reported by Ladd and Foott (1977) for an adjacent site in Amherst, MA based on tests conducted on 102 mm (4 in.) cube laboratory specimens that were rotated 90° to obtain estimates of k_v and k_h. Chan and Kenney (1973) conducted similar tests on 64 mm (2.5 in.) cube specimens of New Liskeard Varved Clay from Ontario, Canada and found the value of r_k for this soil to range from 3 to 4. Casagrande and Poulus (1969) found r_k to range from 4 to 40 with an average value of 10 based on laboratory tests on a varved clay from New Jersey. Differences in the value of r_k for a given varved clay, and among different varved clays, depends on the absolute thickness of the alternating individual "silt" and "clay" varves, the relative thickness of varves and also the hydraulic conductivity of individual varves (Kenney 1963). In general the k_v and k_h profiles at this site are approximately parallel with depth and give higher values in both the vertical and horizontal direction at the top of the deposit and at a depth of approximately 16 m where at both of these locations the silt content is higher than the clay content of the bulk soil (Fig. 1).

Values of horizontal hydraulic conductivity estimated from the oedometer tests tend to give lower values than k_h from the flexible wall tests and the profile is also not parallel to the flexible wall k_h and k_v profiles. Excluding the values at a depth of 1.5 m (5 ft) and 15.2 m (50 ft) the average ratio of k_h (flexible wall) to k_h (oedometer) exhibits a considerable amount of scatter with an average value equal to 2.5 ± 1.8 SD (n = 7). Probable reasons for the differences between both measurements of the horizontal hydraulic conductivity include the fact that the oedometer test is an indirect measurement which has lead some investigators to question the accuracy of using this method in clays (e.g., Tavenas et al. 1983). In addition, because the specimens are trimmed into the oedometer cell in a horizontal orientation, it is not possible to reproduce the in situ K_0 state of stress during consolidation and therefore selecting the appropriate stress level for determining the hydraulic conductivity is difficult and can be subject to errors.

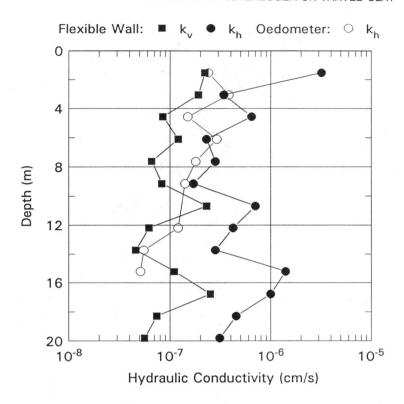

FIG. 3--Results from Laboratory Measurements of Hydraulic Conductivity.

In Situ Tests

Figure 4 presents plots of the normalized excess pore pressure versus log time for the dissipation tests conducted with the CPTU and the DMT. The curves for the DMT do not start at a time less than approximately 60 seconds since this is the amount of time required to take the initial C-reading as previously discussed. In both cases, the results display classical dissipation curves for soft clays. In fact, the position of the curves is not as significant as the fact that the curves all display the same general slope, i.e., with the exception of the 3.05 m (10 ft) tests all the curves obtained for both instruments are essentially parallel at all the different depths. Even though t_{50} may be different for each curve as the depth increases, the stiffness of the clay also increases which offsets the difference in t_{50} in the calculation for k_h.

One of the sources of errors in reducing data from these types of dissipation tests is in the selection of the original in situ pore pressure and therefore it is of use to occasionally allow complete dissipation of excess pore pressures. At this site the in situ pore pressure within the depth of testing is essentially hydrostatic which is

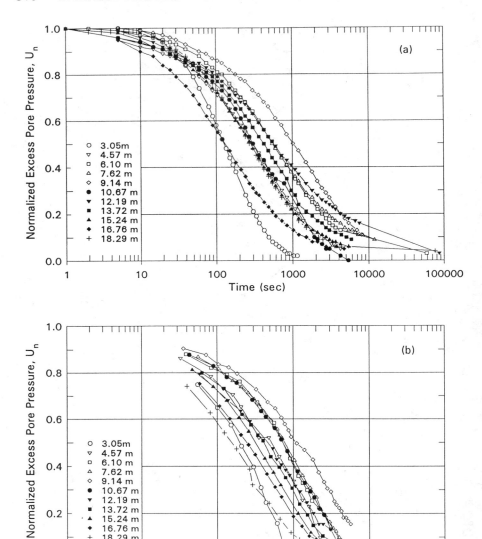

FIG. 4--Excess Pore Pressure Dissipation Curves:
(a) Piezocone; (b) Dilatometer.

confirmed by those tests which were conducted overnight and have a normalized pore pressure that approximately returns to zero. A major advantage of both of these instruments over permanent installations is that if the original in situ pore pressure profile is known or can be reasonably approximated, the tests only need to be conducted to t_{50} which for both devices occurs in less than 20 minutes for this soil. Hence they provide rapid preliminary estimates of the in situ hydraulic conductivity. There is of course an upper bound in which these instruments can be used as for example in granular materials the dissipation of excess pore pressure will occur so rapidly that a reasonable number of readings could not be taken quickly enough to determine t_{50}.

Figure 5 gives estimates of k_h versus depth for both the CPTU and DMT tests and also for the predrilled and push-in piezometers (the laboratory flexible wall k_h values are also plotted for reference). These results indicate that predrilled piezometers give the highest values of horizontal hydraulic conductivity as compared to all the other field and laboratory methods while the push-in piezometers give the lowest values of all the field techniques. The CPTU and DMT tend to give values in between those obtained with the predrilled and push-in piezometers. It is significant that the CPTU and the DMT results, although not equal, are essentially

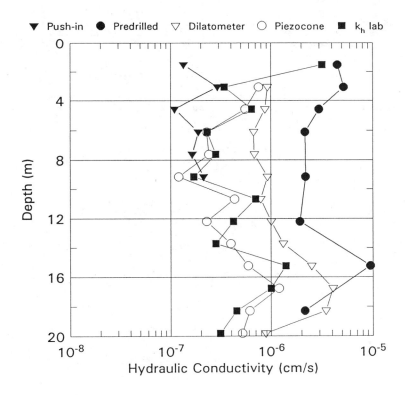

FIG. 5--Results from Field Measurements of Hydraulic Conductivity.

parallel with depth. The ratio of t_{50}(DMT)/t_{50}(CPTU) shows no trend with depth and ranges from 0.9 to 2.3 with an average value of 1.6 ± 0.4 SD (n = 12). Using Eq. 4 and the assumptions for interpreting the DMT and CPTU dissipation data given in the Test Procedures and Interpretation section results in a ratio of c_h(DMT)/c_h(CPTU) that ranges from 1.4 to 3.5 with an average value of 2.3 ± 0.6 SD (n = 12).

The values of c_h required to compute k_h using Eq. 4 should ideally be similar for both devices. However, there is uncertainty associated with the theoretically derived values of T_{50} for both devices and the assumptions used to deal with the fact that the DMT is a flat blade and not a circular device like the CPTU. This latter issue requires two assumptions to be addressed in interpreting the DMT data: (1) is dissipation of excess pore pressures one-dimensional (i.e., radial) or plane strain, and (2) what equivalent radius should be used in Eq. 4 for the DMT? The DMT has an aspect ratio (width/thickness) of 6.8 compared to an aspect ratio of one for the CPTU (Fig. 2). Since no theoretical solution has been published to interpret the DMT dissipation results it is often assumed that the consolidation conditions are one-dimensional rather than plane strain. Using half the DMT blade thickness as an equivalent radius is also an approximation. One could consider using an equivalent radius for an area that is the same as the rectangular cross sectional area of the DMT (96 mm x 14 mm). However, using this assumption would increase the difference between c_h for the two devices rather than making them closer (the average of the ratio of c_h(DMT)/c_h(CPTU) would increase from 2.3 to 20.3; approximately one order of magnitude increase).

The ratio of k_h(DMT)/k_h(CPTU) for the data plotted in Figure 5 ranges from 1.2 to 7.7 with an average of 3.4 ± 1.9 SD (n = 12). As noted previously, although not equal, it is significant that the DMT and CPTU k_h profiles are essentially parallel with depth. Part of the difference is associated with the previously discussed uncertainty in estimates of c_h from both devices but is undoubtedly also due to estimates of the modulus values used in the calculation of k_h (i.e., m_h in Eq. 3). Estimates of M used to compute m_v from both instruments are based on empirical correlations for which there is wide scatter. This could easily result in M values in error by as much as a factor of two.

The CPTU and DMT results show highest values of horizontal hydraulic conductivity at shallow depths and also at approximately 16 m. This trend is nearly identical to the laboratory results as discussed in the previous section because of the fact that the silt content of the soil is highest at the surface and at approximately 16 m (Fig. 1). Similar results have been reported using a CPTU at a site nearby in Amherst, MA by Baligh and Levadoux (1980).

Figure 6 presents the results from a series of both slug and bail tests that were conducted in predrilled piezometers with screens of different length to diameter ratios. Five different piezometers were installed all with a nominal screen diameter of 76 mm (3 in.) and with the center of the screen located at a depth of 3.0 m (10 ft). The results clearly show the influence of length to diameter ratio and suggest the greater influence of macro features as the length of the screen increases. However, for this particular depth there also appears to be an upper limit of L/D for which little change in k_h occurs with increasing L/D. These results also show that no particular trend was found between values measured from slug tests versus bail tests in this soil. In some cases the values obtained from the bail tests were higher (e.g., L/D = 2 and 8) and in the case of L/D = 20 were slightly lower than that obtained from the slug tests. Tremblay and Eriksson (1987) found essentially no difference between measured values of hydraulic conductivity for push-in piezometers in sensitive clays using both bail and slug tests.

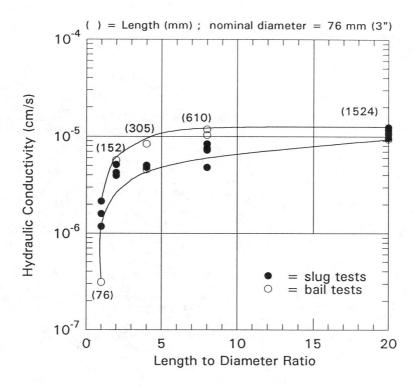

FIG. 6--Influence of Length to Diameter Ratio on Measured Hydraulic Conductivity for Predrilled Slotted Screen Piezometer.

DISCUSSION OF RESULTS

The results presented in the previous section indicate several important aspects regarding the field and laboratory techniques used to measure the hydraulic conductivity of the soil at this site. These include: (1) scale effects, (2) effects of full displacement versus non-displacement in situ techniques, (3) direction of flow during a test relative to the direction of deposition of the soil (i.e., parallel or perpendicular to the varves), and (4) accuracy of methods used to interpret the test results.

The laboratory flexible wall k_h results and the results from the predrilled piezometers with different L/D's (Fig. 6) give a very clear indication of the influence of scale on the measurement of hydraulic conductivity. The data show that as the L/D of the predrilled piezometers increased, the hydraulic conductivity also increased up to a limiting value. Clearly a greater volume of soil is involved in the tests with the larger L/D's and hence the greater the influence of macro features as the length of the screen increases. The piezometer with the smallest L/D = 1 indicates a k_h value approaching that of the laboratory flexible wall k_h value which involves a very small area exposed to flow as compared to the predrilled piezometers. If it were possible to accurately estimate the volume of soil that is influencing the tests conducted in the different size predrilled piezometers then the scale effects could be explicitly quantified. In order to make such an estimate, the radial distance from the piezometer that defines the zone of influence must be determined and also the extent to which the flow remains horizontal (e.g., for a slug test, gravity will cause a downward component of flow and hence involve a greater volume of soil as compared to assuming only horizontal flow in the zone of influence). For these reasons calculating the volume of soil in the zone of influence is very difficult. However, as an indicator of scale effects it is useful to compare the area exposed to flow in the laboratory flexible wall test versus the different sized predrilled piezometers.

The laboratory flexible wall specimen has an area of 0.004 m² in the direction of flow while the predrilled piezometers have circumferential areas of flow ranging from 0.018 m² (L/D = 1) to 0.365 m² (L/D = 20). The predrilled piezometers hence have flow areas that are 4.5 and 90 times greater than the laboratory specimen for the smallest and largest piezometers respectively. The data in Figure 6 indicate that the measured value of k_h in the predrilled piezometers is approximately constant for an L/D greater than 5. This suggests that the scale effect for this particular case has an upper limit beyond which the influence of macro features on the hydraulic conductivity does not increase. For an L/D = 5, the circumferential area is equal to 0.0912 m² which is 23 times greater than that of the laboratory flexible wall specimen (i.e., 0.004 m²). It is interesting to note that the ratio of the average k_h value for the predrilled piezometer estimated from Figure 6 for an L/D = 5 is 22 times greater than that measured in the flexible wall specimen. Hence, in this particular case the ratio of flow areas and hydraulic conductivities are nearly identical.

Further evidence of scale effects is given in the results plotted in Figures 3 and 5 which show that the predrilled piezometers give the highest values of hydraulic conductivity compared to all other field and laboratory methods. The predrilled piezometers used for the data plotted in Figure 5 have an effective screen length of 305 mm and, as noted above, are exposed to a significant volume of soil during a slug test. These piezometers are likely to be more influenced by macro features than the lab tests and thus give a "mass" value of hydraulic conductivity whereas the CPTU and DMT tests are more likely to yield a "matrix" value of horizontal hydraulic conductivity. While some smearing of the borehole does occur during installation of a predrilled piezometer this factor is believed to be significantly less important than the remolding that occurs when using full displacement devices as discussed in the following paragraph.

The screen length of the push-in piezometers is similar to the predrilled piezometer and hence scale effects should not be a factor in comparing results from these two methods. However, the results are significantly different which is most probably due to the fact that the push-in piezometer is a full displacement device which can cause significant smearing/remolding of the soil during penetration. This would be particularly acute in the case of a varved clay since during penetration the push-in piezometer would distort the varves (i.e., the varves get rolled over) and there is a greater influence of the clay on the measured results than the silt compared to if the varves remained intact in a horizontal orientation. It is interesting to note that the push-in piezometers give values similar to the laboratory k_v values which are for flow across the varves (although the trend with depth is not identical). Other problems with the push-in piezometer (as discussed by Lutenegger and DeGroot 1992; 1993) include potential screen clogging although the piezometers were pushed with a positive water head inside the piezometer to minimize this effect.

The CPTU and DMT instruments are full displacement techniques and hence also cause a smearing/remolding of the soil, however, the estimation of hydraulic conductivity is based on the dissipation of excess pore pressure based on the theory of consolidation and hence the compressibility (i.e., modulus) of the soil is used in the calculation. The push-in piezometer on the other hand is a flow test through a remolded/distorted soil and the interpretation of the results does not take into account changes in the compressibility or fabric of the soil. A secondary factor that could be the cause of differences between the DMT and CPTU results is the fact that the DMT is exposed to a larger area of soil since the flexible membrane is 60 mm in diameter (exposed area = 0.0028 m^2) whereas the porous element on the CPTU has a circumference of 112 mm (A = 1000 mm^2) but is only 5 mm thick (exposed area = 0.0006 m^2). Therefore the DMT results would be more prone to be influenced by a larger volume of soil.

The results presented herein clearly show that the hydraulic conductivity of a fine grained soil can be highly dependent on scale effects, the method of measurement, and flow direction. However, the clay tested in this case probably represents an extreme case because of the varved nature of the soil. Selection of values for design should depend on the problem that needs to be solved. If one is conducting fate contaminant transport modelling or predicting pump and treat rates for remediation work using a double porosity model the free water in the formation will be removed quickly (i.e., through cracks, sand lenses, macro features). Once this "free" water is removed then the water held in the matrix has to be removed. Hence one would require estimates of both "mass" and "matrix" hydraulic conductivities in order to reliably evaluate contaminant removal rates. The CPTU and DMT profiling tools provide rapid estimates of the probable matrix hydraulic conductivity. Estimates of the formation hydraulic conductivity should involve a larger volume and hence should probably make use of predrilled piezometers provided that the appropriate L/D ratio is used. For this particular study a L/D of 5 or more is indicated as sufficient to evaluate "mass" flow characteristics. These results may be different for a massive clay deposit that is not as structured as the varved clay investigated in this study. In such a soil the influence of L/D may be less significant and similarly the difference between predrilled, laboratory, CPTU, and DMT results may be much less.

SUMMARY AND CONCLUSIONS

The results of this study show that for a varved clay the measurement of hydraulic conductivity is highly varied depending on scale effects, the test method used, and the direction of flow during testing. Results from predrilled piezometers are dependent on the L/D ratio used; if a large enough value is used (approximately 5 in this case) the predrilled piezometer provides the most likely value of the formation hydraulic conductivity. Unfortunately, these instruments are expensive to install and involve other problems such as disposal of cuttings, etc. The laboratory tests on vertically and horizontally oriented specimens involve small scale estimates of hydraulic conductivity compared to the predrilled piezometers but provide values of the hydraulic anisotropy which cannot normally be determined from in situ tests. The CPTU and DMT methods provide very rapid profiling techniques and probably give estimates of the matrix hydraulic conductivity. For an in situ hydraulic conductivity ranging between 1 x 10^{-7} to 1 x 10^{-6} cm/s a testing time of approximately 20 minutes per test is needed to obtain t_{50} and hence k_h. These methods have the disadvantage of requiring estimates of the coefficient of consolidation and the modulus of the soil which involves a considerable amount of uncertainty. Full displacement push-in piezometers although quick and easy to install give very low values of hydraulic conductivity due to remolding/distortion of the varved soil. Development of a new method for interpreting results from these tests is needed to take into account these factors. Indirect measurements of hydraulic conductivity from oedometer tests on horizontally trimmed specimens involved considerable uncertainty in selecting the appropriate stress range for estimating k. More reasonable estimates may be obtained if direct flow tests are conducted.

ACKNOWLEDGEMENTS

The piezocone dissipation tests were performed by Chris Tonzi. The authors thank Paul Cheever, John DiFini, Mike Lally, Gerry Miller and Anthony Opiatowski for their help in conducting some of the laboratory and field tests. The authors also thank Allen Bennett for his help in preparation of the figures and final manuscript.

REFERENCES

Baligh, M.M. and Levadoux, J.N., 1980, "Pore Pressure Dissipation After Cone Penetration," Massachusetts Institute of Technology, Department of Civil Engineering, Cambridge, MA, Report MITSG 80-13, 368 pp.

Baligh, M.M. and Levadoux, J.N., 1986, "Consolidation After Undrained Piezocone Penetration. II: Interpretation," Journal of Geotechnical Engineering, ASCE, Vol. 112, No. 7, pp. 727-745.

Casagrande, L. and Poulos, S., 1969, "On the Effectiveness of Sand Drains," Canadian Geotechnical Journal, Vol. 6, pp. 287-326.

Chan, H.T. and Kenney, T.C., 1973, "Laboratory Investigation of Permeability Ratio of New Liskeard Varved Soil," Canadian Geotechnical Journal, Vol. 10, pp. 473-488.

Chapius, R.P., 1989, "Shape Factors for Permeability Tests in Boreholes and Piezometers," Ground Water, Vol. 27, No. 5, pp. 647-654.

Daniel, D.E., Trautwein, S.J., Boynton, S.S and Foreman, D.E., 1984, "Permeability Testing with Flexible-Wall Permeameters," Geotechnical Testing Journal, ASTM, Vol. 7, No. 3, pp. 113-122.

Gupta, R.C. and Davidson, J.L., 1986, "Piezoprobe Determined Coefficient of Consolidation," Soils and Foundations Journal, Japanese Society of Soil Mechanics and Foundation Engineering, Vol. 26, No. 3, pp. 12-22.

Hvorslev, M.J., 1951, "Time Lag and Soil Permeability in Groundwater Observations," U.S. Army Waterways Experimental Station, Vicksburg, Miss., Bulletin No. 36.

Kenney, T.C., 1963, "Permeability Ratio of Repeatedly Layered Soils," Geotechnique, Vol. 13, No. 4, pp. 325-333.

Ladd, C.C. and Foott, R., 1977, "Foundation Design of Embankments on Varved Clays," FHWA Report TS-77-214, 234 pp.

Lutenegger, A.J. and Kabir, M.G., 1988, "Dilatometer C-Reading to Help Determine Stratigraphy," Penetration Testing 1988, ISOPT-1, Vol. 1, pp. 549-554.

Lutenegger, A.J. and DeGroot, D.J., 1992, "Measurement of Hydraulic Conductivity in Clay Using Push-In Standpipe Piezometers," Current Practices in Ground Water and Vadose Zone Investigations, ASTM STP 1118, pp. 362-374.

Lutenegger, A.J. and DeGroot, D.J., 1993, Discussion of "Smear Effects of Vertical Drains on Soft Bangkok Clay," by Gergado et al., Journal of Geotechnical Engineering, ASCE, Vol. 119, No. 1, pp. 181-184.

Marchetti, S., 1980, "In Situ Tests by Flat Dilatometer," Journal of the Geotechnical Engineering Division, ASCE, Vol. 106, pp. 299-321.

Mitchel, J.K. and Gardner, W.S., 1975, "In Situ Measurement of Volume Change Characteristics," State-of-the-Art Report, Proceedings of the Conference on In Situ Measurement of Soil Properties, ASCE, Vol. II, pp. 279-345.

Olson, R.E. and Daniel, D.E., 1981, "Measurement of the Hydraulic Conductivity of Fine-Grained Soils," Permeability and Groundwater Contaminant Transport, ASTM STP 746, pp. 18-64.

Robertson, P.K., Campanella, R.G., Gillespie, D. and By, T., 1988, "Excess Pore Pressures and the Flat Dilatometer Test," Penetration Testing 1988, ISOPT-1, Vol. 1, pp. 567-576.

Tavenas, F., Leblond, P., Jean, P. and Leroueil, S., 1983, "The Permeability of Natural Soft Clays. Part I: Methods of Laboratory Measurement," Canadian Geotechnical Journal, Vol. 20, pp. 629-644.

Tortenson, B.A., 1977, "The Pore Pressure Probe," Nordiske Mote, Bergmekanikk, Paper No. 34, Oslo, pp. 34.1-34.15.

Tremblay M. and Eriksson, L., 1987, "Use of Piezometers for In Situ Measurement of Permeability," Ninth European Conference on Soil Mechanics and Foundation Engineering, Vol. 1, Paper No. 23.

Mark A. Phifer[1], Eric C. Drumm[2], and Glenn V. Wilson[3]

EFFECTS OF POST COMPACTION WATER CONTENT VARIATION ON SATURATED CONDUCTIVITY

REFERENCE: Phifer, M. A., Drumm, E. C., and Wilson, G. V., "Effects of Post Compaction Water Content Variation on Saturated Conductivity," <u>Hydraulic Conductivity and Waste Contaminant Transport in Soil, ASTM STP 1142</u>, David E. Daniel and Stephen J. Trautwein, Eds., American Society for Testing and Materials, Philadelphia, 1994.

ABSTRACT: Reductions in water content after construction can produce desiccation cracking and associated aggregation of clay barriers. This produces two flow regimes within the clay barrier. Conduit flow occurs through the desiccation cracks (preferential flow paths), and matrix flow occurs through the aggregates (intact matrix). An investigation of the effect of post compaction water content variations on volumetric strain and saturated hydraulic conductivity of desiccation produced aggregates was conducted. Kaolinite samples for use in flexible wall permeameters were prepared using modified, standard, and reduced proctor methods. The initial volume and hydraulic conductivity of each sample were determined. Each sample was progressively dried back from its molding water content, and its volume and hydraulic conductivity were determined after each dry-back. Correlations between gravimetric water content, dry bulk density, volumetric strain, and saturated hydraulic conductivity were made. Significant volumetric strain occurred, which produced a decrease in the hydraulic conductivity (matrix flow) of the intact aggregate. The strain, however, while resulting in a lower matrix flow in the aggregates, should produce a higher overall flow in barriers due to conduit flow within the resulting desiccation cracks.

KEYWORDS: water content, hydraulic conductivity, kaolinite, landfill, compacted clay layer, desiccation, volumetric strain

INTRODUCTION

EPA has developed the following liquids management strategy (EPA 1989) for RCRA hazardous and mixed waste landfills:

"(1) minimizing the leachate generation by keeping liquids out of the unit, and (2) detecting, collecting, and removing leachate within the unit."

---

[1]Graduate Fellow, Department of Civil and Environmental Engineering, The University of Tennessee, Knoxville, TN 37996.

[2]Associate Professor, Department of Civil and Environmental Engineering, The University of Tennessee, Knoxville, TN 37996.

[3]Assistant Professor, Department of Plant and Soil Science, The University of Tennessee, Knoxville, TN 37901

Furthermore EPA stated the following concerning this strategy (EPA 1989):

> "The Agency considers keeping water out of the unit to be the prime element of the strategy. Thus, the Agency believes that a properly designed and constructed cover becomes, after closure, the most important feature of the landfill structure. The Agency requires that the cover be designed and constructed to provide long-term minimization of the movement of water from the surface into the closed unit."

EPA produced a recommended cover profile which EPA considers to conform to the liquids management strategy. The recommended cover consists of, from top to bottom, a vegetated surface, a soil layer, a drainage layer, and a barrier layer. If the unit has a liner, the barrier layer must be a composite consisting of an upper flexible membrane liner (FML) in direct contact with a compacted clay layer. The FML is optional if the unit does not have a liner. The layers which most directly function to fulfill the liquids management strategy are the drainage and barrier layers. The drainage layer functions to minimize the head (amount and time) of water on the barrier layer, minimize water infiltration into the barrier layer, and provide rapid and efficient removal of water out of the cover profile. The barrier layer functions to provide long-term minimization of water percolation into the waste. If both a FML and clay layer are utilized, they are to provide backup for one another. In all cases, a minimum two foot compacted clay layer or equivalent soil barrier is currently required (EPA 1989).

Much research has been conducted to determine and optimize the construction factors that control production of compacted clay layers with an initially low hydraulic conductivity. The most influential construction factors for production of such a clay layer have been determined to be molding water content, dry density (Daniel and Benson 1990), and fate of clods (Benson and Daniel 1990).

Daniel and Benson (1990) conducted a laboratory investigation to define a zone of acceptability in terms of the molding water content, dry density, and hydraulic conductivity (1 nm/s). The zone essentially encompasses an area from the zero air voids line to a parallel line wet of the line of optimums. This zone defines an area which allows trade offs between molding water content and dry density. The wetter the sample is at compaction, the less dense it is required to be to achieve the required hydraulic conductivity initially.

Benson and Daniel (1990) performed experimentation concerning the influence of clods which indicated that a reduction in clod size, increased compactive effort, and compaction wet of optimum effectively destroy clods and the associated preferential flow paths between clods.

Research on compacted clay layers has indicated that desiccation during construction subsequently increases hydraulic conductivity. Desiccation cracking is typically defined (Kleppe and Olson 1985) as exposure of the clay layer to atmospheric conditions and subsequent air drying of the clay which produces cracking. Several field investigations of constructed clay liners have been conducted. These investigations indicate that high hydraulic conductivities in field installations generally result from hydraulic defects in the clay due to construction methodologies and/or desiccation cracking due to inadequate protection after construction (Daniel 1984; Day and Daniel 1985; Elsbury et al. 1990).

Kleppe and Olson (1985) investigated the desiccation shrinkage of sand/clay mixtures. They concluded that desiccation shrinkage increased linearly with the molding water content but was not affected by density. Samples which were compacted on the dry side of optimum but saturated prior to desiccation behaved essentially identical to samples initially compacted on the wet side. Sand/clay mixtures may be utilized to optimize the crack resistance of barriers while maintaining low hydraulic conductivities (a minimum 25% clay is required).

Boynton and Daniel (1985) performed experimentation on 2.5 in. (6.35 cm) thick slabs of desiccated clay and found that desiccation cracks fully penetrated the slab in several hours. It was demonstrated that hydraulic conductivity of desiccated, cracked clay decreased with increased effective stress. The cracks tended to close when the desiccated soil was moistened, but the hydraulic conductivity remained higher than non-desiccated samples.

Daniel and Wu (1993) stated that these previous studies demonstrate that compacted clay barriers crack as water content decreases and that cover systems are particularly vulnerable. They indicated that two methods are available to address this problem. The compacted clay can be built to have minimum shrinkage potential and/or the cover profile can be designed to maintain a consistent water content within the clay layer. Shrinkage potential is most influenced by the molding water content and coarse particle size fraction. Daniel and Wu (1993) demonstrated the methodology for modification of the hydraulic conductivity zone of acceptability (water content versus dry density) for shrinkage potential. The criteria utilized to determine the overall zone of acceptability included a maximum hydraulic conductivity of 1 nm/s and a maximum shrinkage potential of 4 percent. The 4 percent shrinkage potential represents 10 mm wide crack development from Kleppe and Olson's (1985) study. Its selection is essentially arbitrary, and additional study is needed to determine appropriate criteria. The individual zones of acceptability for hydraulic conductivity and shrinkage potential were determined, the individual zones were overlapped, and the area where all criteria were met was designated as the overall zone of acceptability. This overall zone comprises the portion of the hydraulic conductivity zone of acceptability with low molding water contents and high dry densities.

This research pointed out that the molding water content is probably the most important variable to achieving low hydraulic conductivities during construction since it directly impacts the resulting dry density, destruction of clods, and potential for desiccation cracking. Although desiccation due to air drying of exposed clay barriers during construction has been recognized, it has been generally assumed that desiccation problems are eliminated after placement of the upper protective layers (Daniel and Benson 1990; Kleppe and Olson 1985; Daniel 1984). However, soil water storage is not a static property of a soil that can be described by the presently defined field capacity, wilting point, or residual saturation. It is a time dependent variable based upon the rates of soil water inflow and outflow.

Climatological conditions and the cover profile properties control the post-construction water content variations within compacted clay layers without an overlaying FML. Infiltration into the cover profile which is not completely removed by the processes of lateral drainage through the drainage layer, evapotranspiration, and vapor movement, provide the primary soil water inflow into the clay layer. Redistribution over an extended time period is the primary mechanism contributing to the outflow of water from the clay layers. Redistribution is a continuous movement of water at a progressively decreasing rate that approaches equilibrium (constant soil water potential throughout the profile) only after an extended time. Redistribution will continue to reduce the soil water storage until another precipitation event occurs. (Hillel 1982; Schroeder et al. 1984)

A compacted clay layer with a properly constructed overlaying FML is not part of the water budget of the layers above the FML, since water movement through the FML can occur only due to diffusion and holes in the FML. The clay layer will not receive significant water inflow if the FML is properly installed. However, the process of redistribution will occur between the clay layer and the layers below it. Since the purpose of the liner/cover concept for a landfill is to keep water out

of the waste and remove any leachate which is produced, the extent of
redistribution out of the clay layer will be dependent upon the unit
profile below the FML and the efficiency of leachate removal. (Hillel
1982; Schroeder et al. 1984)

While the post construction variations in water content of covered
compacted clay layers will most likely not decrease to the level of
total air dried desiccation, it is highly likely that the layer's water
content will vary (increase and/or decrease) from its initial molding
water content during its design life. Such reductions in water content
after construction, which result in volumetric strain, have the
potential to produce two flow regimes within the clay barrier. Conduit
flow occurs through the desiccation cracks (preferential flow paths),
and matrix flow occurs through the aggregates (intact matrix).

Based upon past research indicating the importance of water
content, it is evident that the impacts of such water content variations
on the ability of the clay layer to maintain appropriate barrier
properties (both conduit and matrix flow regimes) must be determined.
Such a determination is needed in order to properly assess the long-term
integrity of compacted clay layers, select the materials most resistant
to such variation, determine the impact of cover profile design, and
develop appropriate maintenance measures. This may also mean that a
balance must be maintained between the contradictory goals of minimizing
flow into and through the compacted clay layer and maintaining
sufficient water content within the layer to assure low hydraulic
conductivity.

OBJECTIVES

This paper describes the laboratory investigation of the effects
of post-construction water content variations on volumetric strain and
saturated hydraulic conductivity (matrix flow) of desiccation produced
kaolinite aggregates. The investigation was designed to address the
following aspects of the water content variation on volumetric strain
and hydraulic conductivity:

1) Determine the relationships between post compaction water
content variations and hydraulic conductivity (matrix flow),
volumetric strain (density), and potential for aggregation
(creation of preferential flow paths).
2) Determine the effect of the molding water content and initial
bulk density on such relationships.

EXPERIMENTAL METHOD

Materials

The kaolinite utilized was produced by the J. M. Huber Corporation
under the trade name Barden AG-1. Pertinent material properties are
listed below (Table 1). A 0.01 N calcium sulfate ($CaSO_4$) solution was
utilized as the permeant.

Test Procedure

General test data--All test procedure water content data are
gravimetric water contents expressed as a percent. All density data are
dry bulk densities. All volumetric strain data are ratios expressed as
a percent of the volume change with respect to the original volume. A
negative strain represents shrinkage and a positive strain represents
swelling.

Proctor tests--Modified, standard, and reduced Proctor tests (Fig.
1 and Table 2) respectively were conducted in accordance with ASTM Test

Method for Moisture-Density Relations of Soils and Soil-Aggregate Mixtures Using 10-lb (4.54-kg) Rammer and 18-in. (457-mm) Drop (D 1557), ASTM Test Method for Moisture-Density Relations of Soils and Soil-Aggregate Mixtures Using 5.5-lb (2.49-kg) Rammer and 12-in. (305-mm) Drop (D 698), and ASTM D 698 with 15 drops per layer rather than the specified 25 (Daniel and Benson, 1990).

TABLE 1--Kaolinite properties

Mineral composition[1]	Kaolinite	94.6-97.2%
	Silica, quartz	1.5-3.8%
	Titanium dioxide	1.3-1.6%
Particle size[1]	>45μm	0.0-1.0%
	>5μm	0.3-3.5%
	<2μm	92.0-97.0%
Atterberg limits	Plastic limit	29.9
	Liquid limit	75.0
	Plasticity index	45.1
	Shrinkage limit[2]	17
Miscellaneous[1]	Water content	1.0%
	Specific gravity	2.60

[1](Melton 1992)
[2]Casagrande graphical technique based upon the liquid limit and plasticity index (Holtz and Kovacs 1981)

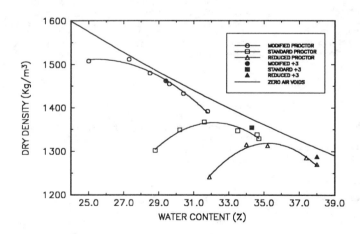

FIG. 1--Moisture-density relationship and initial sample preparation

TABLE 2--Results of Proctor tests

Parameter	Modified	Standard	Reduced
Optimum water content (%)	26.3	31.6	34.6
Maximum dry density (kg/m³)	1520	1370	1320

Sample preparation--Actual samples for the hydraulic conductivity tests were prepared based upon the results of the proctor tests. One sample was prepared using each proctor energy (modified, standard, and reduced) with kaolinite at a water content approximately 3% above optimum for that respective energy. The samples were prepared in 4 in. (10.16 cm) proctor molds, extruded, trimmed to size with a soil lathe, and scarified on the ends. The initial water content, density, and dimensions were determined (Table 3). The samples, as initially prepared, are compared to the Proctor curves (Fig. 1). The Modified +3 designated Mod +3, Standard +3 (Std +3), and Reduced +3 (Red +3) samples were prepared at 3.1%, 2.7%, and 3.3% above the appropriate optimum water content respectively.

TABLE 3--Initial sample preparation

Sample	Dry density (kg/m³)	Water content (%)	Average diameter (cm)	Average length (cm)
Mod +3	1462	29.4	7.0147	8.8400
Std +3	1354	34.3	7.0584	10.8234
Red +3	1288	37.9	7.0739	10.2019

Hydraulic conductivity test procedure--Falling headwater rising tailwater hydraulic conductivity tests were performed in general conformance to ASTM Test Method for Measurement of Hydraulic Conductivity of Saturated Porous Materials Using a Flexible Wall Permeameter (D 5084) Method C. The test procedure was also designed considering other reference sources (Boynton and Daniel 1985; Olson and Daniel 1981; Daniel 1989). The primary pieces of equipment utilized for the hydraulic conductivity tests included a pressure control panel for headwater control, flexible wall permeameters, and burettes for tailwater control. The samples were saturated by applying an approximately 34.5 kPa hydraulic pressure to the headwater and between 67.5 to 84.4 kPa vacuum to the tailwater port for a minimum of 12 hours. The samples were subjected to an average vertical effective stress ranging from 101 to 131 kPa. This effective stress is not representative of typical field conditions where the vertical effective stress might range from 14.5 to 22.5 kPa. The effective stress used was chosen for convenience. Such effective stresses that are higher than field stresses could cause cracks to close thus reducing the tested hydraulic conductivity from that of the field. However no cracking occurred in the samples, thus he effective stress used should not have adversely affected the results. The hydraulic heads and gradients utilized are listed below (Table 4). The temperature varied between 21.0 and 27.0°C during testing. All hydraulic conductivity data has been expressed in units of nm/s (1 nm/s is equivalent to 1×10^{-7} cm/s), and has been corrected for temperature to 20°C and dimensional changes.

TABLE 4--Hydraulic conductivity test parameters

Sample	Headwater hydraulic head range (m)	Tailwater hydraulic head range (m)	Hydraulic gradient range
Mod +3	3.73-3.58	0.69-0.76	32.5-34.9
Std +3	3.72-3.48	0.67-0.77	25.6-28.6
Red +3	3.72-3.48	0.68-0.79	27.2-31.5

Sample dry-back procedure--The samples were permeated immediately after preparation to determine the initial saturated hydraulic conductivity. Then they were removed from the permeameter and allowed to dry-back under atmospheric conditions to a water content lower than the molding water content. After dry-back the samples were again permeated to determine the subsequent hydraulic conductivity. This process of dry-back followed by permeation was continued at subsequently lower levels of dry-back water contents. The sample dimensions and mass were taken immediately after each permeation and at the end of each dry-back. During dry-back the samples remained in the membrane with only the ends exposed, remained in the same orientation as in the permeameter, and were placed on a coarse sand bed overlain by a filter fabric. The temperature ranged from 21.0 to 28.0°C and the relative humidity ranged from 32 to 85% during dry-back. Each dry-back continued until the desired water content was reached as determined by mass measurements.

RESULTS

Mod +3, Std +3, and Red +3 are the sample designations. These designations denote which proctor energy was used (modified, standard, or reduced respectively) and denote that at compaction the water content was approximately 3 percent above the respective proctor energy optimum water content (see Tables 2 and 3). These samples represent desiccation produced intact aggregates (matrix flow) within a compacted clay barrier and not the entire barrier which may include desiccation cracks.

Dry-Back Paths

Mod +3, Std +3, and Red +3 were subjected to cyclic dry-back paths (Fig. 2, 3, and 4). Data point 1 in these figures represents the initial conditions of each sample and subsequent odd numbered data points represent succeeding end of dry-back conditions. The even numbered data points represent end of permeation or rebound conditions with saturated water contents. Actual volumetric measurements of the samples for data points 3 and 5 (end of dry-back conditions) were not taken, however mass and water contents were obtained. Based upon the water contents, previous and succeeding data trends, and identical experiments on other soils, the density and volumetric strains for these data points were assumed.

Mod +3 swelled (density decreased) with the initial permeation (Fig. 2 data point 2). It then shrank (density increased) relative to this initial permeation with succeeding dry-backs. Upon permeation (rebound) after dry-back shrinkage, the sample rebounded toward but did not return to its initial conditions. A limiting dry-back density of approximately 1630 kg/m^3 was observed and a limiting rebound density and rebound water content of approximately 1550 kg/m^3 and 27.2%, respectively, were evident. The limiting densities deviated by about 80 kg/m^3. This indicates that after being reduced to the dry-back water content, which produced the limiting densities, the sample can still

PHIFER ET AL. ON WATER CONTENT 325

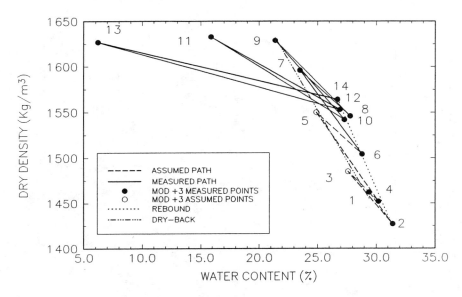

FIG. 2--Modified +3 cyclic dry-back path

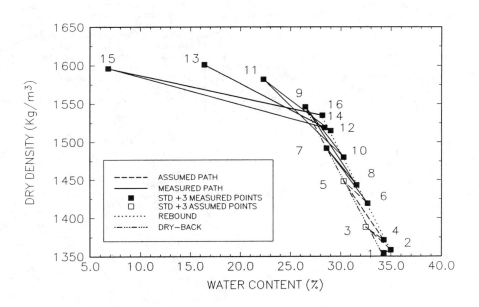

FIG. 3--Standard +3 cyclic dry-back path

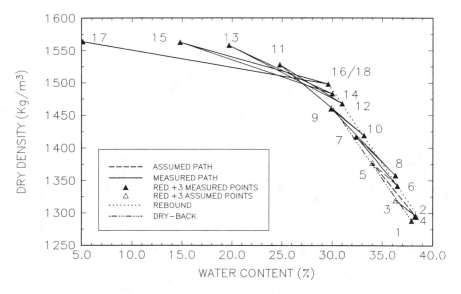

FIG. 4--Reduced +3 cyclic dry-back path

shrink and swell as the water content varies. Linear relationships between rebound (saturated) water content and rebound density, and between dry-back water content and dry-back density were noted. The dry-back linear relationship ends at the limiting dry-back density.

Shrinkage (density increase) was observed for Std +3 and Red +3 for both the initial permeation and succeeding dry-backs relative to the initial conditions. Upon permeation, after dry-back shrinkage, the sample rebounded toward but did not return to its initial conditions. Limiting dry-back densities of approximately 1600 and 1560 kg/m³ were reached, respectively, for Std+3 and Red+3. Limiting rebound densities of approximately 1520 and 1490 kg/m³, and limiting rebound water contents of 28.5 and 29.7% were reached, respectively, for Std+3 and Red+3. The limiting densities deviated by about 80 and 70 kg/m³, respectively, for Std+3 and Red+3. These data findings also indicate that after limiting conditions are reached, the sample can still shrink and swell as the water content varies. Linear relationships between rebound (saturated) water content and rebound density, and between dry-back water content and dry-back density were noted (Fig. 3 and 4). The dry-back linear relationship ends at the limiting dry-back density.

The dry-back density limit of the samples decreases in order of Mod+3, Std+3, and Red+3. The rebound density limit decreases and the rebound (saturated) water content increases in the same order. However, limits of all samples are much closer in value than the initial sample preparation values (Fig. 5). Additionally the magnitude of the saturated water content decrease, and both the rebound and dry-back density increase, follows the order of Red+3, Std+3, and Mod+3 (Table 5). The rate of density change is also greatest in Red+3, followed in order by Std+3 and Mod+3. This indicates that reductions in water content of kaolinite prepared at higher molding water contents and lower densities has the potential to undergo greater changes at a higher rate.

Kaolinite prepared at lower water contents and higher densities undergo less change at a lower rate.

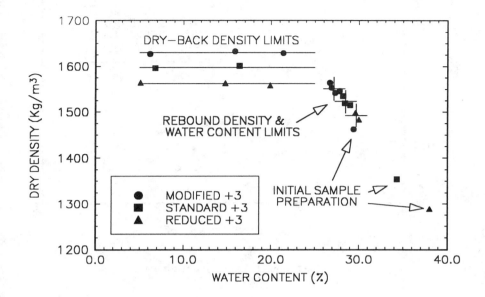

FIG. 5--Dry-back and rebound density and water content limits

TABLE 5--Dry-back effects on water content and density

Sample	Overall sat. w decrease[1]	Overall reb ρ increase[2]	Overall D-B ρ increase[3]	Reb. ρ rate of change[4]
Mod +3	-4.7	102	165	10.0
Std +3	-6.8	181	242	16.2
Red +3	-8.7	210	276	16.5

[1]saturated water content (%); [2]rebound density (kg/m^3); [3]dry-back density (kg/m^3); [4]Rebound density (kg/m^3) increase per 1% dry-back

Hydraulic Conductivity Variations

Density effect--The relationship between rebound density and the associated hydraulic conductivity for each sample was determined (Fig. 6). These measurements correspond to the even numbered data points (rebound points) on the cyclic dry-back paths (Fig. 2, 3, and 4). The density generally increased with the number of dry back cycles. The hydraulic conductivity subsequently decreased linearly as density increased for all samples. Each sample appeared to follow the same linear path and appeared to progress toward similar endpoints even though they started at different points (Table 6).

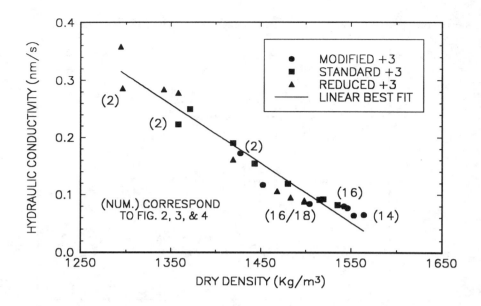

FIG. 6--Rebound density effects on hydraulic conductivity variation

TABLE 6--Hydraulic conductivity

Sample	Initial hydraulic conductivity (nm/s)	Final hydraulic conductivity (nm/s)	Limiting rebound density (kg/m^3)	Limiting sat. water content (%)
Mod+3	0.17	0.07	1550	27.2
Std+3	0.25	0.08	1520	28.5
Red+3	0.36	0.09	1490	29.7

<u>Saturated water content effect</u>--The relationship between saturated water content (rebound) and the associated hydraulic conductivity for each sample was determined (Fig. 7). These measurements correspond to the even numbered data points (rebound points) on the cyclic dry-back paths (Fig. 2, 3, and 4). The saturated water content decreased with the number of dry back cycles. The hydraulic conductivity subsequently decreased linearly as water content decreased for all samples. As previously observed, each sample appeared to follow the same linear path and appeared to progress toward similar endpoints even though they started at different points (Table 6).

<u>Volumetric strain effect</u>--The relationship between dry-back volumetric strain and the associated hydraulic conductivity (measurement following dry-back) for each sample was determined (Fig. 8). These measurements correspond to the odd numbered data points (dry-back points) on the cyclic dry-back paths (Fig. 2, 3, and 4). The

relationship between rebound volumetric strain and the associated hydraulic conductivity for each sample was determined (Fig. 9). These measurements correspond to the even numbered data points (rebound points) on the cyclic dry-back paths (Fig. 2, 3, and 4). In both cases, the volumetric strain increased with the number of dry back cycles, and the hydraulic conductivity linearly decreased for all samples. However, while comparisons of rebound density and saturated water content versus hydraulic conductivity produced one relationship for all three samples, the strain versus hydraulic conductivity produced a separate linear relationship for each sample. Mod+3 underwent both the least magnitude and lowest rate of strain (slope) followed by Std+3 and then Red+3 (Table 7). The dry-back strains were greater than the rebound strains even under final test conditions, and the final conditions appear to represent limiting conditions. The difference in dry-back and rebound limiting conditions is approximately the same for each sample, and it means that the samples are still subject to approximately 3.5% change in volume as the water content varies after limiting conditions have been reached.

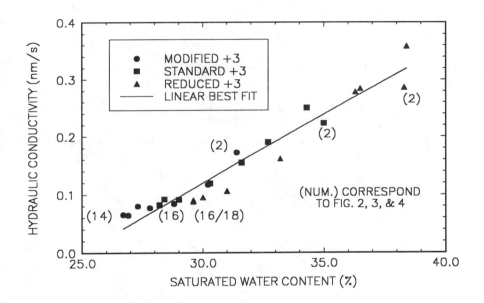

FIG. 7--Saturated water content effects on hydraulic conductivity variation

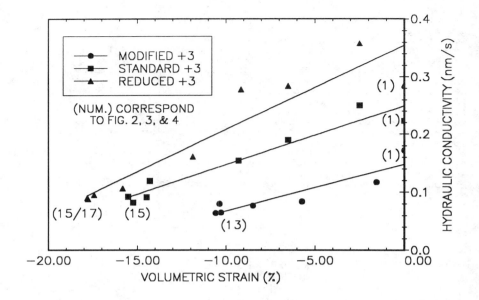

FIG. 8--Dry-back volumetric strain effects on hydraulic conductivity variation

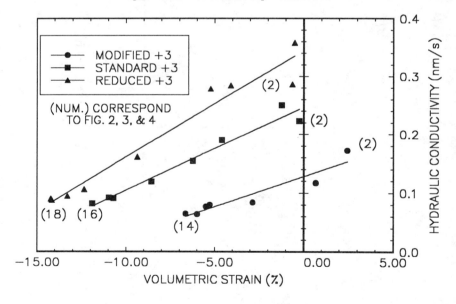

FIG. 9--Rebound volumetric strain effects on hydraulic conductivity variation

TABLE 7--Volumetric strains

Sample	Rebound volumetric strain rate[1]	Final rebound volumetric strain (%)	Final dry-back volumetric strain (%)
Mod+3	-0.66	-6.65	-10.29
Std+3	-1.10	-11.91	-15.24
Red+3	-1.17	-14.23	-17.82

[1]Rebound volumetric strain (%) per 1% dry-back

INTERPRETATION

This was a laboratory investigation of the effects of post compaction water content variation on volumetric strain and saturated hydraulic conductivity (matrix flow) of desiccation produced kaolinite aggregates. As the water content of the samples cycled between saturation and increasingly lower dry-back water contents, the following relationships were observed:

1) As the samples dried back below the molding water content, the dry-back density increased and the sample shrank in a linear relationship with the dry back water content until a limiting dry-back density and strain were reached. These limits are similar to that defined by the ASTM shrinkage limit test, however the conditions of the testing and method of measurement are different, which could result in different values.
2) As the samples were saturated after a dry-back period, a rebound toward the original conditions was observed. However, the original conditions were not replicated and non-recoverable strain and density increases occurred. This resulted in a lower saturated (rebound) water content for the new conditions. A linear relationship between saturated water content and both rebound density and strain was observed. Limiting rebound water contents, densities, and volumetric strains were also observed beyond which increased dry-back water content reductions had no effect.
3) The dry-back and rebound limits were not the same, therefore, shrink and swell of the samples could continue between these limits as water content varied.
4) The hydraulic conductivity of the samples decreased linearly with decreases in the saturated water content, increases in rebound density, and decreases in the sample rebound volume.
5) Samples prepared at higher molding water contents and lower dry densities underwent the greatest changes.

It was evident that the water content variations of the compacted kaolinite controlled changes in density, volumetric strain, subsequent saturated water content, and saturated hydraulic conductivity, and that such changes were not fully recoverable. It was also evident that due to the sample size and/or aspect ratio, the samples behaved as individual clay aggregates rather than a continuous compacted clay layer subject to desiccation. However, the results can be extrapolated to the field, where desiccation cracking occurs by viewing the samples as representing individual clay aggregates (intact matrix) formed within a continuous compacted clay layer during desiccation. As reductions in the water content below the molding water content of continuous compacted clay layers occur, volumetric strain results. At some point, the decreasing water content could result in volumetric strains sufficient to initiate formation of desiccation cracks and associated aggregation of the clay layer. As the water content continues to decrease the aggregates shrink, the individual aggregate hydraulic conductivity decreases, and the desiccation cracks between clay aggregates enlarge and create preferential flow paths. The preferential flow paths can control the hydraulic properties of the clay layer rather than the aggregates. Since such strains are not fully recoverable, these preferential flow path could continue to exist even if rewetting occurs. In fact once preferential flow paths control the clay layer hydraulic properties, the individual aggregates may no longer be able to obtain sufficient water to rebound to the full potential.

Some limits on the interpretation of these results and the subsequent extrapolation include:

1) The kaolinite, which is classified as a low shrink/swell clay mineral, was essentially pure with a very fine, uniform particle size. The results may not be completely applicable to high shrink/swell clay minerals or soils with substantial coarse material.
2) The sample size was small and the aspect ratio (height to width) was high. Therefore, the results can not be directly applied to field conditions were the aspect ratio is extremely low.
3) The use of the flexible wall membrane masks the effects shrinkage has on macroscopic hydraulic properties. It provides only hydraulic properties of intact clay aggregates (matrix flow) without cracks.
4) The sample water content variations were produced under either atmospheric conditions or conditions which promote complete saturation.

CONCLUSION

Water content reductions from the molding water content, much less than that causing total air dried desiccation, can significantly impact the saturated water content, density, volumetric strain, and hydraulic conductivity of compacted kaolinite. The extent of impact is affected by the initial placement conditions of the clay. Initial conditions which are more wet and less dense have the potential for greater change. Due to these impacts, water content reductions may result in the formation of preferential flow paths within clay layers.

Reductions in aggregate (matrix) hydraulic conductivity were observed along with associated volumetric strain. Such reductions in matrix hydraulic conductivity would serve to enhance the preferential flow of water through any desiccation cracks produced by the volumetric strain. This aspect should be incorporated into flow models to properly simulate post compaction field conditions.

Water content variations within a compacted clay layer are dependent upon the climatological conditions, cover profile properties, and the water redistribution process. Variations which may have a negative impact are not necessary eliminated by 2 to 3 feet of cover nor a FML. If a compacted clay layer is to be utilized in a landfill cover the following must be considered:

1) Balance must be maintained between constructability (extent of zone of acceptability for initial clay placement) and shrinkage potential (greater shrinkage potential for clay placed at high water contents and low densities).
2) Balance must be maintained between minimization of flow into and through the layer and maintenance of sufficient water within the layer to assure low hydraulic conductivity.

ACKNOWLEDGEMENTS

The following organizations are recognized for contributions to this research effort:

1) J. M. Huber Corporation, Clay Division, Langley, SC for the donation of one ton of kaolinite.
2) Phillips Fibers Corporation, Greenville, SC for the donation of Supac® 16NP-L17800 filter fabric.

REFERENCES

Daniel, D. E., February 1984, "Predicting Hydraulic Conductivity of Clay Liners," *Journal of Geotechnical Engineering*, ASCE, Vol. 110, No. 2, pp. 285-300.

Daniel, D. E., December 1989, "A Note on Falling Headwater and Rising Tailwater Permeability Tests," *Geotechnical Testing Journal*, Vol. 12, No. 4, pp.308-310.

Daniel, D. E. and Benson, C. H., December 1990, "Water Content-Density Criteria for Compacted Soil Liners," *Journal of Geotechnical Engineering*, ASCE, Vol. 116, No. 12, pp. 1811-1830.

Daniel, D. E. and Wu, Y., February 1993, "Compacted Clay Liners and Covers for Arid Sites," *Journal of Geotechnical Engineering*, ASCE, Vol. 119, No. 2, pp. 223-237.

Day, S. R. and Daniel D. E., August 1985, "Hydraulic Conductivity of Two Prototype Clay Liners," *Journal of Geotechnical Engineering*, ASCE, Vol. 111, No. 8, pp.957-970.

Benson, C. H. and Daniel D. E., August 1990, "Influence of Clods on Hydraulic Conductivity of Compacted Clay," *Journal of Geotechnical Engineering*, ASCE, Vol. 116, No. 8, pp. 1231-1248.

Boynton, S. S. and Daniel, D. E., April 1985, "Hydraulic Conductivity Tests on Compacted Clay," *Journal of Geotechnical Engineering*, ASCE, Vol. 111, No. 4, pp. 465-478.

Elsbury, B. R., Daniel, D. E., Sraders, G. A., and Anderson, D. C., November 1990, "Lessons Learned from Compacted Clay Liner," *Journal of Geotechnical Engineering*, ASCE, Vol. 116, No. 11, pp. 1641-1660.

Environmental Protection Agency (EPA), 1989, *Technical Guidance Document: Final Covers on Hazardous Waste Landfills and Surface Impoundments*, EPA/530-SW-89-047, Office of Solid Waste and Emergency Response, U.S. EPA, Washington, DC.

Hillel, D., 1982, *Introduction to Soil Physics*, Academic Press, Inc., San Diego, California.

Holtz, R. D. and Kovacs, W. D., 1981, *An Introduction to Geotechnical Engineering*, Prentice-Hall, Inc., Englewood Cliffs, New Jersey.

Kleppe, J. H. and Olson, R. E., 1985, "Desiccation Cracking of Soil Barriers," *Hydraulic Barriers in Soil and Rock, ASTM STP 874*, pp. 263-275.

Melton, R. M., 1992, J. M. Huber Corporation, Clay Division, Langley, SC, May 26, 1992 personal correspondence.

Olson, R. E. and Daniel, D. E., 1981, "Measurement of the Hydraulic Conductivity of Fine-Grained Soils," *Permeability and Groundwater Contaminant Transport, ASTM STP 746*, pp. 18-64.

Schroeder, P. R., Gibson, A. C., and Smolen, M. D., 1984, *The Hydrologic Evaluation of Landfill Performance (HELP) Model*, Volume II, Documentation for Version I, EPA/530-SW-84-010, U.S. EPA Office of Solid Waste and Emergency Response.

R. Jeffrey Dunn,[1] and Bert S. Palmer,[2]

LESSONS LEARNED FROM THE APPLICATION OF STANDARD TEST METHODS FOR FIELD AND LABORATORY HYDRAULIC CONDUCTIVITY MEASUREMENT

REFERENCE: Dunn, R. J., and Palmer, B. S., "Lessons Learned from the Application of Standard Test Methods for Field and Laboratory Hydraulic Conductivity Measurement," Hydraulic Conductivity and Waste Contaminant Transport in Soil, ASTM STP 1142, David E. Daniel and Stephen J. Trautwein, Eds., American Society for Testing and Materials, Philadelphia, 1994.

ABSTRACT: Standard ASTM test methods are now available for laboratory and field measurements of hydraulic conductivity of compacted clay soils which are used in liners and covers at waste facilities. They are widely used in waste facility design and construction quality assurance. Many of these methods require input of qualified professionals to select appropriate test parameters. Unless the parameters are carefully selected, conflict may arise over the suitability of test results and thus the regulatory acceptance of a liner or cover. Four case histories are presented which illustrate the use of both laboratory and field test methods and some of the variations which may result. These variations can usually be explained in terms of test parameters, but they emphasize the importance of suitable engineering judgement for careful selection of test conditions, and data reduction methods which are appropriate for to site conditions and allow for comparison of test results from various tests.

KEYWORDS: hydraulic conductivity, field tests, laboratory tests, infiltrometer testing, swelling, soil suction

---

INTRODUCTION

Compacted soil liners, particularly those composed of clayey soils, are often used in containment systems for waste disposal facilities. In many applications, they are now used in conjunction with geomembranes to construct composite liner systems. With federal regulations in 40 CFR Part 258, commonly referred to as Subtitle D, now in effect, construction of composite liner systems for solid waste disposal sites will likely become the "standard" liner system. However, in composite liner systems the compacted soil liner will have to meet rigorous standards for maximum hydraulic conductivity, and the importance of the soil liners, particularly those composed of compacted clay, to containment of wastes will not diminish.

---

[1]Associate, GeoSyntec Consultants, 1600 Riviera Avenue, Suite 420, Walnut Creek, California 94596.

[2]Associate, GeoSyntec Consultants, 16541 Gothard Street, Suite 211, Huntington Beach, California 92647

With the continued use of compacted clay liners, has come increased understanding of the characteristics of the soils used for construction that will result in a liner meeting regulatory standards for hydraulic conductivity of 1×10^{-7} cm/s (Code of Federal Regulations 1991) or less. The importance of compaction moisture content, clod size, compaction equipment, compaction effort , and protection of completed liner sections from environmental degradation such as drying and freezing seems to now be widely acknowledged and taken into consideration.

Standard methods for construction quality assurance (CQA) are also now recognized (Daniel 1990) These CQA techniques are being applied on a much more widespread basis by many landfill operators, and are being required by many states. Construction and testing of liner or cover test sections are also becoming much more routine, and field testing of these sections has become more commonplace.

The recent development of these hydraulic conductivity testing methods and CQA procedures, in the authors experience, has resulted in an improvement in the quality of clay liners, and usually a corresponding reduction in the number of construction defects. This has led to the low hydraulic conductivities measured in the laboratory agreeing more closely with those calculated based on the results of field infiltrometer or borehole tests. This is distinctly different behavior than that highlighted by Daniel (1984), and more recently by Elsbury et al.(1990) It appears that with improvements in construction methods, field prepared clay liner materials now often have a structure much closer to materials prepared in the laboratory.

This paper contains a brief description of recently adopted standard procedures for soil liner testing, and through a review of four case histories of clay liner test section testing, illustrates some of the issues which may arise in the course of construction quality assurance testing and some of lessons learned along the way.

TEST PROCEDURES

Concurrent with improvements to clay liner construction methods and construction quality assurance has come the development of new standardized laboratory and field methods for the evaluation of hydraulic conductivity and infiltration rate such as those by the United States Environmental Protection Agency and ASTM. The use of a number of these standards is now required by state waste regulatory agencies throughout the United States. Most notable has been the adoption of ASTM Test Method for Measurement of Hydraulic Conductivity of Saturated Porous Materials Using a Flexible Wall Permeameter (D 5084) and ASTM Test Method for Field Measurement of Infiltration Rate Using a Double-Ring Infiltrometer with a Sealed-Inner Ring (D 5093). Both of these methods are seeing widespread use in construction quality assurance applications. The authors' experience has been that these two test methods are now often used in conjunction during construction of clay liner test sections, while ASTM D 5084 is more typically used for testing of the production clay liner. Other laboratory and field testing methods such as double ring infiltrometer, air entry permeameter, BAT^{TM} tests and Boutwell permeameters are also used in conjunction with these two methods Daniel (1989).

Consistent with fundamental principles of geotechnical testing practice, ASTM D 5084 was purposely developed to require the input of qualified professionals to select appropriate test parameters, including confining pressure, test permeant, and hydraulic gradient. The method provides some guidance on permeant and hydraulic gradient, but confining pressure must be selected. When measuring hydraulic conductivity of material used in cover sections, confining stresses are clearly quite low and values representative of field stresses can be relatively easily selected and applied in the tests. Selection of test conditions for liner sections is not so straightforward. At the end of construction, stresses imposed on liners are very low, typically in the range of 10 to 15 kPa (1 to 2 psi). As wastes are placed confining pressures are increased and may exceed values of 4000 kPa (580 psi) in deep landfills now being constructed in many areas. Given this variation in confining pressure a question often arises as to what is the representative confining pressure under which to evaluate clay liner materials in laboratory tests, those both for design purposes and for construction quality assurance. A conservative approach is to complete the tests at low confining pressures which represent the stresses that will occur in a soil liner at the end of construction. However, if possible it is highly desirable to complete tests, particularly during the design phase, over a range of confining pressures to evaluate changes in hydraulic conductivity, and thus changes in the performance of a soil liner as wastes are placed.

ASTM D 5093 measures infiltration rate, but regulatory standards for liners and covers are stipulated in terms of hydraulic conductivity. By simply dividing measured infiltration rate by hydraulic gradient, the hydraulic conductivity can be calculated. The mathematics are simple. However, the evaluation of hydraulic gradient is not so easy and there is not yet agreement on a standard method to evaluate gradient accurately.

CASE HISTORIES

Landfill A

The Landfill A expansion project in Southern California includes the construction of a lined base area of approximately 6.2 ha (15 ac) and about 1 ha (2.5 ac) of sideslope lining area. The landfill base was to be lined with composite liner composed of a 0.6-m (2-ft) thick compacted clay liner having an in-place saturated hydraulic conductivity equal to or less than 1×10^{-7} cm/s, overlain by a 1.5 mm (60 mil) thick high density polyethylene (HDPE) geomembrane.

The liner was to be constructed on a prepared subgrade and was to be overlain by a leachate collection and recovery system (LCRS) and operation layer. The liner design for the sidewalls of the disposal area did not include a clay liner, but called for a 2 mm (80 mil) thick HDPE liner placed over approximately 5 m (15 ft) of compacted select soil fill. Liner systems were designed to tie into liners in previously constructed disposal areas surrounding the expansion area.

Borrow Soil Testing—Samples of proposed clay liner material from an off-site source were tested to evaluate their suitability for use in the clay liner test fill. Bulk samples from the off-site source were first composited and thoroughly mixed. Engineering properties were then tested using appropriate ASTM test methods (Table 1). Falling head hydraulic conductivity tests were completed on ten samples compacted to both standard Proctor (ASTM D 698) and modified Proctor (ASTM D 1557) compaction efforts. Corps of Engineers (COE) Method EM 1110-2-1906 (1980) for flexible-wall hydraulic conductivity measurements was used as specified by the project documents. All samples were first back pressure saturated and then consolidated to 308 kPa (45 psi), a confining pressure representing a sizable imposed load of overlying soil and waste. Results of the tests are presented on Figure 1. As shown by the data the measured hydraulic conductivity were quite low ranging from

5×10^{-10} cm/s to 5×10^{-8} cm/s. Low values of hydraulic conductivity, meeting the project requirements, were found even for samples compacted dry of optimum moisture content. This is thought to be primarily due to the elevated confining pressure used in the tests and also the very fine grain size distribution of the borrow soil. No records are currently available on the size of soil aggregates used in the laboratory tests. It is possible that the tests were run with small size aggregates which would help to reduce hydraulic conductivity, even for the samples dry of optimum moisture content.

TABLE 1-- **Landfill A Borrow Soil Characteristics**

Unified Soil Classification	CL
Percent passing #200 sieve, by weight	97%
Atterberg Limits	
Plastic Limit	22
Liquid Limit	37
Plasticity Index	15

Test Pad Construction- A clay liner test pad with dimensions of approximately 12 m by 37 m (40 ft by 110 ft) was constructed to a total thickness of 0.9 m (3 ft), using six 0.15-m (6-in.)thick lifts. Sufficient borrow material for each lift was transported to the pad area using a scraper and placed. It was then wetted and mixed using a pavement reclaimer to the required moisture content of two to four percent above the optimum moisture content as measured by ASTM D 1557. Several types of pad-foot and sheepsfoot compactors were tried, with varying degrees of success to achieve the required minimum density of 90% of maximum dry density as measured by ASTM D 1557. After experimenting on the first and second lifts a compaction procedure consisting of six passes of a REX 350 pad-foot compactor followed by two passes of an Ingersol-Rand SD100D smooth-drum vibratory compactor was found to achieve the required compaction level. Prior to placing each new lift the preceding lift was thoroughly scarified by making two additional passes of the pad-foot compactor to provide for suitable lift bonding. Once construction of the test pad was complete, it was sprayed with water and approximately three-quarters of the pad was covered with 0.1 mm plastic sheeting and a few inches of soil cover to reduce potential for drying. The remaining one-quarter of the pad was left uncovered to allow observations of desiccation cracking with time.

Six shelby tube samples, 7.6 cm (3 in.) O.D. and 7.1 cm (2.8 in.) I.D., were collected from lifts and lift interfaces throughout the thickness of the test section to be used to measure moisture characteristics and laboratory hydraulic conductivity using COE Method EM 1110-2-1906. Samples were the full diameter of the shelby tube. They were back pressure saturated, consolidated to an effective stress of 308 kPa (45 psi), and hydraulic conductivity was measured under a hydraulic gradient of between 1 and 2. Hydraulic conductivity results are listed in Table 2.

Infiltrometer Testing- Two double-ring infiltrometers (DRI) were installed in the test pad following test method ASTM D 3385. The DRIs consisted of round metal inner and outer rings. The outer ring was 0.6 m (2 ft) in diameter while the inner ring was 0.3 m (1 ft) in diameter. Both rings were about 0.5 m (18 in.) high. The outer ring was embedded about 0.15 m (6 in.) into the test section while the inner ring was

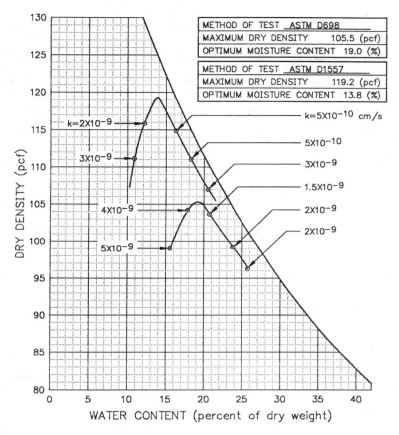

FIG. 1--Compaction characteristics and hydraulic conductivity test results

embedded about 0.1 m (5 in.). Both rings were sealed in place with bentonite grout. They were then filled with water to depths of approximately 0.15 m (6 in.), and were connected to Mariotte tubes to maintain a constant head of water in the rings. Both rings were capped with insulated covers to minimize the potential for evaporation, and soil was mounded around the outer ring to reduce temperature variations.

The two tests were run for 33 days and testing was then terminated as estimates of hydraulic conductivity indicated that the test pad had met the construction requirement of a maximum value of 1 x 10^{-7} cm/s which was below the maximum value of 1 x 10^{-6} cm/s required by regulations in force at the time of the tests. Additional shelby tube samples were collected to evaluate final moisture content below the DRI test locations and the capillary moisture relationship for soils in the test pad.

A comparison of moisture content data from construction and from the shelby tube samples indicated that the wetting front had moved approximately 0.15 m (6 in.) below the surface of the test pad. This depth was verified by measuring pocket penetrometer resistance at approximately 2.5 cm (1 in.) intervals through the depth of the clay liner. At both DRI test locations pocket penetrometer resistances increased from 100 to 400% at depths corresponding to the depth of the wetting front estimated from the moisture content data. The authors have found that measurements of moisture content do not always give a clear indication of the wetting front location. In this case however, there was clear indication of moisture increase in test pad soils over the values measured at the end of compaction. Practitioners are cautioned that moisture content measurements can also be misleading in that compacted clay liners are usually compacted to high degrees of saturation, and moisture content increases to reach field saturation may be only a few percent of moisture by weight. This is within the natural variation which may occur in soils typically used to construct clay liners. Accordingly the authors recommend that moisture content profiles be interpreted conservatively, unless the variation is uniform and clearly shows a distinct wetting front.

The capillary moisture relationship was evaluated using a moisture characteristic cell by increasing moisture content of a sample of the liner test section recovered from the upper 0.15 m (6 in.) of the test pad. The method proposed by Brooks and Corey (1966) and Brackensiek (1977) was used to calculate the water entry value for the liner soil. As shown in Figure 2, which shows effective saturation as a function of suction head, the water entry value was estimated to be 45 kPa (7 psi). Based on the in-situ average moisture content and degree of saturation, the estimated ambient soil suction is approximately equal to 55 kPa (8 psi).

Currently opinions differ on what value of soil suction should be used to calculate the hydraulic gradient, and thus should be utilized to convert infiltration rates into hydraulic conductivity. One view is that the suction at the wetting front depth should be equal to zero. The other view is that the wetting front suction is equal to the ambient suction in the liner at its moisture content during the test. This suction can be measured by tensiometers installed at various depths below the liner surface. Unfortunately, the authors' experience indicates that tensiometers often seem to provide confusing or conflicting data which often does not seem to correspond to the wetting front location as measured by evaluating the moisture content of soil samples collected at approximately 2.5 cm (1 in.) depth intervals from after the tests. It is likely that the true value of suction at the wetting front and hence the hydraulic gradient lies somewhere between these extreme values i.e., zero suction and ambient suction. Bouwer (1966, 1978, 1986) suggests that the soil suction at the wetting front be taken equal to the water entry suction value. Based on the literature reviewed for this paper , it does not seem that there is a simple method to calculate the value of the soil suction that should be used to calculate the gradient in a soil liner. Using the ambient soil suction, the range of hydraulic conductivity was 1.4×10^{-8} cm/s to 4.4×10^{-9} cm/s while using the measured water entry value as the suction the range of hydraulic conductivity was calculated to be 5.1×10^{-9} cm/s to 1.7×10^{-8} cm/s. Assuming zero suction at the wetting front the final hydraulic conductivities from the two tests were calculated to be 8.4×10^{-8} cm/s to 2.2×10^{-7} cm/s. Assuming that the soil suction is equal to zero provides a conservative estimate of the hydraulic conductivity.

<u>Hydraulic Conductivity Summary</u>—A summary of the laboratory and field hydraulic conductivity test results is presented in Table 2. As indicated by a review of the data the results of field tests are higher than the results of laboratory tests on samples from the liner test pad. The variation is about one-half to one and one-half orders of magnitude depending on the suction value assumed at the wetting front. It is assumed that much of this variation is probably due to the elevated confining pressure (310 kPa (45 psi)) used in the laboratory tests as

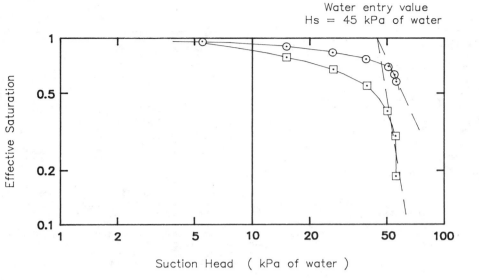

FIG. 2--Suction Characteristics of Test Pad Soil, Landfill A

compared to the low overburden thickness, and thus small confining stress, which occurred in the DRI tests. However, without laboratory tests at low stress levels no further conclusions can be drawn on level of the data agreement, and the actual effect of varying confining pressure on the differences in field and laboratory results.

Infiltration rates measured in both of the DRI tests were below 1 x 10^{-6} cm/s. ASTM Method D 3385 cautions against the use of the DRI method when infiltration rates below this value occur. The authors concur with this recommendation. In the tests described in this paper difficulties were observed in accurately measuring flows and temperature variations had a marked effect on results. Although the authors are confident of the results discussed herein, it is generally their practice to use the SDRI method on all of their projects since this one.

TABLE 2-- Landfill A Hydraulic Conductivity Summary

Laboratory Tests	
(310 kPa consolidation pressure)	
Borrow evaluation	5×10^{-10} to 5×10^{-8} cm/s
Liner test pad	2×10^{-9} to 5×10^{-9} cm/s
Double ring Infiltrometer	
Zero Suction at wetting front	8.4×10^{-8} to 2.2×10^{-7} cm/s
Suction = water entry value	5.1×10^{-9} to 1.7×10^{-8} cm/s
Suction = ambient soil suction	4.4×10^{-9} to 1.2×10^{-8} cm/s

Calabassas and Puente Hills Landfills

The Calabassas and Puente Hills Landfills are discussed together as they are very similar in design, and the testing procedures utilized were essentially the same. Both projects involved construction and testing of clay liner test pads as part of planned expansions of existing landfills. At both sites a composite liner section was planned for use on the landfill expansion base constructed from a 0.3-m (1-ft) thick compacted clay liner, having a hydraulic conductivity equal to or less than 1×10^{-6} cm/s, overlain by a 2-mm (80-mil) thick high density polyethylene (HDPE) geomembrane.

The composite liner was then to be overlain by a coarse sand LCRS layer and 1 m 3 ft) of protective soil. On the sideslopes no clay liner was to be constructed. A 2-mm (80-mil) thick HDPE liner was to be placed above a geotextile cushion layer. The HDPE was then planned to be overlain by 0.6 m (2 ft) of protective soil.

Borrow Soil Testing- Off-site borrow material for use at the Calabassas site had previously been tested by others. Material for use at the Puente Hills Landfill was an on-site siltstone material with some concretions and cobbles in it. This material was tested to evaluate its engineering characteristics including hydraulic conductivity. Triaxial falling head tests (ASTM D 5084) were used to measure hydraulic conductivity at an effective consolidation stress of 38 kPa (6 psi), with a hydraulic gradient of 1 to 2. Results of the test program for both sites are summarized in Table 3.

Test Pad Construction-A single liner test pad was constructed at each landfill. Dimensions were 18 by 30 m (60 by 100 ft) with a total thickness of 0.9 m (3 ft) constructed in six 0.15-m (6-in.) thick lifts. The Calabassas test section was constructed directly on the prepared subgrade while the Puente Hills section was built on a 0.15-m (6-in.) thick drainage layer. At both sites material was first screened to a maximum clod size of 25 mm (1 in.) Calabassas) or 37 mm (1.5 in.) (Puente Hills). Material was moisture conditioned using spray water trucks to attain moisture contents at optimum up to 3 % above optimum, and mixed on the test pad. At Calabassas a towed sheepsfoot roller was used to compact the material, while pad-foot type compactors were used at Puente Hills. Unfortunately, records are not currently available to indicate the operating weight of the compactor at Calabassas. At Puente Hills compactors weighing from 12,247 to 32400 kg (27000 to 71429 lb) were used to construct the test pad. At both sites the first two lifts were utilized to experiment with the number of passes of compaction equipment necessary to attain the required degree of compaction. At both sites the layer thicknesses were approximately 0.2 m (8 in.) prior to compaction and about 0.15 m (6 in.) after compaction. After being proof rolled with a smooth drum roller to provide a smooth surface like

TABLE 3--Calabassas and Puente Hills Landfills Borrow Soil Characteristics

	Calabassas	Puente Hills
Unified Soil Classification	CL	CL
Percent Passing #200 sieve, by weight	43 to 62%	91 to 96%
Atterberg Limits, %		
Plastic Limit	31 to 36	18 to 24
Liquid Limit	53 to 63	43 to 44
Plasticity Index	22 to 31	19 to 25
Hydraulic conductivity, (cm/s)	1.4×10^{-8} to 6.8×10^{-8}	2×10^{-7} to 9×10^{-7}
Test Conditions		
-% Max. dry density	94 to 96	91 to 93
-Molding water content	opt. +4 to 5%	opt.+3 to 5%
-Effective stress, kPa	(1)	38

(1) Test confining pressure not known, tests completed by others.

that required by the project specifications for the production liner, each test pad was then covered with plastic and a thin layer of soil was placed over the plastic to reduce potential for desiccation cracking of the liner test pad.

<u>Laboratory Hydraulic Conductivity Testing</u>- Eight shelby tube samples were collected at Calabassas and six were collected at Puente Hills for laboratory hydraulic conductivity and moisture retention tests. Laboratory hydraulic conductivity tests were completed using ASTM test method D 5084. All samples were back pressure saturated and consolidated before testing. A range of confining pressures from 21 to 345 kPa (3 to 50 psi) were used to evaluate hydraulic conductivity over the range of pressures the liner was expected to be exposed to during landfill operations. Hydraulic gradients of 3 were used at the low end of the effective stress range, while a gradient of 21 was used on the higher effective stress tests. Results of the tests are shown graphically on Figure 3 for Calabassas and Figure 4 for Puente Hills. As shown by the data in these two figures, at both sites hydraulic conductivity values measured for most of the samples tended to group reasonably closely together. However, at both landfill one or two tests showed markedly higher hydraulic conductivity values. This demonstrates the natural variability that occurs even within the small area of a test pad, and in the author's opinion emphasizes the need for measuring hydraulic conductivity values in large scale field tests that represent a suitable amount of material heterogeneity.

<u>Infiltrometer Testing</u>-One sealed double ring infiltrometer (SDRI) was installed at each site following ASTM D 5093. The SDRI consisted of an 3.6 m (12 ft) square outer ring embedded to a depth of 0.45 m (18 in.), with a 1.5 m (5.5 ft) square inner ring which was embedded

approximately 0.13 m (5 in.). A series of tensiometers were installed to various depths up to 0.6 m in the annulus between the inner and outer rings. The rings were filled with water to approximate depths of 0.33 m (13 in.) and reference points were established on the inner ring to allow measurement of liner swell. The tests were run for approximately 60 days.

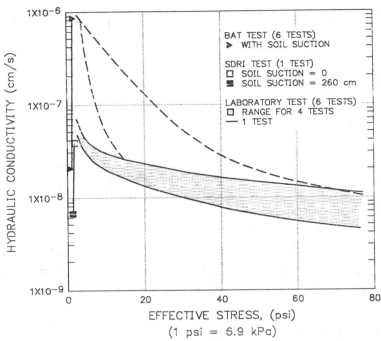

FIG. 3--Summary of hydraulic conductivity data, Calabassas Landfill

Additional shelby tube samples were collected for measurement of moisture contents below the inner ring at the end of the SDRI test. Moisture content measurements of discrete soil samples collected at about 2.5 cm (1 in.) intervals were used in conjunction with the results of tensiometer data to evaluate the depth of the wetting front. At Calabassas, data indicated that the front had advanced to about a depth of 0.53 m (20 in.). At the Puente Hills site the wetting front appeared to have moved to a depth of 0.49 m (19 in.).

Soil suction characteristics were evaluated by similar methods to those used for the Landfill A project. Both liner test sections were compacted to relatively high moisture contents, several percent above optimum moisture contents, and thus the soil suction at ambient moisture content was less than the water entry value. The optimum moisture content, the average as-compacted moisture content, the estimated average ambient soil suction, and water entry value estimated using the method described for Landfill A, for the Puente Hills and Calabassas landfill soils are indicated in Table 4. The ambient value was used to estimate the lower bound of measured hydraulic conductivity, while the upper bound was calculated assuming zero suction at the wetting front. All data were also corrected for swelling of the liner by subtracting the assumed flow equal to swelling from the total measured flow. Swell related flow was attributed to the measured vertical movement of the inner ring multiplied by the area of the inner ring. Swelling was found to occur in the first 30 days of the tests. Results of these calculations indicate a range of in-place hydraulic conductivity at Calabassas of 7×10^{-9} cm/s to 4×10^{-8} cm/s depending on the suction value assumed. At Puente Hills the range was found to be 9×10^{-9} cm/s to 6×10^{-8} cm/s depending on the assumed suction value.

TABLE 4--Selected Characteristics of the Calabassas and Puente Hills Soil Liners

	Calabassas	Puente Hills
Optimum Moisture Content (%)	27.4	16.2
As-compacted average moisture content (%)	34.0	24.9
Estimated Ambient Soil Suction (cm of water)	260	370
Water Entry suction Value (cm of water)*	>800	>400

*Results indicated herein are only indicative. Additional soil suction tests should ideally be performed on these soils to cover a larger range of suction and degree of saturation.

BAT^{TM} Testing-As a supplement to the SDRI and laboratory testing of shelby tube samples, BAT^{TM} tests were completed in six locations adjacent to the SDRI in each of the test pads. Due to space limitations the BAT^{TM} method cannot be described herein, however the reader is referred to Daniel (1989) which provides a description of the method. BAT^{TM} tips were installed to depths of 0.15 to 0.4 m (6 to 16 in.) by hand auguring an initial hole, and then carefully driving the tips into place. Each tip was then allowed to equilibrate for 2 to 12 hours for pore pressure dissipation and a final soil suction reading was taken. Measurements where then completed by the outflow method, where water flows out of the test device into the soil. Hydraulic conductivities were calculated by assuming that the soil suction was that measured by the BAT^{TM} tip at the time of the test. Hydraulic conductivities measured by this method at Calabassas ranged from 2.0×10^{-8} cm/s to 9.0×10^{-7} cm/s. At Puente Hills the range was from 2.2×10^{-8} cm/s to 2.2×10^{-7} cm/s.

Hydraulic Conductivity Summary-Graphical summaries of laboratory and field hydraulic conductivity data for the Calabassas and Puente Hills landfills are presented on Figures 2 and 3, respectively. As shown on these graphs there seems to be reasonable agreement between the different field and laboratory test methods utilized to test the test pad as long as the comparison is based on tests run at similar values of confining stress. All tests completed for both sites met the hydraulic conductivity requirement, and were less than 1×10^{-6} cm/s.

For the Calabassas site, laboratory hydraulic conductivity test results showed a fairly large degree of variability, particularly at the low end of the range of confining pressures used for the tests. The authors think this may be due to the presence of small cracks in the samples caused by sampling stresses, which tend to close as the confining pressure is increased during testing. However, the test programs did not include any microscopic evaluation of samples that might confirm this opinion. Laboratory data agreed fairly well with the range of BAT^{TM} results. Average values for the two test methods also

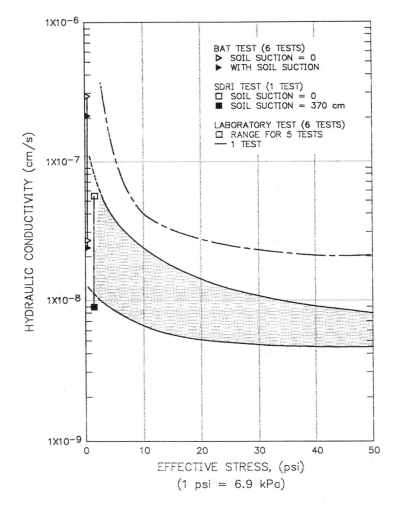

FIG. 4--Summary of hydraulic conductivity data, Puente Hills Landfill

agreed quite well with the average BAT^{TM} value calculated to be 3.8 x 10^{-7} cm/s, while the average laboratory value was 2.8 x 10^{-7} cm/s. This seems to indicate that the BAT™ method may be quite useable as a supplement to laboratory testing of samples collected during production liner construction. The SDRI values at zero suction of 4×10^{-8} cm/s was nearly an order of magnitude below results of the other two methods. This is contrary to most of the author's experience, where SDRI values are usually higher than those measured in the laboratory. This may indicate that the soil sampling or introduction of the BAT™ probe may cause small cracks to develop in the soil, resulting in higher hydraulic conductivity, although no microscopic analyses could be run to confirm this theory.

At Puente Hills laboratory hydraulic conductivity tests also showed variability of several orders of magnitude, which tended to be more pronounced at lower confining stresses as is to be expected. A possible explanation of this phenomenon was provided above. BAT™ results again seemed to agree fairly well with the laboratory tests completed at low confining stresses. Agreement between results of the SDRI and laboratory tests at low stresses is also reasonably good. Comparing average values for the three methods at zero suction values showed agreement within one-half order of magnitude. The average BAT™ value was 1.2×10^{-7} cm/s, 6×10^{-8} cm/s using the SDRI method, and 8.4×10^{-8} cm/s using the laboratory method at a confining pressure of 19 kPa, which is just slightly higher than 3.5 to 10 kPa (0.5 to 1.5 psi) stresses in the field tests.

As indicated by the variability of both the BAT™ and laboratory hydraulic conductivity test results, test methods which both only test limited soil volumes, it is important that sufficient tests be completed to provide an indication of field variability of soil quality, when these methods are used to supplement SDRI tests.

Landfill B

Landfill B, in Northern California, is a new solid waste facility which opened in 1992. The site will have a composite liner consisting of 0.6 m (2 ft) of compacted clay overlain by a 2-mm (80-mil) thick HDPE geomembrane. An LCRS layer overlies the liner system. The liner and LCRS will be constructed over the entire area of the landfill base and sideslopes. On-site soils are used to construct the clay liner at this site. Characteristics of these soils are presented in Table 5.

Prototype Fill-An extensive test fill program was completed at the site to evaluate compaction specifications. This program included compaction at a variety of moisture contents and compaction efforts using a CAT 835 true sheepsfoot compactor rated at an operating weight of 32,400 kg(71,429 lb). Differing liner sections were evaluated by laboratory hydraulic conductivity testing of shelby tube samples recovered from the test liner sections. Compaction criteria that were developed from the test fill included minimum 93% of the maximum dry density, at moisture contents of optimum to optimum plus 4 %, with these compaction criteria as measured by ASTM D 1557. In addition to the requirement for minimum density, the contractor was also required to make a minimum of six passes with the CAT 835 compactor.

A single prototype fill was constructed at the Landfill B site using the compaction specifications, listed in the proceeding paragraph which where developed from the test fill program, to allow field hydraulic conductivity testing using an SDRI with additional laboratory hydraulic conductivity testing as well. The intent of this prototype fill was to verify that the compaction criteria developed during the test fill program would allow construction of a production clay liner which met the hydraulic conductivity requirement of less than 1 x 10^{-6} cm/s applicable to this project.

Infiltrometer Testing–One sealed double ring infiltrometer was installed in the prototype fill, following the methods of ASTM D 5093. Procedures for testing were essentially the same as those used at the Calabassas and Puente Hills sites. The test proceeded in a manner similar to previous tests performed by our firm during the first month,

TABLE 5-- Landfill B Clay Borrow Characteristics

Unified soil classification	CL
Percent Passing #200 sieve, by weight	60 to 65%
Atterberg Limits	
Plastic Limit	15 to 17
Liquid Limit	43 to 48
Plasticity Index	27 to 32
Laboratory Hydraulic Conductivity	
% Max. dry density (D 1557)	92 to 95
Moisture content (D 1557)	opt. 3 to 5%
Confining Stress	
35 kPa	1.6×10^{-8} to 4.8×10^{-8} cm/s
138 kPa	5.4×10^{-9} to 9.3×10^{-9} cm/s
689 kPa	2.3×10^{-10} to 8.8×10^{-10} cm/s

and preliminary indications were that the prototype test section met the hydraulic conductivity requirement of less than 1×10^{-6} cm/s. Readings of liner swell were taken. Inner ring inflow volumes were corrected by subtracting volumes due to swelling based movement of the inner ring. This volume was calculated as equal to the vertical increment of swell multiplied by the plan area of the inner ring. Very early in the test it was noticed that application of this correction for swell over a period of one to three days sometimes resulted in calculated swell volumes which were in excess of the flow measured into the inner ring over a given time period. This type of behavior had never been observed before. It was concluded that this behavior was likely due to inner ring vertical movement monitoring errors. Possible errors were thought to occur due to the measuring reference being located on the test pad and thus being susceptible to swell, or inaccuracies in the measurement technique which was susceptible to temperature variations and human error. Considering these possible errors, it is the authors opinion that swell correction using the techniques described herein are fairly rough. However, the techniques are viable to help to distinguish between a soil that has a high hydraulic conductivity and one that is undergoing swell, but has a low hydraulic conductivity.

Construction was proceeding on other portions of the landfill which had to be built prior to liner construction. Thus time was available to continue the test. SDRI measurements were continued for a period of 137 days. Measurements indicated that the swell of the liner continued until approximately 125 to 130 days, at which time swelling stopped. By that time vertical movement data indicated that the liner had swelled by approximately 3 cm (1 in.). The SDRI was taken down and shelby tubes were pushed through the entire thickness of the prototype liner section.

The surface of the liner in the area between the inner and outer ring was observed to have swelled by at least 3 cm (1 in.) by measuring the level of this area with relation to the level of the prototype fill surface outside the outer ring. The operation of removing the infiltrometer rings led to a large amount of disturbance of the soil within the inner ring area. However, based upon the general appearance of the soils, it appeared that this amount of swell had occurred inside the inner ring as well. Laboratory measurement of moisture content over the depth of the liner indicated that the wetting front had penetrated the entire thickness of the prototype section. Thus suction was assumed to be zero at the wetting front to calculate hydraulic conductivity values based on infiltration rate data obtained at the end of the test.

Although the test data uncorrected for swell were below 1×10^{-6} cm/s, it was the opinion of the engineers supervising the test that hydraulic conductivity values corrected for swelling of the liner would be more accurate and reflective of the "true" field hydraulic conductivity. Various methods were evaluated to calculate the hydraulic conductivity corrected for liner swell. Finally a method in which cumulative flow into the inner ring was plotted along with cumulative flow due to vertical movement of the inner ring. "Saturation-growth" curves, which are mathematically similar to a hyperbolic curve were fit to the data. Curves for both the total flow into the inner ring and that calculated for inner ring vertical movement are presented on Figure 5.

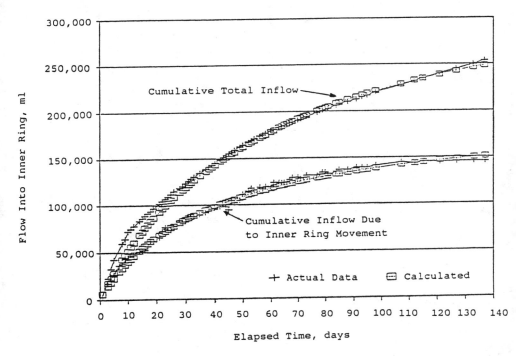

FIG. 5--Cumulative inner ring inflow versus time, SDRI Landfill B

Hydraulic conductivity was then calculated by taking the derivative, with respect to time, of the saturation-growth curves which show cumulative flow into the inner ring or that flow associated with inner ring vertical movement. These values were then divided by the hydraulic gradient and the area of the inner ring. Hydraulic conductivity values uncorrected for inner ring vertical movement were calculated using the total inflow data, while corrected values were calculated by first subtracting the derivative with respect to time for the volume associated with inner ring movement from the total flow volume and then dividing the values by the hydraulic gradient at the time and the inner ring area. Hydraulic conductivity values calculated by this method are shown on Figure 6. As shown on this figure hydraulic conductivity at the end of the test uncorrected for swell was approximately 2×10^{-7} cm/s, while corrected values were about 1×10^{-7} cm/s. This range of values is above that observed in the laboratory hydraulic conductivity tests completed at the lowest confining pressure of 35 kPa (5 psi).

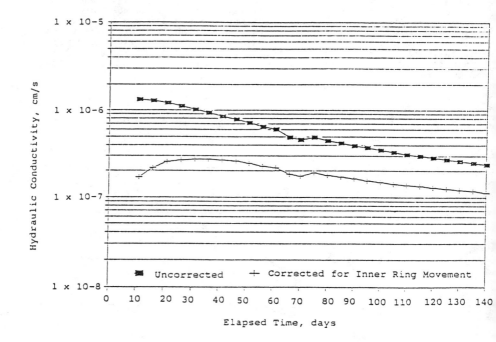

FIG. 6--Hydraulic conductivity versus time, SDRI Landfill B

SUMMARY AND CONCLUSIONS

A variety of lessons were learned from the projects described in this paper. Some of the key points are the following:

1) Construction of low permeability clay liners has become more routine. Provided attention is paid to clod size, proper moisture conditioning to

a wet of optimum moisture content, and appropriate compaction equipment and energy is used to compact the clay, clay liners which meet regulatory performance criteria for hydraulic conductivity can be readily constructed.

2) The results of large scale SDRI field tests sometimes agree quite well with results of small scale laboratory tests completed on shelby tube samples and at or near the low levels of confining stress that occur in SDRI tests. This likely indicates a low level of construction defects. Higher levels of defects would tend to cause field values to exceed those measured in the laboratory. However, in those cases where field hydraulic conductivity values are higher then those measured in the laboratory, a liner is still satisfactory provided the field values meet the applicable hydraulic conductivity standard. Many factors such as presence of cracks in samples, effective stress during testing, and data reduction methodology will impact the degree to which field and laboratory hydraulic conductivity measurements agree.

3) As landfills are filled with wastes, the stresses imposed on clay liners increase. Therefore, the authors feel that evaluation of laboratory hydraulic conductivity over a range of imposed stresses represents good engineering practice. However, it is important that tests also be run at low confining stresses to allow meaningful comparison and correlation to field test results. In our experience there remains much debate over what is the proper stress to run laboratory hydraulic conductivity tests at during production liner construction. While the "correct" stress level varies from site to site, it is our opinion that it should not always be chosen as a low or "worst case" value, but rather should model the stresses the liner will be exposed to as refuse is placed over it, and leachate is generated which could then move through a clay liner. In many cases however, a low stress level should be chosen as the conservative value.

4) There is a definite and strong need for further evaluation and refinement of techniques used to calculate hydraulic gradient in SDRI tests which are then used to calculate hydraulic conductivities from infiltration data. Currently one can assume values of soil suction at the wetting front equal to zero or equal to much higher values, represented by the ambient soil moisture, or more likely the water entry values. These boundary values often result in hydraulic conductivity values which vary over an nearly an order of magnitude in some cases. If both values fall below the required maximum hydraulic conductivity, as was the case in the projects described herein, there is no problem. However, when the calculated range brackets the required hydraulic conductivity, it can be extremely difficult to evaluate when a liner section is acceptable.

5) It is important to correct SDRI data for the effect of liner swell and upward movement of the inner infiltration ring. In most tests the authors have reviewed swell largely stops in about the first month of infiltration, but on occasion swelling continues for a much longer time, and it is rare that a test can be run for as long as the one at the Landfill B which was described herein. Our experience is that swell data can be erratic and a correction technique such as the curve fitting used with the Landfill B test is a good technique for analyzing data. However, until a more reliable technique for measuring swell is utilized practitioners should carefully apply this correction., and always note its use in reporting hydraulic conductivity values corrected for swell.

6) The BAT^{TM} is a technique which may be quite suitable for application during production liner testing, when used to supplement recovery of undisturbed field samples and laboratory hydraulic conductivity testing. The BAT^{TM} method does not have the same boundary conditions as laboratory tests. Thus BAT^{TM} data should be correlated to laboratory

test results using data from a test section prior to production liner testing. The BAT™ test needs to be conducted by qualified personnel who understand the BAT™ system equipment and the general mechanics of unsaturated flow. Like laboratory tests it has the limitation of testing a small volume of soil.

In the past five to ten years there has been a marked improvement in clay liner construction and laboratory and field hydraulic conductivity testing methods which are now routinely utilized as part of design and construction quality assurance practice. Questions remain about the techniques in use, and there is still a need for continued development of testing and data reduction techniques.

REFERENCES

Bouwer, H., " Rapid Field Measurement of Air Entry Value and Hydraulic conductivity of Soil as Significant Parameters in Flow system Analysis," Water Resources Research, Vol.2, pp. 729-732, 1966.

Bouwer, H., Groundwater Hydrology, McGraw-Hill New York , 1978

Bouwer, H., "Intake Rate: Cylinder Infiltrometer," Methods of Soil Analysis, Part 1, Physical and Mineralogical Methods, Agronomy Nomograph No. 9, American Society of Agronomy, Madison, WI pp.825-844, 1986

Brackensiek, D. L., "Estimating the Effective Capillary Pressure in the Green and Ampt Infiltration Equation," Water Resources Research, Vol. 13, No. 3, 1977.

Brooks, R. H., and Corey, A. T., "Properties of Porous Media Affecting Fluid Flow," Journal of the Irrigation and Drainage Division, ASCE, Vol. IR2, 1966.

Daniel, D. E.," Predicting Hydraulic Conductivity of Clay Liners," Journal of Geotechnical Engineering, ASCE, Vol.110, No. 4, 1984, pp.465-478.

Daniel, D. E.,"Summary Review of Construction Quality Control for Earthen Liners," Waste Containment Systems: Construction, Regulation, and Performance, Geotechnical Special Publication No. 26, ASCE, New York, 1990, pp.175-189.

Daniel, D. E.,"In Situ Hydraulic Conductivity for Compacted Clay," Journal of Geotechnical Engineering, ASCE, Vol.115, No. 9, 1989, pp.1205-1226.

Elsbury, B. R., Daniel, D. E., Sraders,, G. A., and Anderson, D. C., "Lessons Learned From Compacted Clay Liner", Journal of Geotechnical Engineering, ASCE, Vol. 116, No.11, 1990, pp.1641-1660

United States Army Corps of Engineers, Engineer Manual EM 1110-2-1906, Laboratory Soils Testing, 1980.

United States Government "Solid Waste Disposal Facility Criteria" Code of Federal Regulations, Title 40. Protection of the Environment, Parts 257 and 258, Subtitle D. , 1991.

Tuncer B. Edil,[1] Jae K. Park,[2] and David P. Heim[3]

LARGE-SIZE TEST FOR TRANSPORT OF ORGANICS THROUGH CLAY LINERS

REFERENCE: Edil, T. B., Park, J. K., and Heim, D. P., "Large-Size Test for Transport of Organics through Clay Liners," Hydraulic Conductivity and Waste Contaminant Transport in Soil, ASTM STP 1142, David E. Daniel and Stephen J. Trautwein, Eds., American Society for Testing and Materials, Philadelphia. 1994

ABSTRACT: Recent investigations indicate that organic chemicals are found in significant concentrations in sanitary landfill leachates. Previous research shows that sorption of organic chemicals is controlled by the amount of organic matter in clay and typical liner clays contain only about 1% organic carbon. Therefore, there is a concern about the fate of organic chemicals found in the leachate. Recent studies indicate the importance of test specimen size on hydraulic conductivity determination and the importance of diffusive transport through low-permeability clay liners. In this paper, development of a large-size liner test is presented. The test was designed to evaluate transport/attenuation behavior of volatile organic compounds through a compacted clay under realistic representation of field conditions. The results of a test that lasted 11 months (9 months of chemical transport) are presented along with the methods for testing and analysis. The results point out to the importance of large-scale, long-term tests and the need to develop other means of attenuation of organic chemicals.

KEYWORDS: Hydraulic conductivity, hydrodynamic dispersion coefficient, partition coefficient, retardation factor, effective porosity, transport, volatile organic compounds, large-scale testing

The guidelines for shallow land disposal facilities for solid wastes require prevention of groundwater contamination by use of compacted clay liners, alone or in combination with flexible membrane liners. Recent investigations indicate that organic chemicals constitute a significant portion of toxics leaching from sanitary landfills (Deville, 1985; Nelson and Book, 1986; Friedman, 1988; Battista and Connelly, 1989). For instance, volatile organic compounds (VOCs) were detected in leachate collected from 18 of the 19 municipal and industrial waste landfills monitored in Wisconsin. Furthermore, some levels of VOC contamination have been found both in the soils and

---

[1]Professor, Department of Civil and Environmental Engineering, University of WIsconsin-Madison, Madison, WI 53706.
[2]Assistant Professor, Department of Civil and Environmental Engineering, University of WIsconsin-Madison, Madison, WI 53706.
[3]Geotechnical Engineer, SEC Donohue, Sheboygan, WI 53082.

the groundwater near the unlined landfill sites (Friedman, 1988). However, landfills constructed during the last decade with modern technology, i.e., with a 1.5-m thick clay liner and a leachate collection system, have not shown any groundwater contamination so far (Friedman, 1988). This may be due to the fact that exfiltration rates have been slow with respect to the 1.5-m thick clay liner so that VOCs in measurable amounts have not yet escaped. Experience shows that clay liners, when properly constructed, are excellent attenuators of inorganic chemicals; however, there is not sufficient basis to reach a similar conclusion with respect to organic chemicals.

Previous research shows that sorption of non-polar organic chemicals is mainly controlled by the amount of organic matter in the clay (Lambert, 1968; Briggs, 1973; Karickhoff et al., 1979). Clays used in Wisconsin in constructing landfill liners contain typically less than 1% organic carbon (Edil et al., 1991a). Accelerated tests on small specimens (102 mm in diameter and 51 mm in height) suggest that the organic chemical transport is not retarded significantly in typical clay liner materials (Edil et al., 1991b). Due to a concern that the current liners may be inadequate in attenuating VOCs in the long term, there is a need to evaluate the rate of transport of VOCs through compacted clay liners and develop means to restrict their release into the environment. The results of recent field studies indicate diffusion as an important transport process in fine-grained soils (Shackelford, 1991). Laboratory column tests for determination of chemical transport parameters typically involve relatively small diameter specimens and commonly use high hydraulic gradients for reasons of cost and convenience. However, it is highly unlikely to encounter a high hydraulic gradient in modern landfills. While tests on small specimens serve a useful purpose, they have been subject to criticism on the grounds that they do not simulate the field conditions of a landfill liner satisfactorily. The importance of test specimen size (i.e., representative elementary volume) for hydraulic conductivity testing of compacted clays have been reported by Trautwein and Williams (1990), Shackelford and Javed (1991), and Benson et al. (1993). The representative volume of liner soil being tested should be large enough to contain macro-fabric features (network of faults, fractures and flow paths) comparable to the field liners. Furthermore, the boundary conditions of the experiment should represent conditions undergone by a compacted clay liner at an active, modern landfill as closely as possible.

In view of these considerations, a large-size test was developed to evaluate transport/retardation behavior of VOCs through a compacted clay liner under realistic representation of field conditions insofar as possible. By its nature, these are long-term tests lasting several years. The methods used and the results from a test terminated at the end of 9 months of chemical transport are presented herein.

DESIGN OF LARGE-SIZE TEST

After reviewing numerous possibilities, a stainless-steel cylindrical tank, 0.6 m in diameter and 0.9 m in height, was constructed to perform the experiment. A sketch of the experimental apparatus is shown in Fig. 1. A clay layer, about 0.3-m thick, could be compacted into this cylinder. Support for the clay was provided by a stamped stainless-steel screen, which rested on a dozen stainless-steel rings 25.4 mm high. The space below the screen served as the effluent reservoir (7.3 L in volume) and slots were filed into the rings to allow for liquid circulation. A layer of wet glass wool was placed over the

screen to act as a filter for soil particles. Two sets of openings were
built as sampling ports, diametrically opposite to each other on the
side of the tank roughly below and above the level of the compacted
liner material in the tank. The space above the liner served as the
influent reservoir (160.5 L in volume). A flange at the top of the tank
allowed for the attachment of a nitrile rubber gasket and a round, flat
lid. The lid had two holes to allow for fluid input and air escape
while filling. Two Teflon bags were connected as shown in Fig. 1 to the
influent and the effluent reservoirs with an elevation difference
(typically 900 mm) between these bags such that the desired hydraulic
gradient across the clay layer could be attained. The bags allowed for
an approximation of the air-fluid interface needed to maintain a
constant-head flow while containing the VOCs.

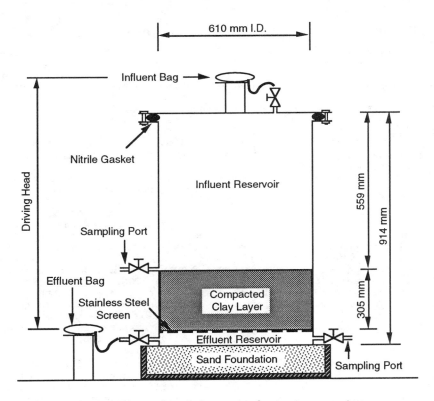

FIG. 1--Schematic of large-scale test apparatus.

It was attempted to maintain the influent concentration as
constant as possible so as to make the data analysis easy. In landfills
that are equipped with a leachate collection system, the hydraulic
gradient is expected to have a low value close to unity. Boundary
conditions at the effluent side of the liner was such that effluent
could be assumed to move freely from the base of the clay layer. This
corresponds to a liner with a secondary leachate collection system or a

stratum with a hydraulic conductivity much higher than that of the liner. All materials used in the test setup, including the fittings, valves and tubes, were selected to be compatible with the organic chemicals to be tested (i.e. metal or Teflon) with negligible capacity for sorption.

The large-size test, having a soil layer of 0.3 m thick with a low hydraulic conductivity ($k \approx 10^{-7}$ cm/s) and hydraulic gradient ($i \approx 3$), necessitates a long-term test extending over one to two years. The performance of the liner soil could be monitored in terms of the effluent quality as a function of time and the distribution of the contaminants with depth in the liner at the termination of the experiment. These data provide a basis also to evaluate the transport and attenuation parameters (diffusion and sorption coefficients) for the VOCs.

MATERIALS

Various landfill liner materials used in Wisconsin were examined to select the soil for the tests. A review of the pertinent properties of five widely used liner clays in Wisconsin indicated that the five soils were comparable in their physical properties and the organic carbon content ranged from 0.27 to 0.91% (Edil et al., 1991a). Of these soils, Kirby Lake till obtained from the Outagamie County Landfill, Wisconsin, was chosen for the experiments. Kirby Lake till (liquid limit: 33%, plasticity index: 16%, percent clay: 47%, activity: 0.3, organic carbon: 0.91%, and cation exchange capacity: 17.4 meq/100g) had a standard Proctor maximum dry unit weight of 17.8 kN/m^3 at an optimum moisture content of 15.5%.

The influent used was Madison tap water spiked with a mixture of three VOCs: methylene chloride (MCL), trichloroethylene (TCE), and toluene (TOL). According to Plumb and Pitchford (1985), these are the three most abundant organic compounds found at waste disposal sites in the United States. Each of the three is commonly used as an industrial solvent. The compounds selected represent a wide range of aqueous solubilities and vapor pressures, and also include species heavier and lighter than water. According to Park et al. (1990), MCL is expected to have a very high mobility, with TCE and TOL having medium mobilities. TCE is expected to be about twice as mobile as TOL. These predicted mobilities are based on soil-water partition coefficients of the compounds. Some properties of these compounds are presented in Table 1. During mixing of the chemicals prior to addition to the influent reservoir of the tanks, some problems were encountered in dissolving these organic compounds in water, especially TCE and TOL, due to their low aqueous solubilities. To resolve this problem, methanol, CH_3OH, was used as a co-solvent. The hydrophobic compounds, TOL and TCE, were dissolved in methanol, MCL was added, and this solution was dissolved in deionized water. Methanol facilitated solution of all of the subject chemicals. The small concentrations of all organic chemicals (mostly less than 20 mg/L) were not expected to alter the hydraulic conductivity of the low activity clays used in the tests (Bowders and Daniel, 1987). Additionally, a nonreactive tracer such as lithium bromide (LiBr) can be added to the influent solution at a concentration of 20 mg/L to determine effective porosity. For this purpose, either the effluent bromide/lithium concentration-time data or the pore fluid bromide/lithium concentration-depth data can be monitored.

TABLE 1--*Physical properties of organic substances used*.

Substance	Methylene Chloride	Trichloroethylene	Toluene
Molecular Formula	CH_2Cl_2	$CHClCCl_2$	$C_6H_6CH_3$
Molecular Weight	84.94	131.29	92.13
Specific Gravity	1.336	1.466	0.866
Vapor Pressure (20°C) (mm Hg)	349	60	22.4
Aqueous Solubility (20°C) (mg/L)	16700	1100	515
log Kow*	1.25	2.53	2.69
Henry's Law Constant	0.093	0.479	0.275

*K_{ow}: Octanol-water partition coefficient.

LINER PREPARATION

Preliminary tests were performed to determine the procedure for preparing the compacted liner in the tank. The soil for these trial tests was air-dried and pulverized to pass a No.4 sieve. The soil was then mixed with water and placed in a plastic bag for at least 18 hours to allow for hydration. Soil was compacted by hand, by dropping a weighted hammer. Table 2 summarizes the conditions of these tests and the measured quantities.

The first four tests resulted in unsatisfactory liners. After compacting the liner in the tank, the influent and effluent reservoirs were filled with water and water was allowed to seep through the liner. The effluent was collected and the seepage rate was determined. An equivalent hydraulic conductivity was calculated from the measured seepage rate assuming a constant head condition and uniform flow across the liner. The equivalent hydraulic conductivities were on the order of 10^{-6} cm/s or more, indicating gross leakage in the first four tests. After each test, three Shelby tube (75 mm in diameter) cores were retrieved from the liner. These cores were sectioned and the moisture content and dry unit weight of each section were determined. Permeability tests were performed on two of these core sections in each trial test in a flexible-wall permeameter at an effective confining pressure of 345 kPa with a back pressure of 276 kPa and a hydraulic gradient of 17. The average values of the measured hydraulic conductivities are also presented in Table 2. The hydraulic conductivities of the core samples were consistently lower than the equivalent hydraulic conductivity of the liner by more than one order of magnitude. A correction applied for the difference in effective confinement pressure did not explain this difference. It is attributed to the presence of macro defects in the liner that could not be represented in 75-mm diameter specimens.

To provide proper sealing, a number of factors were changed successively in each test. For instance, a wetter soil was used in Test 2. It has been shown by Benson and Daniel (1989) that clod size plays a major role in controlling the hydraulic conductivity of compacted clays and clods break down more easily when the soil is wetter. In Test 3, more compactive effort was used resulting in a denser liner, but the seepage rate became even higher than in the previous tests. Test 4 involved a drier clay and, yet a higher compactive effort.

TABLE 2--Liner preparation tests.

Test No.	Hammer Weight[δ] (kg)	No. of Blows per Lift[ξ]	Avg. Water Content (%)	Avg. Dry Unit Weight (kN/m^3)	Equiv. Hyd. Cond. (cm/s)	Core Specimen Properties*		
						Avg. Water Cont. (%)	Avg. Dry Unit Wgt. (kN/m^3)	Avg. Hyd. Cond. (cm/s)
1	7.6	60	16.4	16.2	4×10^{-6}	22±2	15.7±1.4	6×10^{-8}
2	7.6	60	19.9	17.5	3×10^{-6}	23±1	16.0±1.1	9×10^{-8}
3	7.6	120	16.5	17.0	2×10^{-5}	23±3	16.0±0.9	7×10^{-8}
4	12.1	120	14.7	17.6	2×10^{-6}	20±2	16.7±1.1	3×10^{-8}
5	12.1	240	17.6	18.2	3×10^{-9}	18±1	16.7±0.5	2×10^{-7}
6	12.1	60	19.8	17.1	8×10^{-9}	23±2	16.0±1.1	5×10^{-8}

* The cores were taken from the liner after conducting a permeability test in the tank and reflect the soaked condition of the liner.
[δ] Hammer was dropped from 0.46 m.
[ξ] Lift thickness was 5 cm and liner thickness was 15 cm.

During some of the water permeation tests, dye was added to the influent water. When the liner was removed, the interface of the tank and the liner showed no traces of dye, indicating that side flow was not responsible for the excessive seepage. Test 4 provided further verification of this observation. A trench 1 to 2 inches deep and 1 inch wide was excavated along the circumference of the liner and filled with a paste of bentonite (Wyoming natural gel). The seepage test was repeated and yielded an equivalent hydraulic conductivity of 3×10^{-6} cm/s, compared with the value of 2×10^{-6} cm/s that was measured previously in Test 4.

During compaction, it was discovered that deflection of the base plate of the tank at each blow was inducing distress in the liner. The wooden support of the tank was apparently not sufficient to prevent the damage to the liner. In Test 5, a sand bed foundation support was provided under the tank instead of wooden support. In this test, the highest compactive effort was employed on a reasonably wet soil. This resulted in a well sealed liner. It is believed that all factors (higher compactive effort, wetter soil, and sand foundation) contributed to preparing a satisfactorily sealed liner; however, the sand foundation was probably the major factor. This experience has certain implications for compaction of landfill covers over compressible subgrade. Test 6 was performed to confirm the final procedure for sample preparation and to experiment with some of the controlling factors in order to bring the equivalent hydraulic conductivity near the target value of 10^{-8} to 10^{-7} cm/s.

From these trial tests, a compaction methodology was developed to achieve a permeability near the target values of 10^{-8} to 10^{-7} cm/s. This method involved the use of a 12-kg hammer, with 60 blows applied to the lower three lifts, and 75 blows applied to the top three lifts. Lower lifts are subjected to energy transmitted through overlying lifts, and so are not struck as often with the compaction hammer. Six lifts of clay, consisting of 31 kg of clay per lift of approximately 50 mm thick, were compacted in the tank. The surface of each lift was scarified before addition of the subsequent lift, and a very wet slurry of the clay was applied to the walls of the tank before adding each lift of soil. Additional compactive effort was applied to the soil around the edge of the liner. This was accomplished by dropping a 4.5-kg weight from a height of 0.3 m. Water content samples were taken from each lift of material before compaction.

TRANSPORT TEST PROCEDURE

Chemical Transport

After permeation of the liner in the test tank with tap water for approximately two months, the organic chemicals were introduced. This was done by filling the influent reservoir, removing five gallons of water from the tank, and preparing a solution equal to the volume removed from a full tank. This solution was prepared such that a target concentration of each of the VOC species would be in solution in the entire liquid volume above the soil. The graduated cylinder at the outlet of the tank was replaced with a Teflon air sampling bag, connected by Teflon tubing and fitted with a quick-release coupling which permitted disconnection of the bag for weighing it periodically in order to monitor flow. The bag was replaced with an empty one after weighing it. Some air bubbles inside the bags was unavoidable, and this undoubtedly caused some volatilization losses. In addition, there is the possibility that diffusion of the chemicals through the walls of the bag could contribute to losses. For this reason, a "control bag" was filled with a VOC solution at concentrations comparable to those encountered in the tests, and monitored for long-term losses. The results indicated that there are significant long-term losses from the Teflon bags (Heim, 1992). However, the losses were significant after a period of about 20 days. Since the time period between filling and emptying bags was one to two weeks, it is concluded that the losses of organic chemicals from the bags would not affect the test results A Teflon septum was attached to the port on the opposite side of the lower part of the tank for obtaining effluent samples.

A vacuum grease was applied to both sides of the nitrile gasket, and the gasket and tank lid were bolted to the flange welded to the top of the cylinder. As a small amount of head space remained inside the tank, air was allowed to escape the top of the tank while it was filled from the side port with a solution of VOCs. When the tank was filled, as evidenced by the cessation of bubbles emitted from the top hole, an input bag was affixed to the top of the tank. The head difference between the influent and effluent bags was about 0.9 m, providing a hydraulic gradient of approximately 3.

Samples were collected from the bags and the sampling ports on the sides of the reservoirs at regular intervals when the influent bag was nearly empty and the effluent bag was nearly full. This was usually at an interval of 7 to 10 days. Samples were drawn with a glass syringe through Teflon septa located in the outlet of the bags and the sampling ports. At the same time, the influent bag was filled by introducing fresh organic permeant solution into the reservoir through the side port right above the liner and the effluent bag was emptied. These bags were weighed periodically (every 2 to 3 days) to determine the inflow/outflow rates.

Soil Coring

After an operating period of approximately nine months following the initial spiking with the organic permeant, one of the tank tests was terminated. The influent reservoir was emptied and sampling tubes were pushed into the liner to obtain cores of the liner material. The procedure followed is a variation of that used by Patterson et al. (1978). A thin-walled copper tube, 38 mm in diameter and 1.5 mm in thickness, was pushed through the soil. The tube was advanced as far as possible by pushing by hand, and when resistance was met, some torsion was applied with the pressure. When the tube reached the bottom of the soil layer, it was removed by excavating soil around the tube until it could be lifted out easily.

An aluminum cap with a Teflon seal ring was fitted to the ends of the tube, and covered with Teflon tape, polyethylene food wrap, and sealed with a rubber band and masking tape to prevent volatilization losses. Recovery was determined by comparing the depth of the hole created by the sampling tube to the length of soil core inside the tube. Six tubes were driven, of which three, labeled Cores 3, 5, and 6, resulted in recovery sufficient for analysis. Recoveries of these tubes were 80 to 84%.

Analysis of VOCs in the Liner

The VOCs were analyzed both in the pore fluid extracts and in the soil on sections of the core samples. A liquid extractor (Fig. 2) similar to the one described in ASTM Standard Test Method for Pore Water Extraction and Determination of the Soluble Salt Content of Soils by Refractometer (D 4542-85) was used to squeeze a portion of the pore fluid from each section of the cores (38 mm in diameter and 25 mm long) under a pressure of 4,140 kPa.

High applied pressures can affect composition of pore water, so pressures only up to 4,140 kPa were used. This is consistent with pressures used in squeezing techniques described by Patterson et al. (1978) and Johnson et al. (1989). ASTM (1991) recommends 79,500 kPa, but Patterson contends that lower pressures provide greater liquid yields. The applied pressure was generated mechanically, using a universal test machine. The samples were exposed only to copper, brass, aluminum, stainless steel, Teflon and glass during the coring and pore water extraction procedure. An effort was made to minimize the exposure of the samples to air during the procedure. Soil samples were subjected to pressure for 30 minutes, and the pore liquid transferred to a 0.5 mL glass sample vial and capped with a Teflon-lined septum. As with the other liquid samples, these were stored at 4°C. Between uses, the squeezer was disassembled and cleaned with hot soapy water, and rinsed with acetone.

After extraction, a 3-g (± 0.05 g) portion of the soil was immersed in 3 mL of carbon disulfide (CS_2), a nonpolar organic solvent, for extraction of any VOCs not squeezed out with pore water as well as VOCs that sorbed onto soil solids. The soil and carbon disulfide were placed in a centrifuge tube, immersed for at least 12 hours at 4°C, mixed for 10 minutes in a vortex mixer, and placed in a centrifuge. Tubes were spun for 20 minutes at 6,000 rpm. Carbon disulfide-solute samples were then placed into 0.5-mL glass vials with Teflon septa, and returned to the 4°C storage. It was assumed that all of the organic chemicals adsorbed onto the soil solids were desorbed into carbon

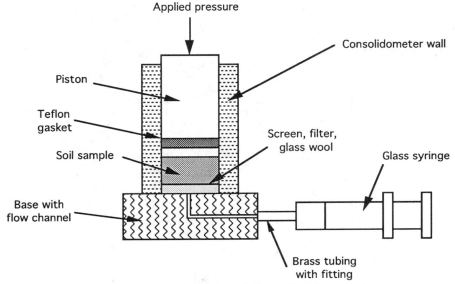

FIG. 2--Schematic of Pore fluid extractor.

disulfide solution. Due to the extreme immiscibility of water and carbon disulfide, it was assumed that no organic chemicals in aqueous pore solution were eluted into the carbon disulfide. Carbon disulfide extraction has been used to desorb VOCs from dry activated carbon by Britton et al. (1989), and is discussed in detail by Clapp et al. (1991). This carbon disulfide extraction procedure was also performed on samples from Core 6 which were not subjected to pore fluid extraction to determine the areal uniformity of solute transport across the tank, and as a check of the squeezing procedure, to determine if any significant volatilization losses occurred.

Organic Chemical Analysis Procedures

Influent and effluent samples were analyzed using a gas chromatograph equipped with a capillary column and a flame ionization detector. Sample vials were shaken for approximately 30 seconds before the cap was removed and 1-µL sample aliquots were drawn into a gas-tight syringe and injected into the gas chromatograph.

It was found through a sensitivity analysis that individual vials were very reproducible, as were multiple injections from the same vial, and two vials collected from the same source were very similar. Only a small amount (about 3%) of VOCs were lost after three weeks of storage at 4°C. Quality assurance measures taken during analysis of samples consisted of running duplicate analyses of effluent samples in which detectable concentrations of VOCs appeared, running a standard, known concentration sample every 10 samples, and mixing new standards and creating calibration curves every two weeks. Calibration curves were created by mixing three standards, injecting each of these three times, and performing linear regression analysis on the results. The resulting equations were used to translate area counts from the integrator attached to the chromatograph into concentrations. R-squared values for regression equations were typically in the range of 0.93-0.99. The detection limits were 1 mg/L for MCL and TCE and 0.5 mg/L for TOL.

DATA ANALYSIS METHODS

Mass Transport

The transport of a reactive solute under hydraulic and concentration gradients in a saturated subsurface environment can be expressed as follows (Freeze and Cherry, 1979):

$$\frac{dC}{dt} = \frac{D}{R_d}\frac{d^2C}{dx^2} - \frac{v}{R_d}\frac{dC}{dx} \qquad (1)$$

where
- C = organic chemical concentration, mg/L,
- t = time elapsed, sec,
- D = hydrodynamic dispersion coefficient, cm^2/s,
- v = seepage (interstitial) velocity, cm/s = $\frac{k \cdot i}{n_e}$,
- k = hydraulic conductivity, cm/s,
- i = hydraulic gradient, cm/cm,
- n_e = effective porosity,
- R_d = retardation factor, dimensionless = $1 + \frac{\rho \cdot K_p}{n}$,
- ρ = dry density of the soil, g/cm^3, and
- n = total porosity.

Partition Coefficient

The retardation factor of a solute shown in Eq. 1 is defined in terms of a soil-water partition coefficient. The distribution of an organic chemical between soil solids and water can be described by an equilibrium expression relating the amount of organic chemical sorbed to the solid surface to the concentration in the water phase as follows:

$$C_s = K_p \cdot C_l \qquad (2)$$

where
- C_s = organic chemical concentration in soil solids, mg/kg,
- K_p = soil-organic chemical partition coefficient, L/kg, and
- C_l = organic chemical concentration in the aqueous phase, mg/L.

For a given type of soil, the sorption behavior of nonionic hydrophobic species is highly correlated to the organic content of the soil (Park et al., 1990). When sorption is closely related to organic fraction, it is convenient to define an organic carbon-normalized soil sorption coeffieicnt, K_{oc}:

$$K_p = f \cdot K_{oc} \qquad (3)$$

where f = organic carbon fraction. There are a number of empirical models relating K_{oc} to K_{ow} and to the aqueous solubility. Hassett et al. (1983) reduced the extensive experimental data of other workers to obtain the following relationships for hydrophobic compounds:

$$\log K_{oc} = 0.088 + 0.909 \log K_{ow} \qquad (4)$$

$$\log K_{oc} = 3.95 - 0.62 \log S \qquad (5)$$

where S = aqueous solubility (mg/L). For a nonreactive solute, the value of K_p is zero (no solute adsorbs to the soil) and the retardation factor has a value of one.

The soil-organic chemical partition coefficient can be calculated using the measured concentrations in the pore water extracted from soil cores and the soil cake eluted by CS_2. The mass of an organic chemical in the pore water, as calculated by the amount of pore water present and the concentrations in the extracted pore water, is subtracted from the total mass of the organic chemical in CS_2 to determine the concentration sorbed on the solid phase on a mass/mass basis.

Effective Porosity

Because of the existence of nonconductive space (such as dead-end pores, pores too small to conduct a significant amount of flow, bound water layers, etc.), the conductive pore volume may be different from the total pore volume, which results in an effective porosity. Effective porosity controls seepage velocity. For nonreactive (conservative) compounds, the effective porosity can be estimated using the following relationship (Freeze and Cherry, 1979):

$$n_e = \frac{k \cdot i \cdot t_{0.5}}{l} \qquad (6)$$

where
  $t_{0.5}$ = time at which the effluent concentration is half of the initial concentration, s, and
  l = specimen thickness, cm.

Hydrodynamic Dispersion Coefficient

The hydrodynamic dispersion coefficient can be determined for an organic chemical once the effective porosity and the retardation factor are known. First, the effective porosity determined from Eq. 6 is adjusted to match the location of the initial breakthrough curve and then the hydrodynamic dispersion coefficient is adjusted to match the shape of the overall breakthrough curve in a solution of Eq. 1 consistent with the boundary conditions of the test. Either an analytical solution (Ogato and Banks, 1991) approximating the test conditions or a numerical solution (Acar et al., 1992) can be used. The data for concentration of the chemical in the pore fluid as a function of depth at a given time provide an independent additional means for determining the hydrodynamic dispersion coefficient in such a model. The effective porosity determined from the bromide breakthrough curve are used in determining the hydrodynamic dispersion coefficient of an organic chemical.

The data collected during the various stages of testing as shown in Table 3 allow more than one basis for evaluation of the needed properties and parameters. For instance, hydraulic conductivity can be determined as a function of time through out the test based on either the effluent or the influent bag weights (Stage 2). Effective porosity can be determined based on either the effluent bromide concentration-time data (Stage 3 and 4) or pore fluid bromide concentration-depth data (Stage 3 and 6). Partition coefficient can be obtained from pore fluid and solid phase VOC concentration-depth data (Stage 6) or independent batch isotherm tests. Finally, dispersion coefficient can be determined from either the effluent VOC concentration-time data (Stage 3 and 4) or the pore fluid VOC concentration-depth data (Stage 3 and 6) once the other parameters are known.

TABLE 3--<u>Primary and redundant samples as data sources.</u>

Test Stages	Primary Sample	Redundant Sample	Data
1 Compaction	Soil Samples /Dimensions	Soil Samples /Dimensions	Water Content, Density (spatial)
2 Permeation	Effluent Bag Weight	Influent Bag Weight	Hydraulic Conductivity (temporal)
3 VOC/Br Spiking	Influent Port Sample	Influent Bag Sample	Entry Concentrations (temporal)
4 Breakthrough	Effluent Port Sample	Effluent Bag Sample	Exit Concentrations (temporal)
5 Liner Coring	Shelby Tube Sample	Shelby Tube Samples	Water Content, Density, Hydraulic Conductivity (spatial)
6 Pore Fluid Extraction	Core Sample	Core Samples	Pore Fluid/Solid Phase Concentrations (spatial)

TEST RESULTS AND ANALYSES

<u>Water Content and Unit Weight Distribution in the Liner</u>

The water content and dry unit weight were nearly constant throughout the liner after it was compacted with the values of 17.2% ± 1.8 and 17.1 kN/m^3 ± 0.5, respectively. These values were determined during the compaction of each of the six lifts from the water content samples, the measured lift thicknesses and the weight of material used. When the tank was dismantled, three Shelby tubes (75 mm in diameter) were also driven into the liner and samples retrieved were sectioned to determine water content, density, and hydraulic conductivity. The water content data collected from sections of at the end of 11 months of permeation showed that the center of the liner was relatively unaffected by the liquid permeant (the water content had increased to 18 to 19%), while the top and the bottom segments near the liquid interfaces had taken up more water with the water contents ranging about 20 to 23%. The top segment of the liner also showed some swelling with a corresponding decrease in dry unit weight to 16 kN/m^3. In general, expansion of compacted clay occurs when soaked without overburden, and some swelling was seen in the soil in the tank at the time of dismantling. The center of the liner, which experienced less compactive energy than the edges and was not influenced by side friction, was noticeably swollen compared to the outer part of the liner.

Hydraulic Conductivity

   After the compaction of the clay liner, the influent and effluent reservoirs of the test tank were filled with tap water and the outflow rate from the bottom of the liner was monitored. Hydraulic conductivity was computed assuming constant head, based on the outflow rate for the period prior to the introduction of the chemical permeant. Volume of flow through the large-scale tank was measured by the periodic detachment and weighing of influent and effluent bags after the introduction of the chemical permeant. As the hydraulic head across the soil was kept constant by the bags, hydraulic conductivity could be calculated based on the influent or outfluent rates determined from the weight change of the bags. Hydraulic conductivity of the liner in the tank is plotted versus time elapsed in Fig. 3. The data include the time before the introduction of organic chemical permeant (tap water only) and after chemical spiking.
   At the initial stage, the hydraulic conductivity of this tank rose to a value of approximately 1.7×10^{-7} cm/s. Some time after the introduction of organic chemical permeant, it dropped to about 8×10^{-8} cm/s. A similar change in hydraulic conductivity was not observed in the other test tanks after the introduction of the organic permeant; therefore, this observation is not attributed to any chemical factors. It is hypothesized that a healing effect occurred in the soil as the liner was soaked.
   Four of the sections obtained from the Shelby tube cores were placed in flexible-wall permeameters. These sections from near the top, center, and bottom of the liner were taken to determine whether there was any significant variation in permeability with depth. Different cores were used to investigate possible areal differences in hydraulic conductivity. The specimens were permeated under an effective confining pressure of 345 kPa with a back pressure of 276 kPa and a hydraulic gradient of 17. The end-of-test core hydraulic conductivities were reasonably uniform across the liner and at different depths with values between 3.6 and 5.1×10^{-9} cm/s. These values are an order of magnitude lower than those calculated from the rate of flow through the liner. This may have been caused by consolidation of the samples due to the applied effective stress, although no measurable contraction of the soil was measured. Dimensions of the sample were measured with an accuracy of 2.5 mm and this may not have been accurate enough to detect sample consolidation. It is more likely that the size of the core samples was too small to include a representative volume of the liner.
   A mass balance on water in the tank system was performed. For this experiment, the conservation of mass principle dictates that:

| Mass of Water in System at Start | + | Mass of Water Added | − | Mass of Water Removed | − | Mass of Water in System at End | = | Loss or Gain |

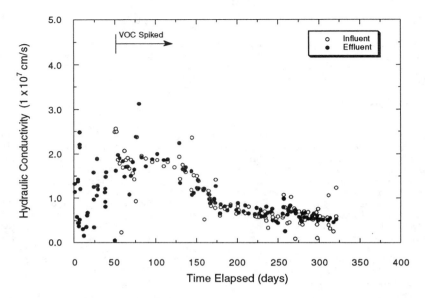

FIG. 3--Hydraulic conductivity of large-size liner.

The results of the water budget indicated a gain of 2.95% water, as related to the total amount of water added to the system. In this test, before spiking with VOCs, the liner was permeated for 2 months with tap water to determine the flow characteristics. Since no data regarding the rate of infiltration of water into the liner during this period were taken, no definitive conclusions can be drawn from the water budget analysis regarding the quality of the test other than attributing the water gain to the water absorbed by the liner. In subsequent tests, this deficiency was corrected.

Organic Chemical Transport

While there was some fluctuation in influent concentrations, the boundary condition of constant inflow concentration would appear to be valid for this experiment. In subsequent tests, a closer control of the influent concentration was achieved by means of the adjusted concentration of the fresh influent fluid added during replacement of the bags. The concentrations of the three chemicals in the influent reservoir were 31 ±5 mg/L for MCL, 31 ±14 for TCE, and 14 ±6 for TOL. There was some difference between the concentrations of solute collected from the influent bags and those taken from the port located at the side of the tanks, near the soil-liquid interface. This appears to have been especially true near the beginning of the test, perhaps before the VOCs were fully mixed into the bulk solution. Reasons for this phenomenon are unknown, but it may have been avoided if there was some way to stir or agitate the solution in the influent reservoir. Very little variation was noted in toluene concentrations between different sampling locations.

The effluent data for replicate analyses of the MCL samples from the sampling port of the effluent reservoir and the effluent bag are shown in Fig. 4. Despite some scatter in the data, the shape of the breakthrough curve is recognizable. The data for TCE were more scattered and a breakthrough curve was not fully established (only about 30% breakthrough) by the end of 9 months of permeation. TOL was at the detection limit of 1 mg/L indicating that significant breakthrough of TOL has not yet occurred. The effluent concentrations given in Fig. 4 are computed based on the mass of MCL delivered (based on either the port or the bag concentration measurements) to the effluent reservoir by the amount of outflow accumulated at a given time.

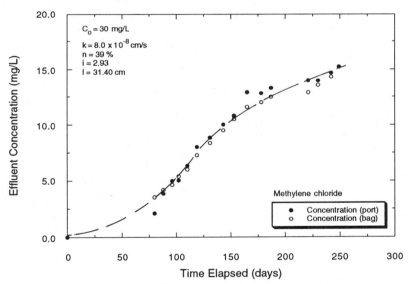

FIG. 4--Effluent concentration vs. time.

Fig. 5 shows MCL and TCE concentrations in the pore water as a function of depth. The concentration of the organic chemicals sorbed on the solid phase of the soil was computed from the mass of the organic chemicals extracted from the soil by carbon disulfide, making a suitable correction for organics in the residual soil moisture. As the partition of VOCs between water and carbon disulfide is not fully understood, it was unknown whether to calculate the concentration of contaminants adsorbed onto soil solids on the basis of dry unit weight of the soil, or after making an adjustment based on the water content and pore water concentration of VOCs. In the former, it is assumed that all of the contaminant mass in the carbon disulfide originates from the soil solids, and none is eluted from the water phase. In the pore water concentration method, the mass of solute in CS_2 is assumed to originate

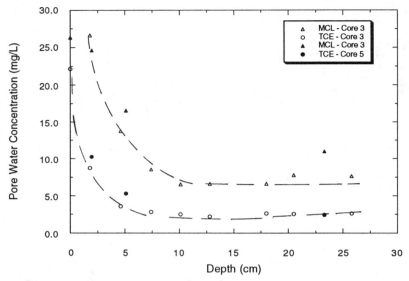

FIG. 5--MCL and TCE concentrations in the pore water vs. depth.

from both the soil solids and the remaining pore water after extraction. The concentrations of VOCs are assumed to correspond to the extracted pore-water concentrations for corresponding depths. A comparison of these two methods was made using Core 3 and Core 6. Pore water was partially extracted from all samples in Core 3, while no extraction was performed on Core 6. The results indicated that the pore water-concentration method is the more appropriate procedure for calculating concentrations of VOCs adsorbed onto soil particles, as the corresponding solid phase concentrations from Cores 3 and 6 more closely corresponded to each other than do the concentrations based on the dry unit weight method. Fig. 6 gives concentrations of MCL and TCE soil solids as a function of depth. The data indicate that a concentration equilibrium between the solid and pore fluid phases was reached for MCL but perhaps not for TCE.

Soil-Organic Chemical Partition Coefficient

The soil-organic chemical partition coefficient (or adsorption coefficient), K_p, was estimated as the ratio of the pore water concentration to the concentration adsorbed on the solid phase. It was calculated using the measured concentrations in the pore water extracted from the cores and the soil cake eluted by CS_2. The mass of an organic chemical in the pore water, as calculated by the amount of pore water present and the concentrations in the extracted pore water, is subtracted from the total mass of the organic chemical in CS_2 to determine the concentration sorbed on the solid phase on a mass/mass basis.

K_p values estimated from the extracted concentrations between the soil and the pore water varied with depth as shown in Fig. 7. Results of the K_p calculations are compared with the data reported by Hassett et al. using Eqs. 4 and 5 in Table 4. The mean values of K_p show reasonable agreement with the values obtained from the empirical relationships described by Hassett et al. (1983).

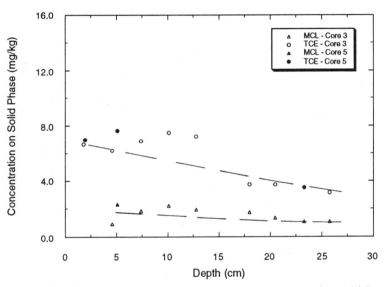

FIG. 6--Concentrations of MCL and TCE on soil solids.

The variation of K_p data with depth resembles the data for dry unit weight (or water content) of the liner at the conclusion of the test. Voice et al. (1983) predicted a relationship between the partition coefficient and the solids concentration of lacustrine sediments. That experiment used low concentrations (10-400 mg/L) of a highly organic soil, with some PCBs and polynuclear hydrocarbons. At higher soil concentrations, K_p values appeared to decrease, with an accompanying increase in turbidity. The microparticles causing the turbidity were not removed by separation processes used in the experiment. Because of this, it was believed that organics sorbed onto this microparticle phase contributed to this phenomenon. Due to the large differences in experimental conditions between this study and those used by Voice et al. (1983), no significant correlations can be made regarding this phenomenon. Soil in the liner model was nearly saturated through the entire depth of soil, so the possibility of volatilization of VOCs within the soil matrix can likely be discounted.

TABLE 4--Results of K_p determination.

Method of Determination	K_p Methylene Chloride	Trichloroethylene
Eq. 4	0.151	2.198
Eq. 5	0.193	1.044
Extraction Procedure:		
Average Value:	0.192	1.804
Std. Deviation:	0.086	0.827

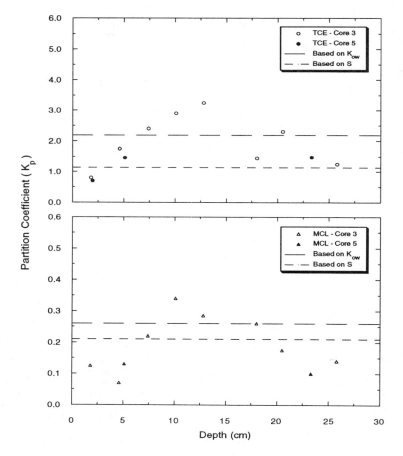

FIG. 7--Variation of soil-water partition coefficient, K_p, with depth: (a) methylene chloride, (b) trichloroethylene.

Analysis of Test Results

In the preceding sections, the development and analysis procedures for a large-size VOC transport test are presented along with the data obtained from the first such test. Three other large-size, long-term tests are currently underway. Some of the shortcomings observed in this first test have been resolved. There is a better control over the test conditions, liner construction, influent concentration, and chemical analysis procedures. However, high variability seems to be the true nature of both the soil behavior and the volatile organic chemicals and becomes more apparent in large-size tests. Therefore, a meaningful interpretation of transport of VOCs requires sufficient number of replicate tests. Until such data become available, modeling of VOCs transport is not undertaken.

CONCLUSIONS

Large-size tests under low hydraulic gradients provide a closer simulation of landfill liner conditions and therefore are considered to be a more appropriate means of determining the transport and attenuation parameters for VOCs through clay liners. The size of test specimen appears to be a very significant factor in conduction and transport investigations. Preliminary data from these and other laboratory tests suggest that the organic chemical transport is not retarded significantly in typical clay liner materials. The test results further suggest that the partition (or sorption) coefficients are reasonably well approximated by the available empirical formulas. If this finding is further substantiated, it would simplify prediction of retardation of a wide variety of organic chemicals in clay. The breakthrough of VOCs through clay liners depends on their solubilities and/or partition coefficients as well as on the organic carbon content of the clay.

ACKNOWLEDGMENT

This research was funded by the Wisconsin Department of Natural Resources. N. H. Severson and P. Fritschel provided support in developing the test equipment and procedures and W. Wambold assisted in preparation of the manuscript.

REFERENCES

Acar, Y. B., Gokmen, C. and Alshawabkeh, A., 1992, Diffuse - 2DFE, A Two Dimensional Finite Element Model for Advective-Dispersive Transport of Reactive Solutes in Porous Media, The Institute of Recyclable Materials, College of Engineering, Louisiana State University, Baton Rouge, Louisiana, Internal report No: CE-GE-05-0992.

Battista, J. R., and C. P. Connelly, 1989, VOC Contamination at Selected Wisconsin Landfills--Sampling Results and Policy Implications, Wisconsin Department of Natural Resources, PUBL-SW-094-89.

Benson, C.H., and Daniel, D.E., 1990, "Influence of Clods on Hydraulic Conductivity of Compacted Clay", *J. Geotech. Engr.*, 116(8): pp. 1231-1248.

Benson, C. H., Hardianto, F. S., and Motan, S., 1993, "Representative Sample Size for Hydraulic Conductivity Assessment of Compacted Soil Liners," *Hydraulic Conductivity and Waste Containment Transport in Soils*, ASTM STP 1142.

Briggs, C. C., 1973, "A Simple Relationship between Soil Adsorption of Organic Chemical and Their Octanol/Water Partition Coefficients," *Proceedings,* 7th British Insecticide and Fungicide Conference, Brighton, England, British Crop Protection Council.

Bowders, J. J., and Daniel, D. E., 1987 "Hydraulic Conductivity of Compacted Clay to Dilute Organic Chemicals," *J. Geotech. Engr.*, 113(12): pp. 1432-1448.

Britton, L. N., Ashman, R.B., Aminabhavi, T.M., and Cassidy, P.E., 1989, "Permeation and Diffusion of Environmental Pollutants through Flexible Polymers," *Journal of Applied Polymer Science*, Vol. 38, pp. 227-236.

Clapp, L.W., Talarczyk, M.R., Park, J.K., and Boyle, W.C., 1992, "Biological Treatability Study," University of Wisconsin-Madison, report submitted to the City of Niagara Falls, New York.

Deville, W. B., 1985, "Superfund Impact on Disposal of Hazardous Waste," *ASTM Standardization News*, March Issue.

Edil, T. B., J. K. Park, and P. M. Berthouex, 1991a, "Effects of Volatile Organic Compounds on Clay Landfill Liner Performance-- Phase 2," *Interim Report to the Wisconsin Department of Natural Resources*, Department of Civil and Environmental Engineering, University of Wisconsin-Madison.

Edil, T. B., P. M. Berthouex, J. K. Park, D. L. Hargett, L. K. Sandstrom, and S. Zelmanowitz, 1991b, "Effects of Volatile Organic Compounds on Clay Landfill Liner Performance," *Waste Management & Research*, Vol. 9, pp 171-187.

Freeze, R.A., and Cherry, J.A., 1979, *Groundwater*, Englewood Cliffs, N.J.: Prentice-Hall, 604 pp.

Friedman, M. I., 1988, "Volatile Organic Compounds in Groundwater and Leachate at Wisconsin Landfills," *Wisconsin Department of Natural Resources*, PUBL-WR-192-88.

Heim, David P., 1992, "Advective and Diffusive Transport of Three Volatile Organic Compounds Through a Compacted Clay," Master of Science Thesis, Civil and Environmental Engineering, University of Wisconsin-Madison.

Hassett, J. J., W. L. Banwart, and R. A. Griffin, 1983, "Correlation of Compound Properties with Sorption Characteristics of Nonpolar Compounds by Soils and Sediments: Concepts and limitations. In *Environment and Solid Wastes*, edited by C. W. Francis and S. I. Auerbach, Ann Arbor Science Book, 161-178.

Johnson, R.L., Cherry, J.A., and Pankow, J.F., 1989, "Diffusive Contaminant Transport in Natural Clay: A Field Example and Implications for Clay-Lined Waste Disposal Sites," *Environ. Sci. Tech*, 23: pp. 340-349.

Lambert, S. M., Omega, 1968, "A Useful Index of Soil Sorption Equilibria," *Journal of Agricultural Food Chemistry*, Vol. 6, pp 572-576.

Karickhoff, S. W., Brown, D. S., and Scott, T. A., 1979, "Sorption of Hydrophobic Pollutants on Natural Sediments," *Water Research*, Vol. 13, pp. 241-248.

Nelson, B. R. and P. R. Book, 1986, "Monitoring for Volatile Organic Hydrocarbons at Minnesota Sanitary Landfills," *8th Annual Madison Waste Conference*, Department of Engineering Professional Development, University of Wisconsin-Madison, pp. 72-77.

Ogata, A., and R. B. Banks, 1961, "A Solution of the Differential Equation of Longitudinal Dispersion in Porous Media," *Geological Survey Professional Paper*, 411-A, A1-A7.

Park, J. K., T. M. Holsen, L. G. Bontoux, D. I. Jenkins, and R. E. Selleck, 1989, "Permeation of Plastic Pipes by Organic Chemicals," Interagency Agreement No. 84/84371 between Department of Health Services, State of California, and the Regents of the University of California, *Sanitary Engineering and Environmental Health Research Laboratory, University of California-Berkeley*.

Park, J. K., Edil, T. B., Berthouex, P. M., 1990, "Effects of Effective porosity, Organic Carbon Contents on Movement of VOCs in a Clay Liner," *13th Annual Madison Waste Conference*, Department of Engineering Professional Development, University of Wisconsin-Madison, Sep. 19-20, pp 438-451.

Patterson, R.J., Frape, S.K., Dykes, L.S., and McLeod, R.A., 1978, "A Coring and Squeezing Technique for the Detailed Study of Subsurface Water Chemistry", *Canadian Journal of Earth Science*, 15: pp. 162-169.

Peyton, G. R., J. P. Givv, M. H. LeFaivre, J. D. Ritchey, S. L. Burch, and M. J. Barcelona, 1985, "Effective Porosity of Geologic Materials," *Proc. of 12th Annual Research Symposium - Land Disposal, Remedial Action, Incineration and Treatment of Hazardous Waste*, EPA/600/9-86/022, pp. 21-28.

Plumb, R. H. and A. M. Pitchford, 1985, "Volatile Organic Scans: Implications for Ground Water Monitoring," NWWA/API, *Proc. of Conference on Petroleum Hydrocarbons and Organic Chemicals in Ground Water-Prevention, Detection and Restoration*, Dublin, Ohio, National Water Well Association, pp. 207-223.

Shackelford, C. D., 1991, "Laboratory Diffusion Testing for Waste Disposal - A Review," *Journal of Contaminant Hydrology*, No. 7, pp. 177-217.

Shackelford, C. D. and Javed, F. 1991, "Large-Scale Laboratory Permeability Testing of a Compacted Clay Soil," *Geotechnical Testing Journal*, Vol. 14, No. 2, pp. 171-179.

Trautwein, S.J. and Williams, C.E. 1990, "Performance Evaluation of Earthen liners", Waste Containment Systems, ASCE, Geotechnical Special Publication No. 26.

Voice, T.C., Rice, C.P, and Weber, W.J., 1983, "Effect of Solids Concentration on the Sorptive Partitioning of Hydrophobic Pollutants in Aquatic Systems", *Environ. Sci. Tech.*, 17: pp. 513-518.

David J. Fallow[1], David E. Elrick[1], W. Dan Reynolds[2], Norbert Baumgartner[1] and Gary W. Parkin[1]

FIELD MEASUREMENT OF HYDRAULIC CONDUCTIVITY IN SLOWLY PERMEABLE MATERIALS USING EARLY-TIME INFILTRATION MEASUREMENTS IN UNSATURATED MEDIA

REFERENCE: Fallow, D. J., Elrick D. E., Reynolds, W. D., Baumgartner, N., and Parkin, G. W., "Field Measurement of Hydraulic Conductivity in Slowly Permeable Materials Using Early-Time Infiltration Measurements in Unsaturated Media," Hydraulic Conductivity and Waste Contaminant Transport in Soil, ASTM STP 1142, David E. Daniel and Stephen J. Trautwein, Eds., American Society for Testing and Materials, Philadelphia, 1994.

ABSTRACT: Most procedures to measure the hydraulic conductivity of slowly permeable materials such as compacted soil liners are based on analyses that assume saturated, one-dimensional flow under steady-state conditions. The overwhelming problem, however, is the very long times, of the order of weeks or months in liner materials, to reach experimentally-measurable steady flow. A new field procedure is proposed for slowly permeable materials that takes advantage of the early transient flow in initially unsaturated soil. Both constant head and falling head techniques are proposed and measurement times are of the order of one half to several hours. The falling head technique has the advantage of requiring only the difference between the field-saturated and initial water contents in addition to the measured position of the falling head above the soil surface as a function of time. An experiment on the experimental soil liner at Champaign, Illinois, gave saturated hydraulic conductivity values using the constant head technique that were in good agreement with previously measured values. A laboratory test demonstrates the advantages of the falling head technique.

KEYWORDS: hydraulic conductivity, compacted soil liners, early-time infiltration, unsaturated flow.

---

[1] Graduate student, professor, technician and graduate student respectively, Department of Land Resource Science, Univ. of Guelph, Guelph, Ontario, Canada, N1G 2W1.

[2] Research scientist, Centre for Land and Biological Resources Research, Agriculture Canada, Ottawa, Ontario, Canada K1A 0C6.

Several techniques have been developed to measure the hydraulic conductivity of slowly permeable materials such as compacted soil liners. However, most of these procedures are based on analyses that assume saturated one-dimensional flow under steady-state conditions (Daniel, 1989). For example, double-ring infiltrometers have been constructed in an attempt to let the outer annular space between the two rings absorb all of the edge and divergence effects. However, this attempt may be futile if the wetting front penetrates below the depth of the embedment of the cylinders. The infiltration rate from the inner cylinder for all but highly permeable material is very much influenced by the divergence and edge effects under these conditions (Bouwer, 1986; Elrick, 1992).

For slowly permeable materials where the field-saturated hydraulic conductivity, K_{fs}, is less than $10^{-9} ms^{-1}$, the overwhelming problem is, however, the very long times required to reach near steady-state flow conditions. Testing times for ring infiltrometers generally take several weeks to several months (Daniel, 1984). Elrick et al. (1990) have shown that steady-state flow conditions are reached much more quickly if three-dimensional flow conditions are established and that the steady flow rate obtained under these three-dimensional conditions depends upon both the saturated and unsaturated components of the hydraulic conductivity. Nevertheless, the very small flow rates measured and testing times of several days to several weeks make even these three-dimensional steady-state flow measurements very time consuming and difficult.

EARLY-TIME TRANSIENT FLOW ANALYSIS

Under field conditions most soils or permeable materials are unsaturated at the surface. If a compacted soil liner or other slowly permeable material is to be field-tested and is initially unsaturated, then advantage can be taken of the early transient flow. The analysis presented below assumes that the soil is incompressible, which may limit the application of the technique under certain conditions; e.g., an initially very dry soil that contains a high percentage of swelling clays. Theoretically, measurements taken over a period of about one to three hours should be sufficient. In Fig. 1 we show a plot of the flow rate Q ($L_w^3 T^{-1}$) vs $t^{-1/2}$ ($T^{-1/2}$) where the subscript w on length L refers to water and T is time. The data in Fig. 1 was obtained from a numerical solution of Richards' equation (Reynolds and Elrick, 1990). For the stereotyped clay soil, the early time behaviour is linear for approximately 17 h and for the stereotyped liner soil, the early time behaviour is linear for approximately 70 hr or 3 days. This data was obtained for a ponded head of zero; for a ponded head of 1m or more the linear portion of the early time behaviour is shortened, perhaps by a factor of 2 or 3, still leaving many hours for obtaining valid measurements.

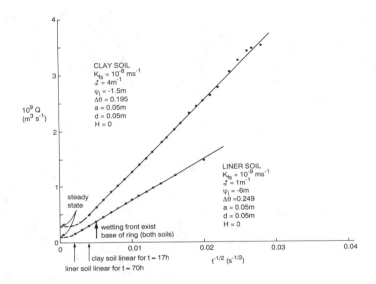

Fig. 1   Flow rate Q, vs $t^{-1/2}$ showing the linear portion of early-time behaviour for the theoretical clay soil (approximately 17h) and the theoretical liner soil (approximately 70 hr). K_{fs} is the field-saturated hydraulic conductivity, α^* is defined in Eq. (9), ψ_i is the initial water potential $\Delta\theta$ the difference in water content between the field-saturated and initial contents, a is the radius of the infiltrometer, d the depth of insertion and H is the ponded head.

It has been shown (Philip, 1957, 1969) that the initial flow of water from a surface ring can be considered to be one-dimensional and free from gravitational effects so that the cumulative infiltration of water I($L_w^3 L_b^{-2}$) or the infiltration rate i ($L_w^3 L_b^{-2} T^{-1}$) for this short early-time period can be described by

$$I = S_H t^{1/2} \qquad (1)$$

or

$$i = dI/dt = 0.5\, S_H t^{-1/2} \qquad (2)$$

where $S_H(L_w^3 L_b^{-2} T^{-1/2})$ is the soil sorptivity for a ponded head H. The subscript b on L refers to bulk length. Note that the flow rate Q in Fig. 1 is given by

$$Q = Ai \qquad (3)$$

where A is the cross-sectional area of the infiltrating surface.

For conditions where water is supplied at a positive pressure head, H, such as occurs with the Guelph Pressure Infiltrometer (Reynolds and Elrick, 1990) or with the Velocity Permeameter (Merva, 1987), then S_H is given by (White and Sully, 1987):

$$S_H = \left[2 \int_{\psi_i}^{H} \frac{(\theta(\psi) - \theta_i) K(\psi)}{F[\theta(\psi)]} d\psi \right]^{1/2} \qquad (4)$$

where θ ($L_w^3 L_b^{-3}$) is the volumetric water content, θ_i, is the initial water content corresponding to the initial water head ψ_i, ψ is the soil water head, $K(\psi)$ ($L_w^3 L_b^{-1} L_h^{-1} T^{-1}$) is the hydraulic conductivity relationship (see Fig. 2) and $F(\psi)$ (no dimensions) is the soil-water flux ratio (Philip, 1973). The subscript h on L refers to head units. If we break the integral in Eq. (4) into two steps where F = 1.0 for $\psi \geq 0$, then Eq. (4) becomes

$$S_H = [2(\Delta\theta) K_{fs} H + S_0^2]^{1/2} \qquad (5)$$

where

$$S_0 = [(\Delta\theta) \phi_m / b]^{1/2} \qquad (6)$$

and where K_{fs} is the field-saturated hydraulic conductivity, S_0 ($L_w^3 L_b^{-2} T^{-1/2}$) is the sorptivity at H = 0, and $\Delta\theta = \theta_{fs} - \theta_i$ where θ_{fs} is the field-saturated water content. Note that the constant b can be set equal to 0.55 with an error of at most of about 10% (White and Sulley, 1987) and that the matric flux potential, ϕ_m ($L_w^3 L_b^{-1} T^{-1}$), is defined (see Fig. 2) as:

$$\phi_m \equiv \int_{\psi_i}^{0} K(\psi) d\psi \qquad (7)$$

In Eq. (5) to (7) we have assumed that there is no tension-saturated zone. Eq. (5) can also be written as

$$S_H = [1 + 2b\alpha^* H]^{1/2} S_0 \qquad (8)$$

where

$$\alpha^* \equiv K_{fs} / \phi_m \qquad (9)$$

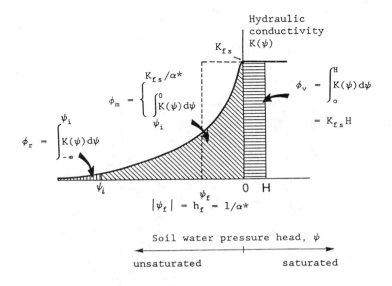

Fig. 2   Hydraulic conductivity, K, as a function of soil water pressure head, ψ, with associated integral properties of residual flux potential ϕ_r, matric flux potential ϕ_m, field saturated hydraulic conductivity K_{fs}, the ratio $K_{fs}/\phi_m = \alpha^*$, and the Green and Ampt suction at the wetting front, $h_f = |\psi_f|$.

A schematic diagram of the constant head and falling head infiltrometers is shown in Fig. (3). For constant head conditions, (Fig.3a) S_H in Eq. (1) is a constant. For falling head conditions, S_H in Eq. (1) is a function of H(t) [see Eq. (5)] and the relationship between H and t is given by (Elrick et al., 1992):

$$t^{1/2} = \frac{a}{A} \frac{(H_0 - H)}{[2(\Delta\theta) K_{fs} H + (\Delta\theta) \phi_m/b]^{1/2}} \qquad (10)$$

where $H_0 = H$ at $t = 0$, A is the cross-sectional area of the infiltrating surface, and a is the cross-sectional area of the falling-head tube (see Fig. 3b).

The cumulative infiltration I is given by

$$I = \left[2(\Delta\theta) K_{fs} H(t) + (\Delta\theta) \phi_m/b\right]^{1/2} t^{1/2} \qquad (11)$$

where the impact of the falling head is given by the term H(t).

For a constant head of H_o we have

$$I = [2(\Delta\theta)K_{fs}H_o + (\Delta\theta)\phi_m/b]^{1/2}t^{1/2} \tag{12}$$

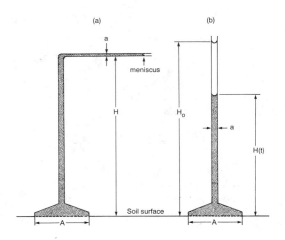

Fig. 3   Schematic diagram of the constant head (a) and falling head (b) infiltrometers, H is head, A is the cross-sectional area of the infiltrometer and a is the cross-sectional area of the falling-head tube.

Early-Time Constant Head Technique

The procedure for determining K_{fs} and ϕ_m using early-time flow data obtained under a constant head proceeds as follows:

(1) Obtain data during the first hour or so of flow of $I(t)$ at a fixed height H, where H is the positive head above the soil surface. A microburette or a tube of known cross-sectional area a may be used to obtain $I(t)$.
(2) Plot I vs $t^{1/2}$ and from the initial straight line portion obtain S_H from the slope:

$$S_H = \frac{I_2 - I_1}{t_2^{1/2} - t_1^{1/2}} = \frac{\Delta I}{\Delta t^{1/2}} \tag{13}$$

(3) Obtain S_o from S_H using Eq. (8) with α^* obtained independently. If no measurements or estimates are

available, then set $\alpha^* = 1$ m^{-1} for clay liner materials (Elrick and Reynolds, 1992).

(4) Measure θ_{fs} and θ_i and calculate ϕ_m from Eq. (6) setting b = 0.55.

(5) Use the value of α^* obtained above and calculate K_{fs} from Eq. (9); i.e.

$$K_{fs} = \alpha^* \phi_m \qquad (14)$$

In this procedure for calculating K_{fs}, $\Delta\theta$ must be measured independently and, more importantly, α^*, in which there may be some uncertainty, must also be known (Reynolds and Elrick, 1992). The following procedure for calculating K_{fs} requires only an independent measurement of $\Delta\theta$ in addition to the flow data.

Early-Time Falling Head Technique

The procedure for determining K_{fs} and ϕ_m using early time data under falling head conditions proceeds as follows:

(1) Select a combination of the cross-sectional area of the infiltrometer, A, and the cross-sectional area of the falling head tube, a, that will give a drop in H of from 1 to 2 m over a period of one or more hours. For slowly permeable soils ($K_{fs} \leq 10^{-9}$ ms^{-1}), H$_0$ should probably be set between 1 and 2 m.

(2) Record H as a function of t, initially taking readings every 30 s or so and then less often as the head falls. Initially H falls very quickly (depending on the choice of A/a), even in slowly permeable soils. If possible record H(t) until H drops to approximately 0.1 m or as low as possible.

(3) The value of H at t = 0 (H$_0$) is difficult to obtain experimentally because the initial wetting of the soil is followed by filling the permeameter and falling head tube with water which may require a few seconds to minutes, depending on the experimental arrangement. Therefore, it is best to obtain a first guess of H$_0$ by a linear extrapolation to t=0 using the first few readings of H vs t$^{1/2}$.

(4) Use a parameter estimation scheme to obtain simultaneous estimates of K_{fs} and ϕ_m

Note that this procedure requires only a measurement of $\Delta\theta$ in addition to the flow data and presents a theoretical advantage over the early-time constant head procedure where

α^* must also be known. Further research is required, however, to test the practicality of this procedure.

EARLY-TIME CONSTANT HEAD EXPERIMENT

A preliminary experiment was carried out at the Illinois State Geological Survey experimental soil liner at Champaign, Illinois (Krapac et al., 1991). The Pressure Infiltrometer attachment to the Guelph Permeameter (see Fig. 4) was driven into the liner by first sprinkling a very small amount of water on the surface and then the 9.6 cm diameter ring was rotated to score the surface. The ring was then pounded carefully using a drop weight procedure, periodically spraying a small amount of water around the inner and outer edges to ease the entry of the ring. The ring was driven in a total depth of 5 cm. The inner surface was inspected for gaps around the ring and, if necessary, were filled with dry bentonite and packed firmly with a spatula. The ring was then filled with water and a stopwatch was started at this time to set t = 0. The acrylic top was then attached to the ring (see Fig. 4), water was added and the system was checked for leaks. The horizontal microburette was then connected to the top of the infiltrometer tube and the position of the meniscus was read. A two-way stopcock on the horizontal burette aided in ensuring that no air was entrapped. A first reading, R_1, was then taken at time t_1, with two sites close to one hour had elapsed before the first reading was taken. The horizontal microburette was then read at two to five minute intervals using a 10 ml burette calibrated to 0.02 ml. Readings were taken for about one to two hours. The procedure outlined previously was then followed. If $\Delta R = R_2 - R_1$ is the change in burette readings in m_w^3 over the time interval $\Delta t = t_2 - t_1$, then the change in cumulative infiltration $\Delta I = I_2 - I_1$ is given by

$$\Delta I = \Delta R / A \tag{15}$$

where A is cross-sectional area of the ring. S_H can then be obtained from Eq. (13) rewritten as

$$S_H = \frac{R_2 - R_1}{A(t_2^{1/2} - t_1^{1/2})} = \frac{\Delta R}{A \Delta t^{1/2}} \tag{16}$$

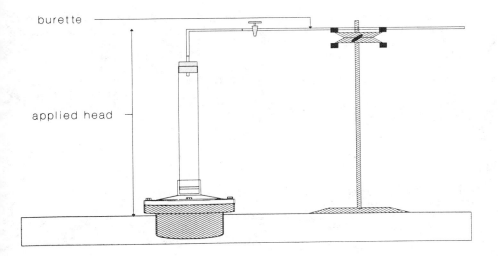

Fig. 4   The schematic of the pressure infiltrometer attachment to the Guelph Permeameter as used at the experimental soil liner at Champaign, Illinois. The head, H, is set by the height of the horizontal burette above the soil surface.

A plot of six separate measurements of R vs $t^{1/2}$ is shown in Fig.5. Rather than use Eq.(16) directly, a linear least squares fit of the early linear portion of the data was used to obtain the slope, $\Delta R/\Delta t^{1/2}$. The geometric mean and standard deviation of the six S_H measurements were 1.4 x 10^{-5} m̶ ̶ ̶ ̶ ̶ x 10^{-6} $ms^{-1/2}$ respectively.

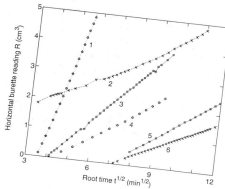

Fig. 5   The horizontal burette reading, R, as a function of $t^{1/2}$ for six separate adjacent sites at the experimental soil liner at Champaign, Illinois.

An estimate of α^* was obtained by fitting the equation of van Genuchten (1980) to the soil moisture characteristic curve (Fig. 18, p. 40 Krapac et al., 1991) of the liner material (Batestown till). A nonlinear least squares fitting technique [MathCAD 3.1 software (Mathsoft, 1992)] gave $\alpha_v = 0.2384$ m^{-1} and $N = 1.4841$ where α_v and N are fitting parameters of the van Genuchten (1980) expression for $\theta(\psi)$ where ψ is the soil water pressure head. The parameter α^* was then calculated from (Elrick et al., 1989)

$$\alpha^* = \left[\int_0^h \Theta^{1/2}[1-(1-\Theta^{1/M})^M]^2 dh \right]^{-1} \tag{17}$$

where the reduced volumetric water content, Θ, is given by van Genuchten (1980) as

$$\Theta = \frac{\theta - \theta_i}{\theta_{fs} - \theta_i} = [1 + |\alpha_v h|^N]^{-M} \tag{18}$$

where $h = -\psi, \ \psi < 0$

and $M = 1 - 1/N$

The upper limit of integration, h, must be sufficiently large as to encompass the entire area under the integrand of Eq. (17). In the above equations θ, θ_{fs} and θ_i are the volumetric water content, the field-saturated water content and the initial water content respectively, and h the soil water suction head. The integration in Eq. (17) gave $\alpha^* = 1.115$ m^{-1}, fortuitously close to our suggested value of 1 m^{-1} for liner materials.

The difference in volumetric water content, $\theta_{fs} - \theta_i = \Delta\theta$ was obtained from initial and final gravimetric samplings of the liner material. The product of the gravimetric water contents and bulk density was used to estimate $\Delta\theta$. The bulk density was assumed to be 1.84 Mg m^{-3} (Krapac et al., 1991) and $\Delta\theta$ was measured as 0.13.

Using the previously calculated value of $\alpha^* = 1.115$ m^{-1}, the field-saturated hydraulic conductivity can be calculated from [Eq. (6), (8) and (9)]:

$$K_{fs} = \frac{\alpha^* S_o^2 b}{\Delta\theta} \tag{19}$$

giving a geometric mean of $K_{fs} = 6.4 \times 10^{-10}$ ms^{-1} using $b = 0.55$. This number compares favourably with the values of 5.3×10^{-10} ms^{-1}, 3.3×10^{-11} ms^{-1} and 6.7×10^{-10} ms^{-1} for the small-ring infiltrometer, large-ring infiltrometer and liner water-balance methods respectively (Krapac et al., 1991). These reported values were obtained using Darcy's law and slightly lower values (about 70% of the above) were obtained using the Green-Ampt model, both based on long-time steady flow.

EARLY-TIME FALLING HEAD EXPERIMENT

Unfortunately no falling head data was obtained at the experimental soil liner site at Champaign, Illinois and therefore a preliminary experiment was carried out in the laboratory using a compacted clay soil (40% clay, 54% silt, 6% sand). The air-dried, sieved material was compacted in a Proctor Density apparatus (A = 8.012 x 10^{-3} m^2) to a bulk density of 1.6 Mg m^{-3}. A 1½ m long vertical tube having a cross-sectional area, a = 8.75 x 10^{-6} m^2, was attached to the top of the apparatus. At t = 0 water was supplied under a positive head and the space above the compacted soil was filled with water. The vertical tube had been filled with water previously, clamped off and then reconnected after the space above the soil had been filled with water. Readings commenced at 30s and were taken every 15s for 3 1/2 min, followed by readings every 30s for the next 3 1/2 min for a total of 7 min. For this material 7 min was sufficient but for more slowly permeable material several hours may be required. Also, a larger diameter falling-head tube would allow more readings over a longer period of time.

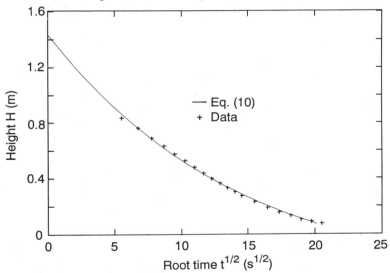

Fig. 6 The measured height of the falling head, H, as a function of $t^{1/2}$. The experimental data and the best-fit Eq. (10) compare favorably.

The difference in water content, $\Delta\theta$, was measured to be 0.32 and the extrapolated H vs $t^{1/2}$ plot for the first 7 data points gave a guess value of $H_0 = 1.193$ m. With b = 0.55, a nonlinear least squares fit of the data using Eq. (10) using the MathCAD 3.1 software (Mathsoft, 1992) gave $H_0 = 1.44$ m, $K_{fs} = 1.44 \times 10^{-8} ms^{-1}$, $\phi_m = 8.2 \times 10^{-9} m^2 s^{-1}$ with $\alpha^* = 1.76 m^{-1}$. Because the fitting procedure is very sensitive to the initial guess value of H_0, the nonlinear least squares program was first applied to the data with H_0 fixed by the extrapolation procedure and only K_{fs} and ϕ_m fitted. With these calculated values of K_{fs} and ϕ_m and with the extrapolation value of H_0 increased by a factor of 1.2 used as guess values the numbers reported above were obtained on fitting for H_0, K_{fs} and ϕ_m. Using this procedure the dependence on the initial guess values was minimized. A final check on the calculations can be obtained by plotting the data vs Eq. (10) using the best estimated values of H_0, K_{fs} and ϕ_m. Such a plot is shown in Fig. 6 and the agreement is excellent. The same data replotted in terms of I vs $t^{1/2}$ [Eq. (11)] is shown in Fig. 7 which also includes the linear I vs. $t^{1/2}$ line for a constant head of H_0 [Eq. (12)].

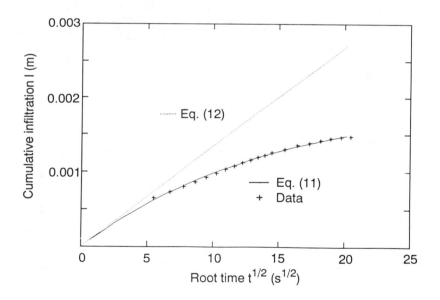

Fig. 7   The same data as in Fig. 6 but with the cumulative infiltration, I, plotted as a function of $t^{1/2}$. In addition to the data and Eq. (11), Eq. (12) is a plot of what the cumulative infiltration would have been if a constant head of H_0 had been maintained.

Note that the estimates of K_{fs} and ϕ_m appear physically reasonable. An assessment of the identifiability, uniqueness and stability of the solution parameters, however, is pending. Perhaps prior information concerning one of the parameters, as in the constant head technique, may be required to aid in insuring that the most appropriate values of K_{fs} and ϕ_m are obtained. (cf. Carrera & Neuman, 1986 and Russo et al., 1991).

CONCLUSIONS

For slowly permeable soils the early-time analysis of flow data appears to offer a promising procedure for obtaining good estimates of the saturated (K_{fs}) and unsaturated (ϕ_m) properties of these materials in a relatively short period of time (several hours). Certainly the time is far less than steady-state techniques which may require weeks or months to attain near-steady flow. The theory is sound for non-swelling soils and the measurements are simple. In particular, the falling head technique is appealing as the analysis only requires a measurement of $\Delta\theta$, the difference between the field-saturated and initial water contents, in addition to the H(t) data.

Both the constant and falling head techniques measure the 1-D vertical flow properties in a relatively thin section, probably < 0.01 m of soil material in slowly permeable soils.

REFERENCES

Bouwer, H., 1986, "Intake Rate: cylinder infiltrometer", IN: *Methods of Soil Analysis, Part I*, Edited by A. Klute, American Society Agronomy, Wisconsin, pp 825-844.

Carrera, J. and Neuman, S.P., 1986, "Estimation of aquifer parameters under transient and steady state conditions: 1. Maximum likelihood method incorporating prior information", *Water Resources Research*, Vol. 22, No. 2, pp 199-210.

Daniel, D.E., 1989, "In situ hydraulic conductivity tests for compacted clay", *Journal of Geotechnical Engineering*, Vol. 115 No. 9, pp 1205-1226.

Daniel, D.E., 1984, "Predicting the hydraulic conductivity of compacted clay liners", *Journal of Geotechnical Engineering*, Vol. 110, No. 2, pp 285-300.

Elrick, D.E., 1992, "Analysis of steady flow from ring infiltrometers", IN: *Proceedings of Twelfth Annual American Geophysical Union Hydrology Days*, Edited by H.J. Morel-Seytoux, Fort Collins, Colorado, pp 95-102.

Elrick, D.E. and Reynolds, W.D., 1992, "Methods for analyzing constant-head well permeameter data", *Soil Science Society of America Journal*, Vol. 56, No. 1, pp 320-323.

Elrick, D.E., Reynolds, W.D., and Tan K.A., 1989. "Hydraulic conductivity measurements in the unsaturated zone using improved well analysis", *Ground Water Monit. Rev.*, 9:184-193.

Elrick, D.E., Reynolds, W.D., Geering, H.R., and Tan, K.-A., 1990, "Estimating steady infiltration rate times for infiltrometers and permeameters", *Water Resources Research*, Vol. 26, No. 4, pp 759-769.

Elrick, D.E., Reynolds, W.D., Fallow, D.J., and Parkin, G.W., 1992, "The analysis of early-time and steady-state infiltration under falling head conditions", IN: *Innovations in measuring unsaturated flow parameters*, American Geophysical Union, San Francisco, Dec. 1992.

Krapac, I.G., Cartwright, K., Hensel, B.R., Herzog, B.L., Larson, T.H., Panno, S.V., Risatti, J.B., Su, W.-J., and Rehfeldt, K.R., 1991, "Construction, monitoring and performance of two soil liners", *Illinois State Geological Survey* Environmental Geology 141.

Math Soft Inc., 1992, *Mathcad*, Cambridge, Mass. USA.

Merva, G.E., 1987, "The velocity permeameter technique for rapid determination of hydraulic conductivity in-situ", <u>Proceedings Third International Workshop on Land Drainage</u>, Ohio State University, Columbus, Ohio, pp G56-G66.

Philip, J.R., 1957, "The theory of infiltration: 1. The infiltration equation and its solution", <u>Soil Science</u>, Vol. 83, pp 345-357.

Philip, J.R., 1969, "Theory of Infiltration", <u>Advances in Hydroscience</u>, Vol. 5, pp 215-296.

Philip, J.R., 1973, "On solving the unsaturated flow equation: 1. The flux-concentration relation", <u>Soil Science</u>, Vol. 116, pp 328-335.

Reynolds, W. D. and Elrick, D.E., 1990, "Ponded infiltration from a single ring: I. Analysis of steady flow", <u>Soil Science Society of America Journal</u>, Vol. 54, pp 1233-1241.

Russo, D., Bresler, E., Shani, U., and Parker, J.C., 1991, "Analyses of infiltration events in relation to determining soil hydraulic properties by inverse problem methodology", <u>Water Resources Research</u>, Vol. 27, No. 6, pp 1361-1373.

van Genuchten, M. Th., 1980, "A closed-form equation for predicting the hydraulic conductivity of unsaturated soils", <u>Soil Science Society of America Journal</u>, Vol. 44, pp 892-898.

White, I. and Sully, M.J., 1987, "Macroscopic and microscopic capillary length and time scales from field infiltration", <u>Water Resources Research</u>, Vol. 23, No. 8, pp 1514-1522.

Moir D. Haug[1], Walter G. Buettner[2], Lionel C. Wong[3]

IMPACT OF LEAKAGE ON PRECISION IN LOW GRADIENT FLEXIBLE WALL PERMEABILITY TESTING

REFERENCE: Haug, M. D., Buettner, W. G., and Wong, L. C., "Impact of Leakage on Precision in Low Gradient Flexible Wall Permeability Testing," Hydraulic Conductivity and Waste Contaminant Transport in Soil, ASTM STP 1142, David E. Daniel and Stephen J. Trautwein, Eds., American Society for Testing and Materials, Philadelphia, 1994.

ABSTRACT: Flow rate measurement errors caused by apparent leakage from the test apparatus during low gradient permeability testing was found to have a significant impact on precision in low gradient hydraulic conductivity test results. A laboratory testing program was conducted on ten similarly configured flexible wall permeameters, using an "impermeable" steel cylindrical block for a test specimen. The fluid pressures used during this test were identical to those used during flexible wall permeability testing. The results of this program showed that all systems had a measurable leakage. This leakage was typically in the range of 5 E-6 ml/sec to 5 E-8 ml/sec. The flow rate measurement error (apparent leakage) was found to have a significant impact on low gradient testing and this impact increased as the hydraulic conductivity measured decreased. Other than lowering the leakage rates through adjustments of the apparatus, increasing the diameter of the test specimen to increase the total flow through the sample was found to be the most effective method of reducing the impact of leakage.

KEYWORDS: hydraulic gradient, hydraulic conductivity, flexible wall permeability testing, soil liners, volume change, leakage analysis

A laboratory testing program was conducted to establish rates and to evaluate the significance of permeameter leakage during hydraulic conductivity tests. Leakage was defined as an apparent measured flow rate not associated with permeant movement through the soil sample during testing. In contrast with other fundamental properties of soil, such as strength and compressibility, hydraulic conductivity can vary widely and testing is required to establish meaningful values. Leakage through the lines, valves and fittings during low gradient testing,

---

[1]Professor of Civil Engineering, University of Saskatchewan, Saskatoon, Sask. S7N 0W0.

[2]Senior Engineer, Golder Associates Inc., 1809 North Mill St., Suite C, Naperville, Il. 60563.

[3]Research Engineer, Department of Civil Engineering, University of Saskatchewan, Saskatoon, Sask. S7N 0W0.

combined with difficulties in measuring low flow rates, has the potential to limit precision. This also represents a barrier below which hydraulic conductivity test results have little meaning. In order to assess the significance of leakage in these tests, a laboratory testing program was established to evaluate the magnitude and significance of flexible wall permeameter leakage. This test program involved conducting a series of flexible wall permeability tests on steel cylinders used to simulate soil test specimens. These tests were conducted on ten similarly configured flexible wall permeameters designed to accommodate specimens 50 to 250 mm in diameter. The leakage tests were conducted for various lengths of time during which the apparent flow in and out of the steel cylinders was continuously recorded. The results of these tests show that leakage is an important consideration in conducting low gradient flexible wall permeability tests. The significance of leakage was found to depend on its magnitude in relation to the amount of flow through the sample. The results of this study also demonstrated that leakage can be reduced through calibration of the test apparatus and its significance minimized with the use of larger diameter test specimens.

BACKGROUND

Hydraulic conductivity tests provide data on which predictions of contaminant migration can be made. It is usually necessary to conduct these relatively short term tests with a high level of precision so that any change in hydraulic conductivity with time can be observed. In some cases, it is also necessary to measure the change in volume of the sample during testing in order to evaluate the effect of chemical or physical processes. These processes may result in the development of a secondary structure in the soil. To accomplish these tasks, increased precision is required during flexible wall permeability tests. There are a number of factors which may influence accuracy and precision in hydraulic conductivity testing. Many of these relate to sample preparation, testing procedure and testing equipment. The focus of this paper is on the leakage associated with the testing apparatus.

Flexible wall permeability tests involve the measurement of fluid flow through confined soil samples under a hydraulic gradient. An equation describing the relationship between the hydraulic gradient used in these tests and discharge velocity was first published by Darcy (1856).

$$v = k_o\, i \tag{1}$$

where

v = discharge velocity

k_o = hydraulic conductivity

i = hydraulic gradient

If the cross sectional area (A) is taken into consideration, the rate of seepage (q) calculated from this relationship is

$$q = k_o \, i \, A \qquad (2)$$

and the total flow (Q) through the sample is given by the relationship

$$Q = k_o \, i \, A \, t \qquad (3)$$

where t = time

Leakage (L), measured in ml/sec, can result in an apparent increase or decrease in the total flow through the sample. This results in an error in the measured hydraulic conductivity (k) which is different from the "true" hydraulic conductivity (ko), as shown in equation (4).

$$k = (q+L) \, / \, (iA) \qquad (4)$$

where k is the measured hydraulic conductivity.

The hydraulic gradients used to drive fluid through the soil samples in flexible wall permeability tests have traditionally been high. High hydraulic gradients shorten the length of time required for flow to pass through the sample and thus, makes the measurement of this flow easier. The use of high effective stress during testing however, may cause consolidation of the sample, lowering its measured hydraulic conductivity. Sample consolidation during permeability testing under high gradients was measured by Kutilek (1972). A decrease in hydraulic conductivity with an increase in hydraulic gradient was also measured by Miller and Low (1963). The movement of fines under high hydraulic gradients through the matrix to locations where they block pores and reduce hydraulic conductivity was reported by Hardcastle and Mitchel (1974) and Frenkel and Rhoades (1978).

The chemical and physical processes which alter soil structure may also induce volume change in a soil liner. A decrease in volume may lead to a development of cracks and fissures, increasing the overall or bulk hydraulic conductivity. Direct laboratory measurements of this increase in hydraulic conductivity are difficult, however, it may be evaluated indirectly from measurements of the percentage decrease in sample volume during permeability testing. Chemical process effects can also be reduced with high hydraulic gradients which mask the impact of contaminants on hydraulic conductivities during testing (Yang and Barbour, 1992 and Fernandez and Quigley, 1991).

In order to assess the effects of these variables during hydraulic conductivity testing, flexible wall permeameter systems now frequently measure flow in and out of the sample, as well as changes in cell fluid volume. Many of these systems are also designed to collect permeant under back pressure exiting the sample. All of these refinements have led to an increase in complexity and have increased the opportunity for leakage. Various attempts have been made to control leakage (Tavenas et al. 1983), however, it still remains a source of potential error.

TEST PROGRAM

A test program was established to measure the range and characteristics of leakage during flexible wall permeability tests. The essential elements of this test program involved measurement of the apparent flow in and out of an "impermeable" test specimen under typical hydraulic conductivity testing conditions.

Description of the Test Apparatus

The permeameter system, including the flexible wall permeameter, regulators and reservoirs, evaluated in this study is shown in Figure 1. This system has been described previously by Buettner (1985), Fernuik and Haug (1990), Yanful et al. (1990), Wong and Haug (1991), and Haug and Wong (1992). The flexible wall permeameter, valves, and supporting fluid tubing lines are constructed of high-grade stainless steel (type 316). Pressure fittings are used for most connections. The flow measuring device consists of a twin tube, double acting burette system. Flow in and out of the sample is measured by the movement of a kerosene - permeant meniscus in these burettes. The most common burette sizes used with this apparatus are 25 ml or 10 ml with resolutions of 0.05 and 0.02 ml respectively, however, 5 ml capacity burettes with resolutions of 0.01 ml can also be used. Any change in the cell fluid volume is also measured on similar volume change indicators. Regulated air pressure is applied to the inflow, outflow and cell reservoirs as well as the leachate collection system. These pressures are monitored by transducers located on the base of the permeameter. The transducers have a measuring range of 0 to 690 kPa, with a resolution of ± 3.45 kPa. Differential pressure transducers, having a range of 0 to 69 kPa and resolutions as low as ±0.6 kPa, monitor head across the sample.

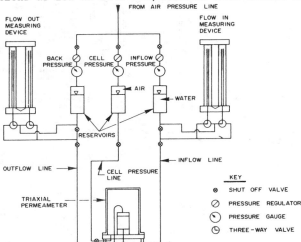

FIG. 1--Schematic illustration of the flexible wall permeameter system.

The dimensions and layout of eight of the ten permeameters are shown in Figure 2. These permeameters have interchangeable base and top caps which permit testing of samples ranging from 50 to 150 mm in diameter. The upper surface of the interchangeable base caps are spirally scrolled to enable flushing when permeants are changed during testing. Samples are placed inside the cell between two coarse corundum stones. The sides of the sample are coated with vacuum grease and surrounded by two rubber membranes. The samples are then attached to the base and top cap and sealed with O-rings. The samples are confined by an "all around" cell pressure. Permeant is introduced through the inflow pressure line at the base of the sample and exits through the back pressure or outflow line. The final two test apparatus examined were similar in configuration, however, these permeameters were capable of measuring samples up to 250 mm in diameter and under a maximum of 7000 kPa confining pressure.

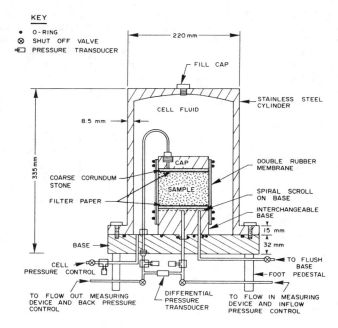

FIG. 2--Section through flexible wall permeameter showing location of inflow lines and "O"-ring seals.

Theoretical Considerations

The time required to make an accurate determination of flow through the sample is a major obstacle in the measurement of apparent permeameter leakage. At low leakage rates, considerable time may be required for the kerosene-permeant meniscus to move one division on the volume change measuring device. This minimum time for an observation is a function of both the magnitude of the leakage rate and the minimum resolution of the burettes. Figure 3 shows on a log-log scale, the minimum time required to measure various leakage rates based on burette resolution. This figure shows that using a 25 ml capacity (0.05 ml resolution burette), approximately 1 day is required for 0.05 ml of flow (movement of one division) with a leakage rate of 1E-6 ml/sec. If the same burette is used, the times required to measure leakage rates of 1E-7, 1E-8 and 1E-9 ml/sec are approximately 8, 80 and 800 days, respectively. If a 10 ml burette is used having a resolution of 0.02 ml, the times required to measure leakage rates of 1E-7, 1E-8 and 1E-9 ml/sec are approximately 5, 50 and 500 days, respectively. Using a very precise 0.01 ml resolution burette, the time required to establish a 1E-9 ml/sec leakage rate is still approximately 100 days. Since it is not always possible to measure leakage rates over an extended period, this figure can also be used to establish a maximum leakage rate. For example, a leakage test conducted for 10 days in which no flow was measured on a 0.05 ml resolution burette would represent a maximum leakage rate of approximately 8E-8 ml/sec.

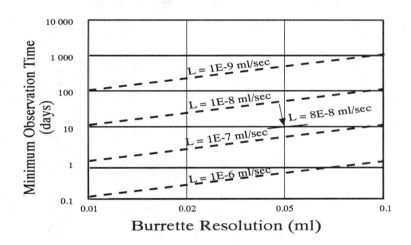

FIG. 3--Theoretical relationship between burette resolution and minimum observation time for various leakage rates.

Potential Apparent Leakage Sources

The flexible wall permeability testing apparatus shown in Figures 1 and 2, contains numerous valves and fittings which may leak. While care is exercised in connecting fittings, leakage can occur from over, or under tightening fittings. Many of these potential sources of leakage are apparent from the photograph shown in Figure 4. This photograph of the back side of the permeameter system shows many of the lines and fittings required to carry out various operations involved in flexible wall permeability testing. This figure also shows a 150 mm diameter sample prior to being tested.

FIG. 4--Photograph showing permeameter lines, valves and fittings representing potential sources of leakage.

An analysis of this system found that there were 31 leakage possibilities on the inflow side of the specimen. Twenty-seven of these sources could result in an apparent higher flow into the sample. Four possible leakage locations could result in an apparent lower flow into the sample. The 27 increased flow sources represent the number of individual connections between the volume change indicators and the sample. These connections are required to turn flow on and off to the sample and re-direct flow from both sides of the volume change indicators. These devices are conventional double tube, double acting volume change indicators (Buettner, 1985). In addition, they enable pressure to be measured across the sample and flushing of the spiralled base to change permeants during testing. The four sources which could result in less flow being measured involve fittings inside the volume change indicators which secure the burettes, and three locations inside the permeameter where O-rings are used to prevent flow from the cell into the inflow lines (Fig. 2).

On the outflow side of the sample there are 23 sources which would contribute to lower measured flow out readings, and eight sources which would contribute to higher flow readings.

Leakage Test Procedures

Two different techniques were examined to measure leakage rates. The first technique involved installing a Teflon tape covered set-screw into the threaded inside of the inflow fluid port on the upper surface of the base pedestal (Fig. 2). A "T" line was installed on the outflow line to enable it to be looped back upon itself so that no flow was permitted. This technique blocked both the inflow and outflow lines.

While this technique eliminated some leakage sources, it also introduced some new ones, such as flow past the O-rings inside the permeameter cell.

The procedure finally adopted for this study involved using a procedure which closely paralleled the flexible wall permeability testing procedure. The only exception was that a solid steel block was substituted for the soil test specimen inside the permeameter. This cylindrical block was placed in the flexible wall permeameters and sandwiched between porous stones. The sides of the steel block were covered with vacuum grease, and confined within two rubber membranes. The membranes were then attached to the top cap and base pedestal and sealed with O-rings.

The first stage of the leakage tests involved vacuum and back pressure saturation. During the vacuum stage, the lines to the inflow, outflow and cell fluid pressures at the permeameter cell were disconnected to allow the vacuum to be applied. Once the vacuum stage was complete, these lines were re-connected to enable the vacuum remaining in the simulated sample to "draw-in" permeant. Inflow and outflow fluid pressures were applied to the block for 24 hours to simulate the saturation stage. The leakage test was then initiated by adjusting the fluid pressures to the levels used during the flexible wall permeability tests.

Leakage tests were conducted on all ten test apparatus for a minimum of five days. More than one leakage test was conducted on six of the apparatus following a period of laboratory use. These tests were designed to evaluate the effect of changing fittings, cleaning volume change indicators and re-calibrating transducers. The final two tests conducted using Apparatus No. 1 were designed to provide information on the impact of changing inflow and cell pressures on the resulting leakage.

LEAKAGE TEST RESULTS

The results of the leakage testing program conducted on each of the permeameters has been summarized (Table 1). This table describes the pressures used during the tests and the duration of these tests. As well, it identifies the burette and its resolution, and lists the average leakage rate. A total of 26 leakage tests were conducted in this program.

Two leakage tests were conducted on Apparatus No. 1. The results of the first test show that the average inflow leakage rate was 1.4E-6 ml/sec and the average outflow leakage rate was 1.6E-6 ml/sec after seven days. Following adjustments and tightening of the fittings, a second leakage test was conducted and it was found that this test showed lower average leakage rates of 6.1E-8 ml/sec and -1.14E-7 ml/sec for the inflow and the outflow, respectively.

TABLE 1--Summary of leakage test results.

Apparatus Number	Test No.	Test Duration (Days)	Inflow Burette Resolution (ml)	Inflow Press. (kPa)	Inflow Avg. Leakage Rate (ml/sec)	Outflow Burette Resolution (ml)	Outflow Press. (kPa)	Outflow Avg. Leakage Rate (ml/sec)	Cell Pressure (kPa)
1	1A	7.0	0.05	220.64	1.40E-06	0.05	206.85	1.60E-06	234.43
	1B	5.7	0.05	220.64	6.10E-08	0.05	206.85	-1.14E-07	234.43
2	2A	8.8	0.05	220.64	7.10E-07	0.05	206.85	3.90E-07	234.43
	2B	8.7	0.05	220.64	9.30E-07	0.05	206.85	6.20E-07	234.43
	2C	5.0	0.05	220.64	1.10E-04	0.05	206.85	5.70E-05	234.43
	2D	1.0	0.05	220.64	0.00E+00	0.05	206.85	0.00E+00	234.43
3	3A	31.2	0.05	220.64	4.10E-07	0.05	206.85	4.10E-07	234.43
	3B	8.8	0.05	220.64	2.60E-08	0.05	206.85	-1.30E-08	234.43
4	4A	4.8	0.05	220.64	5.10E-07	0.05	206.85	9.90E-07	234.43
	4B	21.0	0.05	220.64	3.30E-08	0.05	206.85	2.40E-07	234.43
5	5A	6.0	0.02	220.64	0.00E+00	0.02	206.85	0.00E+00	234.43
	5B	9.7	0.05	220.64	7.00E-07	0.05	206.85	-9.50E-08	234.43
	5C	6.3	0.05	220.64	1.30E-07	0.05	206.85	1.50E-07	234.43
6	6A	89.7	0.02	220.64	1.35E-07	0.02	206.85	5.83E-08	234.43
7	7A	7.0	0.02	220.64	2.00E-07	0.02	206.85	3.50E-07	234.43
8	8A	55.8	0.05	220.64	9.00E-07	0.05	206.85	3.00E-07	234.43
9	9A	8.8	0.02	220.64	1.30E-07	0.02	206.85	6.00E-07	234.43
10	10A	20.9	0.02	220.64	6.10E-08	0.02	206.85	1.70E-08	234.43
	10B	19.5	0.02	220.64	8.90E-08	0.02	206.85	5.80E-08	234.43
1	1.1	5.9	0.05	220.64	2.94E-07	0.05	206.85	2.55E-07	255.12
	1.2	7	0.05	234.43	3.31E-08	0.05	206.85	1.49E-07	268.91
	1.3	7	0.05	262.01	2.65E-07	0.05	206.85	3.31E-07	296.49
	1.4	7	0.05	317.17	3.32E-08	0.05	206.85	3.83E-07	351.65
1	2.1	8	0.05	220.64	1.10E-07	0.05	206.85	5.79E-08	255.12
	2.2	6	0.05	234.43	7.73E-08	0.05	206.85	9.66E-08	268.91
	2.3	8	0.05	262.01	-4.34E-08	0.05	206.85	8.68E-08	296.49

Four tests were conducted on Apparatus No. 2. The results of the first two tests were in the range of 1E-7 ml/sec. This apparatus was then used in the laboratory for a few weeks. A second leakage test was conducted and a leakage rate in the 1E-4 ml/sec range was recorded. All the fittings in the apparatus were tightened and another short leakage test was conducted. This one day test produced no measurable flow, suggesting that the leakage rate was lower than 1E-6 ml/sec based on the relationship shown in Figure 3. The average leakage rate for all other apparatus were found to be in the range of 1E-7 ml/sec.

The last series of tests using Apparatus No. 1 were conducted using different inflow and cell fluid pressures. No clear pattern of increasing or decreasing leakage with change in pressures was apparent from these tests.

Figure 5 shows the results of a long-term test conducted to evaluate the inflow leakage of Apparatus No. 6. This test was conducted for a period of 90 days. Positive values indicate leakage out of the inflow lines. A high initial positive inflow leakage occurred. The leakage rate then decreased with time, over the first 5 days and then fluctuated between positive and negative leakage. Most of these fluctuations were due to the small amount of flow which took place between observations, and small temperature fluctuations. The pattern of high initial leakage rates dropping off with time was typical of all the tests conducted.

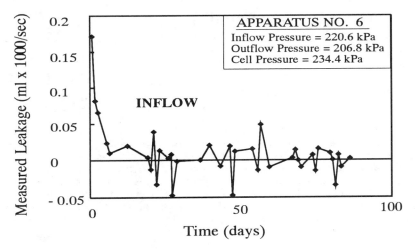

FIG. 5--Leakage rate measured at different stages for Apparatus No. 6.

Figure 6 shows the running average inflow (average value since the test was started) and outflow rates for Apparatus No. 6 as a function of time. A running average was used to level out the fluctuations shown in Figure 5, and establish an average leakage rate. This technique can result however, in some minor variations since the tests were conducted for different lengths of time. Figure 6 shows that for both of the measured flow rates, the running average leakage rates began high, then decreased rapidly over the first week. The values then leveled off. Both the inflow and the outflow leakage rates were positive in this test. The average measured inflow and outflow rates over the duration of the test were 1.37E-7 ml/sec and 5.83 E-8 ml/s, respectively.

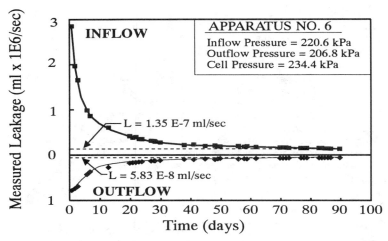

FIG. 6--Running average leakage rate for Apparatus No. 6.

INTERPRETATION OF THE TEST RESULTS

The results of this test program found that all of the apparatus tested had a measurable leakage and that this leakage varied from approximately 5 E-6 ml/sec to 5 E-8 ml/sec. In most cases however, adjustments made to the apparatus were able to reduce leakage rates to 1E-7 ml/sec or below. Some of the test results showed that there was a high initial positive leakage at the start of the tests which dropped rapidly over the first few days. This apparent high initial rate may have been due to some factor other than leakage, such as air bubbles not dissolved in the lines during the saturation stage.

One method of examining the significance of leakage (L) is to calculate the apparent hydraulic conductivity. This leakage hydraulic conductivity (k_L) can be related to the hydraulic gradient and sample diameter by the following relationship

$$L = q = k_L\ F \qquad (5)$$

$$F = i\ A \qquad (6)$$

where F = flow factor

Based on this relationship, and assuming that the steel block used in Apparatus No. 6 was 100 mm high and had a diameter of 100 mm, the flow factor would be 1105. This would give a leakage hydraulic conductivity (k_L), for leakage rates ranging from 5 E-6 ml/sec to 5 E-8 ml/sec, of 4.6 E-09 to 4.6 E-11 cm/s, respectively.

A more valuable approach to examining the significance of leakage is to use the ratio of the leakage influenced hydraulic conductivity (k) divided by the "true" hydraulic conductivity (k/ko), based on equations (3) and (4). Figure 7 shows the relationship between k/ko and hydraulic gradient for a sample 50 mm in diameter and with a leakage of 1E-6 ml/sec. The high end leakage rate combined with the smallest diameter test configurations represent a potential "worst" case situation. This figure shows that k/ko increases as the hydraulic gradient decreases. Using this configuration, a sample having a hydraulic conductivity of 1E-7 cm/s can be tested with a relatively high level of precision (i.e. k/ko < 1.10) down to a hydraulic gradient of approximately six. However, most soil barriers are designed for considerably lower laboratory hydraulic conductivities in order to achieve field (long-term) hydraulic conductivities. For this reason, use of the k = 1E-8 or 1E-9 cm/s curve appears to provides a more valuable measure of hydraulic conductivity limitation in flexible wall permeability testing. Using a k of 1E-8 cm/sec, the minimum hydraulic gradient for a k/ko of 1.1 would be approximately 100.

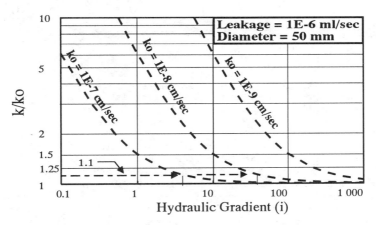

FIG. 7--Relationship between k/ko versus hydraulic gradient for different ko values at a leakage rate of 1E-6 ml/sec and a sample diameter of 50 mm.

The required precision may vary, depending on the application and material being tested, however, a precision associated with a k/ko ratio of 1.25 might represent a typical desired level. Figure 8 shows the minimum reliable hydraulic conductivity (ko) as a function of the flow factor (F) for a k/ko of 1.25. The flow factor governs the total flow through a soil with a particular hydraulic conductivity. The shaded zone in this figure represents the range of average leakage rates measured in the test program. For a particular leakage rate, a larger flow factor is required to reliably measure lower hydraulic conductivity values. Figure 8 also shows four possible ranges of sample diameters and hydraulic gradient combinations which could be used to provide a desired flow factor. Based on this figure, a permeameter with a leakage rate of 1E-7 ml/sec, a diameter of 250 mm and hydraulic

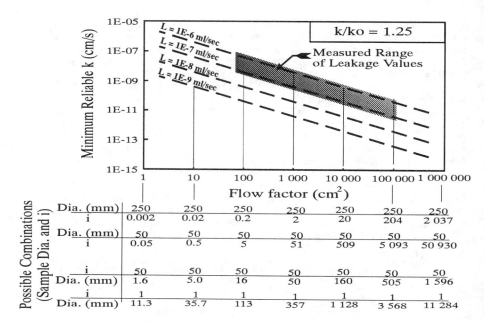

FIG. 8--Relationship between leakage rate, flow factor and minimum reliable hydraulic conductivity for a k/ko of 1.25.

gradient of 20 (F = 10000) could reliably test below the 1E-10 cm/sec hydraulic conductivity range. If a 50 mm diameter sample was used, the hydraulic gradient would have to be increased to 509 to produce the same level of precision in hydraulic conductivity. This clearly illustrates that there is a trade off between area and hydraulic gradient. If the area of the test specimen is decreased, the hydraulic gradient must be increased to obtain the same flow factor.

Figure 8 also shows that testing at low gradients becomes increasingly difficult as the hydraulic conductivity measured drops or the required precision increases. The two most effective methods to improve testing in this range is to attempt to minimize leakage and/or increase the diameter of the test specimen. It was shown, in Figure 3, that establishing lower leakage rates required considerable time. Even with the most accurate burettes in this system (0.01/ml resolution), approximately 100 days was required to measure a leakage rate of 1E-9 ml/sec. Thus, a leakage rate of 1E-9 ml/sec may be the lowest which can be reasonably anticipated since the leakage decreases with time. Figure 8 also shows the difficulty in trying to determine critical or threshold gradients, especially under low gradient conditions.

The influence of leakage on a hydraulic conductivity test is shown conceptually in Figure 9. This figure shows a typical hydraulic conductivity versus time plot for a compacted sand-bentonite test specimen. At the start of the test, the outflow was initially higher than the inflow as the sample consolidated under the increase in effective stress caused by the application of the hydraulic gradient. After a period of time, the rate of inflow surpasses the outflow as the sample swells. At the end of test the hydraulic conductivity measured by the flow in and flow out converge. The zones of uncertainty caused by leakage are shown for the inflow and the outflow hydraulic conductivities. Since all but one of the average leakage rate values was positive, the probabability of the inflow hydraulic conductivity being greater is low. Thus, the upper boundary of the inflow curve is closer to the measured value curve than the lower boundary. These boundaries represent the probable maximum and minimum hydraulic conductivities, and this zone narrows with time as the average leakage rate decreases. On the outflow curve, the spacing of the probability boundaries are reversed, as the chances of a higher outflow hydraulic conductivity are greater than the chances of a lower value. This figure clearly shows the overlap between the two different sets of curves at the end of the test, and why an average of the inflow and outflow hydraulic conductivities (representing range of possible values) provides a realistic value of the "true" hydraulic conductivity. Thus, this average of the flow in and out values is frequently used to report test results. It also shows that in situations where only one volume change indicator is available, less error due to leakage should normally occur if the indicator is placed on the inflow side of the sample. The unusual situation at the start of the test is also illustrated in this figure. This type of deviation at low gradients have been credited to non-Darcian flow (Lutz and Kemper, 1959, and Hansbo, 1960), however Olsen (1962) and others have found little evidence to support this assertion.

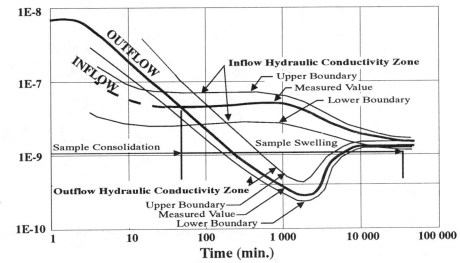

FIG. 9--Conceptual illustration of leakage on hydraulic conductivity test results.

SUMMARY AND CONCLUSIONS

The findings of this research program are based on flexible wall permeameters of a particular configuration. Thus, these leakage rates may not apply to other test apparatus. The results do however provide valuable information on trends and sensitivity which could have applications for all permeameter systems. The conclusions of this work can be summarized as follows:

1. The measured leakage rates were found to be high shortly after the tests were started. This rate then decreased rapidly over the next few days before gradually levelling off. One way to reduce the impact of this apparent high initial leakage rate is to run all hydraulic conductivity tests for longer durations, since the average leakage rates decreased with time to a more steady state value.

2. The average leakage measured on ten similarly configured flexible wall permeameters was found to range from 5E-6 to 5E-8 ml/sec.

3. Most of the leakage rates were in the positive direction (i.e. a loss of fluid from the inflow and outflow lines). This was apparently due to the large number of identified potential leakage sources.

4. In nearly all of the cases, initially high leakage could be reduced by checking and further tightening permeameter fittings.

5. It was not possible to draw a direct relationship between leakage rate and cell pressure in the two series of tests conducted. In one case, increased cell pressure was found to cause a reduction in the leakage rate. In another, increased cell pressure caused a reverse in the direction of the leakage.

6. The significance of the leakage rate was found to be a function of the hydraulic conductivity, area of the test specimen, and hydraulic gradient. An increase in diameter of the test specimen facilitated testing at low hydraulic gradients.

7. The test results showed that in most cases, leakage would cause higher inflow hydraulic conductivity readings and lower outflow hydraulic conductivity values. Based on these findings, measurements of the flow into a specimen with this system tend to provide conservative test results.

ACKNOWLEDGMENTS

The assistance of the National Science and Engineering Research Council of Canada in financially supporting this work is gratefully acknowledged.

References

Buettner, W.G., 1985, "Permeability Testing of Soils with Low Hydraulic Conductivity," Unpublished M.Sc. Thesis, University of Saskatchewan, pp. 203.

Darcy, H., 1856, Les Foutaines Publiques de la Ville de Dijon (The Water of Dijon).

Fernandez, F. and Quigley, R.M., 1991, "Controlling the Destructive Effects of Clay - Organic Liquid Interactions by Application of Effective Stresses," Canadian Geotechnical Journal, Vol. 28, No. 1, pp. 388-398.

Fernuik, N. and Haug, M., 1990, "Evaluation of in situ Permeability Testing Methods," Journal of Geotechnical Engineering, Vol. 116, No. 2, pp. 297-311.

Frenkel, H. and Rhoades, J.D., 1978, "Effects of Dispersion and Swelling on Soil Hydraulic Conductivity," Journal of Testing and Evaluation, Vol. 6, No. 1, pp. 60-65.

Hansbo, S., 1960, "Consolidation of Clay, with special reference to Influence of Vertical Sand Drains", Swedish Geotechnical Institute, Proceedings, No. 18, Stolkholm.

Hardcastle, J.H. and Mitchell, J.K., 1974, "Electrolyte Concentration - Permeability Relationships in Sodium Illite - Silt Mixtures," Clays and Clay Minerals, Vol. 22, pp. 143-154.

Haug, M.D. and Wong, L.C., 1992, "Impact of Molding Water Content on Hydraulic Conductivity of Compacted Sand-Bentonite," Canadian Geotechnical Journal, Vol. 29, No. 2, pp. 253-262.

Kutilek, M., 1972, "Non-Darcian Flow of Water in Soils - Laminar, Region," Fundamentals of Transport Phenomena in Porous Media, International Association for Hydraulic Research, Elsevier Publishing Company, pp. 327-340.

Lutz, J.F., and Kemper, W.D., 1959, "Intrinsic Permeability of Clay as Affected by Clay-Water Interaction," Soil Science, Vol. 88, pp. 83-90.

Miller, R.J. and Low, P.F., 1963, "Threshold Gradient for Water Flow in Clay Systems," Proceedings, Soil Science Society of America, Vol. 27, No. 6, pp.605-609.

Olsen, H.W., 1962, "Hydraulic Flow through Saturated Clays," Clays and Clay Minerals, Vol. 11, pp. 131-61.

Tavenas, F., Leblond, P., Jean, P., and Leroueil, S., 1983, "The Permeability of Natural Soft Clays. Part I: Methods of Laboratory Measurement," Canadian Geotechnical Journal, Vol. 20, No. 2, pp. 629-644.

Wong, L.C. and Haug, M.D., 1991, "Cyclical Closed System Freeze-Thaw flexible wall Permeability Testing of Soil Liner and Cover Materials," Canadian Geotechnical Journal, Vol. 28, No. 6, pp. 748-793.

Yang, N. and Barbour, S.L., 1992, "The Impacts of Soil Structure on the Hydraulic Conductivity of Clays in Brine Environments," Canadian Geotechnical Journal, Vol. 29, No. 5. pp. 730-739.

Yanful, E.K., Haug, M.D., and Wong, L.C., 1990, "Impact of Synthetic Leachate on the Hydraulic Conductivity of a Smectitic Till Underlying a Landfill Near Saskatoon, Saskatchewan," Canadian Geotechnical Journal, Vol. 27, No. 4, pp. 507-519.

Moir D. Haug[1], and Brigitte Boldt-Leppin[2]

INFLUENCE OF POLYMERS ON THE HYDRAULIC CONDUCTIVITY OF MARGINAL QUALITY BENTONITE-SAND MIXTURES

REFERENCE: Haug, M. D., and Boldt-Leppin, B., "Influence of Polymers on the Hydraulic Conductivity of Marginal Quality Bentonite-Sand Mixtures," Hydraulic Conductivity and Waste Contaminant Transport in Soil, ASTM STP 1142, David E. Daniel and Stephen J. Trautwein, Eds., American Society for Testing and Materials, Philadelphia, 1994.

ABSTRACT: The results of a two year laboratory test program demonstrated that the hydraulic conductivity of a marginal quality sodium bentonite could be significantly enhanced with the addition of a small quantity of specially formulated commercial polymer. The first step in the laboratory program involved tests to establish the rheological properties of slurries prepared using 8 % marginal quality (MQ) and 8 % high quality (HQ) bentonite. The marginal bentonite was then modified with various percentages of polymer. A comparison of shear stress versus shear rate characteristics, as a function of polymer addition rate, was used to establish a range of polymer addition rates for hydraulic conductivity tests. Ten low gradient triaxial permeability/volume change tests were conducted on compacted Ottawa sand-8 % bentonite tests samples. The concentration of polymer used in these tests varied from 0 to 0.5 % based on the weight of bentonite. Approximately half of the tests were conducted on newly prepared polymer and the other half on polymer aged for one year prior to being used.

KEY WORDS: hydraulic conductivity, volume change, bentonite, polymers, rheology, triaxial permeability testing

Bentonite is being used increasingly in various applications as a soil barrier material to protect ground water and prevent contaminant migration. By definition, bentonite is a natural occurring rock consisting of a high percentage of expanding lattice smectitic clay minerals. Traditionally, the purest best quality bentonites (those which naturally produce the greatest swelling and slurries with the highest yield) have been used for drilling mud applications while somewhat lower quality materials were used for soil liners and covers. A combination of the anticipated growth in demand for bentonite and gradual depletion of some of the high quality products, has led to the development of special polymers designed to provide more marginal bentonites with the performance of higher quality bentonite. While poly-

---

[1]Professor, Department of Civil Engineering, University of Saskatchewan, Saskatoon, Sask. S7N 0W0

[2]Graduate Student, Department of Civil Engineering, University of Saskatchewan, Saskatoon, Sask. S7N 0W0

mers have been used for many years to enhance the performance of bentonite used in the drilling industry, their use for soil barrier applications has had minimal acceptance. The object of this research was to; 1) examine the influence of polymer concentration on the rheological and hydraulic conductivity characteristics of marginal quality bentonite, 2) to compare these results to those of a typical high quality Wyoming sodium bentonite, and 3) to conduct a preliminary investigation into the impact of polymer aging on the enhancement of marginal quality bentonite.

MATERIAL CHARACTERIZATION

The marginal quality (MQ) and high quality (HQ) bentonites used in this study are both marketed commercially for seepage and contaminant migration control applications. Both of these powdered bentonites were similar in texture and outward appearance. Thus, the first step in this study was to established the difference between the two bentonites in terms of chemical, mineralogy and engineering properties.

A summary of the physio-chemical test results for the MQ bentonite are presented in Table 1, along with the results for the HQ bentonite published previously by Reschke and Haug (1991). In many aspects the two bentonites are quite similar. However, the chemical analysis shows that the MQ bentonite has more silica(SiO_2) and less aluminum

TABLE 1 -- Bentonite Physio-chemical Properties.

Properties	Marginal Quality Bentonite	High Quality Bentonite
Chemical Analysis (%)		
SiO_2	63.63	54.75
Al_2O_3	11.73	19.44
Fe_2O_3	4.02	3.65
MgO	1.22	2.68
CaO	0.94	1.06
Na_2O	1.07	1.69
K_2O	0.60	0.25
Total	83.20	82.76
Specific Surface (m^2/g)	490	668
Exchangable Cations (meq/100g)		
Na	45.6 (65%)	65.9 (62%)
Ca	20.2 (29%)	20.6 (19%)
Mg	4.5 (6%)	20.3 (19%)
K	0.8	1.5
T.E.B.	71	108
CEC	69	90

oxide (Al_2O_3) than the HQ bentonite. The specific surface of the HQ bentonite was 668 m^2/g, or approximately 82 % of the theoretical maximum, compared to only 490 m^2/g, or 62.5 % of the theoretical maximum for the MQ bentonite. The cation exchange capacities (C.E.C.) for the MQ and HQ bentonites were 69 and 90, respectively. The MQ bentonite had plastic and liquid limits of 32.1 and 425, respectively, compared to 31.3 and 590, respectively for the HQ bentonite.

A powdered orientated x-ray diffraction pattern comparison for the HQ and MQ bentonites is shown in Fig. 1. Both of these bentonite samples were glycolated prior to analysis. The smectite basal (001) and basal reflection peak strengths are considerably greater for the HQ bentonite than for the MQ bentonite. The MQ bentonite also contained cristobalite and quartz (6 %), plagioclase (10 %), and illite (7 %). In comparison, the HQ bentonite was relatively pure and contained only a small percentage of other minerals such as plagioclase (3 %) and illite (2 %).

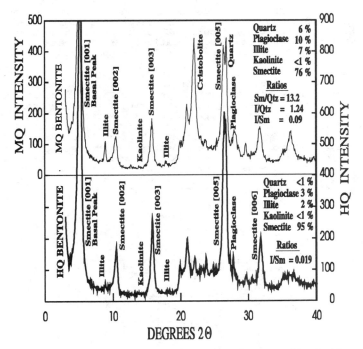

FIG. 1 -- XRD Glycolated Powdered Bentonite Pattern Comparison.

The polymer used for this study was ALCOMER 228, manufactured and distributed by Allied Colloids Ltd., according to International Standard ISO9001-1987 requirements. The polymer is supplied as a free flowing white colored powder which is blended with bentonite prior to its use. Approximately 98 % of the polymer is finer than 0.075 mm. This polymer is an anionic polyacrylamide specifically designed for use in improving the low permeability nature of bentonite. Applications cited by the manufacturer include the sealing of lagoons and tips. The polymer is also described by the manufacturer as being especially effective when inorganic contaminants are encountered (Allied Colloids 1992).

Graded Ottawa sand (ASTM C109) was chosen as the granular constituent in preparation of the hydraulic conductivity test specimens to provide gradation control for testing. Distilled water was used as the permeant in the hydraulic conductivity tests to simulate infiltration of the low ionic rainwater through covers.

RHEOLOGICAL TEST PROGRAM

A preliminary test program was conducted to determine the rheological properties of 8 % slurries of HQ and polymer modified MQ bentonite. This program was designed to provide a measure of the effectiveness of the polymer on the MQ bentonite and assist in establishing a range of polymer addition for triaxial permeability testing of polymer modified MQ bentonite sand samples. Polymer was added to the MQ bentonite in concentrations varying from 0 to 2 %, based on the air-dried weight of bentonite. This test program was conducted using a Fann V-G Model 35A rotational viscometer. The viscometer is commonly used to measures the flow properties of colloidal solutions such as bentonite slurries. This test also provides a relatively precise measure of repulsion and attraction forces between the colloidal particles (Reschke and Haug, 1991).

The relationship between shear rate and shear stress for non-polymerized HQ and MQ bentonite is shown in Fig. 2. The shear stress for the HQ bentonite increases sharply as the shear rate increases and then gradually levels off, increasing uniformly with increasing shear rate. This pattern of sharp initial increase in shear stress followed by a constant uniform increase is typically refered to as a Bingham fluid curve, and is frequently applied to bentonite slurries used in the petroleum field (Xanthakos, 1979). This pattern contrasts with that of a Newtonian fluid, such as water, whose shear stress is directly proportional to shear rate. The nature of the Bingham fluid curve can be characterized by the Bingham Yield and the plastic viscosity. The Bingham Yield being the straight line intercept on the shear stress axis pro-

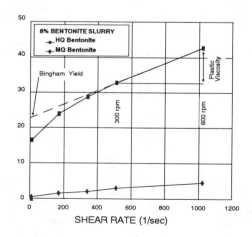

FIG. 2 -- Shear Rate Versus Shear Stress for HQ and MQ Bentonites

jected from the 300 rpm (511/sec.) and 600 rpm (1022/sec.) rotational viscometer readings. In the case of the HQ bentonite the Bingham Yield and plastic viscosity were 23 x 10^{-3} kPa and 19.5 cP. The corresponding values for the non-polymerized MQ bentonite were 2 x 10^{-3} kPa and 3 cP.

The impact of polymer addition on the shear stress versus shear strain characteristics of an 8 % MQ bentonite slurry is illustrated in Fig. 3. This figure shows both the Bingham Yield and plastic viscosity increase, as the percentage of polymer is increased from 0 to 2 %. At the 2 % polymer addition rate the Bingham Yield and plastic viscosity of the MQ bentonite approaches the values of the HQ bentonite shown in Fig. 2. A second series of polymer addition test were also conducted on a different batch of polymer which was aged one year. The test results were similar to those shown in Fig. 3 and no sign of polymer deterioration was apparent.

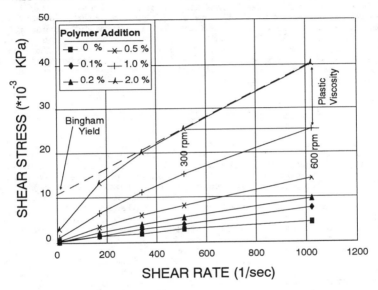

FIG. 3 -- Impact of Polymer Addition of Stress-strain Characteristics of MQ bentonite

HYDRAULIC CONDUCTIVITY EVALUATION

Description of the Test Program

A triaxial test program was established to examine and challenge the sealing capability of the bentonite. This was accomplished by setting the bentonite content at 8 %, using a relatively open graded sand containing no clay size particles, and using distilled water for the permeant. The test program consisted of 10 triaxial permeability/volume change tests conducted on the compacted MQ bentonite and Ottawa sand samples. The initial polymer concentrations were based on the results of the rheological test program. The concentrations were then reduced in steps to determine the minimum concentration which could be used to successfully modify the bentonite without resulting in failure during testing. Three tests were conducted on non-aged (new product) polymer at concentration of 0.05, 0.2 and 0.5 % by weight of bentonite. Four tests were conducted on polymer aged for one year prior to mixing with the bentonite and testing. During this year the powdered polymer was stored in its container at room temperature (approximately 20°C). The aged polymer concentration for aged tests were 0.01, 0.05, 0.2 and 0.5 %. Two additional tests were conducted on samples prepared using non-polymerized MQ bentonite. The final sample tested was prepared using 0.5 % polymer and Ottawa sand with no bentonite.

Triaxial Permeability Sample Preparation

The triaxial permeability tests were conducted on specimens containing 8 % bentonite mixed with Ottawa sand. Polymer was added to the bentonite prior to mixing with the sand. The modified sand-bentonite samples were moisture conditioned to near optimum and compacted at 100 % standard Proctor density (ASTM D698). Optimum water content and the corresponding maximum dry density were 14.8 % and 1.78 Mg/m^3, respectively. The 101.78 mm diameter compacted sand-bentonite samples were extruded from the Proctor molds and trimmed to a height of 43 mm for hydraulic conductivity testing.

Triaxial Permeability Apparatus and Test Procedures

The test apparatus equipment and procedures used in this test program were described previously (Fernuik and Haug 1990; Yanful et al. 1990; Wong and Haug 1991). The essential elements of this system involve the measurement of both flow in and out of the sample during testing. Figure 4 shows a section through the triaxial permeameter and test specimen. The specimens were coated with vacuum grease, sandwiched between filter papers and porous stones, and placed in the triaxial permeameter. The specimens were covered with a top cap and surrounded by two rubber membranes that were sealed to a cap and base pedestal. Saturation during testing was maintained by the application of continuous back pressure, the use of de-aired water, and upward flow of permeant through the sample. Saturation was also confirmed with a pore pressure reaction test before each test started, and the results of this test were evaluated based on the findings of Black and Lee (1973). Prior to the start, and following the test program, each of the triaxial permeameters used were calibrated for leakage. The procedures used in this calibration were previously described (Yanful et al., 1990; Wong and Haug, 1991).

The cell, pore and back pressures used during the test were 234.43, 220.78 and 206.72 kPa. The average effective confining stress on the samples during testing was 17.2 kPa. The hydraulic gradients averaged approximately 30 during these tests. Some fluctuation in the hydraulic gradient occurred during these "constant" head tests as the kerosene water interface moved in the volume change indicators.

HYDRAULIC CONDUCTIVITY TEST RESULTS

The hydraulic conductivity versus time relationship (Fig. 5) for a compacted 8 % HQ bentonite and Ottawa sands sample was described previously by Haug and Wong (1992). This sample was compacted at a water content of 15.8 % and dry density of 1.77 Mg/m^3. The test was conducted using the same average effective sample confining stress (17.2 kPa), test apparatus and procedures as used in this test program. This figure shows that the HQ bentonite had an initially high (near 7 x 10^{-8} cm/s)

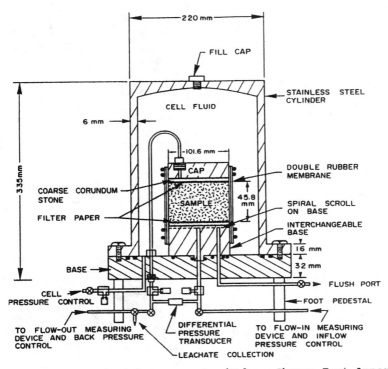

FIG. 4 -- Triaxial Permeameter/Volume Change Test Apparatus.

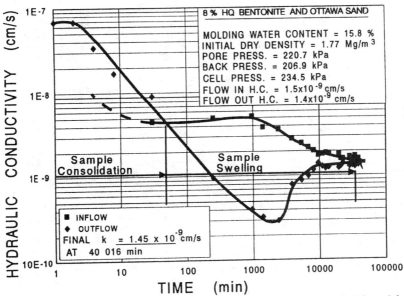

FIG. 5 -- Hydraulic Conductivity Versus Time Relationship for an 8 % HQ bentonite and Ottawa sand.

hydraulic conductivity which decreased rapidly to near 1.45×10^{-9} cm/s after 40,012 minutes of testing. The difference between the flow in and out of this sample shows that it underwent some initial consolidation during the first 80 minutes, followed by a prolonged period of swelling. The swelling resulted in an overall increase in volume of 4 ml or approximately 1 % based on the total sample volume.

A summary of the hydraulic conductivity test results is presented in Table 2. This table also shows the bentonite and polymer contents, the test durations and identifies failures.

TABLE 2 -- Summary of Hydraulic Conductivity Tests Results for Marginal Quality Bentonite

Test No.	Bentonite Content (%)	Polymer Content (%)		Initial Water Content (%)	Dry Density (Mg/cm^3)	Test Duration (min)	Final Hydraulic Conductivity (cm/s)	Comments
1	8	0		14.59	1.77	1,203	2E-5	Failure
2	8	0		14.83	1.78	294	6E-7	Failure
3	8	0.5		14.95	1.77	51,650	1.05E-9	
4	8	0.2		15.01	1.78	40,298	1.2E-9	
5	8	0.05		14.84	1.78	28,298	1.6E-9	
6	0	0.5		14.97	1.77	1,200	8E-6	Failure
7	8	0.2	Aged	14.60	1.78	28,768	1.7E-9	
8	8	0.1	Aged	14.67	1.78	29,586	1.8E-9	
9	8	0.05	Aged	14.67	1.80	28,769	1.5E-9	
10	8	0.01	Aged	14.78	1.82	410	6E-9	Failure

Figure 6 shows the results for test No. 1 conducted on the non-polymerized MQ bentonite and sand samples. The initial water content and dry density of this sample were 14.59 % and 1.77 Mg/m^3, respectively. The hydraulic gradients ranged from 14 to 33 during this test. The hydraulic conductivity of this sample was near 2×10^{-5} cm/s to begin with. The hydraulic conductivity gradually increased over the next few minutes before dropping slightly after 800 minutes and then rapidly moved to failure after 1203 minutes of testing. No significant sample consolidation or swelling was measured during this test. Observations of the failure indicated that a "blow out" of bentonite had occurred, which filled the lucite outflow reservoir and volume change indicator with a brown colored bentonitic slurry.

A second test conducted on the non polymerized moderate quality bentonite also failed. This sample was compacted at a water content and dry density of 14.83 % and 1.78 Mg/m^3, respectively. The initial hydraulic conductivity was also near 1×10^{-7} cm/s. However, the hydraulic conductivity of this sample decreased rapidly for the first 60 minutes, and then increased rapidly to failure after 294 minutes of testing. The water content of the failed sample was 17.48 %. The failure of this sample was closely monitored until it was not possible to manually switch the 25 ml volume change monitoring burettes and therefore the test had to be terminated. However, the failure mechanism appeared to be similar to that for the first test.

Figure 7 shows the result of the triaxial permeability test conducted using MQ bentonite modified with 0.5 % polymer (test No. 3). This sample had an initial water content and dry density of 14.95 and

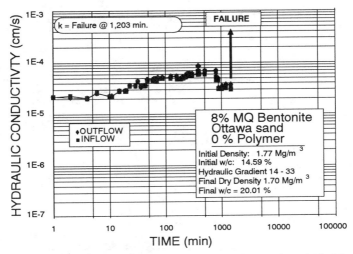

FIG. 6 -- Hydraulic Conductivity Versus Time Relationship for non-polymerized MQ bentonite.

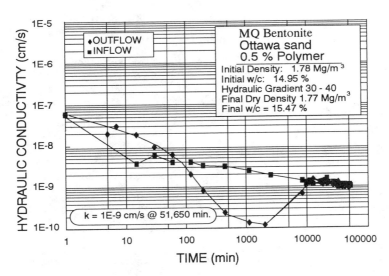

FIG. 7 -- Hydraulic Conductivity Versus Time Relationship for 0.5 % polymer modification of MQ bentonite.

1.77 Mg/m^3, respectively. The initial hydraulic conductivity was near 7 x 10^{-8} cm/s and decreased gradually with time, coming to equilibrium after 51,650 minutes (29 days) of testing. At that time the hydraulic conductivity based on flow into the sample was 1.1 x 10^{-9} cm/s and the hydraulic conductivity based on flow out of the sample was 1.0 x 10^{-9}

cm/s. The final water content and dry density at the end of the test were 17.56 % and 1.79 Mg/m^3. The pattern for this test was similar to that shown for the HQ bentonite shown in Fig. 5.

The change in volume with time for the 0.5 % bentonite polymerized sample is shown in Fig. 8. This sample consolidated during the first 60 minutes of the test, decreasing in volume by 0.05 ml. The sample then swelled for the remainder of the test ending with an overall increase in volume of 2.02 ml. The rate of volume change was near zero (-1.08×10^{-6} ml/s) at the end of the test, indicating that the sample had just quit swelling.

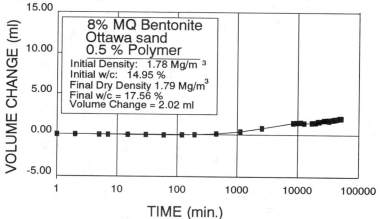

FIG. 8 -- Volume Change Versus Time Relationship for 0.5 % polymer modification of MQ bentonite.

Test No. 4 was conducted on a sample containing 0.2 % polymer addition to the MQ bentonite. The initial water content and dry density for this test specimen were 15.01 % and 1.78 Mg/m^3, respectively. This sample was tested for 40,298 minutes (28 days). The hydraulic conductivity began near 1×10^{-7} cm/s and gradually decreased with time to 1.2×10^{-9} cm/s. At the end of this test the water content and dry density were 17.81 % and 1.79 Mg/m^3, respectively.

The rate of polymer addition was lowered to 0.05 % for test No. 5. The initial water content and dry density for this test specimen was 14.84 % and 1.78 Mg/m^3, respectively. The hydraulic conductivity of this specimen began near 1×10^{-7} cm/s and then decreased to 1.6×10^{-9} cm/s after 28,478 minutes (20 days) of testing. The final water content and dry density following hydraulic conductivity testing were 19.59 % and 1.78 Mg/m^3.

Figure 9 shows the results of a test No. 6 which was designed to evaluate the effectiveness of polymer modification of Ottawa sand containing no bentonite. The initial water content and dry density of this sample was 14.97 % and 1.77 Mg/m^3, respectively. The initial hydraulic conductivity was near 2×10^{-6} cm/s and increased sharply after approximately 15 minutes of testing. The hydraulic conductivity then decreased to approximately 8×10^{-6} cm/s until failure occurred at approximately 1,200 minutes of testing.

The first test conducted on the polymer aged approximately one year prior to use is shown in Fig. 10. This MQ bentonite sample contained 0.2 % polymer. The initial water content and dry density for this sample were 14.6 % and 1.78 Mg/m^3. The hydraulic conductivity began near 1×10^{-7} cm/s and then gradually decreased to 1.7×10^{-9} cm/s after 28,600 minutes (20 days) of testing. The final water content at dry density at the end of this test were 18.34 % and 1.77 Mg/m³, respectively. The sample swelled 1.0 ml during this test.

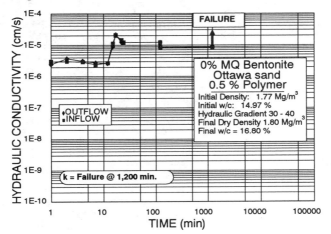

FIG. 9 -- Hydraulic Conductivity Versus Time Relationship for 0.5 % polymer modification of Ottawa sand containing no bentonite.

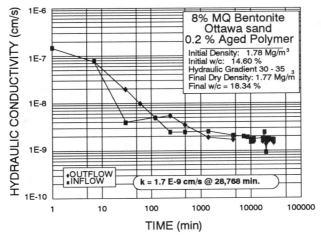

FIG. 10 -- Hydraulic Conductivity Versus Time Relationship for 0.2 % aged polymer modification of 8 % MQ bentonite and Ottawa sand.

Test No. 8 was conducted on 0.1 % aged polymer modified MQ bentonite. This modified bentonite was mixed with Ottawa sand and compacted at a molding water content of 14.67 % and dry density of 1.78 Mg/m³. The initial hydraulic conductivity for this test was near 7×10^{-8} cm/s. The hydraulic conductivity decreased gradually with time to 1.8 x 10⁻⁹ cm/s after approximately 30,000 minutes of testing. At the end of the test the water content and dry density were 18.15 % and 1.78 Mg/m³. This sample swelled 1.33 ml during testing.

The aged polymer concentration was lowered to 0.05 % for test No. 9. This test specimen was prepared at a molding water content of 14.67 % and dry density of 1.80 Mg/m³. The hydraulic conductivity began in the range of 6 x 10⁻⁸ cm/s, decreased sharply and then leveled off at 1.5 x 10⁻⁹ cm/s after 28,769 minutes of testing. At the end of the test the water content had increased to 17.28 % and the density changed to 1.81 Mg/m³.

The final test in this program is shown in Fig. 11. In this test the aged polymer concentration was reduced to 0.01 %, which represents approximately 0.015 g of polymer in a Proctor mold size sample. The initial water content and dry density for this test were 14.78 % and 1.82 Mg/m³, respectively. The initial flow out of the sample gave a hydraulic conductivity of approximately 6 x 10⁻⁸ cm/s. This value decreased sharply and then progressed rapidly to failure after 410 minutes of testing. The hydraulic gradient for this test was relatively low, ranging from 28 to 33. Failure of this sample was characterized by a loss of bentonite from the sample.

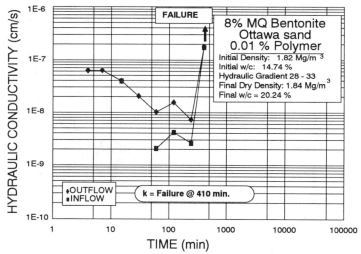

FIG. 11 -- Hydraulic Conductivity Versus Time Relationship for 0.01 % aged polymer modification of 8 % MQ bentonite and Ottawa sand.

DISCUSSION

Figure 12 contains a summary plot which shows the relationship between final hydraulic conductivity and percentage of aged and non-aged polymer added to the bentonite. All of these values apply to identically prepared test specimens containing 8 % MQ bentonite mixed, moisture conditioned and compacted with Ottawa sand. The initial dry densities of these test specimens were tightly grouped near 1.78 Mg/m³, with all of the specimens compacted at optimum. The only variation between these specimens was the percentage of polymer added to the MQ bentonite. This figure shows that a small concentration of polymer has a significant impact on the hydraulic conductivity of a marginal bentonite. The addition of 0.5 % polymer resulted in a decrease in hydraulic conductivity of approximately 4 orders of magnitude (10^{-5} cm/s to 10^{-9} cm/s). The initial impact of a small quantity of polymer is considerable, however, further increases in polymer had a diminishing impact on lowering hydraulic conductivity. In this series of tests, little decrease in hydraulic conductivity occurred with increases in polymer concentration above 0.05 %. However, below that value, all of the test specimens eventually failed, which suggests that a threshold polymer concentration exists to prevent failure. Based on this data and for these test and material conditions, this threshold concentration appears to be near 0.05 %.

Figure 12 also shows that the hydraulic conductivity produced through polymer addition is near that for non-polymerized high quality bentonite. The pattern of change in hydraulic conductivity with time for polymerized MQ bentonite (Fig. 7) is also similar to that for a HQ bentonite (Fig. 5). The sample consolidation and swelling characteristics were also similar for both the 0.5 % polymerized moderate quality bentonite and for the high quality bentonite.

The deterioration of polymers with time is a concern where polymers are used for long-term soil liner applications. No significant variation was found between the shelf aged and non-aged polymer in this test program. However, the behavior of this polymer in an aged field condition was not evaluated. In a three-year study of a field test cover constructed with a different polymerized bentonite-sand mixture, Haug and Wong (1993) found no indication of degradation. Similar field studies are required to confirm the field performance of the polymerized bentonite used in the study.

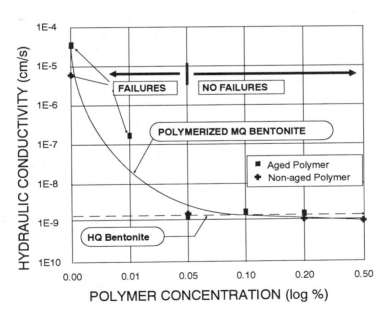

FIG. 12 -- Impact of Polymer Addition on the Final Hydraulic Conductivity of Compacted 8 % MQ bentonite Ottawa sand samples.

CONCLUSIONS

This rheological and triaxial permeability/volume change study found that the low hydraulic conductivity characteristic of a marginal bentonite could be significantly improved with the addition of a small quantity of polymer. The polymer used for the study was ALCOMER 228, which is specifically marketed for use in improving the hydraulic performance of bentonites used for soil liner applications. The addition of polymer concentrations in excess of 0.05 % by air-dried weight of bentonite reduced the hydraulic conductivity of moisture conditioned and compacted 8 % bentonite - Ottawa sand test specimens to near 1×10^{-9} cm/s. The hydraulic conductivity obtained was almost identical to that obtained for a high quality bentonite using no polymer. Polymer additions below 0.05 % had little impact on the final hydraulic conductivity. A test specimen prepared using 0.01 % polymer failed. Triaxial permeability tests conducted on polymer aged one year prior to use and non-aged polymer gave similar results. However, further laboratory and field research is required to establish the longevity of these polymers before their widespread use in soil liners can be recommended. This program also found that the rheological evaluation of the impact of the polymer concentration on MQ bentonite could be used to provide a quick and relatively conservative estimate of the quantity of polymer required for a bentonite modified soil liner application.

Acknowledgments

The assistance of Geoff Gagen, Sales Manager, Allied Colloids (Canada) Ltd. in supplying the polymer used for this study is greatly appreciated. The assistance of Darby Kreitz and Dale Pavier in the conduct of this test program is also acknowledged.

References

Allied Colloids, 1992, ALCOMER 228 Technical and Processing Data, Allied Colloids (Canada) Inc., 11 Automatic Road, Brampton, Ontario.

Black, D.K. and Lee, K.L., 1973, "Saturating laboratory samples by back pressure," ASCE Journal of the Soil Mechanics and Foundation Engineering Division, Vol. 99, pp. 75 - 93.

Fernuik, Neal and Haug, Moir, 1990, "Evaluation of in situ permeability testing methods," Journal of Geotechnical Engineering, Vol. 116, No. 2, pp. 297 - 311.

Haug, Moir D. and Wong, Lionel C., 1992, "Impact of molding water content on hydraulic conductivity of compacted sand-bentonite," Canadian Geotechnical Journal, Vol 29, No. 2, pp. 253 - 262.

Haug, Moir D. and Wong, Lionel C., 1993, "Potash Tails Piles Project, 1992 Progress Report for the Saskatchewan Potash Producers Association," Department of Civil Engineering, University of Saskatchewan.

Reschke, A. and Haug, M.D., 1991, "The physio-chemical properties of bentonites and the performance of sand-bentonite mixtures," *Proceedings*, Canadian Geotechnical Conference, Calgary, Alberta, Vol 2, pp. 62-1 - 62-10.

Wong, Lionel C. and Haug, Moir D., 1991, "Cyclical closed system freeze-thaw permeability testing of soil liner and cover materials," *Canadian Geotechnical Journal*, Vol. 28, No. 6, pp. 784 - 793.

Xanthakos, P. 1979. *Slurry Walls*. McGraw-Hill Inc., pp. 108 - 112.

Yanful, Earnest K. Haug, Moir D. and Wong, Lionel C., 1990, "The impact of synthetic leachate on the hydraulic conductivity of a smectite till underlying a landfill near Saskatoon, Saskatchewan," *Canadian Geotechnical Journal*, Vol. 27, No. 4, pp. 507 - 519.

Irene M.-C. Lo,[1] Howard M. Liljestrand,[2] and David E. Daniel[2]

HYDRAULIC CONDUCTIVITY AND ADSORPTION PARAMETERS FOR POLLUTANT TRANSPORT THROUGH MONTMORILLONITE AND MODIFIED MONTMORILLONITE CLAY LINER MATERIALS

REFERENCE: Lo, I. M.-C., Liljestrand, H. M., and Daniel, D. E., "Hydraulic Conductivity and Adsorption Parameters for Pollutant Transport through Montmorillonite and Modified Montmorillonite Clay Liner Materials," Hydraulic Conductivity and Waste Contaminant Transport in Soil, ASTM STP 1142, David E. Daniel and Stephen J. Trautwein, Eds., American Society for Testing and Materials, Philadelphia, 1994.

ABSTRACT: Montmorillonite clay has been modified by the addition of humic acid (HA) and aluminum hydroxide (AlOH) coatings to enhance the sorption of organics, attenuate their transport, and minimize changes in hydraulic conductivity. Batch sorption experiments and hydraulic conductivity tests have been performed to compare this HA-AlOH-Clay with commercially available Claymax®. Claymax® has an approximately 6.8-19 times greater Langmuir adsorption maximum for Pb^{2+} than that of HA-AlOH-Clay, but the partition coefficients of chlorobenzenes to the HA-AlOH-Clay are 14-25 times greater than that to Claymax®. Hydraulic conductivity tests for Claymax® with and without HA-AlOH-Clay have been performed using tap water, a synthetic leachate, and pure methanol. Synthetic leachate increased the hydraulic conductivity of the HA-AlOH-Clay together with Claymax® to $1.2*10^{-9}$ cm/s or about 3 times over that for tap water. The hydraulic conductivity of HA-AlOH-Clay alone is greater than the minimum required value ($k = 1*10^{-7}$ cm/s) but can be used together with Claymax® to control chemical fluxes.

KEYWORDS: montmorillonite, humic acid, organically modified clay, sorption, lead, partition, chlorobenzenes, hydraulic conductivity.

The present design philosophy for liner systems is to minimize the

---

[1]Lecturer, Civil and Structural Engineering, Hong Kong University of Science and Technology, Clear Water Bay, Kowloon, Hong Kong.

[2]Professor, Civil Engineering, The University of Texas at Austin, Austin, TX 78712.

rate of contaminant transport by reducing the hydraulic transport of leachate and thus the advective flux of contaminants through the liner. However even when advection is minimal, contaminants can migrate through clays by Fickian diffusion at rates that can be significant (Shackelford, et al. 1989). Moreover, the hydraulic conductivity of natural clay barriers can increase upon reaction with organic contaminants (Mitchell and Madsen 1987, Fernandez and Quigley 1988).

A new liner system design (Figure 1) is proposed to maximize the retardation of pollutants, minimize diffusive fluxes, and minimize hydraulic conductivity. A bentonite clay layer (Claymax®) provides cation exchange for the sorption of heavy metal cations and an organically modified clay (clay coated with aluminum hydroxide to form a bridge to a layer of humic acid) for the sorption of the hydrophobic organics. Claymax® is a three layer geotextile clay liner manufactured by James Clem Corp. consisting of an open weave polypropylene geotextile, bentonite with swelling additives affixed with water soluble glue, and open weave polyester geotextile support fabric. The current design for liner systems consists of a permeable material for collection of leachate, a synthetic geomembrane, and 3 m of clay as the primary liner and a redundant set of these three layers as the secondary liner. In Figure 1, the Claymax® plus organically modified clay layers replace the 3 m of clay that is currently recommended to be used in combination with the a geomembrane in a double liner leachate collection system.

The introduction of a new type of clay material coated with humic acid is based on three factors. First, humic coatings are thermodynamically stable (Greenland, 1971, Stevenson, 1982). Second, the material cost is low. Third, dissolved humics generated in disposal sites facilitate the transport of metal ion and neutral hydrophobic organic pollutants. A clay material designed to sorb humics will control this facilitated transport mechanism and, therefore, mitigate the flux of pollutants (Liljestrand et al., 1992).

In this study, column permeameter tests and batch adsorption experiments were used to evaluate the hydraulic conductivity and pollutant attenuation of montmorillonite (as the control case) and the humic acid-aluminum hydroxide-clay (HA-AlOH-Clay). The contaminants chosen were lead, a representative inorganic cation, and a homologous series of chlorobenzenes with a range of octanol-water partition coefficient (K_{ow}). The hydraulic conductivities of the liner materials were determined using flexible wall permeameters permeated with three fluids: tap water, synthetic leachate, and pure methanol.

MATERIALS AND EXPERIMENTAL METHODS

Laboratory Preparation of HA-AlOH-Clay

Synthesis of AlOH-Clay

Hydroxide-aluminum interlayers in expanding layers of Wyoming Montmorillonite SWy-1 obtained from the Source Clay Repository of the Clay Minerals Society available through the University of Missouri were prepared by the cation exchange method (Rengasamy and Oades, 1978, Oades, 1984, Srinivansan, and Fogler, 1990). The cross-linking agents between the expanding layers were hydroxide ions and aluminum ions with OH/Al molar ratios of 1.5, 1.85 and 2.6. The aluminum hydroxide polymers attach to the clay units with a uniform spacing to a pillar

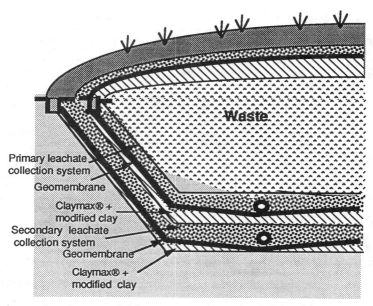

Figure 1. Proposed liner system for hazardous waste landfills

like structures between the clay plates to form what are called cross-linked smectites in the soil science literature (e.g., Lahav et al., 1978). To obtain the aluminum hydroxide polymer cross-linking agents of the desired molar ratios, varying amounts of 1 M NaOH were added to 500 mL of 0.2 M $Al(NO_3)_3$ at a rate of 0.01 eq. NaOH per minute. The solution was mixed continuously for 1 hour and then aged for at least 6 days before use. The cross-linking agents were added dropwise with vigorous stirring to a clay suspension of 200 mg/L. The suspension was then left for 1 day. The supernatant was removed, and the cross-linked clay product was centrifuged, collected, and freeze-dried.

The physical and chemical properties of cross-linked clay products depend on the OH/Al ratio of the aluminum hydroxide solution, as well as the relative amount of aluminum to clay. At least 5.3 mM Al/g clay is required for complete coverage of the surface area of montmorillonite (Turner, 1965), and as much as 16 mM Al/g clay is necessary to provide a uniform distribution of aluminum hydroxide species in the interlayer of the clay (Slaughter and Malne, 1960). In this study, the aluminum provided to clay ratio was set to 5.3 mM Al/g clay in the solutions of OH/Al = 1.5, 1.85, and 2.6, and 16 mM Al/g clay for OH/Al = 2.6.

To characterize the properties of the cross-linked clay products (AlOH-Clays), the cation exchange capacity (CEC), basal spacing, and specific surface area were measured. The results (Table 1) show a slight reduction in CEC of AlOH-Clays with OH/Al = 1.5 and Al/Clay = 5.3 mM/g, and with OH/Al = 1.85 and Al/Clay = 5.3 mM/g, which indicates that the aluminum hydroxide species may not penetrate or may only partially enter the interlayer of clays. This deduction is further substantiated by the measured basal spacing and specific surface area. The aluminum hydroxide species increase the specific surface area of the clay by

Table 1. Physical Characteristics of AlOH-Clay

Samples	Al/clay[2] (mg Al/g Clay) or (mM /g)	Surface[3] Area (m^2/g)	CEC (meq/100g)	d(001)[4] 23°C	d(001) 102°C	d(001) 550°C
Mont.[1]	0	36.47	84.2	12.76	11.28	9.54
OH/Al= 1.5 + Mont.	143 (5.3)	53.78	61.7	14.82	12.22	11.23
OH/Al=1.85 + Mont.	143 (5.3)	38.91	77.7	13.18	ND	ND
OH/Al= 2.6 + Mont.	143 (5.3)	121.42	59.6	ND[5]	ND	ND
OH/Al= 2.6 + Mont.	432 (16.0)	333.69	9.1	19.16	17.94	15.56

1. Mont. = Montmorillonite.
2. Al/clay = The amount of Al in aluminum hydroxide polymer cross-linking agents added to the clay suspension.
3. The specific surface area of pure montmorillonite is measured using N_2, a non-polar molecule unable to enter the interlayer surface of montmorillonite. Any change in the specific surface area between montmorillonite and AlOH-Clay is due to the interlayering of aluminum hydroxide species. Measurements using ethylene glycol monoethyl ether, a polar molecule that penetrates into the clay interlayers yields a higher values around 600 m^2/g (Bohn et al., 1985).
4. d(001) = Basal Spacing of AlOH-Clays.
5. ND = Not Detected.

expanding the interlayers but decrease in CEC of the clay by coating ionic exchange sites.

A comparison of the treated clays with OH/Al = 2.6 and Al/Clay = 5.3 mM/g versus the treated clays with OH/Al = 2.6 and Al/Clay = 16.0 mM/g shows there is a large reduction in CEC as the Al to clay ratio increases. Thus, 5.3 mM Al/g clay is insufficient to cover the clay interlayers. The basal spacing of the AlOH-Clays decreased slightly after heating to 102°C and 550°C. This slight decrease in basal spacing may result from the partial overlap between the hydroxide groups in the aluminum hydroxide polymer cross-linking agents and the oxygen groups of the montmorillonite.

The AlOH-Clay with OH/Al = 2.6 and Al/Clay = 16.0 mM/g has a large specific surface area with a significantly lower cation exchange capacity (indicating the presence of aluminum hydroxide at clay surfaces) and a large basal spacing indicating a higher potential to trap small organic molecules, e.g., those with <10 carbon atoms. The slight change in basal spacing after heating to high temperature indicates a high stability. Therefore, this AlOH-Clay was chosen as the basis for coating with humic acid to form a laboratory synthetic organophilic clay, HA-AlOH-Clay, where humic acid adsorbs onto the aluminum hydroxide polymer coating of the clay surface.

Coating of Humic Acid onto AlOH-Clay

Two grams of freeze-dried AlOH-Clay (OH/Al = 2.6 and Al/Clay = 16.0 mM/g) were added with 1 liter of varying concentrations of Aldrich humic acid. These experiments were performed at constant pH and ionic strength of 6.0 and 0.05 M, respectively. The pH and ionic strength were adjusted using 1 M nitric acid (HNO_3) and 5 M sodium chloride

(NaCl), respectively. Each suspension was transferred from its volumetric flask to an Erlenmeyer flask, which was stoppered and placed in a 20°C constant temperature room. After 24 hours of mixing, the suspensions were centrifuged at 20,000 g for 15 minutes to separate the solids from the solution.

A portion of the supernatant was removed from the centrifuge tube. This extracted solution was diluted and analyzed for its humic acid concentrations using UV-Visible spectroscopy at 254 nm. The adsorption isotherm of humic acid onto AlOH-Clay (Figure 2) is compared with the adsorption isotherm of humic acid to pure montmorillonite at pH 7 (Lo, 1990). The comparison shows that the presence of aluminum hydroxide on the clay surface enhances its affinity for humic acid. This is a result of the ability of the aluminum hydroxide species to neutralize the negative charge of the organic anions/surface sites, which allows the organic matter to approach the clay surfaces. In this study, the modified clays (HA-AlOH-Clay) were synthesized to have 2.5% by weight of humic acid and a weight fraction of organic carbon (f_{OC}) of 0.012 as calculated from a mass balance on humic acid sorbed and the carbon content of Aldrich humic acid, respectively.

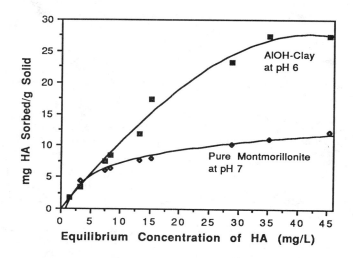

Figure 2. Adsorption of Humic Acid onto AlOH-Clay and Pure Montmorillonite

Batch Sorption Tests

Batch Inorganic Cation Sorption Equilibrium Experiments

For investigation of lead (Pb^{2+}) removal, a known weight of adsorbent (Claymax® or HA-AlOH-Clay) was placed in a 125 mL Erlenmeyer flask. In order to achieve a 10%-50% change in the Pb concentration (i.e., a statistically significant change), 0.5 g/L to 5 g/L of adsorbent was used. A range of fully dissociated Pb concentrations (10 to 100 mg/L) was prepared by serial dilution of a certified atomic absorption standard ($Pb(NO_3)_2$ of concentration 1 mg Pb/mL). During the sorption test, the pH of the suspension was checked and adjusted to the desired pH values (pH = 3, 5, and 7) with either 0.02 M HNO_3 or 0.02 M NaOH every two hours. The volume of acid or base added was recorded and used to recalculate the solution volume and chemical concentrations. After equilibration, the samples were analyzed by inductively coupled argon plasma spectroscopy and compared with standards.

Batch Organic Sorption Equilibrium Experiments

The hydrophobic, non-ionic compounds 1,2-dichlorobenzene (1,2-DCB), 1,2,4-trichlorobenzene (1,2,4-TCB), and 1,2,4,5-tetrachlorobenzene (1,2,4,5-TECB) were used in the organic sorption studies. To measure the organic concentration in an aqueous sample, 2 mL of sample was extracted by 6 mL of chromatographic grade n-hexane by mixing vigorously for 15 minutes on a platform shaker. A 1 µL volume of extractant was injected into a Tracor 550 gas chromatograph equipped with an electron capture detector (sensitive to chlorinated aromatics and insensitive to hexane) and a Supelco SPB-5 chromatographic column (30 m long with 0.53 mm internal diameter and 0.5 µm film thickness). Chromatographic grade nitrogen was used as the carrier gas and purge gas at flow rates of 40 mL/min and 30 mL/min, respectively. The column oven was set at 70°C. The temperatures of the injection port and detector were set at 220°C and 310°C, respectively. Under these operating conditions, the retention times of 1,2-DCB, 1,2,4-TCB, and 1,2,4,5-TECB were approximately 2 minutes, 4 minutes, and 7 minutes, respectively.

Hydraulic Conductivity Tests

To determine the effect of a synthetic leachate containing measured concentrations of 62.2 mg/L of 1,2 DCB, 69.2 mg/L of NaCl, and 100.6 mg/L of Pb^{2+} on the hydraulic conductivity of the modified liner and Claymax® materials, constant-head hydraulic conductivity tests were performed using flexible wall permeameters. In addition to the compounds of interest (1,2-DCB, Pb^{2+}) in the synthetic leachate, 2% by weight of sodium azide and 0.05 N of pH 7 phosphate buffer were added in the leachate to prevent the growth of bacteria and to provide a stable solution pH, respectively. The desired hydraulic gradient was obtained by establishing an elevation difference between the liquid surface of inflow and outflow ends. The hydraulic gradient was adjusted to about 80 in these tests.

The hydraulic conductivity of the modified liner permeated with tap water at 22°C was conducted as a control for the studies using synthetic leachate or pure methanol. (In Austin Texas, drinking water is prepared by lime softening, resulting in a tap water with pH = 10.2±0.2, a specific conductance of 415 µS/cm, and total alkalinity of 67 mg/L as calcium carbonate.) Flexible wall permeameters with a control panel board similar to those used by Shackelford et al. (1989) were used for all tests. All samples were had a diameter of 10 cm and a thickness of 1.5 cm. Porous stones were put on the top and bottom of the test specimen to distribute flow evenly. Filter paper was placed in between the test specimen and the porous stone to prevent clogging of the stones. Latex membranes and O-rings were used to provide a good seal. Double drainage lines to both the top and bottom of the test specimen were installed to flush air bubbles out of the lines. Teflon tape was wrapped around the circumference of the specimen, and the latex membrane was placed over the Teflon tape. Next, aluminum foil was rolled around the membrane, and a second latex membrane was put over the aluminum foil. The HA-AlOH-Clay specimens had a bulk density of 0.715 g/cm^3, a total porosity of 0.579, and a degree of saturation of >95%. The Claymax® specimens had a bulk density of 0.654 g/cm^3, a total porosity of 0.748, and a degree of saturation of 99%. The 1.5 cm thick specimens were saturated using a back pressure of 68.7 kPa (10 psi) for 3 weeks. The effective stress, i.e., the difference between the confining pressure and back pressure, was fixed to be 68.7 kPa.

Falling-head permeability tests were used to measure the hydraulic conductivity of each soil sample permeated with tap water. Then, constant head permeability tests were used to measure the hydraulic conductivity of each soil sample permeated with the synthetic leachate or pure methanol, to avoid contamination of the control panel broad. In the constant head tests, the synthetic leachate or methanol reservoir was one 4.5 L brown bottle instead of several burettes. The desired hydraulic gradient, set to 80, was obtained by establishing an elevation difference between the liquid surface of inflow and outflow ends, and the confining pressure was 135.4 kPa. The flow was considered to be at equilibrium when the variation of inflow and outflow rates was within ±5% and both were constant. The total testing time was 5 months.

EXPERIMENTAL RESULTS

Sorption Results

Inorganic Sorption

The results of Pb^{2+} adsorption onto Claymax® and HA-AlOH-Clay at pH 3, 5 and 7 are presented in Table 2. The Freundlich and Langmuir isotherms described in Stumm and Morgan (1981) both fit well the sorption of Pb^{2+}. The adsorption maximums for Claymax® are 76.76 mg Pb/g (or 74.1 meq/100 g Claymax®) at pH 3 and 81.12 mg Pb/g (or 78.3 meq/100 g Claymax®) at pH 5. These values are comparable to the CEC of Wyoming montmorillonite (84.2 meq/100g clay), which is evidence that Pb removal by Claymax® is primarily an ion exchange - charge attraction adsorption reaction for pH < 7. Since Claymax® consists of 90%-95% montmorillonite, its CEC is estimated to be 72 - 142 meq/100g (Drever, 1982). The adsorption maximums at pH 7 which are greater than the CEC of montmorillonite indicate specific ion adsorption as well as ion exchange.

The solution pH affects the sorption of lead ion onto HA-AlOH-Clay. The sorption maximum decreases 33% as the pH is decreases from 5 to 3. The distribution of the weak acid dissociation constants (K_a) of humics' functional groups can be fit by a discrete ligand model (Fish et al., 1986). The best fit of acid-base titration data using FITEQL, an computer program for fitting equilibrium constants to match experimental results, yielded five discrete functional groups with log K_1 = -10.8, log K_2 = -8.8, log K_3 = -6.5, log K_4 = -4.0 and log K_5 = -3.0 (Fish et al., 1986). The increasing adsorption maximums of Pb^{2+} onto the humic coating between pH 3 and 5 and between pH 5 and 7 reflects the increased ionization of the humic pK_4 = 4.0 and pK_3 = 6.5 weak acid functional groups. Carboxylic (pK_4 = 4.0) and phenolic (pK_3 = 6.5) functional groups are the main adsorption sites for metal ions. At pH > 4.7, the carboxylic acid (-COOH) group is more than 50% dissociated, providing a charge attraction for metal cations and allowing a specific surface

Table 2. Best Fit Pb^{2+} Freundlich and Langmuir Adsorption Isotherms

Sorbent	Parameter*	Freundlich at pH 3	Freundlich at pH 5	Freundlich at pH 7
Claymax®	a	10.38	24.44	7.998
	b	0.571	0.283	0.929
	r^2	0.974	0.963	0.973
HA-AlOH-Clay	a	0.232	0.548	0.775
	b	0.587	0.510	0.830
	r^2	0.986	0.988	0.971

Sorbent	Parameter*	Langmuir at pH 3	Langmuir at pH 5	Langmuir at pH 7
Claymax®	a	76.76	81.12	204.8
	b	0.123	0.246	0.032
	r^2	0.986	0.985	0.978
HA-AlOH-Clay	a	3.948	5.882	30.17
	b	0.029	0.049	0.021
	r^2	0.966	0.964	0.963

* a K_f [mg/g/(mg/L)^b] for Freundlich Isotherm ($q = K_f C^b$) and Q^o (mg/g) for Langmuir Isotherm ($q = Q^o bC/1+bC$)
  r   Least Square Best Fit Correlation Coefficient (n = 12, p < 0.005)

attraction with adjacent phenolic groups. This complexation reaction can be expressed as:

$$\text{[structure]} + Pb^{2+} \longrightarrow \text{[Pb-complex structure]} + H^+$$

Thus, surface complexation between metal ions and humic acid is favored even in weakly acidic solutions.

Organic Sorption

Sorption of 1,2-DCB, 1,2,4-TCB, and 1,2,4,5-TECB onto Claymax® is small, because of the absence of a significant natural organic carbon coating on the clay. While the mineral clay surface provide hydrophilic CEC sites for metal ion sorption, it does not provide sites for hydrophobic partitioning of neutral organics. In this study, values of the distribution coefficient K_d, also known as the linear sorption coefficient of organics to the pure mineral K_m (cm^3/g), for DCB and TCB on Claymax® are found to be 1.44 and 2.2 cm^3/g, respectively. K_m values for DCB and TCB on kaolinite are 1.1 and 2.4 cm^3/g, respectively (Schwarzenbach and Westall, 1981). This small difference between K_m

values for the two clays is unexpected, because the specific surface area of montmorillonite in Claymax® is almost 30 times greater than that of kaolinite. The mineral sorption coefficients may be comparable as a result of a similar number of active sites for neutral organic sorption. The larger specific surface area of montmorillonite corresponds with a larger number of ionic exchange sites. If the number of active sites of Claymax® to adsorb DCB and TCB were small, then the sorption capacity would be low regardless of its specific surface area.

Sorption of 1,2-DCB, 1,2,4-TCB, and 1,2,4,5-TECB onto the humic modified clay was investigated. The K_d distribution coefficients of these chlorobenzenes to HA-AlOH-Clay were 14-25 times greater than that to Claymax® as a result of the humic acid coating. Sorption of neutral hydrophobic organics to soils is generally linear and proportional to the organic carbon content of the solid (Karickhoff et al., 1979). The sorption isotherms are linear, and the distribution coefficients in Figure 3 are the best fit of the form pollutant sorbed (mg pollutant/g soil) = K_d (L/g soil) * equilibrium pollutant concentration (mg pollutant/L). That is, the sorption is assumed to be completely reversible and therefore the linear intercept must be (0,0). The K_d partition coefficients increase with increasing octanol-water partition coefficients K_{ow} of the chlorobenzenes. The partition coefficient K_d normalized to the fraction of soil organic carbon (f_{oc}) is known as K_{oc} = K_d/f_{oc}. The organic carbon normalized sorption of nonionic organic compounds to soils K_{oc} obtained in this study and literature values correlates with K_{ow} for the respective solutes (Figure 4). These soils, although from different sources as indicated in Table 3, are all coated with Aldrich humic acid. Linear regression of this log-log plot (Figure 4) corresponds to the semi-empirical expression of the form given by Karickhoff et al. (1979)

$$\log K_{oc} = a \log K_{ow} + b \qquad (1)$$

where best fit values found in this study are a = 0.878, b = -0.18, and r^2 = 0.89.

Table 3. K_{ow} and K_{oc} Values for Pollutants Sorbed onto Soils Coated with Humic Acid (HA)

Pollutant Sorbate	Clay Base	Organic matter	log K_{ow}	log K_{oc}	Ref.
Trichloro-ethene	Sapsucker soil	Aldrich HA	2.29	1.83	Garbarini and Lion (1986)
Toluene	Sapsucker soil	Aldrich HA	2.69	1.94	
Indole	Bentonite	Aldrich HA	3.08	2.27	Rebhun et al. (1992)
1,4-DCB	Bentonite	Aldrich HA	3.41	2.85	
Fluoranthene	Bentonite	Aldrich HA	5.17	3.90	
1,2-DCB	Montmorillonite	Aldrich HA	3.38	3.22	This Study
1,2,4-TCB	Montmorillonite	Aldrich HA	4.05	3.50	
1,2,4,5-TECB	Montmorillonite	Aldrich HA	4.72	4.33	

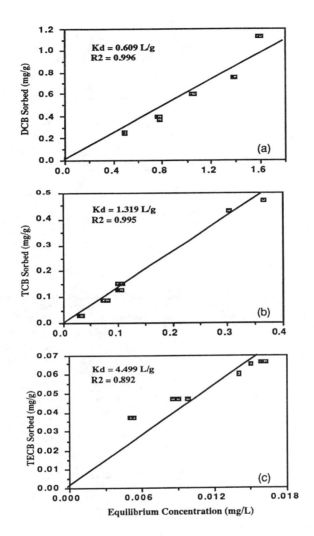

Figure 3. Sorption Isotherms @ 22°C of (a) DCB, (b) TCB, and (c) TECB with HA-AlOH-Clay

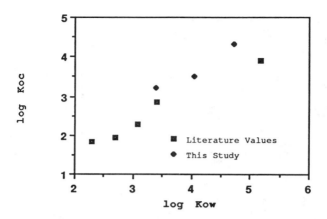

Figure 4. Sorption of Nonionic Organics to Synthetic-Humic Soils

Hydraulic Conductivity

Hydraulic Conductivity of Liners Permeated with Tap Water

The baseline hydraulic conductivity is determined with either tap water or 0.01N CaSO$_4$. This reference hydraulic conductivity of Claymax® to tap water is $4*10^{-10}$ cm/s at 0.7 of a pore volume of flow. The hydraulic conductivity of 0.75 cm of Claymax® layered with 0.75 cm of HA-AlOH-Clay is about $8*10^{-10}$ cm/s at 0.8 of a pore volume of flow. Compared with the US Environmental Protection Agency (1989) standards for landfill containment systems, the hydraulic conductivity of liner samples present with a layer of Claymax® are about 2 order of magnitude lower than that required for a standard composite liner (k $\leq 1*10^{-7}$ cm/s). In contrast, the hydraulic conductivity of HA-AlOH-Clay without Claymax® is unable to achieve the required low permeabilities (k = $8*10^{-6}$ cm/s).

Hydraulic Conductivity of Liners Permeated with Synthetic Leachate

The objective of this experiment was to evaluate the effectiveness of using HA-AlOH-Clay associated with Claymax® to retard the flux of chemicals through the liner and to prevent the attack of chemicals to Claymax®. The hydraulic conductivities of liner samples, which were pre-saturated with de-ionized water and permeated with a synthetic leachate (63 mg organics/L, 90 mg inorganics/L, specific conductance of 630 μmho/cm, and pH = 5.4), are presented in Figures 5-7. The hydraulic conductivities of Claymax® (Figures 5), and Claymax® with HA-AlOH-Clay (i.e., 0.75 cm of Claymax® on the top and 0.75 cm of HA-AlOH-Clay at the bottom as presented in Figure 6) increase slightly with time at the beginning of the permeation with synthetic leachate but stabilize after 1 pore volume of flow with values in the range of $9*10^{-10}$ to $1*10^{-9}$ cm/s. This represents about a 2-5 fold increase as compared with the hydraulic conductivity to tap water.

The hydraulic conductivity of HA-AlOH-Clay with Claymax® (i.e. 0.75 cm of HA-AlOH-Clay on the top and 0.75 cm of Claymax® at the bottom as presented in Figure 7) tended to decrease over the five month testing period from $5*10^{-10}$ cm/s at 0.6 pore volume to $2*10^{-11}$ cm/s at 1.1 pore volume. A possible explanation for this decrease is due to the growth of microorganisms in the permeameter. Black spots were observed around the outer most latex membrane, and hydrogen sulfide odor was noticed. However, the survival of microorganisms in the bulk specimen was expected to be minimal, since sodium azide was added in the synthetic leachate.

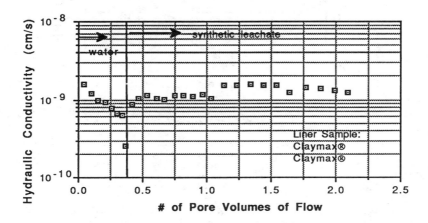

Figure 5. Hydraulic Conductivity of Claymax® to Synthetic Leachate

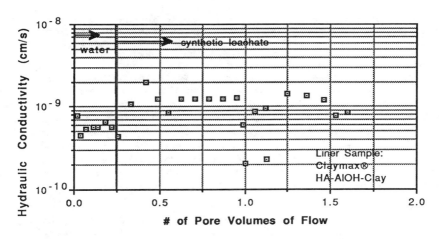

Figure 6. Hydraulic Conductivity of Claymax® with HA-AlOH-Clay to Synthetic Leachate

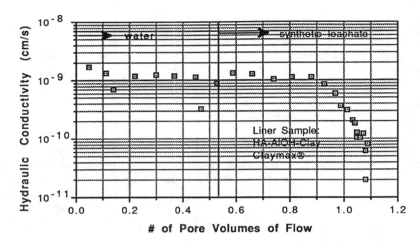

Figure 7. Hydraulic Conductivity of HA-AlOH-Clay with Claymax® to Synthetic Leachate

Volume change measurements of the liner specimens were not made following these hydraulic conductivity tests. However, the specimens were removed and sectioned to determine profiles of water content and pollutant concentrations. Visible cracks were not found and water content did not vary significantly with depth within a material.

Hydraulic Conductivity of Liners Permeated with Pure Methanol

In the hydraulic conductivity test of pure methanol with HA-AlOH-Clay plus Claymax®, the hydraulic conductivity gradually increased to $4*10^{-7}$ cm/s at 8 pore volumes (i.e., the hydraulic conductivity of pure methanol is 3 orders of magnitude greater than that with tap water while the hydraulic conductivity of the synthetic leachate is only 2-5 times greater than that with tap water). In practice, exposure of clay barriers to water-insoluble pure or concentrated organics is an unlikely occurrence except in the case of spills or leaking tanks. Thus, the effect of pure chemicals on liner samples represents a worst case analysis. In contrast, the change of hydraulic conductivity of Claymax® first permeated with water and then pure methanol is not as great (Shan, 1990). The initial hydraulic conductivity of Claymax® to water is $7*10^{-10}$ cm/s, and hydraulic conductivity of this Claymax® presaturated with water to pure methanol ($3*10^{-10}$ cm/s) is comparable to the hydraulic conductivity to tap water ($4*10^{-10}$ cm/s) (Figure 8). The difference in these hydraulic conductivity values indicates the importance of water in the pores of Claymax®. Longer testing to yield more than one pore volume is required to confirm this result.

Breakthrough Curves of Liner Samples

The effluent concentration (C_e) normalized to the influent concentration (C_o) of pollutants to yield C_e/C_o versus permeation time were determined to observe the breakthrough of lead and 1,2-DCB during the column hydraulic conductivity test. After 150 days, the effluent concentrations of lead with the Claymax® and with the Claymax® plus HA-

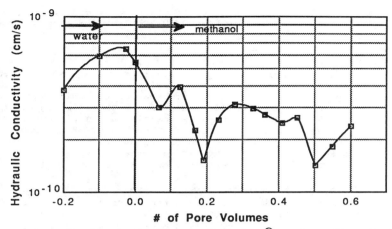

Figure 8. Hydraulic Conductivity of Claymax® to Water/Pure Methanol

AlOH-Clay liner samples were very low as a result of the extremely high sorption capacity of Claymax®. In contrast, the effluent concentration of 1,2-DCB in the Claymax® with Claymax® liner sample has past breakthrough (C_e/C_o = 0.85) after 150 days. However for the same permeation time, 1,2-DCB breakthrough had not occurred in either orientation of Claymax® with HA-AlOH-Clay above or HA-AlOH-Clay with Claymax® above. In both cases, C_e/C_o is less than 0.45. The lower concentration of DCB in the liners containing Claymax® and HA-AlOH-Clay is due to the relatively high sorption capacity of DCB by HA-AlOH-Clay.

CONCLUSIONS

This study examines a new a new material, HA-AlOH-Clay, which may be useful in retarding the advective and diffusive fluxes of pollutants by increasing the sorption of pollutants onto the liner materials as well as controlling the hydraulic conductivity of leachate. To this end, an organically modified clay material for the sorption of organic pollutants was developed based on expanding the clay interlayers with aluminum hydroxide and adsorbing humic acid. To achieve the required hydraulic conductivity with a high sorption capacity for heavy metals, this HA-AlOH-Clay is used together with Claymax® in a two layer liner material system.

The experimental results from this research provide data for the design of liner systems for the containment of the specific pollutants Pb, 1,2-DCB, 1,2,4-TCB, and 1,2,4,5-TECB. The procedures presented would be the same for any cationic or neutral organic pollutant. The species were chosen as being representative of soluble pollutants commonly found in hazardous waste leachates.

The specific conclusions of this study are as follows:

(1) The Langmuir adsorption maximums of Pb^{2+} in Claymax® are 6.8 to 19 times greater than that in the HA-AlOH-Clay. However,

the partition coefficients of chlorobenzenes (1,2-DCB, 1,2,4-TCB, 1,2,4,5-TECB) to HA-AlOH-Clay were 14 - 25 times greater than those to Claymax®.

(2) The removal mechanisms of Pb^{2+} at pH < 7 are probably a result of cation exchange reactions for Claymax® and metal complexation for the organically modified clays (HA-AlOH-Clay). The sorption of chlorobenzenes to humic acid modified clays (e.g., HA-AlOH-Clay) depends on the soil organic carbon content and the octanol-water partition coefficient of the pollutant. The major adsorption mechanism of chlorobenzenes is hydrophobic partitioning to the natural organic carbon, and sorption to the clay mineral phase is more than an order of magnitude less important than that to humic coatings.

(3) The hydraulic conductivity of Claymax® together with organically modified clays to tap water (k = 3 - 8*10^{-10} cm/sec) is at least 2 orders of magnitude lower than the value of 1*10^{-7} cm/s required by the U. S. Environmental Protection Agency (1989). However, the hydraulic conductivity of HA-AlOH-Clay alone is very high, similar to that of sand. These findings imply that two barriers (e.g., Claymax® with HA-AlOH-Clay) are required to reduce the transport of pollutants as well as act as an effective hydraulic barrier.

(4) The hydraulic conductivity of Claymax® changed only slightly when permeated with a synthetic leachate. It increased only 3-fold over that for tap water. The hydraulic conductivity of HA-AlOH-Clay with Claymax® was three orders of magnitude greater for pure methanol than that measured with tap water. This result was probably due to a low water content of the liner materials. The effect of presaturation with water on the hydraulic conductivity of these liner materials should be investigated in future work with pure organic solvents

REFERENCES

Bohn, H. L., McNeal, B. L., and O'Connor, G. A., 1985, Soil Chemistry, 2nd ed., John Wiley and Sons, New York.

Drever, J. I., 1982, The Geochemistry of Natural Waters, Prentice-Hall, Inc., Englewood Cliffs, N.J.

Fernandez, F. and Quigley, R., 1988, "Viscosity and Dielectric Constant Controls on the Hydraulic Conductivity of Clayey Soils Permeated with Water-Soluble Organics," Canadian Geotechnical Journal, Vol. 22, pp. 205-214.

Fish, W., Dzombak, D. A., and Morel, F. M. M., 1986, "Metal-Humate Interactions. 2. Application and Comparison of Models," Environmental Science and Technology, Vol. 20, pp. 676-683.

Garbarini, D. R. and Lion, L. W., 1986, "Influence of the Nature of Soil Organics on the Sorption of Toluene and Trichloroethylene," Environmental Science and Technology, Vol. 20, pp. 1263-1269.

Greenland, D. J., 1971, "Interactions between Humic and Fulvic Acids and Clays," Soil Science, Vol. 3, pp. 34-41.

Karickhoff, S. W., Brown, D. S., and Scott, T. A., 1979, "Sorption of Hydrophobic Pollutants on Natural Sediments," Water Research, Vol. 13, pp. 241-248.

Lahav, N., Shani, U., and Shabtai, J., 1978, "Cross-Link Smectites. I. Synthesis and Properties of Hydroxy-Aluminum-Montmorillonite," Clays and Clay Minerals, Vol. 26, pp. 107-115.

Liljestrand, H. M., Lo, I. M-C., and Shimizu, Y., 1992, "Sorption of Humic Materials onto Inorganic Surfaces for the Mitigation of Facilitated Pollutant Transport Processes," Water Science and Technology, Vol. 26, pp. 1221-1228.

Lo, M. C., 1990, "Sorption of Humic Acid Onto Inorganic Solids and Its Importance in the Transport of Organic Pollutants," M. S. Thesis, The University of Texas at Austin, Austin Texas.

Mitchell, J. K. and Madsen, F. T., 1987, "Chemical Effects on Clay Hydraulic Conductivity," in Geotechnical Practice for Waste Disposal '87, R. E. Woods, ed., ASCE, New York, STP 13, pp 87-107.

Oades, J. M., 1984, "Interactions of Polycations of Aluminum and Iron with Clays," Clays and Clay Minerals, Vol. 32, pp. 49-57.

Rebhun, M., Kalabo, R., Grossman, L., Manka, J., and Rav-Acha, C., 1992, "Sorption of Organics on Clay and Synthetic Humic-Clay Complexes Simulating Aquifer," Water Research, Vol. 26, pp. 79-84.

Rengasamy, P. and Oades, J. M., 1978, "Interaction of Monomeric and Polymeric Species of Metal Ions with Clay Surfaces: III. Aluminium (III) and Chromium (III)," Australian Journal of Soil Research, Vol. 16, pp. 53-66.

Schwarzenbach, R. P. and Westall, J., 1981, "Transport of Nonionic Organic Compounds from Surface Water to Groundwater," Environmental Science and Technology, Vol. 25, pp. 1360-1367.

Shackelford, C. D., Daniel, D. E., and Liljestrand, H. M., 1989, "Diffusion of Inorganic Waste Constituents in Compacted Clay," Journal of Contaminant Hydrology, Vol. 4, pp. 241-273.

Shan, H. Y., 1990, "Laboratory Tests on a Bentonitic Blanket," M. S. Thesis, The University of Texas at Austin, Austin Texas.

Slaughter, M. and Malne, I. H., 1960, "The Formation of Chlorite-Like Structures from Montmorillonite," Clays and Clay Minerals, Vol. 5, pp 114-124.

Srinivansan, K. R. and Fogler, H. S., 1990, "Use of Inorganic-Organo-Clays in the Removal of Priority Pollutants from Industrial

Wastewaters: Structural Aspects," *Clays and Clay Minerals*, Vol. 38, pp. 277-286.

Stevenson, F. J., 1982, *Humus Chemistry - Genesis, Composition, and Reactions*, John Wiley and Sons, New York.

Stumm, W. and Morgan, J. J., 1981, *Aquatic Chemistry*, 2nd ed., John Wiley and Sons, New York.

Turner, R. C., 1965, "Some Properties of Aluminium Hydroxide Precipitated in the Presence of Clays," *Canadian Journal of Soil Science*, Vol. 45, pp. 331-336.

U. S. Environmental Protection Agency, 1989, "Requirements for Hazardous Waste Landfill Design, Construction, and Closure," Seminar Publication Report No.EPA/625/4-89/022, Cincinnati, Ohio.

Alan J. Lutenegger[1] and Don J. DeGroot[2]

HYDRAULIC CONDUCTIVITY OF BOREHOLE SEALANTS

REFERENCE: Lutenegger, A. J., and DeGroot, D. J., "Hydraulic Conductivity of Borehole Sealants," Hydraulic Conductivity and Waste Contaminant Transport in Soil, ASTM STP 1142, David E. Daniel and Stephen J. Trautwein, Eds., American Society for Testing and Materials, Philadelphia, 1994.

ABSTRACT: Geotechnical boreholes provide a direct conduit for the transport of surface and subsurface contaminants if left open or improperly sealed at the end of drilling operations. Current field practices used during the decommissioning of a drilled hole generally rely on a relatively small group of products to act as a sealant or barrier to contaminant transport. Slurries using mixtures of bentonite, portland cement and other materials and compacted barriers composed of native soil and other materials have been used as contaminant transport barriers in applications such as slurry trenches and compacted liners and may be well suited as workable borehole sealants. A literature review provides a summary of hydrologic properties of various materials which might have applications as borehole sealants. This summary is presented in a concise format along with a brief discussion of each of the classes of mixtures to act as permanent seals. In addition to this review, the results of laboratory hydraulic conductivity tests performed on a range of commercially processed bentonite products sold as seal materials are presented. Physical characteristics of various commercial bentonite products are also presented. Results of laboratory hydraulic conductivity tests performed on other typical sealant mixtures are also presented. The results of this study identify a number of alternatives for providing borehole seals to act as contaminant transport barriers in boreholes.

KEYWORDS: borehole sealant, hydraulic barriers, bentonite, hydraulic conductivity, grout

Exploratory boreholes for geotechnical and groundwater investigations create an open conduit for movement of surface or subsurface contaminants through the stratigraphic column. As a result, boreholes pose a threat to the integrity of the natural groundwater system and therefore need to be provided with a hydraulic barrier to eliminate the potential for crosscontamination between aquifers from both advective flow and diffusion. This problem is currently receiving the attention of ASTM Subcommittee D18.21 which is in the process of developing a Standard Guide for decommissioning of groundwater wells and boreholes. This issue is being driven in part by local state and federal agencies which often require some form of permanent abandonment for exploration boreholes, often using practices derived from the oil, gas or water well industry. Because of the wide range in diversity in drilling techniques, groundwater

---

[1]Associate Professor of Civil Engineering, University of Massachusetts, Amherst, MA.

[2]Assistant Professor of Civil Engineering, University of Massachusetts, Amherst, Ma.

conditions, geologic conditions and other local variables, one should expect a wide range in materials and construction procedures used to create a borehole seal to act as a contaminant transport barrier. Examples of seals used for geotechnical or groundwater monitoring instrumentation involving more common sealants have been described by a number of investigators (e.g., Fetzer 1982; Yurchick 1987; Dunnicliff 1988; Riewe 1991; Nielsen 1991). However, in other areas of groundwater protection such as waste encapsulation, a much wider range of materials have been used. This paper presents a review of the potential materials which may have application as hydraulic barriers for seals in borehole to protect the subsurface environment. A description of the recommended criteria which the seal should meet in order to be effective at reducing contaminant transport is also given. The results of an investigation conducted to evaluate hydraulic conductivity of the more viable seals is presented.

SEALING VERSUS BACKFILLING

There is a clear distinction which can be made between the practices of backfilling and sealing an open borehole. Backfilling refers to the practices of placing drill cuttings or other materials into the borehole as a part of clean up and surface restoration. The primary focus is on disposal of cuttings. Sealing is the conscious effort to construct a positive permanent ground water flow barrier to restrict or prohibit the movement of surface and/or subsurface movement of ground water. The term sealing implicitly implies a success has been made in the effort, however, at the present time, there is no reliable means of verifying whether a "seal" really is successful.

BOREHOLE SEALING CRITERIA

Borehole seals should meet a number of specific criteria in order to adequately protect the subsurface environment from contaminant transport. What is really needed is hydrologic restoration of the subsurface stratigraphic column to a quality which is as good as or better than the pre-exploratory condition. In effect this means adequately plugging the conduit created by the test drilling. This plugging must take into account both hydraulic properties and construction procedure in order to be successful. The authors recommend that the following criteria should be considered in the selection of a borehole sealant.

Hydraulic Conductivity

The seal should have a value of hydraulic conductivity lower then the native soil with a practical lower bound value of 10^{-7} cm/s to reduce contaminant movement by advection.

The primary objective of plugging an open hole created by a subsurface investigation is to essentially return the groundwater flow conditions to a state which reflect original behavior. This implies that the most important soil property which must be restored after the investigation is the hydraulic conductivity.

Because hydraulic conductivity varies over at least seven orders of magnitude (from 10^{-2} to 10^{-9} cm/s) in natural geologic deposits, one might expect that seal hydraulic conductivity should also vary over this range, however, this may not necessarily be true. Because contaminant flow will occur in surrounding natural soil of higher hydraulic conductivity, the practical upper value of required hydraulic conductivity of the seal is likely to be on the order of 10^{-4} cm/s which would reflect a typical value for a silty sand. In order to provide for a reasonable margin of safety, it may be desirable to have a seal with a hydraulic conductivity that is lower than the native soil. This will insure flow around the seal and therefore the properties of the seal would not control contaminant movement.

A lower bound value of hydraulic conductivity of 1×10^{-7} cm/s is suggested in order to be compatible with the current upper bound allowable U.S.E.P.A. guidelines for compacted soil liners and final covers for solid waste landfill design.

Stability

The seal material must have sufficient intrinsic structural integrity to prevent seal loss into the native host soil.

The seal material should not be subject to loss into a formation as a result of excessive localized hydraulic gradients or other conditions. Internal erosion from piping, such as occurs with dispersive soils or loss of low solids content bentonite slurries into highly porous formations would render the seal ineffective. In these cases, an alternate seal material that may be more appropriate should be selected.

Compatibility

The seal material should be compatible with the native host soil in order to provide a satisfactory bond at the soil/seal interface.

In addition to eliminating or restricting hydraulic flow through the seal itself, restoration of the vertical column must give consideration to the interface between the native soil and the seal material placed in the hole. This is necessary to prevent hydraulic short circuit or flow of contaminants through an interface zone which may develop between the seal and native material. A short circuit will occur if no bond develops at the interface so that an open channel forms at the interface boundary, such as due to a seal shrinkage.

Compatibility occurs if the same or similar material is used to construct the seal, i.e., if native soil cuttings left over from the test drilling are used to construct the seal. The successful performance of the seal would then be contingent on other factors, but still may include development of a water tight interface. However, practical considerations may still prohibit the use of native soil cuttings from being used as a seal. If drilling proceeds through a zone of contaminated materials and cuttings commingle at the surface, in most cases it would not be possible to use the cuttings back down the hole. Also, since some soil is normally removed during the test drilling, there is usually a net soil volume loss even if the soil is replaced in the hole at the some voids ratio. This requires backfill to be imported onto the site. If compaction or tamping is used to construct the seal then a significant loss of volume may occur, also requiring importing of an additional soil.

Similar consideration should be given to any seal materials used such as Portland Cement or bentonite to form a water tight bond at the borehole walls.

Instrumentation Compatibility

The seal material should be compatible with instrumentation hardware in order to provide a satisfactory bond at the instrument/seal interface.

Holes which are used for installation of instruments for geotechnical or geoenvironmental projects present an additional problem of a potential hydraulic short circuit developing at the interface between the instrumentation material and the seal material. Instrumentation may involve a wide range of construction materials such as teflon, PVC, steel, stainless steel, etc., therefore the seal must be able to form a tight bond with a number of materials at the interface similar to the sidewall interface. By precluding a hydraulic short circuit at either the hole sidewall or instrumentation interface, successful performance of the seal is more heavily dependent on the actual properties of the seal, especially hydraulic conductivity.

Length

The seal should be of sufficient length to reduce movement of contaminants to an acceptable level by advection and reduce contaminant migration by diffusion.

In addition to eliminating movement of potential contaminants by the normal process of advection, diffusion of contaminants may also take place. In order to reduce diffusion it is necessary to design a seal which acts essentially the same as the surrounding native soil. A minimum design seal length is also necessary to provide a sufficient margin of safety in the event that the entire volume of seal does not perform satisfactorily. That is, if one portion of the seal develops a failure along its length, there will be sufficient additional length of seal to preserve the overall integrity.

Longevity

The seal should be considered a permanent repair.

A borehole seal used to protect the subsurface environment from contaminant movement should have an intended useful life of infinity; that is, the repair to the subsurface hydrologic integrity should be considered as permanent.

In order to perform satisfactorily, this implies that a hole seal should have properties which do not degrade with time or under conditions of reasonable environmental change. Deterioration of seal materials by either changing hydrologic conditions, such as internal erosion leading to piping resulting from an elevated hydraulic gradient, or by environmental changes, such as changes in groundwater chemistry, would be undesirable. This means that seal materials must be selected which will be inert to such changes.

Practical Considerations

In addition to the criteria discussed above, there are a number of practical considerations that should be taken into account relative to the selection of a borehole seal.

Constructability--An important practical consideration for normal size and length boreholes used in routine site investigations is a seal which is easy to construct using materials which are easily handled by drillers and drillers helpers and seal "recipes" which allow some degree of flexibility in preparation without compromising the final integrity. Simplicity is preferred over complexity and alternative methods are desirable to allow simple modifications in the field should unforeseen problems occur. The integrity of the seal should not be solely dependant on the quality of the construction.

Seal Location--There is an obvious practical consideration on where the seal should be placed. In most cases seals will need to be placed in stratigraphic units which acted as groundwater protection layers prior to the test drilling. This normally means that seals need to be placed in aquitards or aquicludes above the aquifers which are being protected and not in the aquifer itself. However, as a matter of practicality it is probably easiest to place a seal over the entire length of the borehole rather than attempt to construct a seal in an isolated zone. By using a full length seal, problems associated with basal support of an isolated seal, and diffusion will be eliminated.

Cost--The construction of a positive, permanent hydraulic barrier in a borehole will result in additional costs to an exploration program reflecting both added material and labor costs. A compromise is probably needed to balance material vs. placement costs.

Availability--There is an obvious desire to use seal materials which are readily available under normal drilling constraints or that can be obtained under normal conditions. Even the most technically acceptable seal which more than exceeds minimum criteria will have limited applicability if it can not be obtained with relative ease.
The success of a particular seal system is a function of both the properties of the seal and the quality of the construction practice used in placing the seal. Some seal materials may have sufficiently low intrinsic hydraulic characteristics, but get improperly placed into a borehole such that large voids are present which allow flow around the seal. Similarly, seals which do not adequately create a water tight bond at interfaces, such as at the borehole wall or at an instrument casing will result in a failure. While all of the stated criteria should be evaluated for their appropriateness for a given situation, this paper examines the hydraulic characteristics of different common seal materials.

REVIEW OF CONTAMINANT BARRIERS

Contaminant transport barriers to reduce the movement of liquid contaminants have in the past been constructed of a wide range of materials. While the most common materials used to create barriers of low hydraulic conductivity typically consist in part of either bentonite clay, Portland Cement, or a combination of these materials and native soils, a number of other materials have been used on a more limited basis. This trend may continue as applications are sought for the use of waste by-products as additives for hydraulic barriers. For example, suggestions have been made to use sewage sludge (Shakoor and Schmidt 1989) and paper mill sludge (Aloisi and Atkinson 1991) as cover material for landfills. The most common types of barriers typically consist of compacted materials which are mixed and constructed in-place such as with a compacted soil liner or consist of a slurry which is mixed externally and placed by dumping, such as with a slurry trench. Other types of seals such as bentonite chips or pellets are either dry poured into place and then "activated" by adding water to initiate hydration, premixed with water prior to placing in the hole or are allowed to hydrate in situ from natural ground water.

Compacted Barriers

Compacted barriers for sealing boreholes are constructed in place and are limited to drilling conditions which leave an open borehole and where there is little or no standing water in the hole. Materials which have been used as compacted barriers for both solid waste or lagoon liners and covers may also have applications for sealing boreholes. Such materials range from compacted native soil to soil admixed with a wide range of additives, such as Portland cement, powdered bentonite, lime, fly ash, etc. A summary of reported values of hydraulic conductivity for a number of compacted barriers is given in Table 1.

Compacted Soil--Compacted native soil cuttings may provide adequate hydraulic barriers if conditioned properly and if properly placed. The influence of compactive effort, water content, clod size, soil type, and other variables on the resulting compacted hydraulic conductivity is well documented in the literature. However, the resulting parameters will be site specific and dependent on the native soil characteristics. The main drawback to using compacted native soil cuttings as a borehole sealant is that the integrity of the seal is more dependent on the seal construction than on the seal material. Construction is seriously hindered by length and size of the borehole and the ability to properly place material in "lifts" for proper compaction. The hydraulic conductivity of compacted soil cuttings depends on soil grain-size distribution, plasticity, water content, compaction effort and clod size. Since such a wide range of hydraulic conductivities are possible (10^{-2} - 10^{-10} cm/sec.) the use of compacted soil as a borehole sealant depends primary on the quality of the placement. Normally, significant compaction effort will be required to achieve low hydraulic conductivity ($< 10^{-7}$ cm/sec).

TABLE 1--Reported Hydraulic Conductivity of Compacted Barriers.

Seal	Hydraulic Conductivity (cm/s)	Reference
Soil Cement	$6.5 \times 10^{-8} - 1.5 \times 10^{-7}$	Haxo et al. 1985
Fly Ash	$1.5 \times 10^{-5} - 7.2 \times 10^{-6}$	Bowders et al. 1987
	$1.4 \times 10^{-8} - 7.5 \times 10^{-10}$	Edil et al. 1987
	1.0×10^{-7}	Creek & Shackleford 1992
Fly Ash + Bentonite	$3.2 \times 10^{-8} - 1.9 \times 10^{-9}$	Edil et al. 1987
Soil + Bentonite	$2 \times 10^{-4} - 5 \times 10^{-9}$	Chapuis 1981
	$3 \times 10^{-4} - 1 \times 10^{-9}$	Lundgren 1981
	$3.0 \times 10^{-7} - 1.4 \times 10^{-10}$	Haile 1985
	$1.6 \times 10^{-7} - 1.4 \times 10^{-9}$	Gipson 1985
	$2.5 \times 10^{-9} - 1.2 \times 10^{-10}$	Pusch & Alstermark 1985
	$2 \times 10^{-4} - 8 \times 10^{-9}$	Holopainen 1985
	$2 \times 10^{-6} - 7 \times 10^{-8}$	Jessperger et al. 1985
	$5.0 \times 10^{-7} - 8.0 \times 10^{-10}$	Haxo et al. 1985
	$1.0 \times 10^{-7} - 6.3 \times 10^{-8}$	Haug 1985
	$2.7 \times 10^{-5} - 1.6 \times 10^{-9}$	Garlanger et al. 1987
	$1 \times 10^{-3} - 1 \times 10^{-10}$	Kenney et al. 1992
	$2.7 \times 10^{-6} - 2.9 \times 10^{-7}$	Stockmeyer 1992
Fly Ash + Soil	$5 \times 10^{-4} - 5 \times 10^{-7}$	Parker et al. 1977
	$1.4 \times 10^{-7} - 1.6 \times 10^{-8}$	Edil et al. 1987
	$6.4 \times 10^{-4} - 6.4 \times 10^{-7}$	Creek & Shackleford 1992
Fly Ash + Cement	$1.6 \times 10^{-6} - 6.6 \times 10^{-6}$	Bowders et al. 1987
Fly Ash + Lime	$2.2 \times 10^{-6} - 7.1 \times 10^{-7}$	Bowders et al. 1987
Fly Ash + Sand + Bentonite	3.5×10^{-8}	Haug 1985
	$1.0 \times 10^{-7} - 3.9 \times 10^{-8}$	Edil et al. 1987
	$2.1 \times 10^{-4} - 4.5 \times 10^{-7}$	Creek & Shackleford 1992

Compacted Soil + Additives--Admixtures to native soil to produce hydraulic barriers are used to enhance workability, condition the grain-size in order to produce a material with better packing capabilities and to enhance strength. Since the ultimate goal of a borehole seal is to produce a material of lower hydraulic conductivity, one of the most common additives used in conditioning soil for compacted soil liners is powdered bentonite. Numerous studies have shown that nearly all soils may be conditioned with powdered bentonite to produce a low hydraulic conductivity barrier. The amount of bentonite needed to obtain a desired value of hydraulic conductivity is dependent on the original grain-size distribution of the native soil. The addition of sufficient bentonite can even reduce the need for compaction, in effect producing a seal which is material dependent and not placement dependent. Provided the material can be placed adequately in the hole, and sufficient water is available the seal will perform satisfactorily as a result of in-place hydration of the bentonite. Other additives have been used to condition soil to produce low hydraulic conductivity barriers and include the use of Portland Cement to produce soil/cement; fly ash, to produce soil/fly ash and other combinations such as soil/cement/bentonite or soil/cement/fly ash.

Other Compacted Barriers--A number of other innovative materials have been used to create compacted hydraulic barriers and include: compacted bentonite, fly ash, fly ash + bentonite, fly ash + cement and fly ash + lime. The use of fly ash is primarily a replacement for portland cement. The resulting hydraulic conductivities of these materials, as shown in Table 1, range from about 10^{-5} to 10^{-10} cm/sec.

Slurry Barriers

The use of slurries as hydraulic barriers is a well known and established technology for waste containment, e.g., slurry walls. Slurries may be composed of a number of materials that produce a liquid grout which normally will gel to produce some strength with time. Materials have included native soil mixed with bentonite, bentonite, Portland Cement, fly ash, Portland Cement concrete and combinations of these and other materials. A summary of reported hydraulic conductivities of different slurries is presented in Table 2. In order to adequately perform as a borehole seal, most slurries must be placed full depth in the borehole, and usually will be installed from the bottom of the hole using a tremie pipe.

Soil/Bentonite--Soil/Bentonite slurries are commonly used as hydraulic barriers in cutoff walls (Ryan 1987). The conductivity of soil/bentonite slurries depends on the amount of bentonite in the mix, the initial grain-size distribution and plasticity of the soil and the water content of the mix as clearly illustrated by D'Appolonia (1980) and summarized more recently by Ryan (1987). A number of cases in which slurries made of bentonite with sand as an additive to act as a filler have also been reported.

Bentonite--The hydraulic conductivity of bentonite/water slurries is related to the placement void ratio which may vary over a wide range as the amount of bentonite in the slurry (solids content) varies. The hydraulic conductivity of bentonite slurries can easily vary over 5 orders of magnitude as shown in Table 2. However, since a number of bentonite producers suggest placement of bentonite grout in the range of 20 to 40% solids content, the resulting range in hydraulic conductivity would be expected to be much smaller.

Cement/Bentonite--The use of cement/bentonite slurries has also been shown to be an alternative to soil/bentonite for cutoff walls (Ryan 1987) and may also be suitable as a borehole sealant. Hydraulic conductivities of cement/bentonite slurries tend to be slightly higher than soil/bentonite slurries except for very low bentonite content and show some dependence on the water/cement ratio of the mix (Ryan 1987). The use of neat cement with a small amount of bentonite is a common seal for water wells. The bentonite is used to counteract the potential for shrinkage of the cement.

TABLE 2--Reported Hydraulic Conductivity of Slurries (Grouts).

Seal	Hydraulic Conductivity (cm/s)	Reference
Bentonite	5×10^{-8} - 3×10^{-12}	Mesri & Olson 1971
	1.5×10^{-12} - 8×10^{-12}	Pusch 1982
	3.4×10^{-6} - 4.7×10^{-6}	Alther et al. 1985
	1×10^{-7} - 1×10^{-10}	Kenney et al. 1992
Soil + Bentonite	2×10^{-5} - 5×10^{-9}	D'Appolonia 1980
	2.0×10^{-7} - 2.5×10^{-8}	Schulze et al. 1984
	7×10^{-6} - 8×10^{-9}	Ryan 1987
	2×10^{-3} - 1×10^{-9}	Chapius 1990
	5.5×10^{-6} - 5.0×10^{-7}	Edil & Muhannan 1992
Neat Cement	1.2×10^{-10} - 5.0×10^{-12}	Littlejohn 1982
	3.6×10^{-11} - 8.9×10^{-11}	Banthia & Mindess 1989a
	3.1×10^{-8} - 8.6×10^{-12}	Banthia & Mindess 1989b
Fly Ash	6.5×10^{-5} - 3.0×10^{-6}	Parker et al. 1977
Plastic Concrete	1.8×10^{-6} - 5.0×10^{-8}	Evans et al. 1987
Portland Cement + Bentonite + Fly Ash	5×10^{-6} - 5×10^{-7}	Jefferis 1981
	3.6×10^{-6} - 9.0×10^{-7}	Evans et al. 1987
Cement + Bentonite	1×10^{-7} - 1×10^{-8}	Filho 1976
	5.0×10^{-5} - 6.0×10^{-7}	Logani & Kleimer 1983
	3.2×10^{-6} - 9.0×10^{-7}	Chapius et al. 1984
	9×10^{-6} - 6×10^{-7}	Gill and Christopher 1984
	6×10^{-5} - 2×10^{-7}	Ryan 1987
Cement + Sand + Bentonite	2×10^{-5} - 2×10^{-6}	Gill & Christopher 1984
	1.2×10^{-6}	Chapius et al. 1984

Neat Cement--Neat cement seals have been used for many years to seal water, oil and gas wells. The hydraulic conductivity of neat cement is reported to be very low and is primarily a function of the water/cement ratio and the curing time; both of which control gel formation. One problem which may arise with the use of neat cement is the tendency for shrinkage to occur if free water is removed from around the seal, such as with a fluctuating groundwater table. Kurt and Johnson (1982) indicated hydraulic conductivities on the order of 10^{-5} cm/sec for neat cement seals around PVC casing at low confining pressures. These values are likely influenced by some shrinkage of the grout away from the casing during curing.

Other Slurry Barriers--Other slurries have been used to produce low hydraulic conductivity barriers and include; fly ash, fly ash + cement + bentonite, plastic concrete, Portland Cement concrete, and asphaltic concrete.

LABORATORY INVESTIGATION

An investigation was performed to determine physical, geochemical and hydraulic properties of the more common viable seal materials. Tests were performed primarily on commercial bentonite pellets and chips, bentonite slurries and Portland Cement-based seals but also included compacted soil + bentonite.

Procedures

Characterization--Laboratory characterization tests were conducted on a number of bentonite products currently available on the market and used in the field to provide hydraulic seals for boreholes or isolation seals for piezometers. One of the objectives of this part of the work was to test as many products as possible to determine the typical range of characteristics of "bentonite" which is commercially available. In general the authors found that there was a lack of this information in the literature. Pelletized and granular bentonite products were hand pulverized to pass a No. 40 sieve, while bentonite grouts were tested as supplied. Grain-size distribution was performed using the hydrometer method as described in ASTM D422-63 "Standard Test Method for Particle Size Analysis of Soils" using sodium hexametaphosphate as the dispersing agent with sand content determined by wet sieving through a No. 200 sieve following the hydrometer test. Liquid and Plastic Limits were determined as described in ASTM D4318-84 "Standard Test Method for Liquid Limit Plastic Limit and Plasticity Index of Soils." Carbonate content was measured using the Chittick apparatus described by Dreimanis (1962).

Hydraulic Conductivity--Laboratory hydraulic conductivity tests were performed as constant head tests using clear plexiglass rigid wall permeameters of 101 mm (4 in.) diameter and 101 mm (4 in.) length. Each end of the permeameter was capped with a clear plexiglass end plate which included a sintered bronze porous stone and a piece of Whatman No.5 filter paper to contact the ends of the specimen and prohibit vertical swell while allowing drainage control at both ends of the specimen. Specimens were permeated from the bottom of the specimen, with the outflow line vented to the atmosphere. A standard laboratory permeant consisting of a de-aired solution of 0.005 N calcium sulfate in distilled water was used. A constant head was applied to each specimen with pressurized air to the surface of the inflow permeant column. Because it was anticipated that the hydraulic conductivity of the seals would be very low, tests were performed for a period of approximately 30 days under a hydraulic gradient of 50. While a gradient of 50 is probably unrealistically high in relation to most field conditions this value was chosen to expedite testing and provide a constant value for all tests.

Bentonite chips and pellets were placed into permeameter molds at as-received water content by pouring into the mold. The top cap was then placed on the mold to create a vertical confined swell condition and permeant introduced to allow swelling. Specimens were allowed to hydrate overnight prior to initiation of flow.

An additional test series was performed in which the bentonite chip and pellet products were allowed to hydrate in the permeameter mold under vertical free swell conditions without the top cap in place. After hydration for 24 hours, the amount of bentonite expanded over the top of the mold was cut off and the top cap was then replaced in order to initiate flow. This test series was used to provide a contrasting placement technique to the confined swell tests.

Mixtures of soil and powdered bentonite were mixed by hand to a desired proportion and then brought to a desired water content. Specimens were compacted directly into permeameter molds using a Standard Proctor Hammer. The soil used in this study was a low plasticity (ML) lacustrine varved deposit (PI = 11; < 0.002 mm = 35%) which was air dried, mixed and processed through a No. 4 sieve.

Grouts composed of either bentonite or Portland cement were mixed to a desired proportion with tap water in a mechanical mixer for 20 minutes and then poured directly into permeameter molds after first obtaining a measure of temperature, Marsh Funnel viscosity (API Standard RP 13B-1 (API 1990)) and a measure of mud weight using a mud balance.

RESULTS

Characterization

The results of laboratory characterization tests performed on bentonite products are presented in Table 3a and 3b. A summary of the consistency of the various products is shown on the Casagrande plasticity chart in Figure 1. As indicated, all of the materials show consistency limits which essentially follow the U-line. With only one or two exceptions, all of the processed products (pellets & tablets) fit in a relatively narrow range in the middle of the chart. By contrast, prepackaged "grouts" show much lower plasticity and fall on the lower end of the chart. "Pure" powdered bentonites fall at the top of the chart with the highest plasticity. The results presented in Table 3 indicate that all bentonite products are not the same; i.e., wide variations in plasticity, grain-size distribution and other constituents (e.g. carbonate content) occur throughout the spectrum of products. For example, the clay fraction (< .002 mm) varies from about 55% to 90% while the coarse fraction (> 0.074 mm) varies from about 1% to 9%. This variation in composition is clearly shown in the activity chart presented in Figure 2. While most of the products fall in the general range of activity = 6 which is typical for bentonite, the "grouts" all fall below activity = 4 with some as low as 1.2. The variations in basic compositional properties of these materials is likely related to geologic differences in the formation of a specific bentonite bed resulting from differences in depositional environments. Variations also occur as a result of proprietary processing additives introduced by various manufacturers. These results suggest that in certain situations, it may not be sufficient to simple specify the use of "bentonite" as a component of a seal, but rather rely on a performance criteria as part of the quality control verification. The results presented in Table 3 are consistent with other reported properties of bentonite (e.g., Mesri and Olson 1971; Edil and Erickson 1985; Acar et al. 1985; Kenny et al. 1992.)

Hydraulic Conductivity

<u>Compacted Soil + Bentonite</u>--To illustrate the influence of added powdered bentonite (Pure Gold Grout) content and placement compactive effort on the resulting hydraulic conductivity of a potential borehole seal, the results of laboratory tests using a silty clay (ML) soil are shown in Table 4. As is indicated, low conductivity (< 10^{-7} cm/sec.) seals can be achieved at low bentonite content provided that some compaction is used. Similarly, provided that a sufficient amount of bentonite can be added to the soil, a low hydraulic conductivity seal can be achieved with no compactive effort. These results demonstrate that the performance of seals may be either <u>placement dependent</u> or <u>material dependent</u>. In terms of practicality, an ideal seal should not derive its integrity from placement. That is, since it will be difficult to provide adequate quality control on seal placement, a seal material should be chosen which will perform even if the placement is not properly achieved.

<u>Bentonite Pellets and Chips</u>--Results of laboratory hydraulic conductivity tests conducted on commercial bentonite pellets and chips are presented in Table 5. As can be seen, the void ratio of the confined swell specimens all fall within a relatively narrow range and is on the order of 1.5. Similarly, the results of calculated values of hydraulic conductivity obtained at the end of the 30 day test period all fall within a narrow range and are typically on the order of 5×10^{-10} to 1×10^{-9} cm/sec. These

TABLE 3a--Physical and Chemical Properties of Bentonite Grouts and Gels.

Supplier	Trade Name	LL (%)	PL (%)	PI (%)	Activity	Sand (%)	Silt (%)	Clay (%)	Calcite (%)	Dolomite (%)
ACC	Bentogrout	136	40	96	1.2	3.8	16.2	80.0	1.2	0.7
ACC	High Solids Volclay Grout	330	51	279	3.4	2.7	14.4	82.9	2.9	1.2
ACC	Pure Gold Gel	613	49	564	6.5	1.8	11.0	87.2	1.9	1.4
ACC	Pure Gold Grout	190	46	144	2.5	6.0	35.3	58.7	3.7	1.3
BDF	Aquagel Gold Seal	641	50	591	7.3	2.1	17.2	80.7	8.3	1.5
BDF	Quik-Gel	612	49	563	7.0	2.0	17.6	80.4	3.2	2.0
BHBC	BH Grout	172	44	128	2.3	5.6	39.2	55.2	7.9	3.0
CETC	Super Gel-X	641	48	593	7.1	1.6	15.3	83.1	1.5	0.4
PDSC	PDSC Grout	198	38	160	2.8	5.1	37.4	57.5	2.9	2.9
WB	Enviroplug Grout	233	30	203	2.8	6.6	21.5	71.9	2.6	1.5

TABLE 3b--Physical and Chemical Properties of Bentonite Chips and Pellets.

Supplier	Trade Name	LL (%)	PL (%)	PI (%)	Activity	Sand (%)	Silt (%)	Clay (%)	Calcite (%)	Dolomite (%)
ACC	Pure Gold Tablets 1/4"	458	52	406	4.5	2.3	8.3	89.4	1.8	1.5
ACC	Pure Gold Medium Chips 3/8"	364	34	330	4.3	5.3	17.4	77.3	2.5	1.6
ACC	Volplug Coarse Chips 3/4"	396	48	348	5.4	8.1	27.4	64.5	7.2	5.8
BDF	Benseal Chips 1/8"	566	59	507	5.8	1.3	11.7	87.0	6.7	2.6
BDF	Holeplug Chips 3/8"	425	44	381	5.5	3.7	27.6	68.7	8.5	1.3
BDF	Baroid Pellets 1/4"	625	42	583	6.4	2.0	6.3	91.7	2.5	1.8
BDF	Baroid Pellets 3/8"	573	35	538	6.0	3.7	6.9	89.4	2.5	2.4
BDF	Baroid Pellets 1/2"	620	38	582	6.4	3.6	6.0	90.4	2.3	2.3
BHBC	Bentonite Plug #8	492	34	458	6.5	8.7	21.0	70.3	0.4	2.5

TABLE 3b cont'd.--Physical and Chemical Properties of Bentonite Chips and Pellets.

Supplier	Trade Name	LL (%)	PL (%)	PI (%)	Activity	Sand (%)	Silt (%)	Clay (%)	Calcite (%)	Dolomite (%)
BHBC	Bentonite Plug 3/8"	428	42	386	5.9	3.9	30.6	65.5	2.6	2.8
BHBC	Bentonite Plug 3/4"	452	41	411	5.9	6.1	24.3	69.7	4.2	2.1
BK	Geo Pellets 1/4"	508	47	461	6.1	3.0	21.2	75.8	0.8	2.8
BK	Geo Pellets 3/8"	533	68	465	5.4	3.2	10.6	86.2	2.5	1.6
BK	Geo Pellets 1/2"	555	67	488	6.1	3.1	17.5	79.4	3.2	2.4
PR&D	Bentonite PI Pellets 1/4"	571	54	517	6.8	3.3	20.5	76.2	6.9	1.2
PR&D	Bentonite PI Pellets 3/8"	541	72	469	5.8	2.1	16.6	81.3	6.6	1.8
PR&D	Non-Stick Bentonite PI Pellets 3/8"	623	80	543	7.1	1.1	22.3	76.6	6.2	1.5
PR&D	Bentonite PI Pellets 1/2"	503	65	438	4.7	0.7	5.9	93.4	5.1	2.0
PDSC	Pel Plug Pellets 1/4"	569	62	507	6.6	2.0	19.6	78.4	2.3	2.2
PDSC	Pel Plug Pellets 3/8"	507	54	453	5.2	2.2	10.1	87.7	1.5	1.5
PDSC	Pel Plug 3/8" Pellets with TR30	466	57	409	4.9	4.0	13.3	82.7	2.3	2.9
PDSC	Pel Plug Pellets 1/2"	539	69	470	6.1	2.2	20.3	77.5	2.4	2.8
RT	Peltonite Pellets 3/8"	573	56	517	5.7	2.0	7.8	90.2	1.2	0.3
WB	Enviroplug Medium Chips	443	42	401	4.7	2.7	12.6	84.7	3.7	3.5
WB	Enviroplug Coarse Chips	250	33	217	3.8	7.1	35.7	57.2	1.2	1.7
WB	Enviroplug Tablets 1/4"	495	45	450	5.8	4.7	17.2	78.1	1.4	1.9
WB	Enviroplug Tablets 3/8"	459	47	412	5.4	4.3	20.1	75.6	1.3	1.6
WB	Enviroplug Tablets 1/2"	484	46	438	5.4	4.4	14.8	80.8	1.4	1.5

ACC = American Colloid Co.
BDF = Baroid Drilling Fluids, Inc.
BHBC = Black Hills Bentonite Co.
BK = Brainard-Kilman Geostore
CETC = Colloid Environmental Technologies Co.
PR&D = Piezometer Research and Development
PDSC = Polymer Drilling Systems Co.
RT = Roctest, Inc.
WB = WyoBen, Inc.

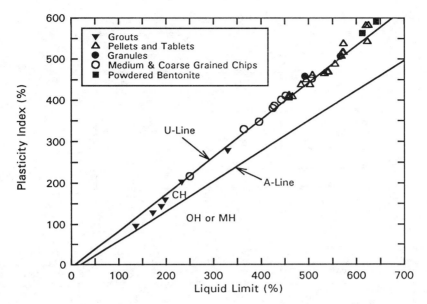

FIG. 1--Plasticity chart of bentonite products.

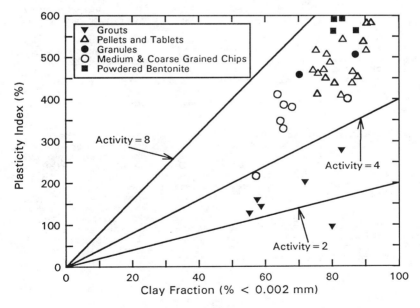

FIG. 2--Activity chart of bentonite products.

TABLE 4--Hydraulic Conductivities of Soil/Bentonite Mixes.

% Bentonite	Compactive Effort (blows/layer)	Hydraulic Conductivity (cm/s)
2	0	4.3×10^{-3}
2	5	5.4×10^{-8}
6	0	3.3×10^{-6}
6	5	4.3×10^{-9}
18	0	1.1×10^{-7}
18	5	1.1×10^{-9}

values are about one order of magnitude higher than values reported by both Mesri & Olson, (1971) and Kenny et al. (1992) for bentonite slurries at similar void ratios. This may be in part related to the fact that not all specimens reached steady-state conditions in the thirty day test period. There is also visual evidence that indicated that because of the "granular" nature of the products, all of the bentonite did not fully hydrate during the test period. These results indicate that provided the bentonite can be placed onto a borehole and remains hydrated, the resulting hydraulic conductivities fall below the recommended criteria of 10^{-7} cm/sec. The results also suggest that under these test conditions there is very little difference in the performance among different suppliers.

The confined swell test results presented in Table 5 represent a best case scenario; i.e., minimum void ratio of these seals in a borehole. The results of free swell tests show that the void ratio increases to about 2.2 and there is a corresponding increase in hydraulic conductivity of about 1 order of magnitude. These results confirm that provided the chips or pellets can be successfully placed in the hole and sufficient free water is available for hydration, a high quality seal may be achieved.

Bentonite Slurries--As previously discussed the hydraulic conductivity of bentonite/water slurries is primarily dependent on the amount of solids in the mixture. In fact, seal performance may actually be dependent on the actual bentonite content of the slurry, which may vary widely for different products based on the results presented in Table 3. Results of tests performed on three commercial bentonite "grouts" prepared using typical manufacturers recommendations are presented in Figure 3. As can be seen, the results show a relatively narrow range in hydraulic conductivity which is proportional to the solids content. Even though these mixes are considered pumpable using common grout pumps encountered on conventional drill rigs, they are generally too viscous to perform a Marsh Funnel viscosity test and therefore, viscosity cannot be used for quality control. The authors have found that a more significant quality control measure is mud weight.

TABLE 5--Hydraulic Conductivity of Bentonite Chips and Pellets.

Supplier	Trade Name	Confined Swell			Free Swell		
		WC (%)	e	Hydraulic Conductivity (cm/s)	WC (%)	e	Hydraulic Conductivity (cm/s)
ACC	Pure Gold Medium Chips 3/8"	69.5	2.14	4.0×10^{-9}	87.8	2.50	4.6×10^{-9}
ACC	Volplug Coarse Chips 3/4"	50.0	1.71	7.6×10^{-10}	76.8	2.13	2.2×10^{-9}
BDF	Holeplug 3/8" Chips	60.5	1.62	8.0×10^{-10}	79.8	2.17	1.4×10^{-9}
BHBC	Bentonite Plug 3/8"	58.0	1.84	9.7×10^{-10}	81.3	2.21	1.7×10^{-9}
BHBC	Bentonite Plug 3/4"	64.4	1.98	9.7×10^{-10}	78.2	2.17	1.4×10^{-9}
BK	Geo Pellets 1/4"	50.2	1.63	2.1×10^{-9}	55.7	1.74	3.6×10^{-9}
BK	Geo Pellets 3/8"	51.9	1.33	5.0×10^{-10}	58.4	1.65	4.9×10^{-9}
BK	Geo Pellets 1/2"	50.9	1.80	2.5×10^{-9}	66.9	2.16	3.4×10^{-9}
PR&D	Bentonite PI Pellets 1/4"	43.3	1.59	5.4×10^{-10}	63.6	2.07	2.9×10^{-9}
PR&D	Bentonite PI Pellets 3/8"	47.4	1.67	4.0×10^{-10}	68.7	1.99	3.2×10^{-9}
PR&D	Non-Stick Bentonite PI Pellets 3/8"	48.7	1.71	5.4×10^{-10}	73.6	2.17	4.5×10^{-9}
PR&D	Bentonite PI Pellets 1/2"	44.6	1.81	1.8×10^{-9}	68.3	2.37	3.3×10^{-9}
PDSC	Pel Plug Size 1/4"	51.6	1.66	7.3×10^{-10}	76.7	2.39	4.0×10^{-9}
PDSC	Pel Plug Size 1/2"	55.4	1.71	7.9×10^{-10}	--	--	--
RT	Peltonite 3/8"	48.3	1.58	1.0×10^{-9}	68.2	2.06	2.8×10^{-9}
WB	Wyo-Ben Env. Coarse 3/4" Chips	61.6	1.82	2.0×10^{-9}	81.0	2.41	4.1×10^{-9}

ACC = American Colloid Co.
BDF = Baroid Drilling Fluids, Inc.
BHBC = Black Hills Bentonite Co.
BK = Brainard-Kilman Geostore

PR&D = Piezometer Research and Development
PDSC = Polymer Drilling Systems Co.
RT = Roctest, Inc.
WB = Wyo-Ben, Inc.

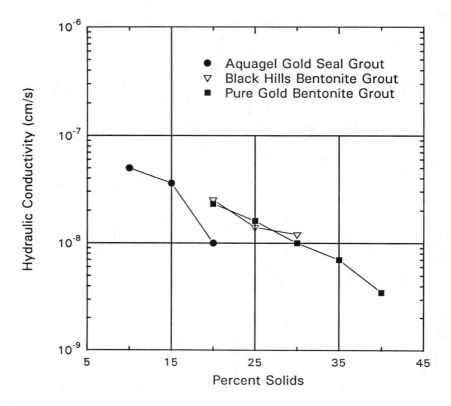

FIG. 3--Hydraulic conductivity of bentonite grouts.

Cement/Bentonite--Results of tests performed on portland cement + bentonite grouts are presented in Table 6. The mixtures cover a wide range of water/cement ratios with a constant bentonite (Pure Gold Grout) content of 5%. With the exception of the test at a water/cement ratio of 0.44 which may have experienced some experimental error, these mixtures result in hydraulic conductivities which are very low and only vary by about half an order of magnitude. No clear trend in hydraulic conductivity was noted with either varying water/cement ratio or mud weight. Even though 5% bentonite was used, the results of Table 3 clearly show that not all bentonites are the same, therefore mixtures using other products may produce different results.

TABLE 6--Hydraulic Conductivity of Cement/Bentonite Grouts.

Water/ Cement Ratio	Mud Weight (lb/gal)	Marsh Funnel Viscosity (sec/qt)	Hydraulic Conductivity (cm/s)
0.35	16.90	--*	2.0×10^{-10}
0.44	16.15	--*	5.3×10^{-9}
0.53	15.30	294	1.2×10^{-10}
0.62	14.60	60	4.2×10^{-11}
0.71	13.75	50	1.7×10^{-10}
0.80	13.50	42	4.7×10^{-11}

*Mixture too thick to perform viscosity test

SUMMARY

Materials used to construct seals in exploration boreholes following the completion of test drilling should meet a number of criteria for consideration as appropriate alternatives. The criteria presented in this paper are related to both the material characteristics and placement techniques and are suggested for consideration. Depending on the given situation the success of a seal will be related directly to some of the criteria more than others. Seals may be broadly categorized into two distinct categories relative to their potential for success as contaminant transport barriers. Material dependent seals are generally insensitive to placement and can perform satisfactorily over a wide range of placement techniques, from very poor to very good. Placement dependent seals require proper placement and generally perform satisfactorily only if great care is taken to construct the seal. In most cases, it is desirable to rely on a material dependent seal so that even under the poorest construction conditions, the seal will essentially "self seal" and perform satisfactorily.

One of the most important criteria a seal material should meet, which has been discussed in this paper, is low hydraulic conductivity. The hydraulic conductivity of potential borehole sealants can vary over several orders of magnitude. A summary of the hydraulic properties of more common seals is presented in Figure 4. While it is clear that there is considerable overlap and all of the seals shown can be considered as low conductivity hydraulic barriers, this does not imply that all seals would perform the same. As previously indicated, hydraulic conductivity is not the sole criteria for evaluating the potential success of a seal. Environmental changes in groundwater conditions or the introduction of various contaminants after seal placement may cause degradation and compromise the integrity of the seal.

CONCLUSIONS

This study has focused on identifying a number of borehole sealants for use in decommissioning exploratory boreholes. For the most part, emphasis has been placed on more commonly used materials, most of which

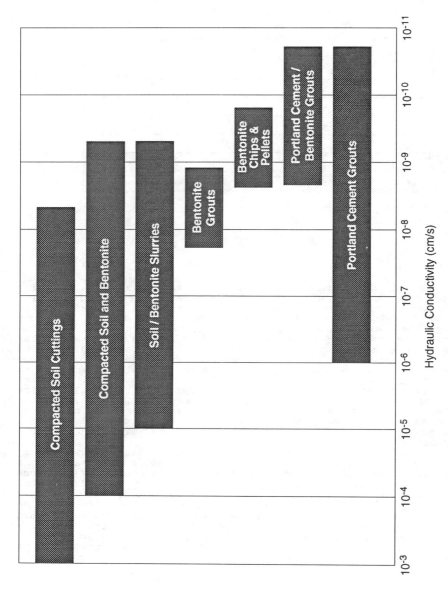

FIG. 4--Summary of hydraulic conductivity of borehole seals.

are readily available and have been used to act as contaminant transport barriers in waste isolation. The results of this work suggest that:

(1) In order to be successful, materials should meet a number of specific criteria to be considered for use as borehole sealants as suggested in this paper.
(2) Borehole seals may be either material dependent or placement dependent. Preference should be given to a seal that is material dependent; that is, the seal should be expected to perform properly even with minimal placement effort.
(3) The characteristics of commercial bentonite products vary widely which may affect their ability to be mixed together and may produce different properties when admixed with other materials.
(4) Native cuttings may be used successfully as a borehole seal provided that sufficient placement effort is used or provided sufficient bentonite is added.
(5) Commercially available bentonite chips and pellets, which are commonly used to seal boreholes have similar hydraulic characteristics and can successfully be used provided they can be placed within the hole and sufficient water is available to ensure hydration.
(6) Bentonite slurries, mixed to proportions in the range recommended by manufacturers (i.e., 20% to 40%), produce low hydraulic conductivity barriers. The hydraulic conductivity is directly related to the solids content of the mix. While viscosity measurements may be helpful in identifying subtle differences in grout mixtures at low solids content, the measurement of viscosity is limited to a relatively narrow range of mixes. Throughout the overall range of mix proportions suggested by a number of manufacturers and currently in use, it appears that mud weight may be a better indication of slurry behavior.

ACKNOWLEDGEMENTS

The work presented in this paper was supported by the National Cooperative Highway Research (NCHRP) program under project 21-4 "Sealing Exploratory Boreholes to Project the Subsurface Environment" which is being conducted in cooperation with Strata Engineering Corp. of Don Mills, Ontario. The authors gratefully acknowledge this support and the input of research team members Cam Mirza and Mike Bozozuk on this phase of the project. Much of the laboratory work presented in this paper was conducted by John Difini and Celeste Artura, graduate students at the University of Massachusetts.

REFERENCES

Acar, Y.B., Hamidon, A., Field, S.D., and Scott, L., 1985, "The Effects of Organic Fluids on the Hydraulic Conductivity of Compacted Kaolinite," Hydraulic Barriers in Soil and Rock, ASTM STP 874, pp. 171-187.

Adaska, W.S., 1985, "Soil-Cement Liners," Hydraulic Barriers in Soil and Rock, ASTM STP 874, pp. 299-313.

Aloisi, W., and Atkinson, D., 1991, "Evaluation of Paper Mill Sludge for Landfill Capping Material," Journal of the New England Water Pollution Control Association, Vol. 25, No. 2, pp. 126-140.

Alther, G., Evans, J.C., Fang, H.-Y., and Witmer, K., 1985, "Influence of Inorganic Permeants upon the Permeability of Bentonite,"

Hydraulic Barriers in Soil and Rock, ASTM STP 874, pp. 64-73.

API Standard RP13B-1: Recommended Practice Standard Procedure for Field Testing Water-Based Drilling Fluids, American Petroleum Institute, 1990.

Banthia, N. and Mindess, S., 1989a, "Water Permeability of Cement Paste," Cement and Concrete Research, Vol. 19, pp. 727-736.

Banthia, N. and Mindess, S. 1989b, "Effects of Early Freezing on Permeability of Cement Paste," Journal of Materials in Civil Engineering, ASCE, Vol. 1, No. 3, pp. 119-132.

Bowders, J.J., Usmen, M., and Gidley, J.S., 1987, "Stabilized Fly Ash for Use as Low-Permeability Barriers," Geotechnical Practice for Waste Disposal 1987, ASCE, pp. 320-333.

Chapius, R.P., 1981, "Permeability Testing of Soil-Bentonite Mixtures," Proceedings of the 10th International Conference on Soil Mechanics and Foundation Engineering, Stockholm, Vol. 4, pp. 744-745.

Chapius, R.P., Pare, J.J., and Loiselle, A.A., 1984, "Laboratory Test Results on Self-Hardening Grouts for Flexible Cutoffs," Canadian Geotechnical Journal, Vol. 21, No. 1, pp. 185-191.

Chapius, R.P., 1990, "Sand-Bentonite Liners: Predicting Permeability from Laboratory Tests," Canadian Geotechnical Journal, Vol. 27, No. 1, pp. 47-57.

Clem, A.G., 1985, "Factors Influencing the Stability of Smectite Sealants," Engineering Geology, Vol. 21, No. 3/4, pp. 273-278.

Creek, D.N. and Schackelford, C.D., 1992, "Permeability and Leaching Characteristics of Fly Ash Liner Materials," Preprint of paper presented at 71st Annual TRB Meeting, Washington, D.C.

D'Appolonia, D.J., 1980, "Soil-Bentonite Slurry Trench Cutoffs," Journal of the Geotechnical Engineering Division, ASCE, Vol. 106, No. GT4, pp. 399-417.

Dreimanis, A., 1962, "Quantitative Gasometric Determination of Calcite and Dolomite by Using Chittick Apparatus," Journal of Sedimentary Petrology, Vol. 32, No. 3, pp. 520-529.

Dunnicliff, J., 1988, Geotechnical Instrumentation for Monitoring Field Performance, John Wiley & Sons, New York, NY.

Edil, T.B. and Erickson, A.E., 1985, "Procedure and Equipment Factors Affecting Permeability Testing of a Bentonite-Sand Liner Material," Hydraulic Barriers in Soil and Rock, ASTM STP 874, pp. 155-170.

Edil, T.B., Berthouex, P.M., and Verperman, K.D., 1987, "Fly Ash as a Potential Waste Liner," Geotechnical Practice for Waste Disposal '87, ASCE, pp. 447-461.

Edil, T.B. and Muhanna, A.S.H., 1992, "Characteristics of a Bentonite Slurry as a Sealant," Geotechnical Testing Journal, Vol. 15, No. 1, pp. 3-13.

Edil, T.B., Chang, M.K., Lan, L.T., and Riewe, T.V., 1992, "Sealing Characteristics of Selected Grouts for Water Wells," Ground Water, Vol. 30, No. 3, pp. 351-361.

Evans, J.C., Stahl, E.D., and Drooff, E., 1987, "Plastic Concrete Cut Off

Walls," *Geotechnical Practice for Waste Disposal '87*, ASCE, pp. 462-472.

Fetzer, C.A., 1982, "Pumped Bentonitic Mixture used to Seal Piezometers," *Journal of the Geotechnical Engineering Division*, ASCE, Vol. 108, No. GT2, pp. 295-299.

Filho, P.R., 1976, "Laboratory Tests on a New Borehole Seal for Piezometers," *Ground Engineering*, Vol. 9, No. 1, pp. 16-18.

Garlanger, J.E., Cheung, F.K., and Tannous, B.S., 1987, "Quality Control Testing for a Sand-Bentonite Liner," *Geotechnical Practice for Waste Disposal '87*, ASCE, pp. 488-499.

Gill, S.A and Christopher, B.R., 1985, "Laboratory Testing of Cement-Bentonite Mix for Proposed Plastic Diaphragm Wall for Complexe LaGrande Reservoir Caniapiscau, James Bay, Canada," *Hydraulic Barriers in Soil and Rock*, ASTM STP 874, pp.75-92.

Gipson, A.H., Jr., 1985, "Permeability Testing on Clayey Soil and Silty Sand-Bentonite Mixtures using Acid Liquor," *Hydraulic Barriers in Soil and Rock*, ASTM STP 874, pp. 140-154.

Haile, J.P., 1985, "Construction of Underseals for the Key Lake Project, Saskatchewan," *Proceedings of the 1st Canadian Technology Engineering Technology Seminar on the Use of Bentonite for Civil Engineering Applications*.

Haug, M., 1985, "Design and Construction of Soil Bentonite Unseal for C-I-L Waste Storage Ponds, Bencancour, Quebec," *Proceedings of the 1st Canadian Engineering Technology Seminar on the Use of Bentonite for Civil Engineering Applications*.

Haxo, H.E., Haxo, R.S., Nelson, N.A., Haxo, P.D., White, R.M., Dakessian, S., and Fong, M.A., 1985, *Liner Materials for Hazardous and Toxic Wastes Leachate*, Noyes Publications, Park Ridge, NJ.

Holopainen, P., 1985, "Crushed Aggregate-Bentonite Mixtures as Backfill Material for Repositories of Low- and Intermediate-Level Radioactive Wastes," *Engineering Geology*, Vol. 21, No. 3/4, pp. 239-245.

Jefferis, S.A., 1981, "Bentonite-Cement Slurries for Hydraulic Cut-Offs," *Proceedings of the 10th International Conference on Soil Mechanics and Foundation Engineering*, Stockholm, Vol. 1, pp. 435-440.

Jessberger, H.L., Ebel, W., and Beine, R.A., 1985, "Bentonite Treated Colliery Spoil for Sealing Waste Disposals," *Proceedings of the 11th International Conference on Soil Mechanics and Foundation Engineering*, San Francisco, Vol. 3, pp. 1193-1198.

Kenny, T.C., van Veen, W.A., Swallow, M.A., and Sungaila, M.A., 1992, "Hydraulic Conductivity of Compacted Bentonite-Sand Mixtures," *Canadian Geotechnical Journal*, Vol. 29, pp. 364-374.

Kurt, C.E. and Johnson, R.C., Jr., 1982, "Permeability of Grout Seals Surrounding Thermoplastic Well Casing," *Groundwater*, Vol. 20. No. 4, pp. 415-419.

Littlejohn, G.S., 1982, "Design of Cement Based Grouts," *Grouting in Geotechnical Engineering*, ASCE, pp. 35-49.

Lundgren, T.A., 1981, "Some Bentonite Sealants in Soil Mixed Blankets," *Proceedings of the 10th International Conference on Soil Mechanics* and Foundation Engineering, Stockholm, Vol. 2, pp. 340-354.

Mesri, G. and Olson, R.E., 1971, "Mechanisms Controlling the Permeability of Clays," Clays and Clay Minerals, Vol. 19, pp. 151-158.

Mitchell, J.K. and Madsen, F.T., 1987, "Chemical Effects on Hydraulic Conductivity," Geotechnical Practice for Waste Disposal, ASCE, pp. 87-116.

Nielson, D.M., 1991, Practical Handbook of Ground-Water Monitoring, Lewis Publishers, Inc., Chelsea, MI, 717 pp.

Ogden, F.L. and Ruff, J.F., 1991, "Setting Time Effects on Bentonite Water-Well Annulus Seals," Journal of Irrigation and Drainage Engineering, ASCE, Vol. 117, No. 4, pp. 354-545.

Parker, D.G., Thornton, S.I., and Cheng, C.W., 1977, "Permeability of Fly Ash Stabilized Soils," Geotechnical Practice for Disposal of Solid Waste Materials, ASCE, pp. 63-70.

Pusch, R., 1982, "Mineral-Water Interactions and Their Influence on the Physical Behavior of Highly Compacted Na-Bentonite," Canadian Geotechnical Journal, Vol. 19, pp. 381-387.

Pusch, R. and Alstermark, G., 1985, "Experience from Preparation and Application of Till/Bentonite Mixtures," Engineering Geology, Vol. 21, No. 3/4, pp. 377-382.

Riewe, T., 1991, "Deep Well Abandonment with Bentonite Chips," Water Well Journal, Vol. 45, No. 11, pp. 32-33.

Ryan, C.R., 1987, "Vertical Barriers in Soil for Pollution Containment," Geotechnical Practice for Waste Disposal, ASCE, pp. 182-204.

Schulze, D., Barvenik, M., and Ayres, J., 1984, "Design of Soil-Bentonite Backfill Mix for the First Environmental Protection Agency Superfund Cutoff Wall," Proceedings of the 4th National Symposium on Aquifer Restoration and Ground Water Monitoring, pp. 8-17.

Shakoor, A. and Schmidt, M.L., 1989, "Lime Sludge Stabilization of Sand for Capping Sanitary Landfills," Bulletin of the Association of Engineering Geologists, Vol. 24, No. 2, pp. 227-239.

Stochmeyer, M.R., 1992, "Organophilic Bentonites for Composite Liner Systems," paper presented at 71st Annual TRB Meeting, Washington DC.

Yurchick, C.W., 1987, "Borehole Grouting With Bentonite: An Easy, Efficient and Cost Effective Hole Abandonment Method," Drill Bits, pp. 7-10.

John J. Bowders, Jr.[1] and Steven McClelland[2]

THE EFFECTS OF FREEZE/THAW CYCLES ON THE PERMEABILITY OF THREE COMPACTED SOILS

REFERENCE: Bowders, J. J., Jr., and McClelland, S., "The Effects of Freeze/Thaw Cycles on the Permeability of Three Compacted Soils," Hydraulic Conductivity and Waste Contaminant Transport in Soil, ASTM STP 1142, David E. Daniel and Stephen J. Trautwein, Eds., American Society for Testing and Materials, Philadelphia, 1994.

ABSTRACT: Three soils: a processed kaolinite, a pulverized shale used as a liner material, and a residual clay soil used as a liner material were compacted using the standard Proctor method and permeated with distilled, deionized water in flexible-wall permeameters. Specimens were subjected to freezing and thawing in an environmental chamber. After being thawed the specimens were again permeated in the flexible-wall permeameters. The freeze/thaw process increased the permeability in all three soils. The kaolinite and residual clay soil showed increases in permeability ranging from one to two orders of magnitude above the pre-freeze/thaw values. The permeability of the pulverized shale increased by less than a factor of three. Post-freeze/thaw permeation at progressively higher effective confining stresses reversed the increase in permeability and resulted in decreasing the permeability from the post-freeze/thaw values. This reduction in permeability was greatest for the kaolinite and the residual clay (one to two orders of magnitude); it was the least for the pulverized shale (maximum of a factor of about two). The results of this study indicated that the magnitude of the change in permeability of specimens subjected to freeze/thaw increased with increasing plasticity of the soils. These findings are in agreement with previous literature that freezing and thawing do increase the permeability of a soil but that the increase can be reduced when effective stress on the soil is increased.

KEYWORDS: compacted soil, landfill liner, permeability, freeze/thaw, clay

---

[1]Assoc. Prof., Dept. of Civil Engrg., West Virginia Univ., Morgantown, WV 26506. (304) 293-3031.
[2]Present Address, Engineering geologist, West Virginia Geological and Economic Survey, Morgantown, WV 26507.

INTRODUCTION

The low permeability of compacted clays combined with their widespread occurrence and relatively low cost make them a candidate for barriers to prevent the migration of contaminants into the groundwater. If seasonal freezing and thawing result in significantly increased permeability of such barriers then compacted clay liners which have been subjected to freezing and thawing prior to being covered with solid waste may not be suitable barriers against groundwater contamination. Hence knowledge of the effects of seasonal freezing and thawing on the permeability of compacted clay is quite important for low permeability barriers being constructed in regions where the ground may freeze.

The objectives of this study (McClelland 1991) were: to investigate the process of freezing and thawing of compacted soils; to provide data on changes, particularly in permeability, which occur as a result of freezing and thawing; and to investigate the effects of effective confining stress on the permeability of compacted clays which have undergone the process of freezing and thawing.

BACKGROUND

Researchers have long investigated the effects of cold environments on geotechnical properties of soils (Jumikis 1977); however, recent attention has specifically addressed the effects of freeze/thaw environments on the permeability of soils (Kim and Daniel 1992, Othman and Benson 1992, LaPlante and Zimmie 1992, Chamberlain et al. 1990). In the cases cited, investigators found that soils, specifically, compacted soils, undergo an increase in permeability after being subjected to freezing and subsequent thawing. Typically the increase has ranged from one to two orders of magnitude (Othman and Benson 1992, LaPlante and Zimmie 1992) (Figure 1).

Kim and Daniel (1992) found that, in the absence of a source of water, the higher the molding water content, the greater the increase in permeability subsequent to freeze/thaw. Specimens compacted at water contents above the optimum showed one to two orders of magnitude increases in permeability after freeze/thaw. Those compacted at water contents below the optimum showed increases by factors of two to six. LaPlante and Zimmie (1992) and Othman and Benson (1992) also examined the effect of molding water content on the permeability of soils subjected to freeze/thaw. All specimens, even those compacted dry of optimum moisture displayed an increase in permeability after freeze/thaw. the ratios of permeability after freeze/thaw to that before (k_{ratio}) for soils from all three studies are shown in Table 1. In general, for small variations in molding water content from that of optimum moisture, there

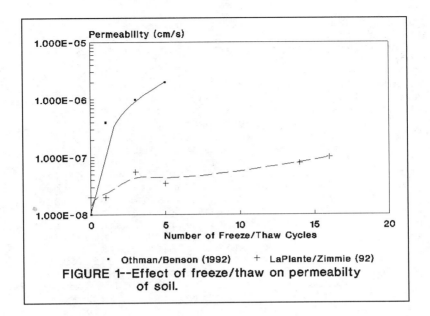

FIGURE 1--Effect of freeze/thaw on permeabilty of soil.

is little difference in resulting permeabilities after freeze/thaw as demonstrated in the k_{ratio}'s shown in Table 1.

The effect of various compaction energies was also evaluated by Kim and Daniel (1992). They found that compaction energies ranging from 60 percent to 455 percent of standard Proctor energy did not appreciably affect the change in permeability induced by the action of freeze/thaw on the soil tested.

The permeability tests performed by Kim and Daniel (1992) were conducted at an effective confining stress of 35 kPa (5psi). Results from Boynton and Daniel (1985) showed that elevated effective confining stress could result in decreasing the measured permeability even for specimens containing cracks and fissures. Thus, if freeze/thaw cycles do result in increased permeability of soil specimens and the increase is due to secondary cracks and fissures resulting from freeze/thaw desiccation (Chamberlain et al. 1990), then it is also important to examine the effects which effective confining stresses might have on the permeability of soils subjected to these conditions.

Chamberlain and Gow (1990) reported the findings of permeability tests on four soils which had been subjected to freeze/thaw conditions. The permeability tests were conducted at various effective stresses subsequent to the freeze/thaw regime. The ratio of post-freeze/thaw permeability to that before freeze/thaw decreased as effective stress in the soil was increased. The decrease was more than an order of magnitude when effective stress was increased to greater than 100 kPa (14 psi). Othman and Benson (1992) measured the permeability of a compacted soil under varying effective confining stresses both before and after it was subjected to freeze/thaw conditions. They found the ratio of permeability after freeze/thaw to that before freeze/thaw (k_{ratio}) decreased as the effective confining stress increased. At a confining stress of 14 kPa (2 psi) the permeability ratio was 300. While at an effective confining stress of 210 kPa (30 psi) the ratio was 2. Additional evidence regarding the effect of confining stress was presented in LaPlante and Zimmie (1992). The authors tested a compacted clay which exhibited permeability ratios (as previously defined) of 30 at an effective stress of 35 kPa (5 psi) and 10 at a stress of 140 kPa (20 psi).

Results from these recent studies indicate that permeability of compacted soils is affected when subjected to freeze/thaw conditions. Furthermore, the effect can be an orders-of-magnitude increase in the permeability above the undisturbed condition. These findings have significant impact when considered in light of performance of clay barriers in regions which experience freezing conditions. Clay liners left exposed to freezing conditions or clay covers lying within the zone of frost penetration could be damaged due to such exposures. However, the findings with respect to the effect of increased effective confining stress indicate that the deleterious effects of freeze/thaw can be reduced by subjecting the soil to increased effective stresses. A working example is given in LaPlante and Zimmie (1992).

TABLE 1-- Ratio of permeability after freeze/thaw to that before freeze/thaw for soils at various molding water contents. All specimens prepared using standard Proctor compaction energy.

SOIL TYPE	LL (%)	PI (%)	MOLDING WATER CONTENT (%)	k_{ratio}	REFERENCE
(1)	(2)	(3)	(4)	(5)	(6)
Glacial Clay	36	19	11.0	2	Kim/Daniel (1992)
			15.1 (opt)	100	
			22.1	84	
Brown Clay	60	34	31 (opt)	4.1	LaPlante/Zimmie (1992)
			35	7.5	
Niagara Clay	39	19	14	50	
			16.5 (opt)	20	
			19	15	
A	34	16	16.5 (opt)	400	Othman/Benson (1992)
			20	400	
B	42	22	18.5 (opt)	250	
			20.5	100	
C	84	60	26 (opt)	67	
			30	750	

MATERIALS AND METHODS

Methodology

In order to accomplish the objectives for this project, three soils were tested in the laboratory. The permeability of compacted soil specimens was measured before and after they were subjected to freeze/thaw cycles. The specimens were tested at various effective confining stresses. In addition, several kaolinite specimens were permeated with a dye and subsequently sectioned and examined.

Soil Specimens

Three different soil types were used in the testing program. One was a prepared commercial kaolinite, Hydrite R^{TM} processed by Georgia Kaolin. The second (Wetzel County) was a pulverized shale which was being used as liner material in an operating landfill. As received, the sample was non-cohesive and behaved like a dusty, coarse sand. On exposure to water it slaked into a low plasticity clay. The third soil (Monongalia) was a clay soil presumed to have developed in place (residual soil) which was being used as liner material in an operating landfill. It is a natural soil and was not processed before being tested. Index properties of the three soils are shown in Table 2.

Preparation of the Specimens

The soil specimens were prepared using ASTM D698 method A (standard Proctor) at a nominal moisture content 1.5% wet of optimum. Optimum moisture contents are shown in Table 2. After the specimens were extracted from the compaction molds they were jacketed with a latex membrane for flexible-wall permeability testing and placed in self-sealing plastic bags. The specimens were kept in a moist room at 22°C (72°F) and 100% relative humidity in the latex membranes and the plastic bags when they were not in the flexible-wall permeameters or not in the environmental chamber.

Freeze/Thaw Procedures

An initial trial consisted of leaving the first set of kaolinite specimens outside and subjecting them to natural freezing and thawing. Results showed that this method would leave any but the most cursory experiment to the vagaries of the weather. The experiment was moved to the controlled freezing of an environmental chamber where the specimens were subjected to 3-dimensional (3-D) freezing. Zimmie and LaPlante (1990) found that 3-D freezing produced similar changes in permeability as compared to the more natural process of one-dimensional (1-D) freezing. Thus for practicality, 3-D freezing was utilized here. The only water in the specimens was that from the molding water and any residual pore water remaining from pre-freezing permeation. Although access to a water source could lead to increased damage to the soil during freeze-thaw cycles, Kim and Daniel (1992) indicate that typically, there is no source of free water for compacted clay barriers in landfill covers. A similar case can be made for soil liners exposed to freezing conditions and as such no source of free water was permitted during the freeze/thaw portion of the experiments.

Table 2-- Properties of soils used in the evaluation of the effects of freeze/thaw on permeability.

Property	Kaolinite	Wetzel County	Monongalia Co.
As-received Water Content (%)	air dry	air dry	17.9
Maximum Dry Unit Weight (kN/m^3) (pcf)	1.38 (86)	1.95 (122)	1.55 (97)
Optimum Moisture Content (%)	31	11	23
Specific Gravity	2.59	2.70	2.68
Percent Finer Than #200 Sieve	90	50	65
Liquid/Plastic Limits	58/34	33/24	60/30
Plasticity Index	24	9	30
Source	Georgia Kaolin	Wetzel Co., W.Va	Monongalia Co., W.Va.
U.S.C.S. Classification	CL-MH	CL-ML	CH
Color	bright white	gray	reddish-brown

Four specimens were tested per set. Two sets of kaolinite (K) specimens were tested. One set of specimens of the Wetzel County (WC) soil was tested and one set of the Monongalia County soil (MC) was tested. The first set of specimens (K1-4) to be placed in the environmental chamber were kaolinite. In the first freezing and thawing experiment the K1-2 specimens were submitted to three cycles per day of -18°C (0°F) to 10°C (50°F). This was continued for three days after which the specimens were remounted in the permeameter and re-permeated. The K3-4 specimens were subjected to five cycles of -18°C to 10°C. Investigations with respect to the number of freeze/thaw cycles and resulting damage have shown that maximum increases in the permeability generally occur within five cycles of freezing and thawing (Othman and Benson 1992, Chamberlain et al. 1990), Zimmie and LaPlante 1990). In the present investigation several additional specimens were subjected to freeze-thaw and sectioned at various times during the process to assess the completeness of the freezing process.

In all cases, the soil specimens were assessed to be completely frozen; however, in some of the earlier tests The second set of kaolinite specimens was permeated with a dye in an attempt to reveal the nature of the flow paths. Specimen K-5 was dyed but not frozen. Specimens K6-8 were frozen for 12 days (24 cycles) of -18°C to 10°C two cycles per day. After they were removed from the environmental chamber there was some question of whether the specimens had thawed during the warming part of the cycle. To make sure that there was both freezing and thawing, the specimens were returned to the environmental chamber for eight days and exposed to 6 cycles per day of -18°C to 43°C (110°F), an extremely severe environment. They were subsequently permeated with a dye.

The Wetzel County specimens (WC1-4) were exposed to one cycle of -18°C to 21°C (70°F) per day for five days. No problems with the specimens freezing and thawing were evident.

The Monongalia County specimens (MC1-4) were tested for 22 days at one cycle per day of -18°C to 21°C. There did not seem to be any problem with the specimens freezing and thawing.

After the specimens were taken out of the environmental chamber they were allowed to thermally equilibrate in the moist room at 22°C (72°F) and 100% relative humidity.

Permeability Testing

After being compacted and extruded from the molds, the specimens were jacketed with a thin, flexible latex membrane and placed in the flexible-wall permeameter. The method of measuring permeability in a flexible-wall permeameter as described in ASTM Method 5084 was followed. Distilled, deionized water was used to permeate the specimens in an attempt to minimize the number of variables in the tests. The available "tap" water varies in chemical constituents and concentration. It was ruled out. Adding a particular solute to the distilled water to make a "standard" water could influence the freezing potential. Therefore, distilled water was deemed the permeant liquid least likely to result in additional parameters affecting the behavior of the soils.

The permeability of each specimen under various effective confining stresses was measured. The specimens and their testing conditions are listed in Tables 3 through 6. It was also necessary to vary the hydraulic gradient in order to allow variation in the effective stress acting in the specimens. Flow was continued at each stress level until the permeability was steady and inflow equaled outflow. Backpressure was used to maintain saturation but varied among the stress states.

Description of Dyeing Procedure

In an attempt to clarify the nature of the flow paths of the permeating liquid in one set of the kaolinite specimens (K5-8), the distilled, deionized water was dyed with FD&C No. 40 dye. The resulting liquid was blood-red and opaque. Specimen K5 was permeated with dye and sectioned for examination of flow paths. This specimen was not frozen. Specimens K6 and K7 were permeated with the dye after being subjected to freeze/thaw conditions. The dye was substituted for the permeant liquid at the initial repermeation stresses (effective confining stress of 23 kPa (3.25 psi)). Specimen K8 was permeated with distilled-deionized water at effective confining stresses of 23 kPa (3.25 psi) and 65 kPa (9.25 psi) after freezing. Dye was inserted into the permeant liquid after the specimen's permeability had reduced to near the pre-freezing value.

RESULTS AND DISCUSSION

The results of four series of permeability tests are presented in this article. Each series consisted of four soil specimens for which permeability was measured before and after they were subjected to freeze/thaw conditions. The testing parameters were described in the previous section of the article. The results of the permeability tests are given in Tables 3 through 6 and Figures 2 through 7. After each series is discussed separately, the commonalities among the series are presented.

Within Tables 3 through 6, the specimen type and number are shown in column 1; the effective confining stress and hydraulic gradient under which the permeability was determined are given in columns 2 and 3 respectively; the cumulative number of pore volumes of flow measured at the point at which the confining parameters were incremented to the next level is given in column 4; the average permeability during the particular stress level is given in column 5; the ratio of the permeability after freeze/thaw (column 5) to that before freeze/thaw is given in column 6; and the number of freeze thaw cycles are shown in column 7.

Series one consisted of kaolinite specimens, K1-4, Table 3. In the first series of tests (K1-4), the post freeze/thaw permeabilities at various effective stresses were compared to the pre-freezing permeability measured at an effective stress of 49 kPa (7.2 psi). The permeabilities of all four specimens increased after freeze/thaw when the specimens were confined with an effective stress of only 25 kPa (3.5 psi). The increase was an order of magnitude for specimen K3 but was less than a factor of four for the other specimens. As effective confining stress was increased to 42 kPa (6 psi), 63 kPa (9 psi) and 115 kPa (16.5 psi), the permeability of all the specimens decreased to nearly the original (pre-freeze/thaw) value as indicated by k_{ratio}'s (Column 6) near unity (Fig. 2). Visual observation of the specimens showed that all four specimens suffered moderate

Table 3-- Permeability test results for specimens of compacted kaolinite subjected to freeze/thaw cycles.

SPEC. NO. (1)	EFFECT. CONFIN. STRESS (kPa) (2)	HYDR. GRAD. (3)	PORE VOLUMES (4)	_k (cm/s) (5)	k_{ratio} (6)	F/T (7)
K1	49	73	5.9	5.0E-8	-	0
	25	20	6.2	7.0E-8	1.4	10
	42	27	6.4	7.0E-8	1.4	-
	42	54	7.4	7.0E-8	1.4	-
K2	49	73	17.1	8.0E-8	-	0
	25	20	19.8	3.0E-7	3.8	10
	42	27	21.7	1.5E-7	1.9	-
	42	54	23.8	1.5E-7	1.9	-
	63	100	26.0	3.0E-7	3.8	-
	115	200	27.4	1.0E-7	1.2	-
K3	49	73	-	6.0E-8	-	0
	25	20	2.7	7.0E-7	11.7	5
	42	27	3.5	8.0E-8	1.3	-
	28	54	4.5	9.0E-8	1.5	-
	63	100	5.6	1.0E-7	1.7	-
	115	200	6.6	7.0E-8	1.2	-
K4	49	73	-	5.0E-8	-	0
	25	20	2.5	8.0E-7	1.6	5
	42	27	2.7	1.0E-7	2.0	-
	28	54	4.0	1.0E-7	2.0	-
	63	100	5.4	8.0E-8	1.6	-
	115	200	6.2	6.0E-8	1.2	-

to severe cracking. Given the appearance of the specimens one might have anticipated much larger increases in permeability than were measured. In addition, the effect of even a relatively small effective confining stress (42 kPa)

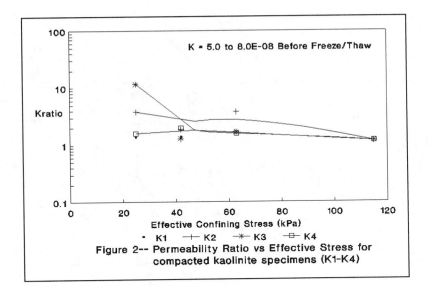

Figure 2-- Permeability Ratio vs Effective Stress for compacted kaolinite specimens (K1-K4)

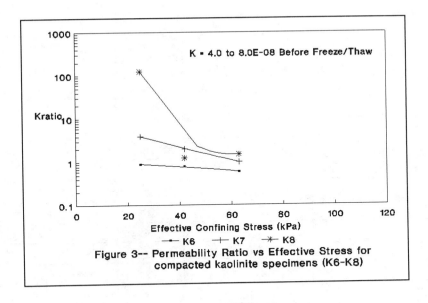

Figure 3-- Permeability Ratio vs Effective Stress for compacted kaolinite specimens (K6-K8)

was sufficient to reduce the k_{ratio}'s to near unity.

Due to the striking impact of the appearance of the first series of specimens, a second series of kaolinite specimens (K5-8, Table 4) were prepared as similarly to the first as possible. In addition to the freeze/thaw regime, these specimens were permeated with dye at various stages during the testing. The dye was intended to highlight preferential flow paths; however, visual results were difficult to document. The dye did permit an informative qualitative assessment of breakthrough times for the various specimens.

TABLE 4 -- Permeability test results for specimens of compacted kaolinite subjected to freeze/thaw cycles and permeated with a dye solution.

SPEC. NO. (1)	EFFECT. CONFIN. STRESS (kPa) (2)	HYDR. GRAD. (3)	PORE VOLUMES (4)	k (cm/s) (5)	k_{ratio} (6)	F/T (7)
K5	42	54	-	1.5E-7	-	0
	63	100	-	7.0E-8	-	0
K6	42	54	-	8.0E-8	-	0
	25	20	4.1	7.0E-8	0.9	72
	63	100	4.5	6.0E-8	0.8	-
	115	200	5.3	5.0E-8	0.6	-
K7	42	54	-	5.0E-8	-	0
	25	20	1.8	2.0E-7	4.0	72
	63	100	2.5	1.0E-7	2.0	-
	115	200	3.3	5.0E-8	1.0	-
K8	42	54	-	4.0E-8	-	0
	25	20	1.2	5.0E-6	125	72
	63	100	1.7	5.0E-8	1.25	-
	115	200	2.5	6.0E-8	1.5	-

Specimen K5 was permeated with the dye and was not frozen. Subsequent sectioning and examination of the specimen showed a uniformly colored cross-section. As expected, no preferential flow paths were macroscopically discernable. Specimens K6-8 showed post freeze/thaw k_{ratio}'s ranging from unity to more than two orders of magnitude indicating a likewise increase in permeability under a confining stress of 25 kPa. Similar to the first series of specimens, permeability decreased with increasing confining stress (Fig. 3). All three specimens showed a k_{ratio} near unity at the final confining stress of 115 kPa.

Dye was inserted into specimens K6-7 upon permeation after the freeze/thaw regime. Within minutes, the dye emanated from the effluent ends of the specimens. Both specimens exhibited marked physical disturbance due to the freeze/thaw process. Specimen K8 was permeated with distilled water (undyed) at effective confining stresses of 25 kPa and 49 kPa. When the confining stress was increased to 63 kPa, dye was introduced into the permeating liquid. There was no rapid breakthrough of the dye and k_{ratio}'s were near unity.

The introduction of the dye into the soil specimens did not visually accent any preferred flow channels which might have existed in the specimens as a result of the freeze/thaw process; however, it did accentuate the observations of the specimens being remolded upon application of increased confining stress after being subjected to a freeze/thaw regime.

The third series of specimens consisted of the Wetzel County (WC1-4) soil shown in Table 5. This soil has a low plasticity index, 9, and a low optimum moisture content of 11 percent. Examination of the k_{ratio}'s shown in Table 5 indicate that the greatest increase in post freeze/thaw permeability was by an approximate factor of 2.5. All four specimens exhibited a small decrease in the k_{ratio} as the effective confining stress was increased. Final k_{ratio}'s were near unity (Fig. 4).

A soil containing the highly plastic Monongalia County (MC1-4) clay, was used in the fourth series of specimens (Table 6). The soil had a liquid limit of 60%, a plasticity index of 30% and an optimum moisture content of 23 percent. Permeabilities were measured at identical effective confining stresses before and after the specimens were subjected to freeze/thaw. The k_{ratio}'s for the MC soil (Table 6) exhibit large increases in permeability at the low effective confining stresses. As the confining stress increased, the k_{ratio} and associated permeability decreased as much as an order of magnitude (Fig. 5). However, at the largest confining stress employed (115 kPa), there remained an increased permeability for all of the specimens compared to their pre-freezing values. The increase ranged from a factor of 5 to an order of magnitude.

TABLE 5-- Permeability test results for specimens of compacted Wetzel County (WC) soil.

SPEC. NO. (1)	EFFECT. CONFIN. STRESS (kPa) (2)	HYDR. GRAD. (3)	PORE VOLUMES (4)	k (cm/s) (5)	k_{ratio} (6)	F/T (7)
WC-1	42	54	-	1.0E-7	-	0
	25	20	2.3	2.0E-7	2.0	5
	42	54	3.2	2.5E-7	2.5	-
	115	200	5.3	1.5E-7	1.5	-
WC-2	42	54	-	1.0E-7	-	0
	25	20	2.0	2.0E-7	2.0	5
	42	54	2.7	2.0E-7	2.0	-
	115	200	4.3	1.0E-7	1.0	-
WC-3	42	54	-	9.0E-8	-	0
	25	20	1.9	2.0E-7	2.2	5
	42	54	2.9	2.5E-7	2.8	-
	115	200	5.0	1.5E-7	1.7	-
WC-4	42	54	-	1.5E-7	-	0
	25	20	3.29	1.5E-7	1.0	5
	42	54	3.77	1.5E-7	1.0	-
	115	200	5.0	8.0E-8	0.5	-

The permeability results from series 3 and 4 indicate a relationship might exist between the plasticity of a soil and the relative degree to which the permeability is altered when the soil is subjected to freeze/thaw conditions. The results of those two series indicate that a greater degree of disturbance might be expected for soils having larger plasticity indices. However, since plasticity is an approximate measure of the the volume of bound water in a soil, it would be more accurate to attribute the degree of disturbance to the volume of water available during the freeze/thaw process. The greater the volume of water, the larger the ice lenses which form in the voids of the soil leading to larger pores and cracks in the thawed soil. In

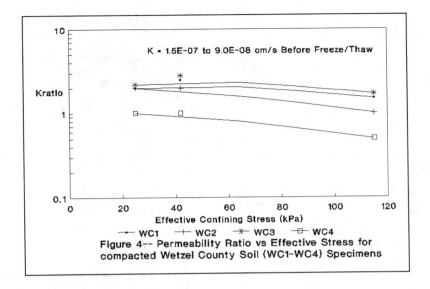

Figure 4-- Permeability Ratio vs Effective Stress for compacted Wetzel County Soil (WC1-WC4) Specimens

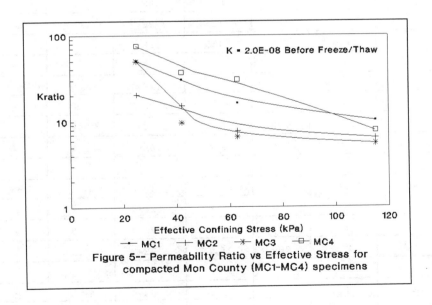

Figure 5-- Permeability Ratio vs Effective Stress for compacted Mon County (MC1-MC4) specimens

this study the only water available during freezing was that remaining in the voids from the molding water and prior permeation. Calculation of the porosity of the soils used in the study indicated that the kaolinite and Mon County soils had similar porosities, 0.47 and 0.42 respectively, or consequently, similar volume of voids at maximum dry density. The Wetzel County soil had a porosity of 0.27 at maximum dry density. Thus, assuming saturated conditions just prior to placing the specimens in the environmental chamber, one concludes that the kaolinite and Monongalia County soils would undergo similar and the greatest increase in permeability. This is born out by the results shown in Fig. 6.

TABLE 6 -- Permeability test results for specimens of compacted Mon County (MC) soil.

SPEC. NO. (1)	EFFECT. CONFIN. STRESS (kPa) (2)	HYDR. GRAD. (3)	PORE VOLUMES (4)	k (cm/s) (5)	k_{ratio} (6)	F/T (7)
MC-1	25	20	0.3	1.3E-6	50.0	22
	42	54	4.5	6.8E-7	30.5	-
	63	100	11.5	3.8E-7	16.2	-
	115	200	14.5	1.6E-7	10.1	-
MC-2	25	20	0.4	3.2E-7	20.4	22
	42	54	1.7	3.0E-7	15	-
	63	100	4.1	1.7E-7	7.7	-
	115	200	6.2	1.0E-7	6.4	-
MC-3	25	20	0.5	1.1E-6	49.6	22
	42	54	4.8	6.2E-7	9.6	-
	63	100	11.4	3.2E-7	6.6	-
	115	200	15.1	1.7E-7	5.5	-
MC-4	25	20	0.5	2.6E-6	75.2	22
	42	54	9.1	1.9E-6	37.1	-
	63	100	28.6	1.3E-6	30.4	-
	115	200	39.3	2.1E-7	7.7	-

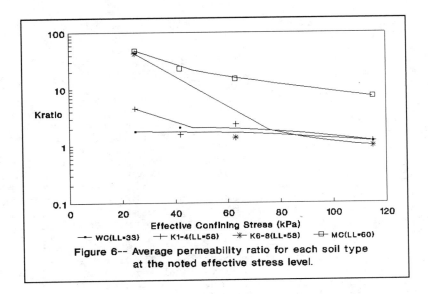

Figure 6-- Average permeability ratio for each soil type at the noted effective stress level.

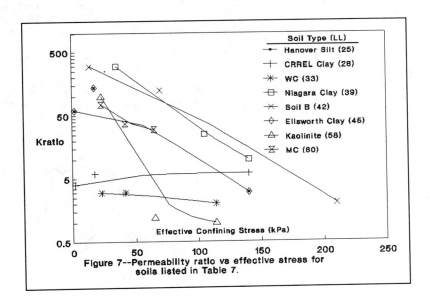

Figure 7--Permeability ratio vs effective stress for soils listed in Table 7.

It is also important to examine the significance of effective stress on the permeability of the soils after they have been subjected to freeze/thaw. In general, all cases shown in Tables 3 through 6 illustrate a decrease in the k_{ratio} as the effective confining stress is increased. More importantly, it is significant to note the magnitude of the decrease in the k_{ratio}'s. Specimens which had a one or two order increase in permeability, showed one to two orders of decrease in permeability when the maximum effective stress was applied. Although not fully reversible in all cases, the increase in permeability brought about by the freeze/thaw process was, in most cases, minimized with the application of a 115 kPa confining stress.

Thus, information from this study indicates that volume of water available during freezing and effective confining stress are important variables in assessing the hydraulic properties of soil subjected to freeze/thaw conditions. A search through the literature revealed a limited amount of data on this topic; however, the existing data appear to support the findings of this study. The literature data along with the findings of this study are summarized in Table 7. The plasticity of the soils (Column 3) increases as one reads from top to bottom from Table 7. The k_{ratio}'s for the soils are shown in Columns 4, 5 and 6. Directly beneath the K_{ratio}'s are the effective confining stresses (shown in parentheses) at which the post freeze/thaw permeabilities were measured.

The data shown in Table 7 support the findings from the study reported herein. Namely, the permeability of soils subjected to freeze/thaw processes does increase and, increased effective stress applied to the soil after freeze/thaw results in a permeability decrease (Fig. 7). To a lesser degree, the data show a trend of increasing permeability (post freeze/thaw) as plasticity of the soil increases; however, this is more likely a function of the volume of water available during freezing as previously noted. The subsequent decrease in permeability when effective stress is increased showed values approaching unity. This indicates that permeabilities in soils subjected to freeze/thaw processes might be remediated with the application of increased effective stress. In some cases the k_{ratio}'s of these soils were less than unity, a case in which the post freeze/thaw permeability ends up being lower than that prior to freezing.

Table 7-- Summary of available data showing the effect of freeze/thaw behavior on the permeability of soil, including the effects of effective confinement stress and plasticity. (Note: Chamberlain and Gow 1979 tests are on sedimented soil specimens)

REFERENCE (1)	SOIL TYPE (2)	LL/PL (%) (3)	K_{ratio} (σ_3) kPa (4)	K_{ratio} (σ_3) kPa (5)	K_{ratio} (σ_3) kPa (6)
Chamberlain/Gow (1979)	Hanover Silt	25/NP	60 (1.75)	140 (17.5)	3 (140)
Chamberlain/Gow (1979)	Morin Clay	26/7	--	8 (17.5)	1.2 (140)
Chamberlain/Gow (1979)	CRREL CLAY	28/5	4 (1.75)	6 (17.5)	6 (140)
McClelland (1991)	WC	33/9	1-2 (23)	1-3 (42)	.5-2 (114)
Kim/Daniel (1992)	Glacial Clay	36/19	160 (5)	--	--
LaPlante/Zimmie (1992)	Niagara Clay	39/19	300 (35)	25 (105)	10 (140)
Othman/Benson (1992)	Soil B (CL)	42/22	300 (14)	125 (70)	2 (210)
Chamberlain/Gow (1979)	Ellsworth Clay	45/20	60 (1.75)	140 (17.5)	3 (140)
McClelland (1991)	Kaolinite	58/24	.9-100 (23)	.6-1.2 (65)	.1-1 (114)
McClelland (1991)	MC	60/30	20-75 (23)	9-37 (42)	7-30 (65)

CONCLUSIONS

Three soils: a processed kaolinite, a pulverized shale and a residual clay soil were compacted using the standard Proctor method and permeated with distilled, deionized water in flexible-wall permeameters. Specimens were subjected to freezing and thawing in an environmental chamber. After they had thawed they were again permeated in the flexible-wall permeameters.

The freeze/thaw process increased the permeability in all three soils. The kaolinite and residual clay soil showed increases in permeability ranging from one to two orders of magnitude above the pre-freeze/thaw values. The permeability of the pulverized shale increased less than a factor of three.

Post-freeze/thaw permeation at progressively higher effective confining stresses resulted in decreasing the permeability from the post-freeze/thaw values. This reduction in permeability was greatest for the kaolinite and the residual clay (one to two orders of magnitude); it was the least for the pulverized shale (maximum of a factor of about two).

The results of this study indicated that the magnitude of the change in permeability of specimens subjected to freeze/thaw increased with increasing plasticity of the soils; however, other investigators have not found a strong correlation. It is more probable that the magnitude is a function of the volume of water available during freezing as shown earlier. These findings are in agreement with those in the literature which indicate that freezing and thawing do increase the permeability of a soil; however, the increase is at least partially reversible when effective stress in the soil is increased.

ACKNOWLEDGEMENTS

The authors are appreciative of the support extended by Dr. Sam Kiger, Chairman, WVU Civil Engineering; the West Virginia Solid Waste Management Board and Mr. David C. Brown, P.E., engineer for the Wetzel County landfill for soil samples; and Drs. L.K. Moulton, J.C. Sencindiver and M.A. Gabr for technical review of the project. A special thank you is extended to Patricia G. Schnitzer for selflessly sharing her editorial expertise.

REFERENCES

Boynton S.S. and Daniel, D.E., "Hydraulic Conductivity Tests on Compacted Clay," *Journal of Geotechnical Engineering*, 111(4) April, 1985, 465-478.

Chamberlain, E.J. and Gow, A.J.,"Effects of Freezing and Thawing on the Permeability and Structure of Soil," *Engineering Geology*, 13(1), 1979, 73-92.

Chamberlain, E.J., Iskander, I. and Hunsicker, S.E., "Effect of Freeze/Thaw Cycles on the Permeability and Microstructure of Soils," *Proc. Int'l Symp. on Frozen Soil Impacts on Agricultural, Range, and Forest Lands*, March 21-22, 1990, Spokane, WA, 145-155.

Jumikis, A.R. *Thermal Geotechnics*, Rutgers Univ. Press, New Brunswick, N.J., 1977.

Kim, W.H. and Daniel, D.E., "Effects of Freezing on the Hydraulic Conductivity of Compacted Clay," *Journal of Geotechnical Engineering*, 118(7) July, 1992, 1083-1097.

LaPlante, C.M. and Zimmie, T.F., "Freeze/Thaw Effects on the Hydraulic Conductivity of Compacted Clays," Paper presented at the 71st Annual Meeting of the Transportation Research Board, January, 1992. To be published in the Transportation Research Record, In Press.

McClelland, S. "The Effects of Freeze/Thaw Cycles on the Permeability of Three Compacted Soils," M.S. Thesis, Dept of Civil Engineering, West Virginia Univ., Morgantown, WV, 1991.

Othman M.A. and Benson, C.H. "Effect of Freeze/Thaw on the Hydraulic Conductivity of Three Compacted Clays From Wisconsin," Paper presented at the 71st Annual Meeting of the Transportation Research Board, January, 1992. To be published in the Transportation Research Record, In Press.

Zimmie, T.F. and Laplante, C.M.,"The Effects of Freeze/Thaw Cycles on the Permeability of a Fine-Grained Soil," Proc. 22nd Mid-Atlantic Industrial Waste Conference, Drexel University, Philadelphia, 1990, 580-593.

Harold W. Olsen,[1] Arthur T. Willden,[2] Nicholas J. Kiusalaas,[3] Karl R. Nelson,[4] and Eileen P. Poeter[5]

VOLUME-CONTROLLED HYDROLOGIC PROPERTY MEASUREMENTS IN TRIAXIAL SYSTEMS

REFERENCE: Olsen, H. W., Willden, A. T., Kiusalaas, N. J., Nelson, K. R., and Poeter, E. P., "Volume-Controlled Hydrologic Property Measurements in Triaxial Systems," Hydraulic Conductivity and Waste Contaminant Transport in Soil, ASTM STP 1142, David E. Daniel and Stephen J. Trautwein, Eds., American Society for Testing and Materials, Philadelphia, 1994.

ABSTRACT: New capabilities for hydrologic property measurements in triaxial systems include: (1) volume-controlled and simultaneous measurements of hydraulic conductivity and one-dimensional consolidation (or specific storage) of a saturated test specimen; and (2) volume-controlled measurements of hydraulic conductivity, matric potential, and the variation of these properties with the moisture content of an unsaturated test specimen.
Data on saturated kaolinite demonstrate simultaneous hydraulic-conductivity and one-dimensional consolidation tests with continuous monitoring of both vertical and horizontal effective stresses. Data on well-graded silty sand demonstrate the feasibility of concurrent constant-flow hydraulic conductivity and matric potential measurements, and the variation of these properties with moisture content, for undisturbed and unsaturated specimens mounted in triaxial cells. Refinements needed to realize the full potential of these capabilities include a more rigid triaxial cell to minimize compliance, and an improved technique for measuring hydraulic-head differences within an unsaturated test specimen.

KEYWORDS: compressibility, consolidation, hydraulic conductivity, matric potential, permeability, specific storage, capillary pressure, flow pumps, volume control.

---

[1]Research Geotechnical Engineer, U. S. Geological Survey, Box 25046, Mail Stop 966, Denver, CO 80225.
[2]Geological Engineer, Geotrans Inc., 3300 Mitchell Lane, Boulder, CO 80301.
[3]Geological Engineer, Advanced Sciences Inc., 405 Urban Street, Lakewood, CO 80228.
[4]Associate Professor, Department of Engineering, Colorado School of Mines, Golden, CO 80401.
[5]Associate Professor, Department of Geology and Geological Engineering, Colorado School of Mines, Golden, CO 80401.

INTRODUCTION

Since the late 1970's, applications of the constant-flow method for hydraulic conductivity measurements have been increasing in both research and practice. This method, which involves generating a constant flow rate through a test specimen with a flow pump and measuring the pressure difference induced thereby with a transducer, has recently been accepted in the ASTM Standard Test Method for Measurement of Hydraulic Conductivity of Saturated Porous Materials Using a Flexible Wall Permeameter (D5084-90). The fundamental advantage of this volume-controlled method, compared with constant-head and falling-head methods, is that hydraulic conductivity measurements can be obtained much more rapidly and with substantially smaller hydraulic gradients (Olsen et al. 1985, 1988, 1991; Aiban and Znidarcic 1989).

In recent years, applications of flow pumps in laboratory testing have been expanding for additional hydrologic properties of both saturated and unsaturated test specimens. These applications involve volume control not only for hydraulic conductivity measurements, but also for changing the volume and/or the moisture content of test specimens in one-dimensional consolidometers and in triaxial cells. For saturated specimens in a one-dimensional consolidometer, Gill, Olsen, and Nelson (1991) introduced a volume-controlled method for superimposing direct constant-flow hydraulic conductivity measurements on constant-rate-of-deformation (CRD) consolidation tests. For saturated specimens in triaxial cells, Olsen et al. (1991) introduced a volume-controlled method for obtaining hydraulic conductivity versus effective stress and/or void ratio relationships on specimens under isotropic consolidation and rebound. For unsaturated specimens in triaxial cells, Znidarcic, Illangasekare, and Manna (1991) introduced a volume-controlled method for measuring matric potential (also known as suction and capillary pressure) versus degree of saturation relationships.

This paper presents improvements to the above-mentioned volume-controlled method for saturated specimens under isotropic consolidation and rebound in a triaxial cell. In addition, new applications of volume control are introduced for: (1) simultaneous measurements of hydraulic conductivity and consolidation (or specific storage) on saturated specimens under one-dimensional deformation in a triaxial cell; and (2) simultaneous measurements of hydraulic conductivity, matric potential, and the variation of these properties with the degree of saturation of an unsaturated specimen in a triaxial cell.

SATURATED SPECIMENS UNDER ISOTROPIC DEFORMATION

Figure 1 illustrates a triaxial system for measuring hydraulic conductivity versus effective stress on specimens under isotropic consolidation and rebound. A single-carriage flow pump equipped with an infuse-withdraw actuator [P] generates identical flow rates through opposite ends of the test specimen [S] while one transducer [M] monitors the pressure difference across the specimen and a second transducer [N] monitors the pressure difference between the pore fluid in the specimen and chamber fluid in the triaxial cell. A dual-carriage flow pump, equipped with identical infuse-withdraw actuators [CC], extracts (or infuses) pore fluid from (or to) the specimen, and thereby changes its volume at a constant rate. A high-pressure bellowfram [B] allows the chamber pressure to be elevated and maintained constant at high levels, on the order of 2760 kPa (400 psi).

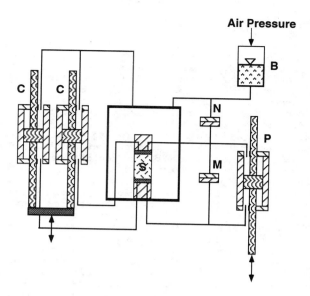

Figure 1--Triaxial system for saturated specimens under isotropic deformation.

One improvement in this system, compared with previous systems (Olsen et al. 1991), is the use of infuse-withdraw actuators not only for generating identical flow rates through opposite ends of a test specimen, but also for extracting or infusing pore water to or from the ends of the specimen. The use of infuse-withdraw actuators for this latter purpose, instead of conventional stainless-steel syringes in previously described systems, provides a means to minimize the reaction force on the flow pump from the shafts connected to the pistons in the actuators. In figure 1, this force is proportional to the pressure difference between the pore fluid in the specimen and the chamber fluid in the triaxial cell. Alternatively, separate pressure sources could be used to control the chamber pressure and to counteract the force generated by the specimen pore water pressure in the infuse-withdraw actuators. The other improvement in this system is the use of a high-pressure bellowfram, whose design is documented elsewhere (Willden 1991). These improvements were motivated by our experience that response times needed to reach steady state for very low hydraulic conductivity measurements can be minimized by elevating pore fluid and chamber pressures above 690-1035 kPa (100-150 psi).

Data obtained with the system in figure 1 on 5.08-cm diameter by 2.54-cm thick specimens of shale from Oklahoma are shown in figures 2 and 3. In these figures, the bottom plots show the time history of the effective stresses induced in the specimens during consolidation and rebound. The middle plots show the hydraulic conductivity measurements that were superimposed during consolidation and rebound. The upper plots show the hydraulic conductivity versus effective stress relationships derived from the lower and middle plots.

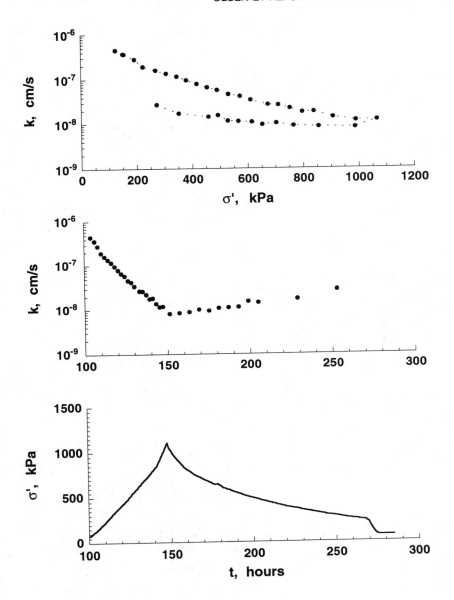

Figure 2--Elapsed time (t) versus effective stress (σ') [bottom plot] and hydraulic conductivity (k) [middle plot] on an Oklahoma shale specimen (from 20.4-m depth) in the triaxial system illustrated in figure 1; σ' versus k relationship from data in the middle and lower plots [top plot].

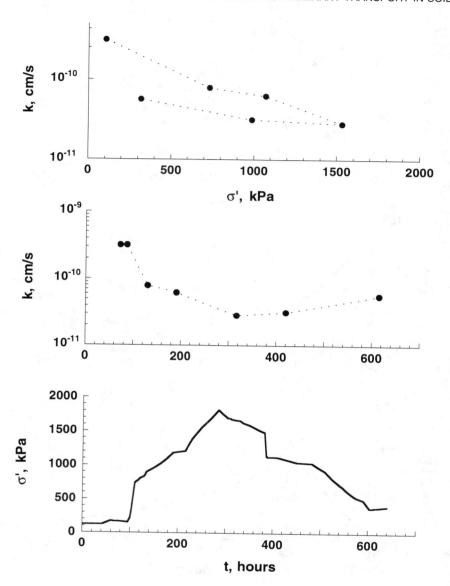

Figure 3--Elapsed time (t) versus effective stress (σ') [bottom plot] and hydraulic conductivity (k) [middle plot] on an Oklahoma shale specimen (from 36.2-m depth) in the triaxial system illustrated in figure 1; σ' versus k relationship from data in the middle and lower plots [top plot].

The specimens were first equilibrated for about 100 hours at low effective stresses. Thereafter, the specimens were consolidated and rebounded continuously, although in figure 3, the rate of loading was modified at elapsed times of about 105, 390, and 490 hours. Figure 2 shows 35 hydraulic conductivity measurements in the range of 10^{-6} to 10^{-8} cm/s were obtained between elapsed times of 100 to 270 hours, or an interval of about seven days. In contrast, figure 3 shows seven hydraulic conductivity measurements in the range of 10^{-10} to 10^{-11} cm/s were obtained before, during, and following consolidation and rebound, between elapsed times of 100 to 600 hours, or an interval of about 21 days. These data include the lowest values of hydraulic conductivity that we have measured to date.

Figures 2 and 3 illustrate that the elapsed time needed for obtaining hydraulic conductivity versus effective stress relationships varies with the hydraulic conductivity of the material. The governing factor is the response time needed for reaching steady state during each hydraulic conductivity measurement, which depends on the ratio of the compressibility (or specific storage) divided by the hydraulic conductivity of the material. This ratio, which is commonly referred to as either the coefficient of consolidation or the hydraulic diffusivity, is independent of the hydraulic gradient induced during a constant-flow hydraulic conductivity test. Our experience in this and previous studies (Olsen et al. 1985, 1988, 1991) shows response times for constant-flow hydraulic conductivity (k) measurements in triaxial systems to be on the order of: (a) a few minutes or less for materials with k values greater than 10^{-6} cm/s, (b) a few hours for materials with k values in the range of 10^{-8} to 10^{-9} cm/s; and (c) a few days for materials with extremely low k values, in the range of 10^{-10} to 10^{-11} cm/s.

SATURATED SPECIMENS UNDER ONE-DIMENSIONAL DEFORMATION

Figure 4 shows an experimental system Willden (1991) developed for superimposing constant-flow hydraulic-conductivity measurements with one-dimensional constant-rate-of-deformation (CRD) consolidation of a test specimen in a triaxial system. This system includes the components of the system in figure 1. In addition, (a) it is equipped with a hydraulic loading system [H] for controlling the axial deformation of the specimen, (b) the upper ends of the pair of identical infuse-withdraw actuators [CC] transmit fluid to or from the hydraulic loading chamber, and (c) an additional transducer [O] measures the pressure difference between the hydraulic loading chamber and the cell chamber.

One-dimensional consolidation or rebound of the specimen is accomplished with two identical infuse-withdraw actuators [CC], driven by a dual-carriage flow pump. For one-dimensional deformation with double drainage, these actuators simultaneously withdraw (or infuse) pore fluid from (or into) opposite ends of the specimen, and transmit an identical volume of fluid to (or from) the piston loading chamber. The total volume change of the specimen equals the volume extracted from (or infused into) the top and bottom of the specimen. Alternatively, for one-dimensional deformation with single drainage, both actuators can withdraw (or infuse) pore fluid from (or into) one end of the specimen. This requires some additional valves and tubing not shown in figure 1. With either double or single drainage, the total volume change generated by extracting pore fluid from the specimen equals the volume change displaced by axial deformation because the piston loading chamber has a

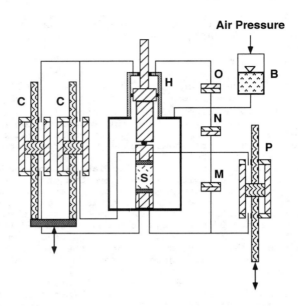

Figure 4--Triaxial system for saturated specimens under one-dimensional deformation.

net cross-sectional area equal to that of the test specimen. This maintains the average diameter of the specimen constant during consolidation and rebound. Radial (lateral) deformations are limited to those associated with moisture migration, such as can be induced by hydraulic gradients that cause swelling and consolidation at opposite ends of the specimen.

Data obtained with this system on a reconstituted specimen of kaolinite are presented in figure 5. This specimen was prepared from Standard Air Floated (SAF) Kaolinite from the Georgia Kaolin Company. Dry powdered clay was mixed with distilled water to form a slurry with an initial moisture content of 110%. The slurry was deaired under vacuum, poured into a rigid 7.62-cm diameter cylinder, and consolidated one-dimensionally between porous discs to a consistency with suffcient stiffness to allow a 5.08-cm diameter specimen to be trimmed and mounted in the triaxial system. The specimen was maintained under a very small effective stress while the chamber fluid pressure was elevated in three steps from 0-690 kPa (0-100 psi), 690-1380 kPa (100-200 psi), and 1380-2760 kPa (200-400 psi), and allowed to equilibrate for 24 hours following each step. B-bar values measured at the end of the successive equilibration periods were 98%, 99%, and >99.5%.

In figure 5, the elapsed time scale begins at the onset of CRD consolidation with single drainage, following equilibration of the specimen at pore fluid and chamber fluid pressures of approximately 2760 kPa (400 psi) and 2795 kPa (405 psi) respectively. The lower plot shows the time history of the void ratio during consolidation and rebound at a constant rate of deformation of 10^{-6} s^{-1}. The middle plot shows the corresponding time history of the induced vertical and horizontal

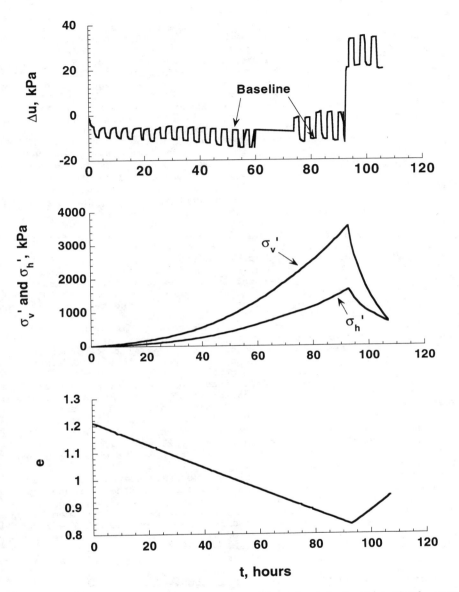

Figure 5--Elapsed time (t) versus void ratio (e) [bottom plot], pore-pressure difference across the specimen (Δu) [middle plot], and the vertical (σ_v') and horizontal (σ_h') effective stresses [top plot] on a kaolinite specimen in the triaxial system illustrated in figure 4.

effective stresses. The upper plot shows the time history of the pore-pressure difference, Δu, across the specimen. The deformation-induced component of this pressure difference, designated as the baseline, is caused by CRD deformation with single drainage. The additional periodic changes in this pressure difference were induced by generating a flow rate through the specimen with the infuse-withdraw actuator driven by a single-carriage flow pump. Approximately 20 constant-flow hydraulic conductivity measurements were thus obtained during consolidation for about 93 hours and subsequent rebound for an additional 12 hours.

Constitutive relations derived from these data are shown in figure 6. The lower plot shows: the void ratio (e) versus the vertical effective stress (σ_v'); the void ratio vs. the horizontal effective stress, σ_h'; and the void ratio versus K_O, where $K_O = \sigma_h'/\sigma_v'$. The upper plot shows the void ratio versus hydraulic conductivity (k), and also the void ratio versus specific storage (S_S) relationships. The specific storage is calculated from the compressibility, (a_v), which is the slope of the void ratio versus vertical effective stress relationship (de/dσ_v').

A limitation of the triaxial system in figure 4 arises from three sources of compliance that tend to cause deviations from the zero-radial-strain condition: (a) compressibility of the pore water in the test specimen and the interconnected water in the tubing and infuse-withdraw actuators; (b) compressibility of the water in the hydraulic loading chamber and the interconnected water in the tubing and infuse-withdraw actuators; and (c) vertical elongation of the test cell induced by axial force in the specimen. To minimize errors from these factors, very high fluid pressures were used in the test specimen, the triaxial chamber, and the hydraulic loading system. In addition, the vertical elongation of the test cell in response to the axial force generated by hydraulic loading was calibrated and taken into account in determining the thickness of the specimen throughout the test.

Willden (1991) examined deviations from the zero-lateral-strain condition by comparing direct measurements of the thickness and diameter of the specimens before and after CRD consolidation and rebound, and also by comparing the K_O values obtained during virgin consolidation with those compiled from the literature by Mayne and Kulhawy (1982). These comparisons are presented in Tables 1, 2, and 3 for four test specimens of the same kaolinite, including specimen T6 whose data were presented in figures 5 and 6.

Table 1 shows the radial strains were small compared with the vertical strains. For three of the specimens, the final diameters were approximately 0.5% larger than the initial diameters, after the specimens had been consolidated from about 12-18%. For specimen T5, the larger radial strain appears to be unreasonable, due to a value of the initial diameter that is low compared with the initial diameters of the other specimens. Comparison of Tables 2 and 3 shows the range and the average values of K_O obtained from the four kaolinite specimens (Table 2) are reasonable because they are within the range of values for kaolinite that Mayne and Kulhawy (1982) compiled from the literature (Table 3).

UNSATURATED SPECIMENS UNDER STRESS CONTROL

Figure 7 shows an experimental system for measuring hydraulic conductivity, matric potential, and the variation of these properties with the moisture content of a unsaturated specimen. Our initial goal for this system was to determine whether constant-flow hydraulic

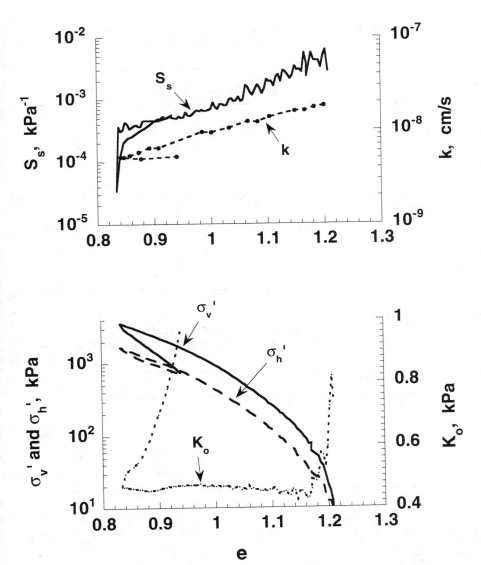

Figure 6--Constitutive relations derived from data in figure 5. Upper plot: void ratio (e) versus specific storage (S_S) and hydraulic conductivity (k). Lower plot: void ratio (e) versus vertical effective stress (σ_v'), horizontal effective stress (σ_h'), and the coefficient of lateral earth pressure ($K_o = \sigma_h'/\sigma_v'$).

TABLE 1--Initial and final dimensions and strains for kaolinite specimens tested by Willden, 1991

Sample number	T3	T4	T5	T6
Sample height				
Initial, cm	2.236	2.103	2.280	2.201
Final, cm	1.846	1.790	1.850	1.930
Vertical strain	17.4%	14.9%	18.9%	12.3%
Sample diameter				
Initial, cm	5.130	5.110	5.060	5.130
Final, cm	5.155	5.144	5.150	5.156
Radial strain	0.5%	0.7%	1.8%	0.5%

TABLE 2--K_o values measured by Willden, 1982 on Standard Air Floated Kaolinite from Georgia Kaolin Company

Specimen	Range	Average
T3	0.46-0.50	0.49
T4	0.42-0.49	0.45
T5	0.46-0.48	0.47
T6	0.44-0.47	0.46

conductivity measurements could be performed on an unsaturated specimen by infusing and withdrawing identical flow rates at opposite ends of the specimen with a flow pump equipped with an infuse-withdraw actuator [P].
The pedestal and top cap have 5-bar (high-air-entry) ceramic discs through which water is transmitted to or from the specimen. The upper disc has a center hole, and is sealed with O-rings in a doughnut shaped cavity in the top cap. A hole through the center of the top cap allows pore-air pressure in the specimen to be controlled by an external pressure regulator [PR]. The moisture content of the specimen is varied by using a second flow pump [W] to infuse or withdraw water at the base of the specimen. Insitu stresses on the specimen can be simulated using a differential pressure regulator [DPR] to maintain a constant pressure difference between the chamber pressure and the pore-air pressure. Alternatively the specimen volume can be controlled (to a degree limited

TABLE 3--K_o values compiled by Mayne and Kulhawy, 1982

Reference	Soil Name	K_o Value
Burland, 1967	kaolinite	0.69
Sketchley and Bransby, 1973	Spestone kaolinite	0.66
Parry and Wroth, 1976	kaolinite	0.64
Parry and Nadarajah, 1973	Spestone kaolinite	0.64
Moore and Cole, 1977	Australian kaolinite	0.56
Singh, 1971	kaolinite	0.51
Poulos, 1978	Sydney kaolinite	0.48
Moore and Cole, 1977	Australian kaolinite	0.44

by compliance of the system) by infusing or withdrawing water into/from the triaxial chamber with a third flow pump [V].

One side of each of three differential pressure transducers [M, N, and Q] are connected to one of the pore water lines into the base pedestal, where they monitor the water pressure below the 5-bar ceramic disc at the base of the specimen (P_b). The other side of transducer M monitors the water pressure in the top cap above the 5-bar ceramic disc on top of the specimen (P_t). The other side of transducer Q monitors the air pressure in the center hole of the top cap that connects with the pore-air pressure at the top of the specimen (P_a). The other side of transducer N monitors the chamber pressure (P_c).

The data presented below were obtained on an "undisturbed" sample of well-graded silty sand, having 14.5% finer than #200 sieve, from a recharge research site 3 miles north of Golden, CO. The sample was extracted from the bottom of a hole drilled with a 8.3-cm (3.25-inch) diameter hand auger, at a depth of 1.5 meters in soil overlying the approximate contact of the Denver and Arapahoe Formations (Kiusalaas 1992). The sample was obtained by using a 35.6-N (8-lb) slide hammer to advance a hollow-tube sampler equipped with a 5.08-cm diameter by 15.2-cm long brass liner. Subsequent moisture changes were minimized by placing end caps on the sample-filled liner immediately after extraction, and storing the sample in a humid room.

The test specimen was prepared by cutting a 5.08-cm section of the liner-covered sample with a diamond rock saw. The liner was cut lengthwise to eliminate frictional resistance while removing the specimen from its liner. The specimen was enclosed in a cylindrical latex membrane, mounted in the triaxial cell, and sealed between the base pedestal and top cap with O-rings. The triaxial cell chamber was filled with distilled and deaired water from an elevated reservoir (not shown). Prior to mounting the specimen, the hydraulic components of the system were saturated with distilled and deaired water. After mounting, the specimen was saturated by evacuating the air in its pore space with a vacuum pump attached to the pore-air line leading to the center hole in the top cap over the specimen. Distilled and deaired water, at atmospheric pressure in the elevated reservoir, was then allowed to flow

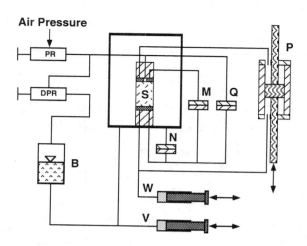

Figure 7--Triaxial system for unsaturated specimens under isotropic stress control

upwards through the specimen until it appeared in the pore-air line connected to the vacuum pump. The vacuum pump was then disconnected, the pore-air line was equilibrated with the atmosphere, and the differential pressure regulator was set to maintain the pressure in the chamber fluid 69 kPa (10 psi) greater than the pressure in the pore-air line at the top of the specimen. It appears reasonable to suppose that this procedure removes most of the pore air from the specimen; however, the efficiency of this procedure for saturating test specimens has yet to be determined.

The time history of a test on the above specimen is illustrated in figures 8, 9, and 10. ΔP_{a-b} is the difference between the pore-air pressure at the top of the specimen and the water pressure below the porous stone in the base pedestal. When there is no flow through the system, and hence no head loss through the ceramic discs, this pressure difference equals that between the pore-air pressure and the pore-water pressure within the specimen, P_m, which is commonly referred to as matric potential, or suction. ΔP_{t-b} is the difference between the water pressure above the ceramic disc in the top cap, and the water pressure below the ceramic disc in the base pedestal. The response of this pressure difference to an externally generated flow of water through the specimen arises from resistance to water flow, both through the specimen and also through the ceramic discs above and below the specimen. When there is no externally generated flow through the system, finite values of ΔP_{t-b}, which are generally very small, may arise from sources within a test specimen, such as gradients in the composition and properties of the materials and/or chemical reactions among the constituents (Olsen, Nichols, and Rice 1985). In our experience, significant magnitudes of ΔP_{t-b} from sources within test specimens arise primarily in materials

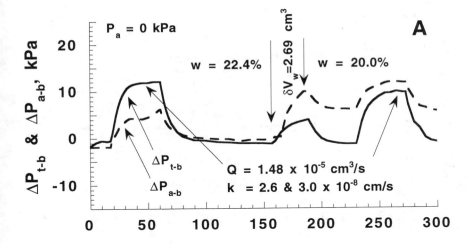
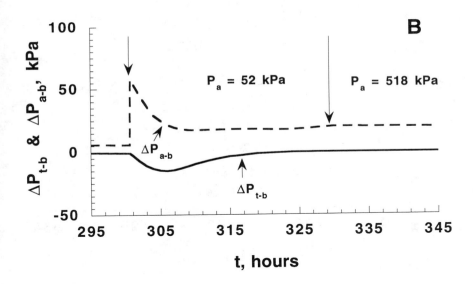

Figure 8--Data obtained during elapsed time (t) from 0 to 345 hours for a specimen in the triaxial system illustrated in figure 7. ΔP_{a-b} is difference between pore-air pressure at top of specimen and water pressure below the porous stone in the base pedestal. ΔP_{t-b} is difference between water pressures above and below the porous stones in the top cap and the base pedestal.

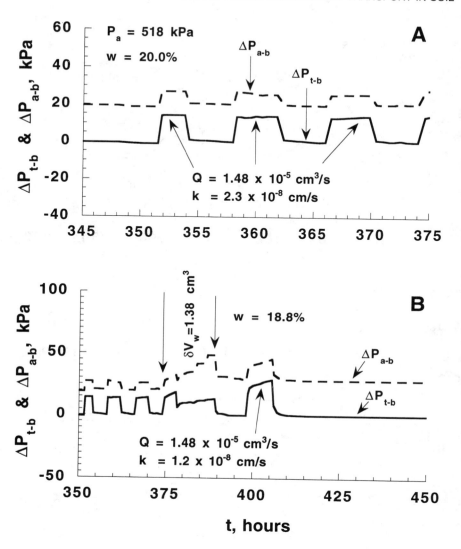

Figure 9--Data obtained during elapsed time (t) from 345 to 450 hours for a specimen in the triaxial system illustrated in figure 7. ΔP_{a-b} is difference between pore-air pressure at top of specimen and water pressure below the porous stone in the base pedestal. ΔP_{t-b} is difference between water pressures above and below the porous stones in the top cap and the base pedestal.

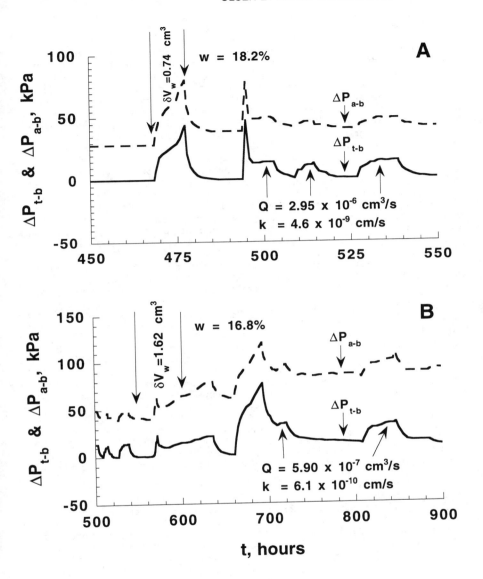

Figure 10--Data obtained during elapsed time (t) from 450 to 900 hours for a specimen in the triaxial system illustrated in figure 7. ΔP_{a-b} is difference between pore-air pressure at top of specimen and water pressure below the porous stone in the base pedestal. ΔP_{t-b} is difference between water pressures above and below the porous stones in the top cap and the base pedestal.

with extremely low hydraulic conductivities, on the order of 10^{-9} cm/s or less.

Figure 8A shows the first 300 hours of testing when the pore-air pressure was maintained at atmospheric pressure. About halfway through this period, the moisture content of the specimen (in % of dry soil weight) was reduced from 22.4% to 20.0% by extracting 2.69 cm^3 of pore water from the specimen. Before and after this change in moisture content, hydraulic-conductivity measurements was conducted by generating a flow rate of 1.48×10^{-5} cm^3/s through the specimen with a single-carriage flow pump equipped with an infuse-withdraw actuator. The hydraulic conductivity values shown were calculated from the magnitudes of ΔP_{t-b} induced by the externally imposed flow rate. Recall that the induced values of ΔP_{t-b} include resistance to flow through both the specimen and the ceramic discs. The hydraulic conductivity did not change appreciably after reducing the moisture content, probably because the magnitude of ΔP_{t-b} was dominated by resistance to flow through the ceramic discs. Low hydraulic conductivity of the ceramic discs is further reflected in the induced response of ΔP_{a-b} that parallels the induced ΔP_{t-b} during the hydraulic-conductivity measurements. However, the matric potential, P_m, which equals ΔP_{a-b} when there is no flow through the specimen, increased from nearly zero to about 6 kPa after reducing the moisture content from 22.4% to 20.0%. Finally, note that the response times for these hydraulic conductivity measurements are on the order of 20-30 hours.

Figure 8B shows the period from 300 to 345 hours when the pore-air pressure was elevated, first to 52 kPa (7.5 psi), and second, to 518 kPa (75 psi). During the first step, ΔP_{a-b} shows the increase in pore air pressure generated a pressure difference across the lower ceramic disc that dissipated at a decreasing rate over a period of about 10 hours. Similarly, ΔP_{a-b} shows that the pore air pressure increase generated a transient difference between the fluid pressure below the lower ceramic disc and the fluid pressure above the upper ceramic disc that dissipated in about 20 hours. These transient responses were avoided in the second step by elevating the pore-air pressure and the water pressures above and below the porous stones simultaneously. The data show the matric potential, P_m, which equals ΔP_{a-b} when there is no flow through the specimen, was increased from about 6 kPa to about 16 kPa during the first step, and from about 16 kPa to about 18 kPa psi during the second step.

Figure 9A shows the period from 345 to 375 hours when three hydraulic conductivity measurements were obtained while the pore-air pressure and the moisture content were maintained constant at 518 kPa and 20.0%, respectively. Note that the matric potential remained constant at 18 kPa during this time interval, and further note that this 30-hour time interval is one-tenth of the 300-hour time interval in figure 8A. The increase in pore-air pressure from 0 to 518 kPa decreased the response time for hydraulic-conductivity measurements from more than 20 hours to less than 1 hour. This change in response time is not accompanied by a significant change in the hydraulic conductivity.

Figure 9B shows the period from 350 to 450 hours. The first part of this period includes the three hydraulic conductivity tests in figure 9A, at a substantially condensed time scale. Thereafter, 1.38 cm^3 of porewater was extracted which decreased the moisture content from 20.0% to 18.8%, increased the matric potential from about 18 kPa to about 30 kPa, and reduced the hydraulic conductivity from 2.0×10^{-8} to 1.03×10^{-8} cm/s.

Figure 10A shows the period from 450 to 550 hours where the moisture content was decreased to 18.2%, which increased the matric potential to about 40 kPa and reduced the hydraulic conductivity to 1.03×10^{-9} cm/s. Figure 10B shows the period from 500 to 900 hours where the decrease in the moisture content to 16.8% increased the matric potential to about 90 kPa, and reduced the hydraulic conductivity to 5.19×10^{-10} cm/s. Comparison of the response times for hydraulic conductivity measurements in figure 9 and 10 show they increased substantially with decreasing hydraulic conductivity.

One complication in the behavior of ΔP_{t-b} first arose in figure 9B, and became more pronounced in figures 10A and 10B. In figure 9B during the hydraulic conductivity test between 395 and 410 hours, ΔP_{t-b} continued creeping upward rather than reaching a constant steady state value. In figure 10A during the hydraulic conductivity tests between 490 and 505 hours, ΔP_{t-b} initially climbed at a rate that did not decrease with time towards a constant steady state; however, a steady state was reached by reducing the externally imposed flow rate by a factor of 3, from 14.8 to 2.95×10^{-6} cm^3/s. This behavior occurs again in figure 10B during the hydraulic conductivity test between 630 and 720 hours. ΔP_{t-b} initially climbed at a rate that did not decrease with time towards a constant steady state; however, an approximate steady state was reached by reducing the externally imposed flow rate by a factor of 5, from 2.95×10^{-6} to 5.90×10^{-7} cm^3/s. A second complication in the behavior of ΔP_{t-b} first appears in figure 10B, following the hydraulic conductivity test between 630 and 720 hours and again following the test between 810 and 860 hours. ΔP_{t-b} does not return to zero after terminating the externally imposed flow rate. A pressure difference on the order of 10 kPa persists across the specimen.

Figure 11 summarizes the variations of hydraulic conductivity and matric potential that were obtained concurrently at successively lower moisture contents. The nearly constant hydraulic conductivity for moisture contents above 20% probably reflects the upper limit controlled by the hydraulic conductivity of the 5-bar ceramic discs.

Figure 12 shows, in the upper plot, the interrelationship between the hydraulic conductivity and matric potential values in figure 11. The lower plot in this figure shows how the response time (RT) varies with pore-air pressure and decreasing magnitudes of the hydraulic conductivity. Note that the variation of response time with hydraulic conductivity, for elevated pore-air pressures at 518 kPa, are comparable to the response times for hydraulic-conductivity measurements in saturated specimens that were summarized at the end of the previous section of the paper concerning saturated specimens under isotropic deformation.

DISCUSSION

Our experience in this and previous studies (Olsen et al 1985, 1988, 1991) shows response times for constant-flow hydraulic conductivity (k) measurements in triaxial systems to be on the order of: (a) a few minutes or less for materials with k values greater than 10^{-6} cm/s, (b) a few hours for materials with k values in the range of 10^{-8} to 10^{-9} cm/s; and (c) a few days for materials with extremely low k values, in the range of 10^{-10} to 10^{-11} cm/s. In contrast, Benson and Daniel (1990) report the following time intervals for falling-head hydraulic conductivity tests in rigid-wall and flexible-wall

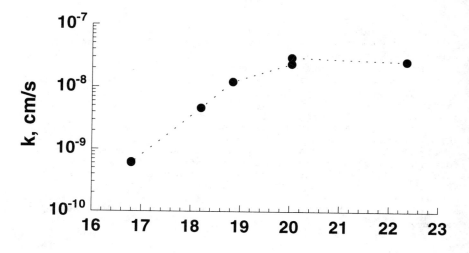
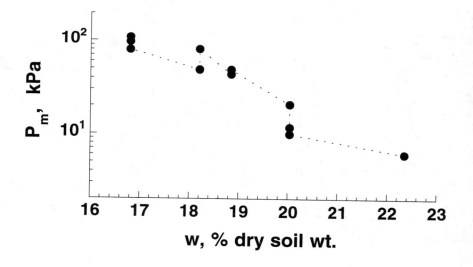

Figure 11--Moisture content (w) in percent of dry soil weight versus hydraulic conductivity (k) [upper plot] and matric potential (P_m) [lower plot].

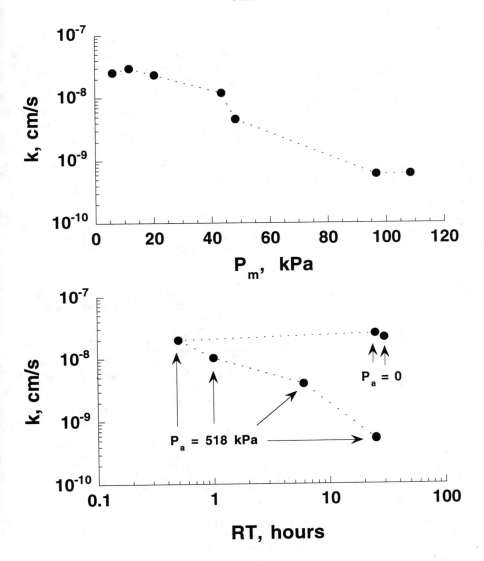

Figure 12--Hydraulic conductivity (k) versus matric potential (P_m) [upper plot] and response time (RT) [lower plot].

permeameters: "Tests on soil with $k > 10^{-7}$ cm/s typically lasted for about 2 weeks; soils with $k < 10^{-7}$ cm/s were permeated for several months." This comparison illustrates the previously reported advantage of the constant-flow method, compared with conventional constant-head and falling-head methods; namely, that hydraulic conductivity measurements can be obtained much more rapidly and at substantially smaller hydraulic gradients.

The development of the triaxial system in figure 4 for saturated specimens under one-dimensional deformation was preceded by a previously described system for superimposed constant-flow hydraulic conductivity and CRD-deformation measurements in an Anteus backpressured one-dimensional consolidometer (Gill, Olsen, and Nelson 1991; Olsen et al. 1991). One advantage of the triaxial system is that it provides continuous data on the horizontal effective stress, which is unknown in conventional one-dimensional consolidometers. Another advantage is that many earth materials can be tested in triaxial systems that cannot be properly trimmed for mounting in one-dimensional consolidometers. Such materials include very stiff materials like sandstones and shales. They also include irregularly-shaped materials which cannot be trimmed to the exact and uniform diameter needed to avoid sidewall leakage in one-dimensional consolidometers.

The data obtained with the triaxial system in figure 4 illustrate the infuse-withdraw actuator is a useful tool, not only for hydraulic conductivity measurements, but also for strain-path control in triaxial systems. In this study, strain-path control is limited to the special case of one-dimensional deformation, commonly referred to as K_o consolidation or rebound, wherein the volumetric and axial deformation rates in the specimen are maintained equal. By adding an additional flow pump for infusing or withdrawing hydraulic fluid to or from the hydraulic loading chamber, the system would be capable of controlling arbitrary combinations of vertical and volumetric deformation rates for a wide range of strain paths during either consolidation or shear phases of a triaxial test.

This volumetric approach to strain-path control in triaxial systems is attractive in that it avoids the need for instrumentation to monitor radial deformations and to counteract these deformations with feedback-induced chamber-pressure adjustments. However, the evidence presented in this paper suggests small deviations occurred from the requirement of zero radial strain for one-dimensional consolidation and rebound. It remains to be determined whether, and under what conditions, these deviations can be considered negligible for practice and research, and also whether these deviations can be minimized with a more rigid triaxial cell. Nevertheless, the radial strain is small compared with the vertical strain, and therefore the system in figure 4 provides a very substantial improvement over conventional triaxial systems that allow only isotropic consolidation and rebound for simulating insitu stress conditions.

The data obtained with the triaxial system in figure 7 demonstrates the feasibility of concurrent constant-flow hydraulic conductivity and matric potential measurements, and the variation of these properties with the degree of saturation, for specimens of undisturbed and unsaturated soils mounted in triaxial cells. However, the data show there are limitations to the applicability of this approach when hydraulic conductivity values become extremely low, on the order of 10^{-10} cm/s. Also the system described has limitations when the hydraulic conductivity of a specimen exceeds that of the porous stones above and below the specimen. Steps we are pursuing to minimize these limitations include: (a) adding the capability of measuring the water pressure difference across the specimen at the interfaces between the specimen surfaces and the high-air-entry porous stones, (b) adding the capability of measuring air permeability on either the same specimen or a replicate specimen, and (c) developing a more rigid triaxial cell that allows volume control of the specimen by infusing or withdrawing liquid to or from the triaxial cell chamber.

SUMMARY AND CONCLUSIONS

New capabilities for hydrologic property measurements in triaxial systems include: (1) volume-controlled and simultaneous measurements of hydraulic conductivity and one-dimensional consolidation (or specific storage) of a saturated test specimen; and (2) volume-controlled measurements of hydraulic conductivity, matric potential, and the variation of these properties with the moisture content of an unsaturated test specimen.

For volume control of hydraulic conductivity measurements in both saturated and unsaturated test specimens, a single-carriage flow pump equipped with an infuse-withdraw actuator generates identical flow rates into and out of opposite ends of a test specimen. For volume control of one-dimensional deformation of saturated specimens, a dual-carriage flow pump equipped with two identical infuse-withdraw actuators generates identical flow rates of water from (or into) the ends of a test specimen and into (or from) a hydraulic piston that deforms the specimen vertically. By dimensioning the hydraulic piston to maintain the volumetric and vertical deformations equal, radial deformation of the specimen is prevented. An additional flow pump provides volume control of the moisture content of an unsaturated specimen.

Data on saturated kaolinite demonstrate integrated hydraulic-conductivity and one-dimensional consolidation tests with continuous monitoring of both vertical and horizontal effective stresses. For many applications, this capability provides a means to conduct similar tests on a wide variety of earth materials including very stiff claystones, shales, and sandstones that cannot be properly trimmed for mounting in conventional consolidometers. For some applications, refinements may be needed to minimize or eliminate small deviations from the requirement of zero-lateral-strain that arise from compliance in the equipment.

Data on well-graded silty sand demonstrate the feasibility of concurrent constant-flow hydraulic conductivity and matric potential measurements, and the variation of these properties with moisture content, for undisturbed specimens mounted in triaxial cells. Refinements needed to realize the full potential of this approach include an improved technique for measuring hydraulic-head differences within the specimen and a more rigid triaxial cell that allows volume control of the specimen by infusing or withdrawing liquid to or from the triaxial cell chamber.

REFERENCES

Aiban, S. A. and Znidarcic, D., 1989, "Evaluation Of The Flow Pump And Constant Head Techniques For Permeability Measurements," Geotechnique, Vol. 39, No. 4, pp. 655-666.

Benson, C. H., and Daniel, D. E., 1990, "Influence Of Clods On Hydraulic Conductivity Of Compacted Clay," Journal Of Geotechnical Engineering, Vol. 116, No. 8, pp. 1231-1248.

Gill, J. D., Olsen, H. W., and Nelson, K. R., 1991, "Volume-Controlled Approach For Direct Measurements Of Compressibility And Hydraulic Conductivity," Geotechnical Engineering Congress 1991, Geotechnical Special Publication No. 27, Vol. 2, ASCE, pp. 1100-1111.

Kiusalaas, Nickolas J., 1992, "Estimation of Groundwater Recharge Using Neutron Probe Moisture Readings Near Golden, CO," Master of Science Thesis, Colorado School of Mines, Golden, CO 80401.

Mayne, P. W. and Kulhawy, F. H., 1982, "K_0-OCR Relationships In Soil," Journal Of The Geotechnical Engineering Division, Proceedings Of The American Society of Civil Engineers, Vol. 108, No. GT6, pp. 851-872.

Olsen, H. W., Nichols, R. W., and Rice, T. L., 1985, "Low-Gradient Permeability Measurements In A Triaxial System," Geotechnique, V. 35, No. 2, pp. 145-157.

Olsen, H.W., Morin, R.H., and Nichols, R. W., 1988, "Flow Pump Applications In Triaxial Testing," in Robert T. Donaghe, Ronald C. Chaney, and Marshall L. Silver, eds., Advanced Triaxial Testing of Soil and Rock, ASTM STP 977, American Society for Testing and Materials, Philadelphia, pp. 68-81.

Olsen, H. W., Gill, J. D., Willden, A. T., and Nelson, K. R., 1991, "Innovations In Hydraulic-Conductivity Measurements," Geotechnical Engineering 1991, Transportation Research Record 1309, Transportation Research Board, National Research Council, pp. 9-17.

Willden, Arthur T., 1991, "K_0 Consolidation With Simultaneous Hydraulic-Conductivity Measurement In A Modified Triaxial Cell Using Flow-Pump Methods," Master of Science Thesis, Colorado School of Mines, Golden, CO 80401.

Znidarcic, Dobroslov, Illangesekare, Tissa, and Manna, Marilena, 1991, "Laboratory Testing and Parameter Estimation For Two-Phase Flow Problems," Geotechnical Engineering Congress 1991, Geotechnical Special Publication No. 27, Vol. 2, ASCE, pp. 1089-1099.

Sibel Pamukcu[1], Ilker B. Topcu[2], and Cengiz Guven[3].

HYDRAULIC CONDUCTIVITY OF SOLIDIFIED RESIDUE MIXTURES USED AS A HYDRAULIC BARRIER

REFERENCE: Pamukcu, S., Topcu, I. B., and Guven, C., "Hydraulic Conductivity of Solidified Residue Mixtures used as a Hydraulic Barrier," Hydraulic Conductivity and Waste Contaminant Transport in Soil, ASTM STP 1142, David E. Daniel and Stephen J. Trautwein, Eds., American Society for Testing and Materials, Philadelphia, 1994.

ABSTRACT: The feasibility of using a solidified product of steel process residue in construction of a barrier system was investigated in the laboratory and in-situ. A solidified product of the dewatered sludge residue, filter cake, which exhibited high strength and low hydraulic conductivity was considered as a barrier cap for an existing lagoon. Variations in the placement and curing methods in the field, along with additive composition of the base mixture, influenced the strength development, hydraulic conductivity and durability of the resulting material. Strength of the material did not correlate well with hydraulic conductivity. However, high material strength appeared to increase resistance to freeze-thaw effects, thus demonstrating resistance to deterioration of the desired low permeability. Compositional variations such as proportion of surface reactive agents, water to cement ratio and the proportion of cement influenced strength as well as the measured hydraulic conductivity in the laboratory. Finally, a comparative analysis of fraction of three heavy metals removed by water permeation of three different composite samples demonstrated that chemical processes during the solidification may play an active role in the performance of the final product. Three slightly different composites of fairly high strength and low hydraulic conductivity showed selective release of cadmium from the solidified material.

KEYWORDS: Solidification, residual materials, hydraulic conductivity, strength, durability, freeze-thaw, leaching of heavy metals

---

1 Associate Professor, Department of Civil Engineering, Lehigh University, Fritz Engineering Laboratory, Bethlehem, Pa 18015

2 Assistant Professor, Department of Civil Engineering, Anadolu University, Eskisehir, Turkey

3 Research Engineer, Environmental Studies Center, Lehigh University, Bethlehem, Pa 18015

INTRODUCTION

Reuse of residual materials in constructed or engineered facilities has gained much attention over the last decade (Ahmed and Lowell 1992). Present-day need for feasible ways of resource recovery and minimization of waste and residual materials necessitate further work in this area. One process which has been developed over the last 30 years is the 'Chemical Fixation and Solidification' (CFS) technology for liquid wastes and sludges (Conner 1990; Cullinane and Jones 1986; Jones 1990). Chemical fixation or stabilization and solidification technologies have been used by geotechnical engineers in ground improvement (e.g. soil stabilization, soil and rock grouting) prior to that. The types of materials used as stabilizing agents have ranged from earthen (lime, gypsum) and siliceous pozzolin (cement, fly ash) materials to various resins and other polymerizing agents (Winterkorn and Pamukcu 1991). ASTM standards have been developed for mixing, testing and applying most of these composites of soil and cementitious materials in construction of earthen structures. A comprehensive set of standards are not yet available for development of stabilized or solidified products of waste or residual materials. The best studied materials among the industrial residues are furnace flyashes and slags (Ahmed and Lowell 1992). Standardization of CFS procedures for waste materials is perhaps difficult owing to the heterogeneity and inconsistency of many of these materials. Such variations in the chemistry and physical properties of these materials dictate that each process be compatible with the specific waste constituents. Therefore each process must be developed independently to optimize selected physical and chemical properties in the finished product. Selection of those properties to optimize - such as hydraulic conductivity, leachability, strength, rigidity and durability - should depend on the intended function of the final product.

Stabilized (chemically fixed) and solidified waste or residual materials, especially those considered for reuse, are often required to possess the following properties: (a) minimized solubility and mobility (leachability) of toxic substances, (b) minimized permeability and thus leachability and volume change, (c) high physical durability, strength and stability. Stabilization of a residue or waste is the chemical modification of the material to detoxify it or reduce the mobility of its toxic constituents. This process may not necessarily improve the physical properties, such as strength. Solidification, on the other hand, refers to the encapsulation of the waste material in a solid matrix with high structural integrity and often high strength. In United States, compliance with the requirement of minimized solubility and mobility of toxic substances are assessed by the EPA Extraction Procedure Toxicity (EPT) or Toxicity Characteristic Leaching Procedure (TCLP) tests. The hydraulic conductivity of the product is sometimes used as an indicator of leachability, since reduced hydraulic conductivity is expected to result in reduced rate of solubility and transport of toxic substances through the porous matrix by advection.

The feasibility of using a solidified product of a steel process residue in construction of a barrier or capping layer was investigated in the laboratory and *in-situ*. The residue, a dewatered sludge of fine particulate material (filter cake), is mixed with pozzolanic additives and fillers which then produce a concrete-like material when solidified. The intended function of this recycled material was to place it as a capping layer over an existing lagoon to reduce infiltration of surface waters to the layers below. This required minimization of hydraulic conductivity and maximization of the durability of the capping material. This has been a long-term study involving various phases of laboratory and field work. Earlier results and discussions pertaining to strength development, durability and quality control in field application, and subsequent performance of the solidified product, have been presented elsewhere (Pamukcu et al. 1989, 1991). This work presents and discusses the effect of mix design on hydraulic conductivity and short-term leachability of the solidified residue composites. Also, the effects of freeze-thaw on hydraulic conductivity of a few solidified mixtures were investigated in the laboratory. A few of the laboratory findings were compared with field measurement of hydraulic conductivity in trial applications of pilot scale capping.

BACKGROUND

A cement-based solidification process was used to treat a steel industry water quality control station sludge bearing heavy metals (Fe($\approx$ 50%); Cr($\approx$2%);Zn ($\approx$1%), Pb, Ni, Cu (less than 1%)),calcium and magnesium oxides($\approx$10%), silicate and aluminum oxide ($\approx$5%), free carbon ($\approx$2%), oils and greases ($\approx$1%), and other trace materials. The sludge was first mechanically dewatered, producing a moist, light weight fine particulate material or a 'filter cake', designated as WQCS-7. The specific gravity of the filter cake was measured at 1.56 and its particle size was relatively fine and uniform, on the order of a few microns ($\approx$80% passing No. 200 sieve). The filter cake, owing to a high affinity for water, contained 40 to 45% water by total weight, even in the dewatered state. When mixed with water the iron rich filter cake set up fairly rapidly, demonstrating cementitious character. The raw filter cake, as well as its solidified product, were subjected to TCLP and EPT tests by the material producer. The results of these tests showed that the materials were non-toxic and represented no threat to the environment.

The initial solidification treatment work started in 1988 with the laboratory investigations of trial composites of solidified filter cake for durability and permeability. A mix formula, designated as **T-14**, was selected based on the laboratory findings of its low hydraulic conductivity (on the order of 10^{-7} cm/s) and good durability with respect to material loss (10% material loss upon 12 cycles of freeze-thaw). Subsequently, by late Fall 1989, a cap of 15 cm thickness was constructed over a 16,200 m² desiccated sludge lagoon using the T-14 mixture. The mixture was a thick slurry composed of <u>37% filter cake</u>, <u>16% cement</u> (type II), <u>35% slag</u> (between 6.35 mm and 38.1 mm sizes) and <u>12% water</u>, by weight. In this construction, the lagoon surface was first graded and a 2.5 mm thick ($\approx$ 100 mil) polyethylene plastic liner was placed over the surface to prevent migration of slurry water into the desiccated zone below. Slurry mixture was poured onto the plastic liner in 15 cm thick slab sections of 7.6 m by 61 m. Control and expansion joints were cut into the freshly poured material and they were then sealed with a rubber silicone sealant. To control the surface moisture evaporation, a thin coating of rubberized sealer and 2.5 mm plastic liner was applied over each section immediately after finishing the surface. In Spring of 1990, after the first freeze-thaw, extensive surface cracking and spalling occurred on the cap, followed by complete deterioration of a number of the slabs. These failures were observed to be random in frequency and location. This observation suggested that variations in the conditions at the time of each pour were responsible for the random failures. Such factors may be curing conditions and field mixture composition such as water content. The results of the earlier laboratory investigations and those on the probable causes of random cracking and spalling have been presented elsewhere (Pamukcu et al. 1989, 1991). Based on the information gathered from the field application and analysis, two pilot scale sections were constructed in the field to observe long-term durability. The first section was intended to investigate the effects of curing conditions on durability and permeability of the original T-14 mixture. The second section was constructed a year later to investigate the effects of variations in the basic mixture composition on durability, strength and permeability.

Field Slabs of T-14 Composites

The first section was constructed on a relatively stable corner of the original cap. It was composed of 6 slabs of 4.9 m x 7.3 m with expansion joints along the larger dimension. The thickness of these slabs were all 15 cm. The composition of T-14 was slightly changed in each slab, one being the base mixture and the other five containing either fine slag (less than 6.35 mm sizes) or coarse slag (between 6.35 mm and 38.1 mm sizes) and variable quantities of mixed water, designated as high (12%), medium (11%) or low (10%) by weight. The slag characterized as furnace slag with specific gravity of 3.57, and moisture content of 9% by dry weight. The water/cement ratio was kept approximately constant for all the slabs, and the percentages of ingredients were redistributed according to each composition. One extra slab was constructed with reduced cement content and added fly ash. Each slab was subdivided into 4 subsections of 2.45 m x 3.65 m by control joints. These subsections were each treated with a different curing cover: 75-mm and 150-mm thick wet sand, wet burlap, rubberized sealer and plastic liner. A selected number of the slabs were instrumented with infiltrometers to assess the magnitude and time variation of their field hydraulic conductivity. As these slabs were left to cure, laboratory specimens of the solidified WQCS-7 filter cake were prepared with minor variations of the T-14 compositions. These variations were water/cement ratio, surface active reagents percentages, and slag size. Sufficient number of replicate specimens of various compositions were prepared and tested for strength, durability and permeability after 28 days of curing in the humidity chamber.

Field Slabs of New Composites

The second section was constructed on prepared foundations on the cap. The existing surface was wetted and compacted initially, then over laid with either plastic liner, compacted slag, or compacted fill materials. Two sets of slabs were poured which were identical in composition but varied in thickness as 15 cm and 30 cm. Each set consisted of 5 different compositions of the solidified product. In these compositions the percentage as well as the type of base ingredients were changed. In each of the mixtures the main ingredient was the WQCS-7 filter cake with the highest percentage (32-37%). Some of the compositions contained minor quantities of high strength developing agents such as microsilica. These 5 different mixtures were designated as **L18, E7, RI, KB and KW**. The L18 and E7 mixtures were proposed by the Lehigh University group. Each of the other three mixtures were proposed by a commercial group. Two additional slabs were constructed using standard concrete mix and slag concrete mix - in which the aggregate of standard concrete was replaced by slag of comparable gradation. At the time of the field construction, large (30.4 cm length x 15.2 cm diameter) and small (15.2 cm length x 7.6 cm diameter) specimens of each composition were poured into waxed heavy cardboard cylinders. Once the slurry mixture set up, the cylinders were removed and exposed specimens were left at the site to cure under the same conditions as did the slabs. At the end of the four-week curing period, the cylindrical samples were transported to the laboratory for strength and permeability testing. A sufficient number of replicate specimens of each composition were poured at the site to facilitate tests on multiple specimens and assess repeatability.

INVESTIGATION

Laboratory and Field Test Results of T-14 Composites

The composite referred to as T-14 was the base mixture originally applied at the lagoon site. In the pilot scale field applications this base mixture was supplemented with surface reactive agents such as air entrainment and superplasticizer reagents. These were added at rates of 2 to 4% by weight of cement. The main contribution of these agents expected were improved durability of the composite against freeze-thaw (air entrainment reagent) and improved flowability and thus better mixing before placement (superplasticizer reagent).

When preparing laboratory specimens of T-14, the composition was varied slightly to investigate the effect of the additives, slag size and water/cement ratio on strength, durability and hydraulic conductivity of the solidified products. In these tests, replicate specimens of each composition were prepared and cured in a humidity chamber (25°C and 98% humidity) for 28 days. Subsequently, they were divided into 5 sets, each set often having more than one replicate specimen. One set of specimens was tested for unconfined compressive strength immediately. A second set of specimens was subjected to 12 cycles of freeze-thaw following the standard procedure of ASTM D4842. A third set was used as the control group according to D4842; another set was dried to constant weight at 65°C. The last set of specimens was tested for hydraulic conductivity according to ASTM D5084 *Constant Rate of Flow Test* method with backpressure. The initial degree of saturation for these specimens varied from about 84% to 91%. Near saturation condition was therefore achieved without large backpressures. A few of the specimens that underwent freeze-thaw were tested for hydraulic conductivity afterward to assess the influence on hydraulic conductivity.

Table 1 presents the results of strength, durability and hydraulic conductivity testing for these T-14 specimens. As observed from this Table, the hydraulic conductivity ranged between 10^{-8} and 10^{-6} cm/s. The lowest hydraulic conductivity was achieved when the proportion of cement was increased. The addition of surface reactive reagents appeared to influence hydraulic conductivity as well as strength. In the sample group designated as S7-1, the water/cement ratio was kept constant while the other compositional parameters were varied. It appeared that there may be an optimum percentage of the surface reactive agents for the highest strength and lowest conductivity of the material. When no air entrainment agent was added, the specimen exhibited the highest permeability. The permeability reduced an order of magnitude when 4% air entrainment agent (by weight of the cement) was mixed in. No clear trends were observed between hydraulic conductivity and strength of the materials. The ratios of the compressive strength after freeze thaw, $q_{f/t}$, to compressive strength of the control specimen, q_c, were greater, in general, than one. This suggested that reduced rate of curing at the reduced temperature perhaps promoted the strength increase. Reduced rate of curing may have restrained formation of matrix cracks and fissures owing to lowered rate of water evaporation. A similar situation was also observed when a number of specimens cured outside under ambient temperature and humidity exhibited higher strengths than those cured in the humidity chamber. The average temperature (averaged over day and night over 28 days) was lower than the humidity chamber temperature of 25°C. The lowest permeability in this group was achieved when the cement proportion was increased to 20%. The change in the gradation of the slag did not influence the permeability as observed by the specimen S7-2/1. The effect of increased cement content in reducing hydraulic conductivity of the composite, even at a higher water/cement ratio, was also observed in the second group of samples (S7-2 series).

Two specimens designated as S2 and S6 were mixed using dewatered sludge material from two existing lagoons on the site. These materials were aged and found to contain a higher concentration of iron oxides as also confirmed by the higher specific gravity of the material (Table 2). These two specimens exhibited significantly higher strengths and lower hydraulic conductivity with 16% cement content. Compressive strength was comparable to that of standard concrete. Although no consistent relationship was observed between strength and hydraulic conductivity of the composites, it appeared that specimens exhibiting strengths above a threshold value of around 3.00 MPa also showed reduced hydraulic conductivity.

A number of specimens were tested for hydraulic conductivity immediately after the 12th freeze-thaw cycle. Since a majority of the specimens were intact and did not experience significant weight loss (Table 1), replicate specimens that were not subjected to unconfined compression could be used for hydraulic conductivity testing. Table 2 shows the results of these tests. As observed, the specimens exhibiting

TABLE 1--Strength, durability and hydraulic conductivity of T-14 samples

SAMPLE GROUP	NO of SAMP.	COMPOSITIONAL VARIATION FROM BASE MIXTURE T-14				UNCONFINED COMPRESSIVE STRENGTH (MPa)					FREEZE THAW DURABILITY		HYDRAULIC CONDUCT. (cm/s)
		Slag size (cm) [1]	R1-R2 (%) [2]	W/C [3]	C(%) [4]	Cured Ambient Cond. 28 days	Cured Hum. Ch. 28 days	After Freeze Thaw, [5] $q_{f/t}$	Control of Freeze Thaw, q_c	Oven Dried @ 65°	$q_{f/t}/q_c$	Weight Loss (%)	
S7-1/1	1	-2.54	0-4	0.78	16	...	2.38	2.96	2.81	1.65	1.05	1.3	6.60×10^{-7}
S7-1/2	1	-2.54	2-2	0.78	16	...	1.81	2.22	1.90	1.32	1.17	0.1	3.20×10^{-7}
S7-1/3	1	-2.54	7-7	0.78	16	...	1.11	1.74	1.21	0.95	1.44	0.0	1.14×10^{-7}
S7-1/4	4	-2.54	4-4	0.78	16	...	2.66	2.97	2.47	1.84	1.05	1.2	6.45×10^{-8}
S7-1/5	1	-1.9,+0.9	4-2	0.78	20	4.94	3.65	4.84	3.66	2.26	1.32	0.0	1.5×10^{-8}
S7-2/1	2	-1.9,+0.9	4-2	1.0	16	2.48	1.82	2.04	2.47	1.10	0.83	0.0	4.01×10^{-7}
S7-2/2	1	-0.31	4-2	1.0	16	3.11	2.79	2.33	1.96	1.16	1.18	0.0	3.06×10^{-7}
S7-2/3	1	-2.54	4-2	1.1	16	1.72	2.52	2.53	2.34	1.33	1.08	0.0	4.90×10^{-7}
S7-2/4	3	-2.54	4-2	1.2	16	...	0.92	0.99	1.01	0.51	1.10	1.0	4.75×10^{-7}
S7-2/5	2	-2.54	4-0	1.2	16	...	2.30	2.64	2.07	1.87	1.27	0.0	1.7×10^{-7}
S7-2/6	2	-2.54	4-2	1.0	20	3.17	3.40	3.01	2.80	3.10	1.07	0.0	8.7×10^{-8}
S2*-1/1	1	-2.54	4-2	0.75	16	7.41	5.94	7.04	8.11	7.03	0.86	0.0	1.8×10^{-8}
S6*-1/1	1	-1.9,+0.9	4-2	0.93	16	5.96	5.87	12.29	13.97	7.34	0.88	0.0	2.4×10^{-8}
Slag Concrete ✧	1	-2.54	4-2	0.75	16	0.76	0.90	1.25	1.54	1.01	0.81	0.0	1.18×10^{-7}
Agg. Concrete ✧	1	...	...	0.33	24	7.03	7.78	12.60	13.30	11.09	0.94	0.0	3.14×10^{-7}

[1] (-) smaller than, (+) larger than the indicated size
[2] R1: air entrainment reagent; R2: superplasticizer; by weight of cement
[3] W: water; C: cement; water/cement ratio
[4] C: cement percentage by total weight
[5] The freeze thaw is performed after 28 day humidity chamber curing
★ S2 and S6 are filter cakes from a different source in which more iron oxide products were found than there were in WQCS-7 (S7)
✧ No filter cake was used in these mixtures

low initial compressive strengths showed an order of magnitude increase in hydraulic conductivity after freeze-thaw. In contrast, essentially no change in hydraulic conductivity was observed for the high strength composite S6-1/1. The strength maker in these specimens was their high iron oxide content. It is not clear, for this particular case, if the constancy of the hydraulic permeability after freeze-thaw is simply due to the iron oxide content, or the initial strength developed. Although not confirmed by visual inspection, the increase in permeability was attributed to micro-cracks and fissures that may have formed within the matrix, however, not evident on the surface of the specimens. The high compressive strength measurements of these specimens after the freeze-thaw (Table 1) also suggest that these small faults do not necessarily influence their compressive strength, but probably decrease their stiffness. These findings suggest that durability strength measurements may not be indicative of conduction potential of the solidified composites. However, initial high strength, perhaps above a threshold value, may suggest potentially high resistance to fissures and micro cracks in the matrix that lead to increased rate of water conduction through the composites.

TABLE 2--Hydraulic conductivity before and after freeze-thaw

SAMPLE GROUP	k[1] before Freeze-thaw, cm/s	k after Freeze-thaw cm/s	Unconfined Compressive Strength, MPa	Specific Gravity of Composite, Gs
S7-2/5	1.70×10^{-7}	2.49×10^{-6}	2.07	2.68
S7-1/4	6.45×10^{-8}	1.24×10^{-6}	2.47	2.81
S7-2/6	8.70×10^{-8}	1.36×10^{-6}	2.80	2.83
S6-1/1	2.4×10^{-8}	1.9×10^{-8}	13.97	3.31

[1] Coefficient of hydraulic conductivity

In the field, a few of the constructed slabs were instrumented with infiltrometers to measure field permeability. The infiltrometers were installed during the placement of the slabs, and were embedded in the top half of the 15 cm thick slab. These instruments were basically single wall steel tube casings of 3.6 cm ID with an attached polypropylene graduated cylinder and a glass burette for finer incremental measurements. Each infiltrometer was capped and vented to minimize evaporation during the test period. On the 7th day of curing, the casings were filled with water to a given head of about 85.5 cm, and the drop in head was recorded in time for 24 hours. The infiltration rates were calculated using these readings. A calculated wetting front depth was used to compute hydraulic gradient. The test was repeated at 60 days following initial placement. In the second test, data was recorded over 48 hours, assuming a 24 hour pre-wetting period. This quick and approximate method was selected over the standard double ring infiltration method, primarily due to time and economic considerations. The calculated hydraulic conductivities based on the infiltration rates are given in Table 3. As observed, fairly good agreement was achieved between the laboratory measured k and the estimated k in the field. The field hydraulic conductivity decreased with time suggesting continuing pozzolanic and cementitious binding reaction in solidified matrix or unsteady state flow condition. An added reason for the reduced hydraulic conductivity might be due to possible shrinking of the material around the infiltrometer where it is embedded.

TABLE 3--Approximate field hydraulic conductivity of T-14

SLAB NO	Comparable Laboratory Sample No	Field k, cm/s (7 day)	Field k, cm/sec (60 day)	Laboratory k, cm/s (28 day)
T-14-2	S7-2/1	...	4.09×10^{-7}	4.01×10^{-7}
T-14-7	S7-1/4	1.26×10^{-5}	2.04×10^{-7}	6.45×10^{-8}
T-14-16	S7-2/2	2.16×10^{-5}	8.11×10^{-7}	3.16×10^{-7}
T-14-25①	None	3.59×10^{-6}	1.92×10^{-7}	...

① Contains 6% Flyash and 10% Cement

Laboratory Test Results - New Composites with WQCS-7

As discussed in the Background section, five new solidified products of the WQCS-7 filter cake were produced. The intended goal was to achieve improved strength and durability characteristics for the composite. Two of the formulas were developed at Lehigh University and three others were developed by commercial groups. Table 4 presents the approximate compositions of two Lehigh formulas, designated as L18 and E7, and one commercial formula designated as KB. The compositions of the other two commercial formulas, which will be referred to as the RI and KW formulas are proprietary and unavailable. However, the measured properties of each will be presented and compared.

TABLE 4--New composite formulas and properties

COMPONENT	E7 (%)	L18 (%)	KB (%)
Filter Cake	37	33	32.6
Cement (Type I)	24	24	22
Flyash	6	0	11.4
Slag (-1.9, +0.63cm)	7	27	18.8
Sand	2	0	0
Stone	5	0	0
Water	19	16	15.2

Slump	7.5	...	...
Bulk density, pcf	114	125	125
Specific Gravity	2.54	2.82	2.74

The L18 composition was similar to T-14, the main difference being the increased cement content from 16% to 24%. In the E7 formula most of the slag was replaced by graded aggregate, sand, stone, slag and flyash for the silt size particle. The KB mixture contained a moderate amount of calcium bearing (class C) flyash. All of the mixtures contained surface reactive reagents at varying percentages by weight of cement(1 to 4% air entrainment and superplasticizer), and some of them contained minor quantities of strength making agents (e.g. microsilica). One set of E7 specimens were prepared without the air entrainment reagent (E7-NA) to assess effects of this reagent on long-term permeability. Table 5 presents the average 28 day unconfined compressive strength results and hydraulic conductivities of the new composites. As observed from Table 5, higher strengths were achieved with the new composites than with the T-14 composites. However, the hydraulic conductivity did not improve significantly over the former. The absence of air entrainment in E7-NA did not appear to influence hydraulic conductivity but its strength.

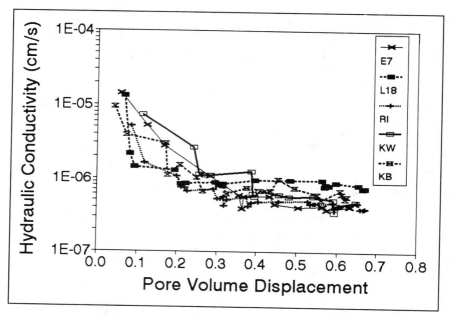

FIG. 1--Average hydraulic conductivity for the five new composites

Figure 1 presents the typical range of <u>average k</u> values achieved for each of the five composites. The range appears to be between 9×10^{-7} and 3×10^{-7} cm/s as also given in Table 5. For all the different composites, k became constant after about 0.4 pore volume displacement of fluid. There is good agreement between the hydraulic conductivity trends measured for the five replicate specimens of L18 composition, as shown in Figure 2. This suggests that homogeneous specimens with repeatable results could be mixed using the L18 formula. The E7 formula produced larger variability of k values from about 2×10^{-7} to 1×10^{-6} cm/s, as observed in Figure 3 and Table 5. The number of specimens tested for hydraulic conductivity and the coefficient of variance for each composite type are also given in Table 5. The larger variance observed in E7 specimens could be attributed to the heterogeneity of the

Table 5--Average strength and hydraulic conductivity of new composites

Sample	Unconfined Compressive Strength (MPa) (28 day)					Average Strength (MPa) (28 day)	Hydraulic Conductivity Tests		
	Sample Size 7.6 D x 15.2L (cm)		Sample Size 15.2 D x 30.4 L (cm)				No of samples	Average Hydraulic Conduct. (cm/s)	Coef. of Variance
	Humidity Chamber	Ambient Condition	Humidity Chamber	Ambient Condition					
L18	5.70	5.72	6.20	7.73		6.35	4	9.42×10^{-7}	0.29
E7	7.40	8.87	6.93	8.75		7.99	11	4.14×10^{-7}	0.51
E7-NA[1]	...	6.05	7.71	5.85		6.62	2	4.84×10^{-7}	0.99
KB	11.62	9.50	11.05	12.36		11.13	7	5.97×10^{-7}	0.27
RI	9.87	13.90	8.75	13.17		11.46	3	3.84×10^{-7}	0.29
KW	7.03	10.16	6.87	7.85		7.98	4	3.66×10^{-7}	0.40

[1] No air entrainment reagent

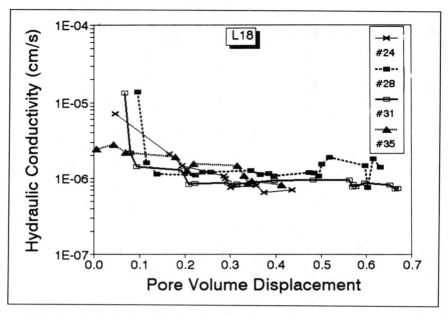

FIG. 2--Variation of hydraulic conductivity for L18 specimens

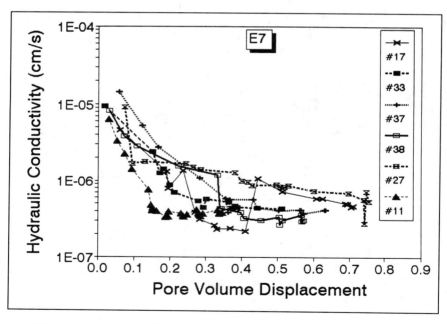

FIG. 3--Variation of hydraulic conductivity for E7 specimens

aggregate component (sand, slag, stone and flyash) of this mixture. The improved gradation is expected to result in better mixing and thus less chance to segregate. However, variability in specific gravities of these materials might have caused localized accumulation of different types of aggregates in the specimens. The resulting spread in hydraulic conductivity suggests that field quality control of proper mixing may become an important factor in achieving the targeted hydraulic conductivity when using a mixture such as E7. Commercial products also showed consistent trends in hydraulic conductivity, which suggested similar bulk compositions for all three composites. However, the RI and the KB mixtures exhibited significantly higher strengths than that of the KW mixture (Table 5). The strength of KW mixture is comparable to that of the E7 mixture.

Finally, one set of specimens from L18, KW, and KB groups were permeated with local tap water for different durations and the tailwaters, or the leachates were collected for chemical analysis. Three heavy metals, cadmium, lead and zinc were selected for quantitative analysis. Hydraulic conductivity as related to pore volume of water displacement for the specimens of the L18, KW and KB composites are given in Figure 4.

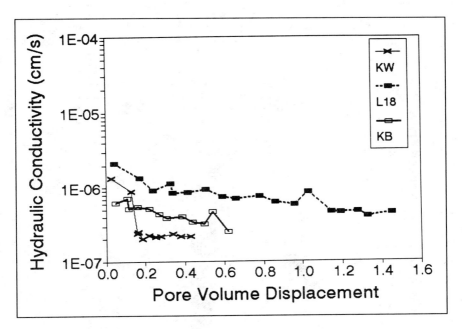

FIG. 4--Hydraulic Conductivity versus pore volume displacement

The KW samples exhibited the lowest hydraulic conductivity, whereas L18 samples showed the highest. The fraction of each metal removed by leaching was computed as the ratio of the mass (milligrams) of metal found in the leachate to the mass of metal originally found in the solidified composite. The fraction of each metal removed per pore volume of flow for the three composites are presented in Figure 5. In this comparison, the general assumptions made were that the contaminant migration is due to advection only and the rates of metal removal per pore volume of flow are constant for the duration of leaching tests performed. The fraction of removal per pore volume of flow was fairly low and consistent for all three specimens. The closeness of the values obtained indicated that leaching behavior of the three composites were similar. In all three composites, cadmium removal rate was highest, while zinc removal rate was lowest. The average percent removal per pore volume of flow for zinc, lead and cadmium were 0.0003±0.0003, 0.285± 0.039 and 0.519±0.055, respectively.

The significant conclusion from these limited observations is that solidification process may not encapsulate or bind all the metals equally in the final composite product. In the three slightly different solidification processes discussed above, cadmium was selectively more available for leaching than the other two metals analyzed. Therefore, although solidification is often viewed as a process by which the substance is rendered stable physically, the chemical and surface reactions that take place during solidification may play an important role in the chemical characteristics of the final product.

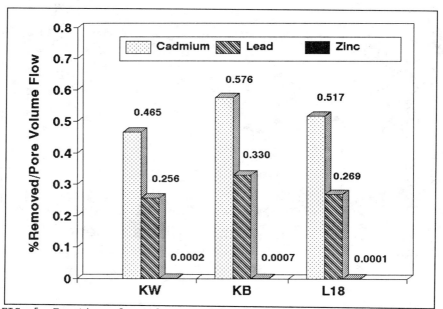

FIG. 5--Fraction of metal removed in the leachate per pore volume of flow in three different solidified composites of WQCS-7 filter cake

CONCLUSIONS

The outcome of this study increased the understanding about mechanical behavior of a solidified mixed residue of steel industry. The inconsistency and chemical heterogeneity of the residue made it difficult to apply a standard technique that would render a solidified product of predictable properties. The methods applied in here, although based on common practices of solidification and stabilization of wet granular and powder materials, had an element of trial and error in it. Therefore some of the results presented in this paper may be tentative due to insufficient number of replicate testing.

In addition to vast practical information gained that would help in selection of the final solidified product to function as a barrier material, the following specific conclusions were derived from this work:

1) Hydraulic conductivity and strength of the solidified material were not necessarily related.
2) Strength above a threshold value appeared to be indicative of resistance to micro cracking upon freeze-thaw and thus indicative of resistance to increased permeability.
3) Surface reactive agents promoted low permeability when used at an optimum percentage, that would vary by the bulk composition formula of the solidified product.
4) Solidification may not encapsulate or bind equally all the components (e.g. heavy metals) of the final product.

ACKNOWLEDGMENTS

This study was supported financially through a liaison program between the Environmental Studies Center of Lehigh University and the Bethlehem Steel Corporation.

REFERENCES

Ahmed, I. and Lowell, C.W., 1992, "Use of Waste Materials in Highway Construction: State of the Practice and Evaluation of the Selected Waste Products," <u>Transportation Research Record No. 1345</u>, TRB, National Academy Press, Washington, D.C., pp. 1-9.

Conner, J.R., 1990, <u>Chemical Fixation and Solidification of Hazardous Wastes</u>, Van Nostrand Reinhold Co., New York.

Cullinane, M.J. and Jones, L.W., 1986, "Stabilization/Solidification of Hazardous Waste," <u>Report EPA/600/D-86/028</u>, Hazardous Waste Engineering Research Laboratory, USEPA.

Jones, L.W., 1990, "Interferences Mechanisms in Waste Stabilization /Solidification Process," <u>Report EPA/600/2/89/067</u>, Hazardous Waste Engineering Research Laboratory, USEPA.

Pamukcu, S., Kugelman, I.J. and Lynn, J.D., 1989, "Solidification and Re-Use of Steel Industry Waste," *Proceedings* of 21st Mid-Atlantic Industrial Waste Conference, Technomic Publishing, Lancaster, Pa., pp. 3-15.

Pamukcu, S., Topcu, I.B., Lynn, J.D. and Jablonski, C.E., 1991, "Reuse of Solidified Steel Industry Sludge Waste for Transportation Facilities," *Transportation Research Record*, No. 1310, TRB, National Academy Press, Washington, D.C., pp. 93-105

Winterkorn, H.F., and Pamukcu, S., 1991, "Soil Stabilization and Grouting," Chapter 9, in *Foundation Engineering Handbook*, 2nd ed., H.Y. Fang, Editor, pp. 317-378, Van Nostrand Reinhold Co., New York.

Charles D. Shackelford[1] and Michael J. Glade[2]

CONSTANT-FLOW AND CONSTANT-GRADIENT PERMEABILITY TESTS ON SAND-BENTONITE-FLY ASH MIXTURES

REFERENCE: Shackelford, C. D., and Glade, M. J., "**Constant-Flow and Constant-Gradient Permeability Tests on Sand-Bentonite-Fly Ash Mixtures,**" Hydraulic Conductivity and Waste Contaminant Transport in Soil, ASTM STP 1142, David E. Daniel and Stephen J. Trautwein, Eds., American Society for Testing and Materials, Philadelphia, 1994.

ABSTRACT: Flexible-wall permeability (hydraulic conductivity) tests are conducted on compacted specimens consisting of 50 percent Class F fly ash with a 50 percent soil admixture (sand with 0, 6.25, 12.5, 18.75, and 25 percent bentonite). All test specimens are back-pressured before permeation with distilled water. Three specimens from each sand-bentonite-fly ash mixture are tested under high (72 or 73) and low (6.1) constant-gradient conditions as well as under constant-flow conditions for a total of 15 tests. Induced hydraulic gradients from the constant-flow tests range from 3.5 to 41. A decrease in permeability with an increase in bentonite content is evident regardless of the test method. However, the permeability is minimized at a bentonite content of 18.75 percent regardless of the test method. The overall lowest permeability value is 1.6×10^{-9} m/s. An increase in hydraulic gradient of 11.9X resulted in a 3.1X to 6.2X decrease in measured permeability values for all test specimens. A significant portion of the difference in measured permeability values is attributed to differences in applied or induced hydraulic gradients regardless of the test method.

KEYWORDS: permeability, hydraulic conductivity, fly ash, soil admixture, constant gradient, flow pump, flexible wall permeameter, laboratory testing

INTRODUCTION

Fly ash is a by-product of the coal combustion process in industrial power facilities. Over 60 percent of the electric power produced in the United States comes from the coal combustion process. Fly ash is considered a "waste material", and an estimated 70 million

[1]Associate Professor, Department of Civil Engineering, Colorado State University, Fort Collins, CO 80523.
[2]Environmental Engineer, Engineering-Science, Inc., 1700 Broadway, Suite 900, Denver, CO 80290.

metric tons is produced every year (Edil et al. 1987). Over a billion tons of coal fly ash have been stockpiled (McLaren and DiGioia 1987).

Fly ash is collected by electro-static precipitators in fabric filter baghouses and hoppers during the coal burning process. Fly ash consists mostly of glassy spheres less than 0.075 mm in diameter, with some crystalline matter and carbon. Fly ash samples demonstrate a significant variability in engineering properties depending on the coal source and the operational and collection methods of the power facility. Fly ashes typically are classified as either Class C or Class F. Class C fly ash usually is derived from sub bituminous coal sources and generally demonstrates a significant pozzolanic behavior when mixed with water. Class F fly ash, usually from bituminous coal, generally requires a lime or cement additive before significant pozzolanic behavior is evident.

Several studies have evaluated the use of fly ash, admixed fly ash, and fly ash stabilized soil for waste containment barriers (e.g., Bergstrom and Gray 1989; Bowders et al. 1990; Creek and Shackelford 1992; Gray et al. 1991; Edil et al. 1987; and Vesperman et al. 1985). In some cases, permeability values of $\leq 1 \times 10^{-9}$ m/s have been measured in the laboratory on specimens of fly ash, admixed fly ash, and stabilized fly ash (e.g., Bergstrom and Gray 1989; Bowders et al. 1990; Edil et al. 1985; Usmen et al. 1992; and Vesperman et al. 1985). However, other studies have indicated that permeability values $\geq 1 \times 10^{-9}$ m/s may be more common (e.g., Bowders et al. 1987; Creek and Shackelford 1992; McLaren and DiGioia 1987; and Parker and Thornton 1976). The wide range of reported permeability values usually is attributed to the type and amount of fly ash (Class C or Class F), admixture, and/or stabilizer, among other factors. However, there has been little mention of the potential effects of the laboratory test conditions on the reported test results, including the practice of utilizing relatively high applied hydraulic gradients (≥ 100). In addition, there have been little, if any, reported test results for test specimens containing fly ash in which the constant-flow test method (ASTM D 5084 - Standard Test Method for Measurement of Hydraulic Conductivity of Saturated Porous Materials Using a Flexible Wall Permeameter) has been used to measure the permeability. Therefore, the purposes of this study are to evaluate differences between constant-gradient and constant-flow test methods for measuring permeability of compacted specimens of soil admixed fly ash, and to quantify the potential influence of the hydraulic gradient, applied or induced, on the test results.

MATERIALS AND METHODS

Materials

The pertinent physical properties of the fly ash, sand, and bentonite are summarized in Table 1. The fly ash used in this study is a light gray, well-graded material collected from the Western Ash Company in Denver, Colorado. The fly ash is referred to as Nixon Fly Ash since the original source for the fly ash is the Nixon Power Plant in Colorado Springs, Colorado. As indicated in Table 1, the fly ash is a fine-grained material classified as a ML according to ASTM D 2487 (Standard Test Method for Classification of Soils for Engineering Purposes). Based on the data in Table 2, Nixon fly ash is classified as a Class F fly ash, as evident by the relatively low CaO content.

TABLE 1--Physical properties of mixture materials.

Property	Standard	Mixture Materials		
		Fly Ash	Sand	Bentonite
Liquid Limit, g/g	ASTM D 4318	NP	NP	461%
Plastic Index, g/g	ASTM D 4318	NP	NP	427%
Particle Sizes:	ASTM D 422			
Sand, g/g		10%	100%	0
Silt, g/g		75%	0	22%
Clay, g/g		15%	0	78%
Specific Gravity, G_S	ASTM D 854	2.20	2.65	2.82
Classification	ASTM D 2487	ML	SP	CH

NP = non plastic

A clean, medium sand was chosen for this study because previous results have indicated that a lower permeability value is achieved when a medium sand is used as a fly ash admixture as opposed to a fine sand (Gray et al. 1991). The poorly-graded sand used in this study is from the Colorado Lien Company in LaPorte, Colorado. The sand is referred to as a 20-40 silica sand since the majority ($\approx$ 95 percent) of the sand particles are between the No. 20 (0.085 mm) and the No. 40 (0.425 mm) sieve sizes.

Bentonite was used as an admixture in this study since previous studies have indicated that bentonite can reduce significantly the permeability of fly ashes when mixed with the fly ashes in small quantities (e.g., Bowders et al. 1990; Creek and Shackelford 1992; and

TABLE 2--Properties of Nixon fly ash relative to ASTM requirements for Class F fly ash.

Property	Fly Ash[1]	ASTM C 618
Silicon Dioxide, SiO_2, %	56.1	--------
Aluminum Oxide, Al_2O_3, %	29.6	--------
Iron Oxide, Fe_2O_3, %	4.60	--------
Sum of SiO_2 , Al_2O_3 , and Fe_2O_3 , %	90.3	70 min.
Calcium Oxide, CaO, %	6.60	--------
Magnesium Oxide, MgO, %	1.50	--------
Sulfur Trioxide, SO_3, %	0.29	5 max.
Loss on Ignition, %	0.85	6 max.
% Retained on No. 325 Sieve	26.3	34 max.
Pozzolanic Activity Index (PAI):		ASTM C 311
Portland Cement at 7 days, % of control	69	--------
Portland Cement at 28 days, % of control	84	75 min.
Lime at 7 days, psi	1098	800 min

[1]Tests performed by Resource Materials Testing, Inc., of Lakewood, Colorado.

Gray et al. 1991). The bentonite used in this study is a sodium bentonite from Black Hills, Wyoming. The bentonite was purchased as "Ecco Gel" from the Eisenman Chemical Company in Greeley, Colorado. A cation exchange capacity (CEC) of 76 meq/100g was measured for the bentonite. Further details of the physical and chemical properties of the bentonite can be found in Shackelford (1994).

Permeant

The permeant was prepared by passing tap water through a Barnstead ion exchange column. The ion exchange reactions in the column reduced the specific conductance of the tap water from 27.9 $\mu S/cm$ (1 $\mu S/cm$ = 1 $\mu mho/cm$) to 0.57 $\mu S/cm$ at 25°C. Based on Standard Methods (1985), water with a specific conductance value less than 1.0 $\mu S/cm$ is classified as distilled water (DW) whereas a specific conductance value of less than 0.20 $\mu S/cm$ is required for deionized water. Therefore, the permeant for this study is considered to be DW, or Type II reagent water (ASTM D 1193 - Standard Specification for Reagent Water). The chemical properties of the permeant are summarized in Table 3.

The use of distilled water (DW) as a permeant for permeability tests generally is not recommended because the lack of a significant ionic content can result in permeability values which are unconservative

TABLE 3--Properties of distilled water (DW) used as permeant.

Property	Value
pH	5.8
Specific Conductance @ 25°C ($\mu S/cm$)	0.57
Temperature	22°C
Metal Element Concentrations (mg/l):	
Al	<0.01
B	0.01
Ba	<0.01
Ca	0.20
Cd	<0.01
Cr	0.01
Cu	0.01
Fe	0.01
K	1.80
Mg	<0.01
Mn	<0.01
Mo	0.01
Na	<0.01
Ni	0.01
P	<0.01
Pb	<0.05
Sr	<0.01
Zn	<0.01
Nonmetal Anion Concentrations (mg/l):	
CO_3^{2-}	<0.01
HCO_3^-	3.90
Cl^-	0.70
SO_4^{2-}	0.10
NO_3^-	<0.01

(too low) with respect to the desired application (Olson and Daniel 1981; and ASTM D 5084). However, an additional objective of the present study (not presented herein) was to evaluate the effect of the bentonite content on the leaching of metals from the test specimens (Glade 1992). As a result, DW was used to minimize the background concentrations of metals and other chemical constituents in the effluent of the permeability tests. The DW was used as the permeant for the permeability tests as well as for preparing and back pressuring the compacted test specimens. The water quality was checked before the start of each test sequence.

Mixtures

Five mixtures consisting of 50 percent fly ash with 50 percent sand-bentonite admixtures were evaluated in this study. Three test specimens were prepared from each mixture for the permeability tests. The test specimens were compacted prior to permeability testing to simulate the use of soil admixed fly ash as a waste containment liner material. The compaction procedure followed the standard Proctor procedure (ASTM D 698 - Method A - Standard Test Methods for Moisture-Density Relations of Soils and Soil-Aggregate Mixtures Using a 5.5-lb (2.49-kg) Rammer and 12-in (305-mm) Drop) except the molds (specimens) were half size (i.e., 5.82-cm (2.29-in) height and 10.16-cm (4-in) ID). The percentages, by dry weight, of each of the three constituents - fly ash, sand, and bentonite - in the mixtures as well as the maximum dry unit weights and optimum moisture contents resulting from compaction of the mixtures are summarized in Table 4. The data in Table 4 indicate that the maximum dry unit weight decreased and the optimum moisture content increased as the percentage of bentonite in the mixture increased.

Specimen Preparation

Each test specimen was compacted at moisture contents from one to three percent on the wet side of optimum moisture content and within ± five percent of the maximum dry unit weight for the mixture. The test specimens were wrapped in plastic and sealed in double plastic bags for curing after compaction. The specimens were cured for seven days at room temperatures ranging from 19.3°C to 22.3°C (67°F to 72°F) in a humidity storage chamber. The storage chamber consisted of a styrofoam box and lid with standing water at depths ranging from 2.5 to 5 cm in the bottom of the box. Baffles were placed at the bottom of the box to prevent

TABLE 4--Specimen mixtures for permeability tests.

Mixture No.	Fly Ash Content[1], %	Sand Content[1], %	Bentonite Content[1], %	Maximum Dry Unit Weight[2,3], kN/m^3	Optimum Moisture Content[2], %
1	50	50	0	17.17	11.8
2	50	43.75	6.25	16.58	13.7
3	50	37.5	12.5	15.99	14.6
4	50	31.25	18.75	15.27	16.4
5	50	25	25	14.94	18.2

[1]By dry weight.
[2]ASTM D 698 - Method A.
[3]1 kN/m^3 = 6.371 lb/ft^3.

contact between the test specimens and the standing water. In all cases, less than 0.1 percent change in weight of the test specimens was measured for the curing period. The test specimens were extruded from the compaction molds after curing for permeability testing in flexible-wall permeameters. If either excessive cracking or non-homogeneity of the extruded specimens was observed, the specimen was discarded and another specimen from the same mixture was prepared.

Hydraulic Conductivity Tests

The prepared test specimens were placed in flexible-wall permeability cells, the cells were assembled, and the test specimens were back-pressured before permeation. The specimens were back-pressured from both top and bottom under the same pressure. The cell pressure and back pressure were raised in 34.5 kPa (5 psi) increments every 24 hours such that a 34.5 kPa (5 psi) difference between the cell and back pressures was maintained at all times. The back-pressure stage lasted 14 days resulting in B-values (B = $\Delta u/\Delta\sigma$) ≥ 0.90 for all test specimens.

Both constant-gradient (constant-head) and constant-flow test methods were used to determine the permeability values of the test specimens. Two constant-gradient tests and one constant-flow test were performed for each of the five material mixtures for a total of 15 flexible-wall permeability tests. For each material mixture, one constant-gradient test was performed at a relatively high hydraulic gradient of 72 or 73 whereas the other constant-gradient test was performed at a relatively low hydraulic gradient of 6.1. The constant-flow tests were performed at Darcian velocities, or fluid fluxes, ranging from 1.28×10^{-7} m/s to 1.38×10^{-7} m/s resulting in induced hydraulic gradients at steady-state flow ranging from 3.5 to 41. All tests were performed at ambient laboratory temperatures which ranged from 19.3°C to 22.3°C (67°F to 72°F). No attempt was made to control specimen porosity or to measure specimen volume change during permeation.

A schematic of the constant-gradient test apparatus is shown in Fig. 1. The apparatus is essentially the same as depicted in ASTM D 5084 except the outflow head is maintained constant because of the elevated exit tube contained within the tailwater reservoir. As water flows through the test specimen, the elevation head in the headwater reservoir decreases and, therefore, the test conditions actually model a falling-head test method. However, the headwater reservoirs were refilled at or before each quarter pore volume of flow to minimize gradient fluctuations. As a result, the effect of the change in elevation head in the headwater reservoir was considered to be negligible, particularly when considering the relatively high air pressures applied to the water in the headwater reservoir. For example, the hydraulic gradient changed by less than two to three percent for the tests performed at the high hydraulic gradients, and from four to five percent for the tests performed at the low hydraulic gradients. As a result, a constant-gradient was assumed to exist in the tests.

Hydraulic gradients can be applied across the test specimens by increasing the headwater pressure and/or decreasing the tailwater pressure. However, the increase in the headwater pressure is limited to a value which is somewhat less than the difference between the back pressure and the cell pressure, and an excessive decrease in the tailwater pressure can result in release of air from solution and excessively high effective stresses in the test specimen. The high hydraulic gradients (72,73) in this study were achieved by both increasing the headwater pressure and decreasing the tailwater pressure

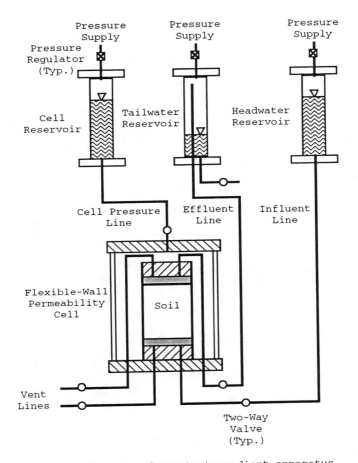

FIG. 1--Schematic of constant-gradient apparatus.

such that the average effective stress in the test specimen was maintained at 34.5 kPa (5 psi) in four of the five tests and at 41.3 kPa (6 psi) for the test specimen containing 12.5 percent bentonite. In all cases, the low hydraulic gradient (6.1) was achieved by decreasing the tailwater pressure 3.45 kPa (0.5 psi) resulting in an average effective stress in the test specimens of 36.2 kPa (5.25 psi).

The permeability (hydraulic conductivity) was calculated periodically for each constant-gradient test in accordance with the following form of Darcy's law:

$$k = \frac{\Delta V}{Ai\Delta t} \quad (1)$$

where
k = the coefficient of permeability, ms^{-1},

ΔV = the increment of effluent volume collected in the tailwater reservoir during the interval between successive readings, m^3,
A = the cross-sectional area of the test specimen perpendicular to flow, m^2,
i = the applied hydraulic gradient, and
Δt = the elapsed time between successive readings, s.

All constant-gradient tests were terminated only after the permeability values had stabilized to a constant value.

The test apparatus for the constant-flow tests is shown schematically in Fig. 2. The apparatus is the same as shown in Fig. 1 for the constant-gradient test except that the volumetric flow rate is maintained constant through the use of a Harvard Apparatus flow pump (model No. 944) and a stainless-steel syringe system. The flow rate is controlled via mechanical displacement of a piston arm through a syringe of constant cross-sectional area. The flow rate is controlled by setting a constant speed for the displacement of the piston arm and calibrating the displacement speed with the resulting flow rates. The differential pressure loss, or head loss, across the test specimen resulting from flow is measured using a differential pressure transducer and a strip chart recorder (not shown) which indicates the transient response time; i.e., the time required to achieve a steady-state hydraulic gradient. As a result, hydraulic gradients in the constant-flow tests are induced, not applied.

The flow-pump system utilized in this study is similar to systems described by Olsen (1966), Olsen et al. (1985), and Aiban and Znidarcic (1989) except two syringes in parallel are utilized instead of only one syringe. Flow can be induced through the test specimen by either infusion or withdrawal. Infusion results from outward displacement of the piston arm whereas withdrawal results from inward displacement of the piston arm (Morin and Olsen 1987). However, all constant-flow tests in this study were of the infusion type. The advantages of the flow-pump system for constant-flow permeability tests are described by Olsen et al. (1985). Further details of the flow-pump system used in this study are described elsewhere (Glade 1992).

Since the constant-flow test results in a continuous readout of the differential pressure loss across the test specimen, the permeability at any time is calculated as follows:

$$k = \frac{C}{\Delta u} \quad (2)$$

where
   Δu = the differential pressure loss across the test specimen, Pa, and
   C = a constant given by

$$C = \frac{QL\gamma_w}{A} \quad (3)$$

where
   Q = the volumetric flow rate, m^{3}s^{-1},
   L = the length of the test specimen, m, and
   γ_w = the unit weight of the water, Nm^{-3}.

At steady-state, Δu and, therefore, k are constants.

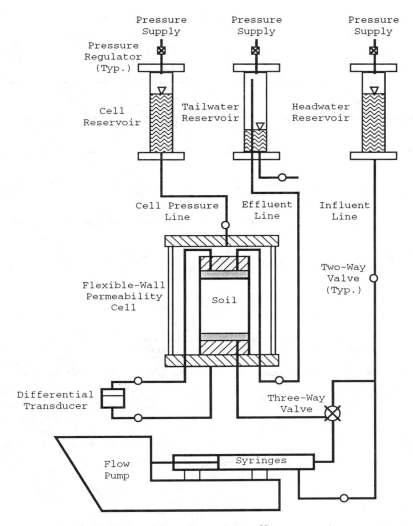

FIG. 2--Schematic of constant-flow apparatus.

RESULTS AND DISCUSSION

Test Specimen Properties

The initial properties of each of the 15 test specimens before permeation are summarized in Table 5. In general, an increase in bentonite content resulted in an increase in specimen porosity at compaction (before back pressure).

TABLE 5--Initial properties of test specimens.

Mixture Constituents (by dry weight)	Mixture Specific Gravity[1], G_S	Test No.	Dry Unit Weight[2,3], kN/m^3	Molding Water Content, %	Mixture Porosity	B Value
50% Fly Ash - 50% Sand	2.40	1	17.09	13.1	0.274	0.94
		2	16.72	13.6	0.290	0.91
		3	16.67	13.3	0.292	0.91
50% Fly Ash - 43.75% Sand - 6.25% Bentonite	2.41	4	16.34	15.8	0.310	0.90
		5	16.36	16.0	0.309	0.97
		6	16.31	16.0	0.311	0.97
50% Fly Ash - 37.5% Sand - 12.5% Bentonite	2.42	7	15.66	17.2	0.341	0.98
		8	15.66	17.1	0.341	0.90
		9	15.70	17.1	0.340	0.90
50% Fly Ash - 31.25% Sand - 18.75% Bentonite	2.43	10	15.04	18.5	0.371	0.90
		11	15.01	18.7	0.372	0.94
		12	15.08	18.7	0.369	0.94
50% Fly Ash - 25% Sand - 25% Bentonite	2.44	13	14.66	20.7	0.390	0.96
		14	14.60	20.9	0.393	0.93
		15	14.53	20.9	0.395	0.93

[1]Calculated from constituent specific gravity values in Table 1
[2]ASTM D 698-Method A
[3]1 kN/m^3 = 6.371 lb/ft^3

Hydraulic Conductivity Tests

Plots of permeability versus pore volumes of flow for the constant-gradient and constant-flow tests are shown in Figs. 3 and 4, respectively. The test results are summarized in Table 6. The difference between the final water content and the molding water content generally increased as the bentonite content in the test specimens increased.

The plots in Figs. 3 and 4 generally indicate a decrease in permeability with time (i.e., with pore volumes of flow). This is confirmed by the ratios of the initial permeability value for a given test to the final, or steady-state, hydraulic conductivity value, k_i/k_f, in Table 6. A k_i/k_f value greater than one indicates an overall decrease in k during the test.

A decrease in k with time typically is attributed, in part, to compression of the test specimens due to seepage forces (induced consolidation). Therefore, one would expect that the greater the compressibility of the test specimen and/or the greater the applied hydraulic gradient, the larger the k_i/k_f value. In addition, one might expect that the greater the bentonite content in the test specimen, the greater the compressibility of the test specimen. Based on the values of k_i/k_f in Table 6, the expected trends are not evident in all cases although the test specimens with relatively low bentonite contents ($\leq$ 6.25%) generally appear to be less compressible (lower k_i/k_f values) than the test specimens with the higher bentonite contents ($\geq$ 12.5%). Also, for a given mixture, the values of k_i/k_f for the constant gradient tests performed at a relatively high hydraulic gradient (72, 73) are greater, in all cases, than the values of k_i/k_f for the constant gradient tests performed at a relatively low hydraulic gradient (6.1). This trend is consistent with the expected effect of an increase in applied hydraulic gradient. Since the infusion type of constant-flow test results in an increase in headwater pressure and a decrease in

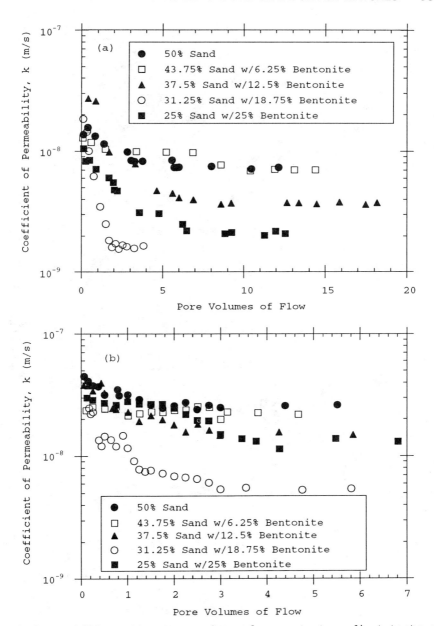

FIG. 3--Permeability versus pore volumes for constant-gradient tests on specimens containing 50% fly ash with 50% sand-bentonite admixtures: (a) high (72,73) hydraulic gradients; and (b) low (6.1) hydraulic gradients.

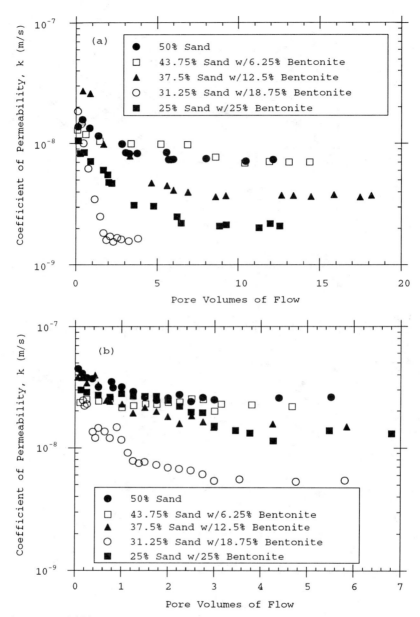

FIG. 3--Permeability versus pore volumes for constant gradient tests on specimens containing 50% fly ash with 50% sand-bentonite admixtures: (a) high (72,73) hydraulic gradients; and (b) low (6.1) hydraulic gradients.

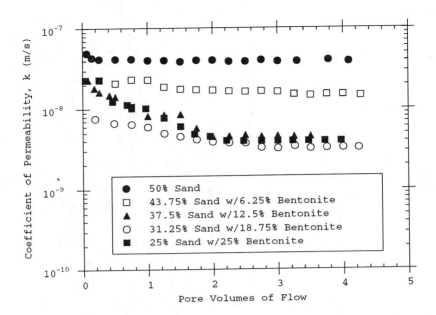

FIG. 4--Permeability versus pore volumes for constant-flow tests on specimens containing 50% fly ash with 50% sand-bentonite admixtures.

effective stress, a comparison of the k_i/k_f values from the constant-flow tests with the k_i/k_f values from the constant-gradient tests is not appropriate, i.e., due to differences in applied stress conditions. The failure to follow the expected trend may be due to the slight pozzolanic character of the fly ash in the test specimens. Observation of the extruded specimens after preparation revealed that the compacted specimens were stiff.

Effect of Bentonite Content

The effect of the bentonite content on the permeability is illustrated in Fig. 5. The test results indicate a general decrease in permeability with an increase in bentonite content regardless of the test method. However, as shown in Fig. 5a, the specimens containing 18.75 percent bentonite (Test Nos. 10-12) exhibited the lowest permeability values for all three series of tests. As shown in Fig. 5b, this trend does not change when the test results are plotted in terms of the relative magnitude of the applied or induced hydraulic gradient.

The overall lowest permeability value was 1.6 x 10^{-9} m/s. This seemingly high value resulted even though distilled water was used as the permeant. Higher values would be expected for the test specimens containing bentonite if tap water or standard water (ASTM D 5084) had been used since the higher ionic content of these waters likely would have resulted in flocculation of the bentonite. Several reasons may be offered for the seemingly high k values.

TABLE 6—Permeability (k) test results.

Mixture Constituents (by dry weight)	Sand: Bentonite (by dry weight)	Test No.	Final Water Content, %	Final Degree of Saturation, %	Type of Test[1]	Hydraulic Gradient	Total Pore Volumes of Flow	Final Permeability k_f, m/s	k_i/k_f Ratio[2]
50% Fly Ash – 50% Sand	NA	1	14.8	95	CG	72	12.1	7.4×10^{-9}	2.1
		2	16.4	97	CF	3.5	4.1	3.9×10^{-8}	1.1
		3	16.9	99	CG	6.1	5.5	2.6×10^{-8}	1.5
50% Fly Ash – 43.75% Sand – 6.25% Bentonite	7:1	4	19.5	100	CG	73	14.4	7.0×10^{-9}	2.1
		5	19.6	100	CF	8.4	4.3	1.5×10^{-8}	1.6
		6	19.8	100	CG	6.1	4.7	2.2×10^{-8}	1.2
50% Fly Ash – 37.5% Sand – 12.5% Bentonite	3:1	7	20.0	95	CG	72	18.2	3.8×10^{-9}	7.2
		8	20.2	95	CF	29	3.5	4.5×10^{-9}	3.6
		9	20.4	96	CG	6.1	5.8	1.5×10^{-8}	2.3
50% Fly Ash – 31.25% Sand – 18.75% Bentonite	1.67:1	10	27.5	100	CG	72	3.8	1.6×10^{-9}	8.8
		11	26.2	100	CF	41	4.2	3.3×10^{-9}	2.3
		12	25.9	100	CG	6.1	5.8	5.4×10^{-9}	4.2
50% Fly Ash – 25% Sand – 25% Bentonite	1:1	13	25.5	98	CG	73	12.6	2.1×10^{-9}	4.0
		14	25.4	96	CF	34	4.0	4.0×10^{-9}	5.8
		15	27.1	100	CG	6.1	6.8	1.3×10^{-8}	2.2

[1]CG = constant gradient test; CF = constant flow test
[2]k_i = initial permeability value at 0.25±0.10 pore volumes of flow; k_f = final permeability value

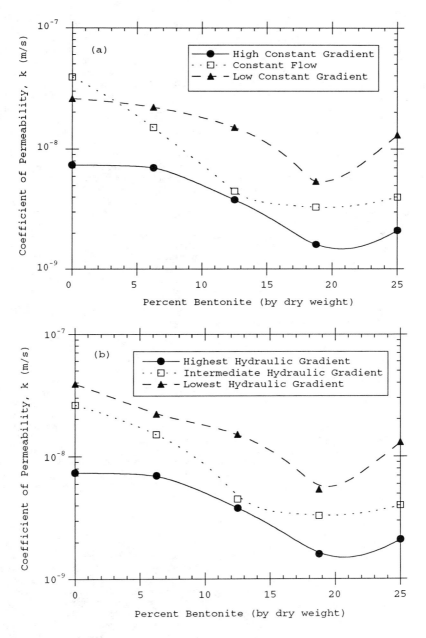

FIG. 5--Permeability versus percent bentonite in terms of (a) test method, and (b) relative hydraulic gradient.

First, it simply may not be possible to achieve k values less than 1×10^{-9} m/s with the constituent materials used in this study. For example, Creek and Shackelford (1992) performed constant-gradient, rigid-wall permeability tests using tap water as the permeant and applied hydraulic gradients ranging from 24 to 124 on 32 test specimens consisting of various mixtures of the same constituent materials used in this study. The lowest k value reported by Creek and Shackelford (1992) is 1×10^{-9} m/s for a 100-percent fly ash specimen tested at a hydraulic gradient of 100, and even this test result was considered to be in error on the low side. Therefore, the constituent materials and material mixtures used in this study may not be appropriate for use as low-permeability (k ≤ 1×10^{-9} m/s) barriers.

Second, the mixtures used in this study may not have represented optimum mixtures with respect to minimization of the specimen permeability. For example, Bergstrom and Gray (1989) and Gray et al. (1991) tested mixtures consisting of fly ash (100% < No. 200 sieve), a medium sand, and bentonite. They found that the value of k increased at fly ash contents greater than 45 percent and 25 percent for mixtures containing two and four percent (dry weight) bentonite, respectively. In the present study, all test specimens consist of 50 percent fly ash of which 90 percent is considered fine-grained material (i.e., < No. 200 sieve). Therefore, the amount of fine-grained material (fly ash plus bentonite) in the test specimens ranged from 45 to 70 percent which may have been greater than the amount required to optimize the mixtures for minimization of the specimen permeability.

Third, the curing period of seven days used in this study may not have been sufficient for minimization of the specimen permeability. Bowders et al. (1987, 1990) and Usmen et al. (1992) have shown that lower k values result when specimens containing fly ash are allowed to cure longer. For example, Bowders et al. (1987) show that a 28-day curing period resulted in lower permeability values relative to a seven-day curing period for lime and cement stabilized fly ash specimens. Although their materials and test procedures were different than the present study, a seven-day curing period in this study may not have been sufficiently long with respect to minimization of the specimen permeability. However, longer curing periods may not be practical for field applications.

Finally, lower permeability values probably could have been achieved if higher hydraulic gradients were used. For example, a k value ≤ 1×10^{-9} m/s was reported by Bowders et al. (1987, 1990) for a Class F fly ash specimen with only 10 percent bentonite, but the applied hydraulic gradient was 100. The lowest k value in this study is for an applied hydraulic gradient of 72.

Effect of Hydraulic Gradient

Except for the test specimens containing only sand and fly ash (Test Nos. 1-3), the permeability values from the constant-flow tests are consistently higher than the values from the high constant-gradient tests and lower than the values from the low constant-gradient tests (Fig. 5a). The apparent discrepancy at zero percent bentonite content is removed when the test results are plotted in terms of the relative magnitude of the hydraulic gradient, as shown in Fig. 5b. As a result, consideration of the hydraulic gradient results in a more consistent trend in permeability values for a given mixture than does consideration of type of test method.

The ratio of permeability values in terms of the lowest and highest gradients is plotted as a function of bentonite content in Fig.

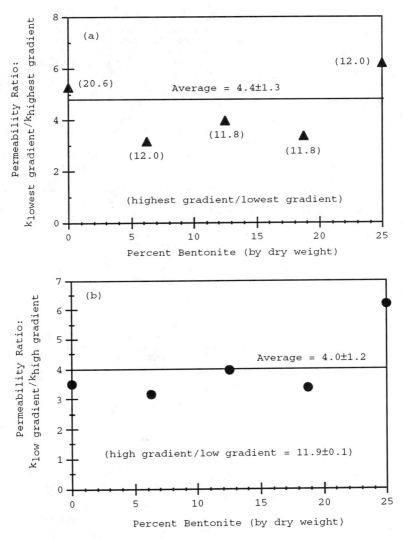

FIG. 6--Permeability ratio (high-to-low k values) versus percent bentonite: (a) all tests; and (b) constant-gradient tests.

6a. For all mixtures, the permeability ratios range from 3.1 to 6.2 with an average value of 4.4. However, the ratio of the highest to lowest hydraulic gradients for the test specimens with zero percent bentonite is not close to the same ratio for the test specimens for the other mixtures.

The discrepancy in hydraulic gradient ratios in Fig. 6a is removed in Fig. 6b where the ratios of permeability values for only the constant-gradient tests are considered. In this case, the ratio of high-to-low hydraulic gradients is relatively constant at 11.9±0.1, the range in ratios of high-to-low k values is still 3.1 to 6.2, but the average value of the high-to-low k ratios is lower at 4.0. Therefore, the effect of an increase in hydraulic gradient of approximately 11.9X resulted in a decrease in permeability ranging from 3.1X to 6.2X for the entire range of bentonite contents. This difference has significant ramifications with respect to field application of laboratory permeability test results; i.e., a test specimen achieving a regulatory limit on k (e.g., $k \leq 1 \times 10^{-9}$ m/s) in a laboratory test in which a high hydraulic gradient (e.g., $i \geq 100$) is applied may not achieve the regulatory limit in the field under lower hydraulic gradients.

Also, a constant high-to-low k ratio for all test specimens in which all other factors except the hydraulic gradients are constant would indicate that the differences in measured k values are due to differences in applied or induced hydraulic gradients as opposed to the test method. Therefore, the relatively narrow range in the ratios of high-to-low k values for the test specimens in this study suggests that a significant portion of the observed differences in the permeability values for test specimens in each mixture can be attributed to differences in the applied or induced hydraulic gradients. Aiban and Znidarcic (1989) showed that the constant-gradient and constant-flow test methods yield essentially the same permeability values for fine-grained soil specimens when the specimens are tested under similar conditions.

The influence of the hydraulic gradient is illustrated further in Fig. 7 where the permeability is plotted versus the applied or induced hydraulic gradient for all test specimens. For a given mixture, the permeability values decrease as the hydraulic gradient increases. This trend is consistent with the data in Figs. 5b and 6. However, the controlling influence on the permeability actually is the effective stress in the test specimen, not the hydraulic gradient.

For example, consider the stress conditions for flexible-wall permeability tests schematically illustrated in Fig. 8. At the end of the back-pressure saturation stage (Fig. 8a), the effective stress in the test specimen is uniform, and the hydraulic gradient is zero. As previously mentioned, flow can be induced through the specimen by increasing the headwater pressure (Fig. 8b) and/or decreasing the tailwater pressure (Fig. 8c). If the headwater pressure is increased or the tailwater pressure is decreased by the same amount, Δu, the hydraulic gradient across the specimen is the same, but the effective stress distribution in the specimen is different. Due to the higher effective stresses in the test specimen in Fig. 8c, one would expect the permeability of the specimen in Fig. 8c to be lower than the permeability for the specimen in Fig. 8b, all other factors being equal. The infusion type of constant-flow test results in a specimen stress distribution analogous to that shown in Fig. 8b (see Morin and Olsen 1987).

The effect of the specimen effective stress in flexible-wall tests previously has been shown by Boynton and Daniel (1985) and Daniel et al. (1985). In their tests, the permeability of compacted clay specimens decreased from one to three orders of magnitude as the average effective stress in the specimen increased from 13.8 to 103.4 kPa (2 to 15 psi). The greater decreases in k were associated with test specimens which were desiccated. As a result of the above considerations, the

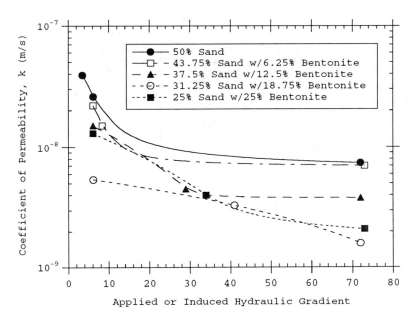

FIG. 7--Permeability versus hydraulic gradient for 50% fly ash with 50% sand-bentonite admixtures.

permeability values for the present study are plotted versus the average effective stress in Fig. 9a. The data do not illustrate the expected trend. At least three reasons for this discrepancy appear plausible. First, the range in average effective stresses in Fig. 9a is much narrower than the 13.8 to 103.4 kPa (2 to 15 psi) range reported by Boynton and Daniel (1985) and Daniel et al. (1985). As a result, the data in Fig. 9a actually may represent scatter over a narrow stress range in which no trend is evident. Second, the pozzolanic nature of the fly ash may have resulted in a cementation effect and relatively stiff specimens which do not exhibit the same stress behavior as specimens consisting only of uncemented soil. Third, the flow conditions in flexible-wall tests actually represent two-dimensional flow, not one-dimensional flow as typically assumed. As a result, the maximum effective stress may be a more appropriate indicator of the effect of effective stress on the permeability of flexible-wall specimens.

With respect to the last reason, consider the schematic diagram in Fig. 10 illustrating the difference between one-dimensional and two-dimensional flow in flexible-wall permeameters. For the one-dimensional flow situation, all parameters in Darcy's law are constants; i.e., Q, k, i, and $A \neq f(x,t)$, where x is the direction of flow. However, in the two-dimensional flow situation, none of the parameters are constant. For example, at steady-state flow, $Q = Q_{in} = Q_{out}$, but due to the stress conditions on the flexible-wall specimens, $A_{in} > A_{out}$. Therefore, $Q/A_{out} > Q/A_{in}$. Since the void ratio at the effluent end of the specimen is smaller than at the influent end of the specimen, one would expect $k_{out} < k_{in}$ and, therefore, $i_{out} > i_{in}$. The typical assumption of one-dimensional flow in which $i_{out} = i_{in}$ and k is homogeneous really is only

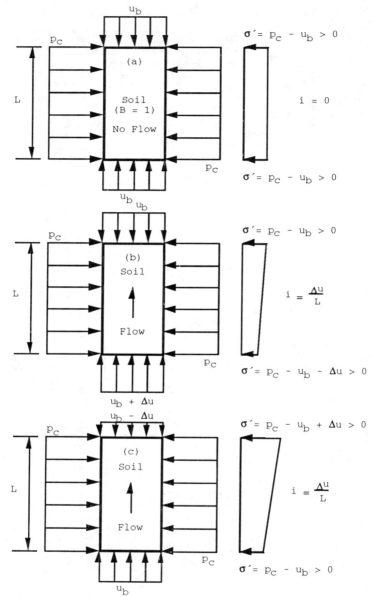

FIG. 8--Stress conditions in flexible-wall permeameters: (a) back-pressure stage; (b) increase headwater pressure; and (c) decrease tailwater pressure [u_b=backpressure; p_c=cell pressure; σ'=effective stress; Δu=differential pressure].

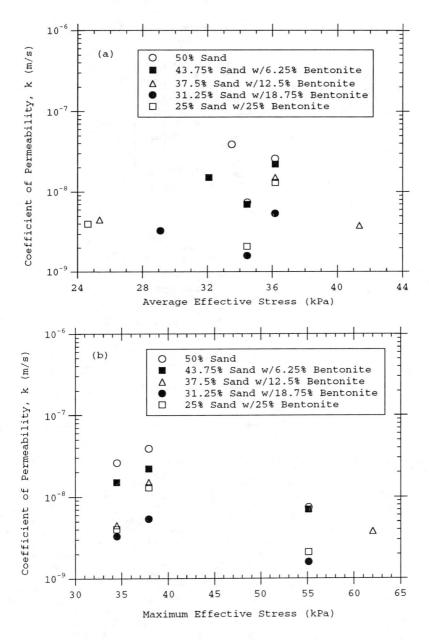

FIG. 9--Permeability versus (a) average effective stress and (b) maximum effective stress in 50% fly ash with 50% sand-bentonite admixtures.

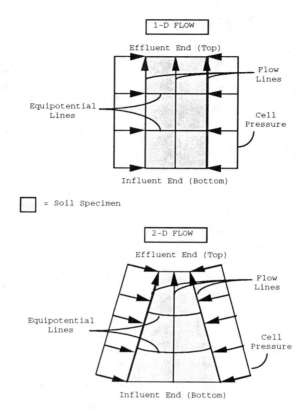

FIG. 10--Schematic of one-dimensional versus two-dimensional flow through flexible-wall test specimens.

an approximation in flexible-wall tests. The approximation becomes more significant as the compressibility of the test specimens and the applied total stresses (induced effective stresses) increase.

Based on this analysis, the measured permeability values have been plotted versus the maximum effective stress in Fig. 9b instead of the average effective stress. The first effect from plotting the data versus the maximum effective stress is that the overall stress range has been increased. In addition, if the variation in the expected trend occurring at the low maximum effective stresses of 34.5 kPa and 37.9 kPa (5.0 psi and 5.5 psi, respectively) is accepted as scatter, then the expected trend of a decrease in k versus applied effective stress is more evident than when the data are plotted in terms of average effective stress. However, there is not sufficient data to make a definite conclusion in this regard, particularly since the test specimens were observed to be relatively stiff and may have exhibited a cementitious behavior due to the presence of the fly ash. Nonetheless, the analysis is correct in principle.

SUMMARY AND CONCLUSIONS

Flexible-wall permeability tests were performed on compacted specimens consisting of 50 percent Class F fly ash mixed with 50 percent soil admixtures. The soil admixtures contained sand with 0, 6.25, 12.5, 18.75, and 25 percent bentonite by dry weight. Three specimens from each sand-bentonite-fly ash mixture were tested using either constant-gradient or constant-flow test methods. The constant-gradient tests were performed either at relatively high (72 or 73) or relatively low (6.1) applied hydraulic gradients. The constant flow tests resulted in induced hydraulic gradients ranging from 3.5 to 41.

In general, the permeability of the test specimens decreased as the bentonite content in the soil admixture increased. However, the permeability of the test specimens was minimized at a bentonite content of 18.75 percent. Both of these trends were followed regardless of whether the comparison of data was based on the results from the constant-gradient tests at the same hydraulic gradient or the constant-flow tests. Therefore, the observed trends are independent of the test methods.

The overall lowest permeability (k) value measured was 1.6×10^{-9} m/s. Four possible reasons are offered for the inability to achieve a k value $\leq 1 \times 10^{-9}$ m/s: (1) the materials are not suitable; (2) the optimum mixture proportions of the constituent materials for minimizing k were not achieved; (3) the curing time of seven days was too short; and (4) the hydraulic gradient (72) was too low.

For a given mixture, a consistent decrease in permeability with increase in applied or induced hydraulic gradient was evident regardless of test method. For the constant-gradient tests in which the ratio of high-to-low applied hydraulic gradients was maintained relatively constant at 11.9 ± 0.1, the corresponding ratios of high-to-low k values for the two specimens within a given mixture ranged from 3.1 to 6.2 for all mixtures, with an average value of 4.0. Therefore, although different permeability values resulted from the different test methods, the relatively narrow range in ratio of high-to-low k values suggests that a significant portion of the observed difference can be attributed to the applied or induced hydraulic gradients regardless of test method.

The expected trend of a decrease in permeability with increase in average effective stress in the specimen was not observed. This contradiction is attributed, in part, to the possibility of scatter in the data over the relatively narrow range in average effective stresses occurring in the test specimens and the possible influence of cementing of the test specimens due to the pozzolanic nature of fly ash. When the test data are plotted in terms of maximum effective stresses, the expected trend is more evident, but not conclusive. The use of maximum effective stresses instead of average effective stresses is consistent with the two-dimensional flow conditions typically occurring in flexible-wall permeability tests as opposed to the commonly assumed one-dimensional flow conditions.

ACKNOWLEDGMENT

This material is based upon work supported by the National Science Foundation under Grant No. MSS-8908201. The Government has certain rights in this material. Any opinions, findings, and conclusions or recommendations expressed in this material are those of the authors and do not necessarily reflect the views of the National Science Foundation.

REFERENCES

Aiban, S. A. and Znidarcic, D., 1989, "Evaluation of the Flow Pump and Constant Head Techniques for Permeability Measurements," Geotechnique, Vol. 39, No. 4, pp. 655-666.

Bergstrom, W. R. and Gray, D. H., 1989, "Fly Ash Utilization in Soil-Bentonite Slurry Trench Cutoff Walls," Proceedings, 12th Annual Madison Waste Conference, University of Wisconsin, Madison, pp. 444-458.

Bowders, Jr., J. J., Gidley, J. S., and Usmen, M. A., 1990, "Permeability and Leachate Characteristics of Stabilized Class F Flay Ash," Transportation Research Record No. 1288, Soils, Geology, and Foundations, Geotechnical Engineering 1990, Transportation Research Board, Washington, D. C., pp. 70-77.

Bowders, Jr., J. J., Usmen, M. A., and Gidley, J. S., 1987, "Stabilized Fly Ash for Use as Low-Permeability Barriers," Geotechnical Practice for Waste Disposal '87, ASCE Geotechnical Special Publication No. 13, R. D. Woods, Ed., American Society of Civil Engineers, New York, pp. 320-333.

Boynton, S. S. and Daniel, D. E., 1985, "Hydraulic Conductivity Tests on Compacted Clay," Journal of Geotechnical Engineering, ASCE, Vol. III, No. 4, pp. 465-478.

Creek, D. N. and Shackelford, C. D., 1992, "Permeability and Leaching Characteristics of Fly Ash Liner Materials," Symposium, Geo-Environmental Properties of Soil and Rock, 71st Annual Meeting of the Transportation Research Board, Washington, D. C.

Daniel, D. E., Anderson, D. C., and Boynton, S. S., 1985, "Fixed-Wall Versus Flexible-Wall Permeameters," Hydraulic Barriers in Soil and Rock, ASTM STP 874, A. I. Johnson, R. K. Frobel, N. J. Cavalli, and C. B. Pettersson, Eds., American Society for Testing and Materials, Philadelphia, pp. 107-126.

Edil, T. B., Berthouex, P. M., and Vesperman, K. D., 1987, "Fly Ash as a Potential Waste Liner," Geotechnical Practice for Waste Disposal '87, ASCE Geotechnical Special Publication No. 13, R. D. Woods, Ed., American Society of Civil Engineers, New York, pp. 447-461.

Glade, M. J., 1992, "The Influence of Cation Exchange Capacity on Metals Leaching from Soil Amended Fly Ash," M. S. Thesis, Colorado State University, Fort Collins, CO.

Gray, D. H., Bergstrom, W. R., Mott, H. V., and Weber, W. J., 1991, "Fly Ash Utilization in Cut-Off Wall Backfill Mixes," Proceedings, 9th International Coal Ash Utilization Symposium, Orlando, FL.

McLaren, R. J. and DiGioia, Jr., A. M., 1987, "The Typical Engineering Properties of Fly Ash," Geotechnical Practice for Waste Disposal '87, ASCE Geotechnical Special Publication No. 13, R. D. Woods, Ed., American Society of Civil Engineers, New York, pp. 683-697.

Morin, R. H. and Olsen, H. W., 1987, "Theoretical Analysis of the Transient pressure response from Constant Flow rate Hydraulic Conductivity Test," *Water Resources Research*, Vol. 23, No. 8, pp. 1461-1470.

Olsen, H. W., 1966, "Darcy's Law in Saturated Kaolinite," *Water Resources Research*, Vol. 2, No. 2, pp. 287-295.

Olsen, H. W., Nichols, R. W., and Rice, T. L., 1985, "Low Gradient Permeability Measurements in a Triaxial System," *Geotechnique*, Vol. 35, No. 2, pp. 145-157.

Olson, R. E. and Daniel, D. E., 1981, "Measurement of Hydraulic Conductivity of Fine Grain Soils," *Permeability and Groundwater Contaminant Transport*, *ASTM STP 746*, T. F. Zimmie and C. O. Riggs, Eds., American Society for Testing and Materials, Philadelphia, pp. 18-64.

Parker, D. G. and Thornton, S. I., 1976, "Permeability of Fly Ash and Fly Ash Stabilized Soils," *Final Report*, Highway Research Project 47, Arkansas State Highway Department, Little Rock, AK.

Shackelford, C. D., 1994, "Waste-Soil Interactions that Alter Hydraulic Conductivity," *Hydraulic Conductivity and Waste Contaminant Transport in Soils*, *ASTM STP 1142*, David E. Daniel and Stephen J. Trautwein, Eds., American Society for Testing and Materials, Philadelphia, 1994.

Standard Methods for the Examination of Water and Wastewater, 1985, 16th Edition, Arnold E. Greenberg, R. Rhodes Trussell, and Lenore S. Clesceri, Eds., American Public Health Association, Washington, D.C.

Usmen, M. A., Baradan, B., and Yazici, S., 1992, "Geotechnical and Geoenvironmental Properties of Stabilized Lignite Fly Ash," *Proceedings*, Mediterranean Conference on Environmental Geotechnology, Cesme, Turkey, May 25-27, 1992, A. A. Balkema Publ., Rotterdam, pp. 419-427.

Vesperman, K. D., Edil, T. B., and Berthouex, P. M., 1985, "Permeability of Fly Ash and Fly Ash-Sand Mixtures," *Hydraulic Barriers in Soil and Rock*, *ASTM STP 874*, A. I. Johnson, R. K. Frobel, N. J. Cavalli, and C. B. Pettersson, Eds., American Society for Testing and Materials, Philadelphia, pp. 289-298.

Van Maltby[1] and Laurel K. Eppstein[1]

A FIELD-SCALE STUDY OF THE USE OF PAPER INDUSTRY SLUDGES AS HYDRAULIC BARRIERS IN LANDFILL COVER SYSTEMS

REFERENCE: Maltby, V., and Eppstein, L. K., "A Field-Scale Study of the Use of Paper Industry Sludges as Hydraulic Barriers in Landfill Cover Systems," Hydraulic Conductivity and Waste Contaminant Transport in Soil, ASTM STP 1142, David E. Daniel and Stephen J. Trautwein, Eds., Amercian Society for Testing and Materials, Philadelphia, 1994.

ABSTRACT: Four field-scale landfill cover test cells have been constructed in order to facilitate a comparison of the field performance of sludge and clay hydraulic barriers. Two covers contain clay as the barrier material, one contains a paper mill primary sludge, and one contains a paper mill combined (primary and biological) sludge. One of the clay cells is designed to allow for a field verification study of the Hydrologic Evaluation of Landfill Performance (HELP) model.

Both sludge cells have produced smaller volumes of seepage and greater volumes of runoff than the clay cells to date. Calculated barrier hydraulic conductivities for the paper mill sludges are somewhat lower than those values calculated for the clay.

KEYWORDS: Hydraulic conductivity, runoff, seepage, HELP model, paper mill sludge, evapotranspiration, landfill cover, water balance

---

Widespread awareness of the need to protect groundwater quality is prompting closer scrutiny of the design and operation of solid and hazardous waste landfills. Municipalities and other landfill owners are experiencing growing pressure to upgrade landfilling practices, including closure. Some of the changes, such as more stringent liner and cover requirements, may be financially burdensome, particularly when clay soils used for construction purposes are not locally available. It may be possible, however, to reduce

---

[1]Research Scientist and Research Assistant, respectively, National Council of the Paper Industry for Air and Stream Improvement, Western Michigan University, Kalamazoo, MI 49008.

the cost of compliance by the utilization of unconventional yet readily available materials in landfill construction.

BACKGROUND

The United States pulp and paper industry generates considerable quantities of (a) sludge from the treatment of wastewater and (b) fly ash from the combustion of bark and other wood waste, and coal (NCASI 1991, 1992). Disposal in landfills remains the principal means of managing these wastes, although the industry is pursuing alternatives which emphasize waste materials as a resource (NCASI 1984, 1985, 1989, 1990, 1991, 1992; Schroeder et al. 1983a). One possible alternative is the use of sludge, fly ash, or their combination as a hydraulic barrier material (used to minimize infiltration) in landfill covers. Hydraulic and other mechanical properties of wastewater sludges from the paper industry, and coal fly ashes from the power industry, have been evaluated to some extent, but there is limited experience in the use of these materials as hydraulic barriers (NCASI 1989).

This paper describes ongoing field-scale research conducted as part of a USEPA cooperative agreement entitled "The Use of Pulp and Paper Mill Sludge and Fly Ash as Barrier Material in Covers for Municipal, Industrial, and Hazardous Waste." This research was funded in part by the USEPA under Cooperative Agreement No. CR811878-01-1. Research was organized into a laboratory phase and a field phase. A brief description of each phase follows.

Laboratory Study

The laboratory study included two main objectives. The first objective was to review existing information and industry experience related to the use of sludge and fly ash as landfill cover materials. The second objective was to generate new information on physical and chemical properties of sludges and fly ashes. The physical property of principal interest was hydraulic conductivity, which was measured by fixed-wall permeameter.

Thirteen fresh sludges from wastewater treatment systems of mills that encompass major pulp and paper production categories were tested. The hydraulic conductivity results are summarized in **Table 1**. A complete description of the laboratory study may be found in NCASI Technical Bulletin No. 559 (NCASI 1989). The laboratory study yielded the conclusion that pulp and paper mill sludges represent a potential alternative to clay for use as hydraulic barrier material in landfill covers.

Table 1--**Results of Laboratory Hydraulic Conductivity Tests**

MATERIAL	HYDRAULIC CONDUCTIVITY, cm/sec		
	Minimum	Maximum	Geometric Mean*
Sludge	5.8×10^{-8}	4.2×10^{-4}	1.8×10^{-6}
Fly Ash	2.8×10^{-7}	1.7×10^{-5}	2.8×10^{-6}

* The lognormal distribution adequately characterizes hydraulic conductivity observations for both sludge and fly ash.

Field Study

Two sludges which exhibited particularly low hydraulic conductivity were selected for investigation in the field. The field study has been in progress for approximately six years. Results from the first two years of the field study were published as NCASI Technical Bulletin No. 595 (NCASI 1990).

The field study has two main objectives. The first objective is to compare, under field conditions, the performance of paper mill sludge as a landfill hydraulic barrier material with the performance of a typical clay soil barrier material. The comparison is based upon monitoring components of the water balance (e.g., runoff and seepage) and hydraulic conductivity of the barrier material, for test cells with cover barrier layers constructed from either clay or sludge. Some chemical characterization of seepage water from sludge cells is also being performed.

The second objective is to collect data useful for verifying the Hydrologic Evaluation of Landfill Performance (HELP) Model (Schroeder et al. 1983a, 1983b). This computer program, which essentially models the water balance of a landfill, was developed by the U.S. Army Corps of Engineers Waterways Experiment Station for the USEPA. One of the test cells was specially constructed to facilitate verification of the model.

TEST CELL DESIGN AND CONSTRUCTION

Construction of four landfill cover test cells was completed in November of 1987. These test cells were designed as "typical" covers based on consultation with experts in the field. The clay cells contain a clay soil sold locally expressly for use in landfill construction. The sludge cells contain primary sludge and combined sludge (primary and biological). The paper mill sludges are from treatment of wastewater generated in nonintegrated fine papermaking processes. The particular sludges were selected for several reasons: (1) they come from mills which represent large sectors of the industry, (2) the hydraulic conductivity values determined for these sludges in the laboratory portion of the project indicated that they might be suitable barrier materials, and (3) the proximity of the sources to the field site minimized hauling. Technical Bulletin No. 595 (NCASI 1990) provides complete characterization of the clay and sludge materials and their placement in the cells.

Figure 1 is a schematic cross-section of the landfill cover test cells. In plan view each cell is approximately 9.5 meters square. At the cell top is a 15 cm (0.5 ft) layer of topsoil with vegetation. Below this is a 45 cm (1.5 ft) layer of the surface soil from the site, which is primarily a sand containing small amounts of silt and clay. Below this is a 61 cm (2.0 ft) hydraulic barrier layer. Below the barrier is a 61 cm (2.0 ft) layer of graded and compacted clean sand. At the bottom of each cell is a 30-mil PVC flexible membrane liner (FML) which was installed with no field seams other than those needed for various penetrations. A perforated 10 cm (4 in) schedule 40 PVC

FIG. 1--Cross Section of Typical Landfill Cell

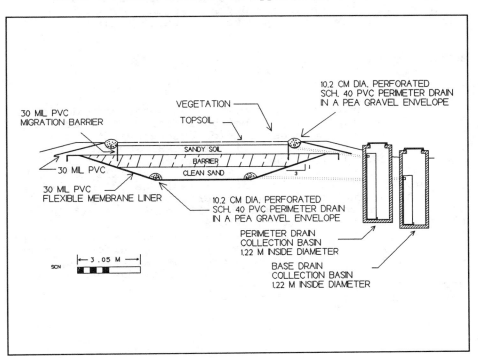

pipe collection system embedded in a pea gravel envelope lies on the perimeter of the cell top. The outside of the trench in which this runoff collection pipe is located defines runoff record areas ranging between 91.3 and 92.5 square meters (983 and 996 square feet). Another perforated PVC pipe collection system lies at the bottom of the cell embedded in a gravel envelope. This system collects water that has seeped through the barrier layer. Each collection system directs flow to a separate collection basin where the volume of water collected may be measured. The vertical FML migration barrier insures vertical percolation through the cell and precludes wall effects. This barrier defines vertical infiltration record areas ranging between 64.8 and 69.5 square meters (697 and 748 square feet).

In order to meet the objectives of the field study, the collection of meteorological and water balance data is necessary. Meteorological data along with runoff, seepage, and lateral drainage quantities are collected either manually or by an electronic data logger, or a combination of both. Backup data for precipitation, temperature, runoff, and seepage, are manually collected three times each week. Saturated thickness and soil moisture data are collected three times a week. Other data are collected on a seasonal basis or as needed. Four slotted tube piezometers placed to a depth of two feet on the top of each cell provide an indication of the saturated depth on each barrier layer. Data presented in this paper were manually collected.

RESULTS AND DISCUSSION

Consolidation

Consolidation of paper mill sludges has been described by Charlie et al. (1979) as being similar to that of soils. A description of the consolidation observed in the test cells has been presented earlier in NCASI Technical Bulletin No. 595 (NCASI 1990). The consolidation observed in the test cells is shown in **Figure 2**. The clay cells have undergone very little consolidation. The sludge cells still appear to be in the secondary compression stage of consolidation, in which the movement of clay particles within the sludge and/or decomposition of organic matter results in further compaction. The primary sludge has experienced approximately 17 cm (6.7 in) consolidation or almost 28 percent of the original thickness over the last six years. Over the same time period, the combined sludge has experienced approximately 20 cm (7.9 in) consolidation or 33 percent of the original thickness.

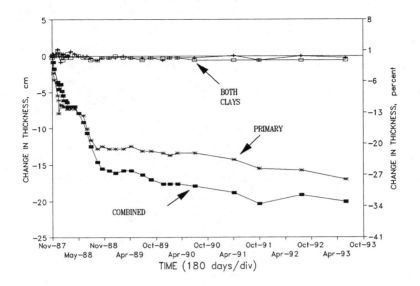

FIG. 2--Barrier layer consolidation

Runoff

Figure 3 presents the cumulative equivalent depth of surface runoff collected as a function of time. The cumulative equivalent depth is calculated by dividing the total accumulated volume of surface runoff by the runoff record area of the cell. The rate of runoff is indicated by the slope of the curve.

Examination of Figure 3 reveals that both sludge cells have produced substantially greater amounts of runoff than either of the clay cells. The combined sludge and primary sludge cells have produced approximately 200 cm and 170 cm (6.6 ft and 5.6 ft) of runoff, respectively. Clay cells 2 and 4 have produced approximately 120 cm and 78 cm (3.9 ft and 2.6 ft) of runoff, respectively. The combined sludge cell has continuously produced a slightly greater amount of runoff than has the primary sludge cell.

The difference in performance between the cells may be related to the hydraulic conductivities of the barrier layers as discussed later.

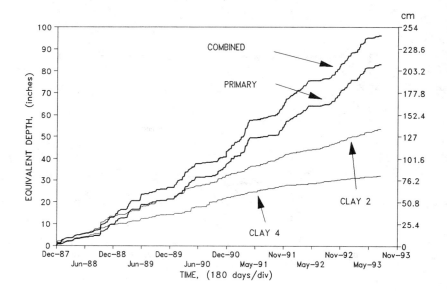

FIG. 3--Runoff cumulative equivalent depth

Seepage

Figure 4 is a plot of the cumulative equivalent depth of seepage collected as a function of time. The cumulative equivalent depth of seepage is calculated by dividing the total accumulated volume of seepage by the infiltration record area of the cell. The rate of seepage is indicated by the slope of the curve. Both clay cells have produced approximately 190 cm (6.2 ft) of seepage. The combined sludge and primary sludge cells have produced approximately 42 cm and 78 cm (1.4 ft and 2.6 ft) of seepage, respectively. Using the volume of seepage produced as a performance indicator, the sludge barriers to date have outperformed the clay barriers. Both of the clay cells have produced more than twice the volume of seepage produced from either of the sludge cells.

Saturated Depth

Figure 5 presents the average saturated depth (thickness of saturated soil above the barrier layer) of the overburden on top of the hydraulic barriers for two cells. Curves are presented for one sludge cell and one clay cell.

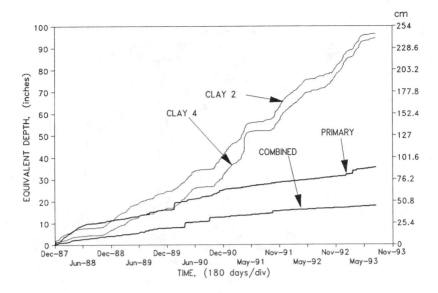

FIG. 4--Seepage cumulative equivalent depth

The curves for the cells not shown are nearly identical to those shown. Visual examination of these data along with a plot of precipitation (not shown) indicates that very little relationship exists between precipitation and saturated depth.

Examination of <u>Figure 5</u> reveals a distinct fluctuation of saturated depth over the course of a year for the combined sludge cell, whereas that of the clay cell is relatively constant throughout the year. Examination of the same data for the primary sludge cell and the other clay cell reveals almost identical trends. For the combined sludge cell, the growing season (approximately April - September) average saturated depth is 9 cm (3.5 in). During the season when vegetation is dormant (approximately October - March), the average saturated depth for the same cell is 48 cm (19 in). The clay cell 2 average saturated depth for growing and dormant seasons is 0 cm and 2 cm (0 in and 0.8 in), respectively. During the dormant months, the sludge cell has approximately 24 times the saturated thickness on top of the barrier as that of the clay cell. Averaged over the entire timespan, the saturated depth of both sludge cells is 15 times that of the clay cells. The difference in the amount of saturated depth between the two types of

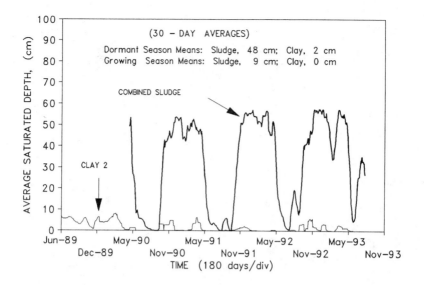

FIG. 5--Saturated depth -vs- Time

barriers is further evidence of the greater ability of the sludge barriers to minimize the rate of infiltration.

The likely mechanism for the reduction in sludge cell saturated depth during the growing season is evapotranspiration. The vegetative growth on the sludge cells rapidly reduces the accumulated water on top of the barrier during the period of growth, as shown in the curves by the sharp seasonal dips in saturated thickness. At the start of the dormant season in the fall, the sludge cell saturated depth rapidly increases to about 50 cm (20 in). Because the clay cells allow for a larger component of precipitation to migrate through the barrier, less of this water is available to plants so less is lost to evapotranspiration. Thus, the cyclical change in saturated depth in the clay cells is much less apparent.

Qualitatively, both sludge cells have a denser vegetative cover than do the clay cells; an indication that plants on top of the sludge cells have access to more water. One quantitative measure of vegetative cover is the leaf area index (LAI), an expression of leaf area per unit ground area. In July of 1989, the LAI for the combined sludge cell was more than three times that of clay cell 2 when measurements were made simultaneously on the two cells.

Hydraulic Conductivity

Comparison of the relative performance of the materials as hydraulic barriers may be made by computing hydraulic conductivity values using Darcy's Law for flow of water through saturated porous media. Yearly average hydraulic conductivity is presented for each cell in **Figure 6**. The variables for Darcy's Law - flow, area, and hydraulic gradient - were calculated as twelve-month averages. It should be noted that hydraulic conductivity values calculated in this manner are averages over the entire infiltration record area. The computations were made using adjusted barrier layer thickness based on the consolidation values presented in **Figure 2**.

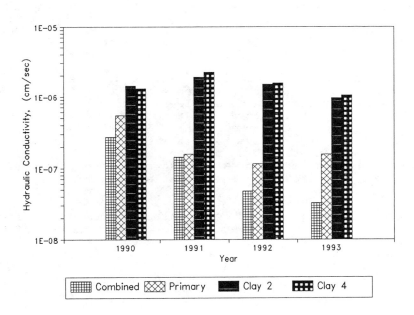

FIG. 6--Averaged yearly hydraulic conductivity

Overall long-term average hydraulic conductivity values for the primary and combined sludge barriers are 1.24×10^{-7} and 2.43×10^{-7} cm/sec, respectively. Overall long-term hydraulic conductivity values for clay cells 2 and 4 are 1.44×10^{-6} and 1.54×10^{-6} cm/sec, respectively. Examination of **Figure 6** indicates approximately one order of magnitude difference in performance between the sludge cells and the clay cells.

Several reasons exist that may explain the relatively poor performance of the clay barriers when compared to the hydraulic conductivity of the same clay determined in the laboratory. First, due to time constraints and equipment availability during construction, the clay barriers could not be compacted using conventional compaction equipment. The compaction equipment used, however, did result in design density objectives being met, and shelby tube hydraulic conductivities of core samples determined in the laboratory were reported to be 1.9×10^{-8} and 4.5×10^{-9} cm/sec respectively (NCASI 1990). Since the time of construction, several researchers have demonstrated that clod size can have a large influence on the hydraulic conductivity of compacted soils (Benson and Daniel 1990; Elsbury et al. 1990). Data from these studies suggest that if clods survive the compaction process, hydraulic conductivities may be increased by as much as two orders of magnitude due to preferential flow.

The effect of test specimen size may also account for differences in hydraulic conductivity data generated in the laboratory and hydraulic conductivity data generated in the field. Benson et al. (1993) demonstrated recently that for laboratory hydraulic conductivity testing of liner block specimens, 30 cm diameters were necessary to yield the same results as tests from sealed double-ring infiltrometers. A similar study by Shackelford and Javed (1991) indicated that hydraulic conductivity for a given soil fraction was higher in large-scale permeameters than it was in small-scale permeameters. An assumption made in both of these studies is that as sample size increases, macroscopic defects (e.g. clods) are incorporated, allowing for preferential flow through pathways.

Since hydraulic conductivity calculations for the four test cells are made from overall water balances, any effect from macroscopic defects would be included in the calculation. At the end of this research project, a diagnostic investigation is planned to determine the reasons for the differences in the performance between the clay barriers and the sludge barriers. Potential approaches include visual examination, enhanced by the use of dyes, of the components of the clay cells.

SUMMARY

The use of pulp and paper mill sludge as hydraulic barrier material in landfill covers is being examined at the field scale in four landfill cover test cells. Primary and combined paper mill sludge are being compared to clay.

After approximately six years of operation, the paper mill sludge barriers have consolidated as much as 33 percent, while the clay barriers have consolidated little, if any. Either of the sludge cells has produced almost twice the volume of runoff and less than one-half the volume of seepage as either of the clay cells. Saturated depth is high enough in the sludge cells for them to be affected by evapotranspiration, but too low in the clay cells for them to experience depletion due to evapotranspiration.

Calculated hydraulic conductivities for the sludge cells are approximately one order of magnitude lower than those calculated for the clay cells.

REFERENCES

Benson, C.H., and Daniel, D.E., August 1990, "Influence of Clods on Hydraulic Conductivity of Compacted Clay," Journal of Geotechnical Engineering, Vol. 118, No. 8.

Benson, C.H., Hardianto, F.S., and Motan, E.S., 1993, "Representative Sample Size for Hydraulic Conductivity Assessment of Compacted Soil Liners," Hydraulic Conductivity and Waste Contaminant Transport in Soils, ASTM STP 1142, David E. Daniel and Stephen J. Trautwein, Eds., American Society for Testing and Materials, Philadelphia.

Charlie, W.A., Wardwell, R.E., and Andersland, O.B., 1979, "Leachate Generation from Sludge Disposal Area," ASCE Journal of the Environmental Engineering Division, Vol. 105, No. EE5, pp. 947.

Elsbury, B.R., Daniel, D.E., Sraders, G.A., and Anderson, D.C., November 1990, "Lessons Learned from Compacted Clay Liner," Journal of Geotechnical Engineering, Vol. 116, No. 11.

NCASI, August 1984, "The Land Application and Related Utilization of Pulp and Paper Mill Sludges," Technical Bulletin No. 439.

NCASI, December 1985, "Recent Studies and Experience with the Use of Sludge and Fly Ash on Farm Crop and Forest Lands," Technical Bulletin No. 478.

NCASI, January 1989, "Experience with and Laboratory Studies of the Use of Pulp and Paper Mill Solid Wastes in Landfill Cover Systems," Technical Bulletin No. 559.

NCASI, September 1990, "A Field-Scale Study of the Use of Paper Industry Sludges in Landfill Cover Systems: First Progress Report," Technical Bulletin No. 595.

NCASI, February 1991, "Progress in Reducing Water Use and Wastewater Loads in the U.S. Paper Industry," Technical Bulletin No. 603.

NCASI, September 1992, "Solid Waste Management and Disposal Practices in the U.S. Paper Industry," Technical Bulletin No. 641.

Schroeder, P.R., Morgan, J.M., Walski, T.M., and Gibson, A.C., 1983a, U.S. Environmental Protection Agency, "Hydrologic Evaluation of Landfill Performance (HELP) Model," Vol. I, User's Guide for Version 1, EPA/530-SW-84-009.

Schroeder, P.R., Morgan, J.M., Walski, T.M., and Gibson, A.C., 1983b, U.S. Environmental Protection Agency, "Hydrologic Evaluation of Landfill Performance (HELP) Model," Vol. I, Documentation for Version 1, EPA/530-SW-84-010.

Shackelford, C.D., and Javed, F., June 1991, "Large-Scale Laboratory Permeability Testing of a Compacted Clay Soil," ASTM Geotechnical Testing Journal, Vol. 14, No. 2.

John F. Wallace[1], Ralph R. Sacrison[2] and Erick E. Rosik[1]

Two Case Histories: Field Sealed Double Ring Infiltrometer(SDRI) and Laboratory Hydraulic Conductivity Comparison Test Programs

---

REFERENCE: Wallace, J. F., Sacrison, R. R., and Rosik, E. E., "Two Case Histories: Field Sealed Double Ring Infiltrometer (SDRI) and Laboratory Hydraulic Conductivity Comparison Test Programs," Hydraulic Conductivity and Waste Contaminant Transport in Soil, ASTM STP 1142, David E. Daniel and Stephen J. Trautwein, Eds., American Society for Testing and Materials, Philadelphia, 1994.

ABSTRACT: Two case histories are discussed presenting the results of Sealed Double Ring Infiltrometer (SDRI) tests conducted in conjunction with laboratory test programs. Testing was performed on liner demonstration test fills and on liner materials obtained from within the SDRI test cells. One series of tests was conducted on a compacted weathered shale being used as a primary liner in a gold mine tailings facility. The second site was a test fill constructed as a demonstration for material suitability to be used as a tertiary liner in a low level mixed waste commercial landfill. In the second test, series of both fixed and flexible wall constant head permeability tests were performed. In both case histories, SDRI measured hydraulic conductivities were well within the same order of magnitude as those measured in the laboratory. In one case the field measured values were within a factor of 2 of those determined in the laboratory. Good agreement was also obtained for hydraulic conductivity between values measured using fixed wall and flexible wall testing procedures. These case histories present significant additions to the small but growing body of published data showing reasonable agreement between hydraulic conductivity measured in situ in the field and those values measured in the laboratory.

KEYWORDS: hydraulic conductivity, case histories, SDRI, landfill, liner

---

The widespread use of low-permeability compacted soil liners for secondary and in some cases primary means to control and collect leachate in waste management facilities has propagated a significant amount of interest in the regulatory and research community over the past ten years. Numerous authors (Elsbury et al, 1988; Boynton and Daniel, 1985; Daniel, 1984; Edil and Erickson, 1985) have acknowledged the importance of 1) compaction equipment selection, 2) moisture content, 3) lift thickness, 4) compactive effort and 5) material macro particle characteristics or "clod size" as critical aspects to a constructed liners hydraulic performance. Several authors (Elsbury et al, 1988 and Daniel and Trautwein, 1986 and Daniel, 1984) have suggested that actual field performance of compacted soil liners may differ by between 1 and 3

---

[1] Managing Principal and Lab Manager respectively, Dames & Moore, Salt Lake City, UT
[2] Mining Engineer, American Barrick Resources, Mercur, UT

orders of magnitude when compared to performance based on laboratory determined hydraulic conductivity values. Elsbury et al, 1988 concluded that laboratory measured hydraulic conductivity underpredicts field performance by "about 20,000 times." This was concluded even though the "field liner for this project did not meet the basic objectives, because: 1) the roller was too light for the soil moisture content.....lifts were too thick.... number of passes of the roller was probably too few....."

The case histories presented in this paper suggest that better correlation can be achieved between large scale field and laboratory measured values of hydraulic conductivity if the five critical aspects enumerated above are properly addressed during construction.

Case 1 - Mercur Tailing Impoundment Liner

Liner Material Characteristics

The Reservation Canyon Tailing Impoundment at the Mercur Mine utilizes an earthen liner to retard seepage of tail water. The primary concern is controlling a dilute (1-10 ppm) sodium cyanide (NaCN) solution. The liner consists of 46 cm (18 in) of earth, placed in two layers upon a prepared foundation. The lower layer consists of 30 cm (12 in) of screened clay, with a maximum particle size not exceeding 2 inches. This is a carbonaceous clay, with a liquid limit ranging from 33 to 40 and a plasticity index of 18 to 24. The mineralogy of the clay is summarized in Table 1. (Chatwin, 1989). Though having a significant amount of carbon, the material exhibits sufficient plasticity to warrant a CL classification. This lower layer is the hydraulic barrier within the liner system.

TABLE 1

MINERALOGY OF THE LONG TRAIL SHALE
BARRICK MERCUR

QUARTZ	-	24 Percent
PLAGIOCLASE	-	3 Percent
ORTHOCLASE	-	5 Percent
CALCITE	-	7 Percent
DOLOMITE	-	4 Percent
SIDERITE	-	3 Percent
GYPSUM	-	5 Percent
PYRITE	-	3 Percent
CHLORITE (CHL)	-	5 Percent
CHL/SMECTITE (mixed layer clay)	-	2 Percent
ILLITE (IL)	-	16 Percent
IL/SMECTITE (mixed layer clay)	-	18 Percent
KAOLIN	-	6 Percent

The upper layer consists of 15 cm (6 in) of weathered shale, with a maximum particle size of 15 cm (6 in). This material provides erosion and frost protection for the lower layer. The liner system material specifications are provided in Table 2. Table 3 presents the actual values for both the 1991 liner construction and the 1991 SDRI test pad.

TABLE 2

LINER MATERIAL SPECIFICATIONS

Gradation	BOTTOM, CLAY		TOP, EROSION CAP	
	Sieve Size	Percent Passing	Sieve Size	Percent Passing
	2 inches	100	6 inches	100
	3/4 inches	74		
Density	95% MDD		None	
Moisture	-3 to +3%		None	
Atterberg (PI)	10%		None	
Permeability	1×10^{-7} cm/sec		None	

TABLE 3

LINER MATERIAL CHARACTERISTICS

Gradation (% Finer)	Ave.	St. Dev.	Min.	Max.	Number of Tests
1"	99.78	0.875	96.5	100.0	19
3/4"	97.91	1.656	94.3	100.0	
1/2"	92.44	2.290	86.6	95.8	
3/8"	89.47	2.226	84.6	93.0	
#4	82.46	2.625	76.6	86.9	
#8	76.78	2.636	71.3	81.4	
#16	71.41	2.603	66.3	76.7	
#30	66.65	2.550	61.9	72.3	
#50	62.04	2.604	57.5	68.4	
#100	58.39	2.786	53.5	64.9	
#200	55.05	2.860	50.2	61.7	
Density -Kg/m³ (pcf)	1896 (118.3)	32 (1.99)	1838 (114.7)	1947 (121.5)	20
Moisture - %	14.7	1.76	11.5	18.3	20
Plasticity Index	22.56	1.525	18.7	24.8	16
Lab Hydraulic Conductivity - cm/sec	$3.82 \times 10^{-8(1)}$	5.60×10^{-8}	4.61×10^{-9}	1.99×10^{-7}	20

(1)-arithmetic mean

Liner Construction

Prior to placement of the first lift, the foundation surface was prepared by smoothing and contouring to a final 3:1 slope. This was accomplished with dozing and grading equipment. The tracked dozers were D7 models, the grader was a Cat 14G. In select places, a 3/4 CY trackhoe (Komatsu 220LC) with a contouring blade attachment was used to shape the surface. The surface was then moistened and compacted with a minimum of three passes of a sheepsfoot soil compactor. The unit used was a Caterpillar 815B, with a nominal operating weight of 20mT (22 st). This four-wheel unit has sixty feet(segmented pads) per wheel in five rows. The end area per foot is 116 cm^2 (18 in^2).

A minimum of 40 cm (16 in) of loose clay was then placed on the foundation. The placement equipment was primarily the dozers and graders noted above. When expedient, a Cat 996D 4-1/4 CY wheel loader and/or the Komatsu trackhoe assisted in this function.

Compaction of the liner was effected by use of a Cat CS563 smooth drum vibratory roller. The unit has a nominal operating weight of 11.1 Mt (12 St). The drum width is 213 cm (84 in), with a diameter of 152 cm (60 in). The vibratory system uses a hydraulic-driven eccentric weight at a frequency of 1,800 VPM. The two amplitude settings deliver a centrifugal force of 35,000 to 50,000 lb. The clay was compacted to not less than 95 percent of maximum dry unit weight defined by the Standard Proctor test (ASTM D-698-78). The acceptable moisture range was specified as -3 to +3 percent of optimum.

A four year history during which over 19 ha (47 ac) of liner was constructed using this clay established the acceptable moisture range required to meet hydraulic conductivity performance criteria. An investigation of the moisture-hydraulic conductivity relationship for this clay indicates a value of 5.1×10^{-8} cm/sec at a moisture content as low as 7 percent below optimum. Hydraulic conductivity was determined in general accordance with ASTM D-5093 procedures on specimens compacted to a minimum of 95% of the maximum dry unit weight defined by the ASTM D-698 compaction test procedure.

Moisture was added primarily at the clay stockpile. The stockpile was bladed and tilled to achieve a uniform moisture. Minor spraying occurred when the clay was delivered to the liner, but care was taken to prevent overwetting. This would have been counterproductive on sloped areas of the impoundment. Constructed liner hydraulic conductivity was confirmed using drive tube samples tested by the ASTM D-5084 procedure. A statistical summary of these test results is presented in Table 3.

Following confirmation of the 1×10^{-7} cm/sec objective, the erosion and frost cap were placed over the liner. Compaction of this upper layer commenced when a minimum of 23 cm (9 in) inches of loose fill were placed. The placement and compaction equipment were the same as for the lower layer. Compaction consisted of four passes of the smooth drum vibratory roller.

Hydraulic Conductivity Measurements

A test pad was constructed concurrently with the liner installation. The pad was nominally 9.1 m (30 ft) on each side. This allowed construction using the same equipment as the actual liner.

The test liner included the erosion cap used on the actual liner. This necessitated some alteration to the normal ASTM D-5093 procedures:

1. The inner and outer rings penetrated the test pad an additional 15 cm (6 in) to account for the erosion cap.

2. The duration of the test was increased to account for the time required to stabilize flow into and through the erosion cap.

3. Temperature corrections for the calculated K values could not be applied. This was due to the inability to quantify the total volume within the inner ring. That inability arose from the relative uncertainty of the actual porosity of the erosion cap.

In order to control the temperature of the test, an insulated building was erected around the pad. Heat was provided and regulated, providing some measure of uniformity through the late fall and early winter. The temperature fluctuations are shown in conjunction with the calculated hydraulic conductivities, on Figure 1.

FIGURE 1

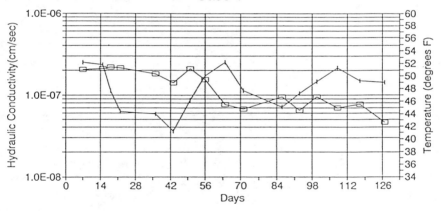

The test ran for 126 days. Stable measurements of hydraulic conductivity were achieved in 77 days from initiating the test. These values stabilized following temperature stabilization, and the test was concluded at a calculated hydraulic conductivity of 6.99×10^{-8} cm/sec.

As an index of expected performance, a laboratory hydraulic conductivity test was performed on a test fill specimen obtained prior to initiating the SDRI test. Test results showed a hydraulic conductivity of 9.3×10^{-8} cm/sec. An additional test was conducted on the hydrated liner material following completion of the SDRI using the ASTM D-5084 procedure. Calculated hydraulic conductivity from this test was 2.16×10^{-8}. Drive tube samples having a 6 cm(2.4 inch) diameter were taken for quality control laboratory hydraulic conductivity measurements for the production liner. Using ASTM D-5084 test methods an average hydraulic conductivity of 8.09×10^{-8} cm/sec was calculated with a standard deviation on the of 2.69×10^{-7} for the 183 specimens tested. Twenty three specimens remolded to a minimum 95% of the maximum dry density (ASTM D-698 basis) and subjected to the same testing procedure yielded an average hydraulic conductivity of 6.51×10^{-7} cm/sec with a standard deviation of 1.67×10^{-6}.

Case 2 - Mixed Waste Commercial Landfill Site

Liner Material Characteristics

The compacted material in Case 2 was used as secondary containment for development of a commercial mixed waste landfill cell. The site is located in a portion of the Basin and Range more commonly referred to as the West Desert approximately 80 miles west of Salt Lake City, Utah. Indigenous clays were used for the test fill construction. These materials typically had less than 5 percent fine sand, a liquid limit of 38 and a plasticity index of 14, thus classifying them as CL in the Unified Soil Classification System. A complete particle-size analysis for the liner material is presented as Figure 2. Compacted characteristics for this soil include a maximum dry unit weight of 1604 Kg/m³ (100.1 pcf) and an optimum moisture content equal to 23.5 % based on ASTM D-698, Method A.

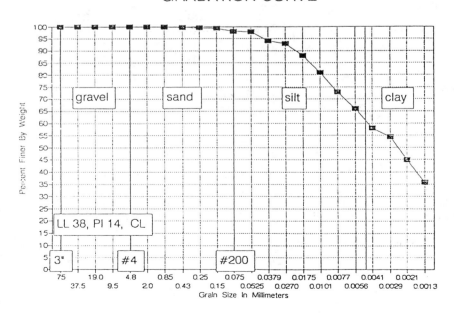

Figure 2

GRADATION CURVE

Liner Construction

The liner test fill measuring approximately 12 x 12 m (40 ft by 40 ft) was constructed to a total depth of 61 cm (24 in). Liner material was spread in loose lift thickness approximately 20 cm (8 in) in thickness and compacted using a CAT 815 segmented pad compactor. Successive layers were placed in a similar fashion and compacted to achieve the full 61 cm (24 in) depth of the test fill area. Densities measured using drive tube sampling methods (ASTM D-2937) indicated a compacted dry unit weight ranging between 90 and 97 percent of the maximum. Compacted moisture contents ranged between 21.4 and 26.0 percent or minus 2 to plus 2.5 percent relative to optimum. Prior to installation of the SDRI test unit the upper 5 to 7.5 cm (2 to 3 in) was removed using a motorgrader for leveling and to remove a majority of the cleat depressions caused during compaction.

Hydraulic Conductivity Measurements

In Situ Measurements in the Field

Field hydraulic conductivity of the test fill was calculated based upon infiltration rates measured using a Trautwein SDRI device following the procedures outlined in ASTM D-5093-90. The test was conducted over a 30 day period with relatively stable readings being obtained over the final 18 days. Calculated hydraulic conductivity of the liner test fill ranged between 4.0×10^{-8} and 5.0×10^{-8} cm/sec yielding an average value equal to 4.5×10^{-8} cm/sec. The calculated hydraulic conductivity over the course of the testing is presented in Figure 3.

An example calculation of the hydraulic conductivity for data obtained on day 11 of the test is presented below:

Depth of Water - dw - 26.1 cm(10.3 in)
Saturated Liner Thickness - slt - 15.9 cm(6.28 in)
Time of Flow - T - 511,200 sec
Volume of Flow - V - 1562 ml
Area of Flow - A - 23,225.8 cm² (25 ft²)

Calculated Hydraulic Gradient - $i = \frac{dw + slt}{slt}$

Calculate Infiltration Rate - I = V/(AT) = 1562/(23,225.8*511,200)cm/sec
   = 1.31×10^{-7} cm/sec

Calculate Hydraulic Conductivity - k = I/i = $1.31 \times 10^{-8}/2.64 = 4.98 \times 10^{-8}$

A moisture profile was developed for the testfill at the conclusion of testing to confirm the depth of saturation. Moisture contents determined in accordance with ASTM D-2216-80 ranged between 35.9 percent at the surface of the fill to 24.2 percent at a depth of 16.5 cm (6.5 in). Table 4 presents a summary of the moisture profile data. These data show moisture contents

FIGURE 3

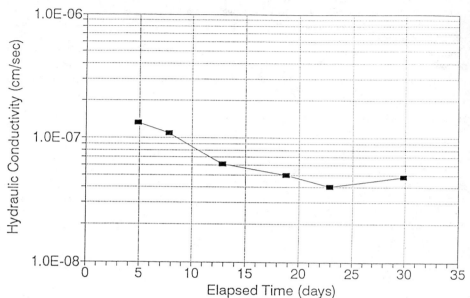

at or very near saturation through the depth measured in the testfill when compared to the optimum moisture and its relationship to the zero air voids relationship. For the purpose of hydraulic conductivity calculations, full depth saturation was conservatively assumed.

Table 4
Moisture Content vs. Depth
In Case 2 Test Fill

Depth-cm	Moisture Content - %
0-5	35.9
7-10	28.2
15-18	24.2
23-25	24.6
30-32	25.7
35-37	25.6

Laboratory Measured Properties

A series of hydraulic conductivity measurements were made on testfill specimens taken at the conclusion of the in situ testing from within the area defined by the inner ring of the SDRI apparatus. Fifteen cm (6 in) long by 6 cm (2.4 in) diameter drive tube methods (ASTM D-2937) were used to obtain relatively undisturbed specimens. Three specimens 5 cm(2 inches) in length were tested using flexible wall (ASTM D-5084) methods and three additional specimens were tested using fixed wall methods. Net confining pressure applied in the flexible wall tests was approximately 6.9 kPa(144 psf). Applied hydraulic gradients ranged between 79 and 83. Fixed ring tests were conducted under a 4.8 Kpa (100 psf) surcharge with applied hydraulic gradients ranging between 23 and 24. A summary of the test results is presented in Table 5 and Table 6.

Table 5
Flexible Wall
Hydraulic Conductivity Test Summary

Test No.	Dry Unit Weight - Kg/m3 (pcf)	Moisture - %	Hydraulic Conductivity - cm/sec
1	1470(91.7)	32.8	1.57×10^{-8}
2	1585(98.9)	28.8	1.66×10^{-8}
3	1534(95.7)	26.8	1.62×10^{-8}

Table 6
Fixed Wall
Hydraulic Conductivity Test Summary

Test No.	Dry Unit Weight - Kg/m3 (pcf)	Moisture - %	Hydraulic Conductivity - cm/sec
1	1452(90.6)	33.4	6.99×10^{-8}
2	1502(93.7)	30.2	3.44×10^{-8}
3	1558(97.2)	27.7	2.38×10^{-8}

Discussion and Conclusions

The two case histories presented above represent somewhat of a departure from like data presented previously in the literature. The most significant difference may be that of using large compactors for test fill construction in conjunction with adequate moisture conditioning of materials. Procedures and equipment were essentially those in use for liner construction in Case 1 and that proposed for use in Case 2.

Based upon the results of the test results obtained for the two case histories presented above the following conclusions can be drawn.

1. Hydraulic conductivity values determined through laboratory measurement were, in both case studies, lower those values measured in situ using the SDRI procedure.

2. The hydraulic conductivity values determined by either fixed wall or flexible wall methods are generally in good agreement with one another, i.e. within a half order of magnitude of one another.

3. Hydraulic conductivity determined using the SDRI were less than one half order of magnitude greater than those values determined using either laboratory method.

4. The effects of macro structural defects associated with "clods", desiccation cracks, non-homogeneity and poor compaction previously reported as resulting in a several order of magnitude disparity between laboratory and field measured hydraulic conductivity are significantly reduced through the use of adequately sized compaction equipment, and proper moisture conditioning and placement procedures.

REFERENCES

Boynton, S. S. and Daniel, D. E., 1984, "Hydraulic Conductivity Tests on Compacted Clay," Journal of Geotechnical Engineering, ASCE, New York, Vol. 111, No. 4, pp. 465-478.

Chatwin, T.D., Aug. 1989, Final Report on Cyanide Attenuation Characteristics of the Long Trail Shale, Submitted to Barrick Resources by Resource Recovery and Conservation Consultants, Salt Lake City, UT.

Daniel, D. E., 1984, "Predicting Hydraulic Conductivity of Clay Liners," Journal of Geotechnical Engineering, ASCE, New York, Vol. 110, No. 2, pp. 285-300.

Daniel, D. E. and Trautwein, S. J., 1986, "Field Permeability Test for Earthen Liners," Proceedings, Use of In Situ Tests in Geotechnical Engineering, ASCE, New York, pp. 146-160.

Daniel, D. E., 1987, "Earthen Liners for Land Disposal Facilities," Proceedings, Geotechnical Practices for Waste Disposal '87, ASCE, New York, pp. 21-39.

Edil, T. E. and Erickson, A. E., 1985, "Procedures and Equipment Factors Affecting Permeability Testing of a Bentonite-Sand Liner Material," ASTM STP 874, ASTM, Philadelphia, pp. 155-170.

Elsbury, B. R., Sraders, G. A., Anderson, D. C., Rehage, J. A., Sai, J. O. and Daniel, D. E., "Field Testing of a Compacted Soil Liner," Risk Reduction Engineering Laboratory, Office of Research and Development, U. S. EPA, Cincinnati, OH, EPA/600/2-88/067, November 1988.

Albert T. Yeung[1]

EFFECTS OF ELECTRO-KINETIC COUPLING ON THE MEASUREMENT OF HYDRAULIC CONDUCTIVITY

REFERENCE: Yeung, A. T., "Effects of Electro-Kinetic Coupling on the Measurement of Hydraulic Conductivity," *Hydraulic Conductivity and Waste Contaminant Transport in Soil, ASTM STP 1142*, David E. Daniel and Stephen J. Trautwein, Eds., American Society for Testing and Materials, Philadelphia, 1994.

ABSTRACT: Hydraulic conductivity is the proportionality constant relating the volume flow rate of a fluid through a porous medium to the imposed hydraulic gradient. Hence, there should not be any gradient or driving force other than a hydraulic gradient imposed on the sample during the measurement of the parameter. However, no precaution is taken to prevent the development of a streaming potential during the experiment. As a result, an electrical gradient is always generated across the sample and the measured hydraulic flow volumes are actually induced by the combined hydraulic and electrical gradients. The effects of electro-kinetic coupling are more pronounced in fine-grained soil. Moreover, the measured conductivity coefficient cannot be used in coupled flow equations to quantify the hydraulic volume flow rate contributed by the hydraulic gradient without appropriate corrections. This paper attempts to present the results of the theoretical analysis and experimental evaluation of the effects of electro-kinetic coupling in the measurement of hydraulic conductivity.

KEYWORDS: coupled flow, electro-kinetics, electro-osmosis, fine-grained soils, hydraulic conductivity, coefficient of electro-osmotic conductivity, laboratory measurement, streaming potential, thermodynamics of irreversible processes

Hydraulic conductivity is the proportionality constant relating the volume flow rate of a fluid through a porous medium to the applied hydraulic gradient across the medium and the gross total area perpendicular to the flow direction through which the flow occurs (Darcy 1856). If there is no other driving force existing in the system, it can be determined experimentally *in-situ* or in the laboratory to quantify the resistance provided by a porous medium to a hydraulic flow through the medium driven by an imposed hydraulic gradient according to

$$Q = k_h i_h A \qquad (1)$$

where
Q = volume flow rate (m³/s)
k_h = hydraulic conductivity (m/s)

---

[1]Assistant Professor of Civil Engineering, Texas A&M University, College Station, TX 77843-3136

i_h = hydraulic gradient (dimensionless)
A = gross total flow area (m^2)

As the flow is driven by the potential gradient of the same type, it is classified as a direct flow (Mitchell 1993). There exists a large number of similar experimental relations describing irreversible processes in the form of proportionalities, e.g. Fourier's law between heat flow and thermal gradient, Ohm's law between electrical current density and imposed electrical gradient, Fick's law between flow of a chemical component in a mixture and its chemical gradient. Each of these fluxes J_i is linearly related to its corresponding gradient or driving force X_i according to the conjugated flow equation,

$$J_i = L_{ii} X_i \qquad (2)$$

where
J_i = flux or flow per unit area of type i
L_{ii} = conductivity coefficient for type i flow
X_i = gradient or driving force of type i

Thus, Eq. (1) is a special case of the generalized flow equation in the form of Eq. (2).

When two or more of these phenomena occur simultaneously, they interfere with each other and produce cross effects. Examples include electro-osmosis (hydraulic flow driven by an electrical gradient), the Peltier effect (evolution or absorption of heat at junctions of metals resulting from the flow of an electrical current), thermoelectric force (electromotive force resulting from the maintenance of the junctions at different temperatures), the Soret effect (chemical gradient developed as a result of an applied thermal gradient), the Dufour effect (thermal gradient generated due to the existence of a chemical gradient), and so on. As a flow of one type is driven by a potential gradient of another in these examples, it is classified as a coupled flow (Mitchell 1993). The types of interrelated coupled flows that can occur under the influences of hydraulic, electrical, thermal and chemical gradients are listed in Table 1 (Yeung and Mitchell 1993). These effects can be described mathematically by adding terms proportional to the coupling driving forces to the corresponding conjugated flow equations. That is, any flux is related to the gradients or driving forces by

$$J_i = \sum_{j=1}^{n} L_{ij} X_j \qquad (i \leq n) \qquad (3)$$

where
J_i = flux or flow per unit area of type i
L_{ij} = coupling coefficients relating type i flow and type j force
X_j = gradient or driving force of type j

Supporting evidences for the validity of Eq. (3) in soil systems have been given by Abd-El-Aziz and Taylor (1965), Kemper and Letey (1968), and Letey and Kemper (1969) for the simultaneous flows of water and salt; Gray and Mitchell (1967), and Olsen (1969) for the simultaneous flows of water and electricity; and Hadas and Taylor (1968) for the simultaneous flows of vapor and heat and the simultaneous flows of heat and water.

Eq. (2) is a special case of Eq. (3) when all coupling driving forces vanish. If the hydraulic conductivity of a soil defined by Eq. (1) is measured without the interference of any gradients or driving forces other than a hydraulic gradient, the conductivity coefficient L_{ii} in Eqs. (2) and (3) should be a linear function of hydraulic conductivity defined in Eq. (1) when J_i is the hydraulic volume flux and X_i is the imposed hydraulic gradient. However, state-of-the-art hydraulic

TABLE 1 -- Coupled and direct flow phenomena (after Yeung and Mitchell 1993)

Flow J (1)	Gradient X			
	Hydraulic (2)	Electrical (3)	Thermal (4)	Chemical (5)
Fluid	Hydraulic conduction (Darcy's Law)	Electro-osmosis	Thermo-osmosis	Normal osmosis
Electric current	Streaming potential	Electric conduction (Ohm's Law)	Seebeck effect	Diffusion and membrane potentials
Heat	Isothermal heat transfer	Peltier effect	Thermal conduction (Fourier's Law)	Dufour effect
Ion	Streaming current	Electro-phoresis	Soret effect	Diffusion (Fick's Law)

conductivity measurement apparatuses, techniques and procedures do not take the effects of electro-kinetic coupling generated by the applied hydraulic gradient into consideration. Thus the hydraulic volume flow rate measured in a hydraulic conductivity measurement is always generated by the combination of a hydraulic and an electrical gradient. The coupling effects are more pronounced in fine-grained soils typical of those used in the construction of the compacted clay liner of an engineered waste repository facility. Recently, the potential applications of electro-kinetics in hazardous waste site remediation are being investigated. Most applications involve the coupled flows of fluid, electricity and contaminants. If Eq. (3) is used to describe these flows and predict the concentrations of contaminants as a function of time and space, the error induced by expressing the conductivity coefficient as a linear function of hydraulic conductivity alone may be significant. This paper attempts to present the results of the theoretical analysis and experimental evaluation of the significance of the these electro-kinetic coupling effects on the measurement of hydraulic conductivity. However, the analysis and evaluation of the coupling effects of chemical gradients are beyond the scope of this paper.

POTENTIAL APPLICATIONS OF ELECTRO-KINETIC FLOW PROCESSES IN HAZARDOUS WASTE SITE REMEDIATION

Several electro-kinetic phenomena arise in a wet plug of clay when there are couplings between hydraulic and electrical driving forces and flows. These phenomena can broadly be classified into two groups by the driving forces causing the relative movement between the liquid and the solid phases. The first group consists of *electro-osmosis* and *electrophoresis*, in which the liquid or the solid phase moves relative to the other under the influence of an externally applied electrical potential. The second group consists of *streaming potential* and *migration* or *sedimentation potential*, in which the liquid or the solid phase moves relative to the other under the influence of a hydraulic or gravity force, thus inducing an electrical potential difference across the medium. These four electro-kinetic phenomena are schematically shown in Fig. 1 and reviewed in detail by Yeung (1993).

There are many conceivable potential uses of electro-kinetic flow processes for hazardous waste site remediation. These potential uses of electro-kinetics were assessed by a forum of research engineers and scientists experienced in electro-kinetics in a two day workshop held in Seattle, Washington. The workshop was sponsored by the Hazardous Waste Engineering Laboratory of the United States Environmental Protection Agency and hosted by the Department of Civil Engineering at the University of Washington (United States Environmental Protection Agency 1986). As a result of the outcome of the workshop (Mitchell 1986) and recent developments in the area, possible uses of electro-kinetics for hazardous waste site remediation may include:

(1) concentration, dewatering and consolidation of wastewater sludges, slimes, coal washeries, mine tailings or dredged materials;
(2) injection of grouts to control ground water flow;
(3) injection of cleaning agents into contaminated soils;
(4) injection of vital nutrients for the growth of micro-organisms essential to the biodegradation of specific wastes;
(5) electro-kinetic barriers to contaminant transport through compacted clay landfill liners or slurry encapsulation walls;
(6) electro-kinetic extraction of contaminants from polluted soil;
(7) modification of flow pattern of ground water, and manipulation of the movement and size of a contaminant plume;
(8) rapid and reliable *in-situ* determination of hydraulic conductivity of compacted clay landfill liners;
(9) *in-situ* characterization of contaminants in soil pore fluid;
(10) *in-situ* generation of reactants for cleanup and/or electrolysis of contaminants;
(11) Retrofitting of leaking in-service geomembrane liner.

Detailed review of the current status of these innovative technologies is given by Yeung (1993). Since these processes involve the simultaneous flows of fluid, electricity, chemicals and heat under hydraulic, electrical, chemical and thermal gradients, the net transport of fluid and chemicals depends on a number of complex interactions. Quantitative evaluations and predictions of the flows require that these interactions be accounted for appropriately. If the measured hydraulic conductivity is used directly as a conductivity coefficient in Eq. (3) to quantify the volume of hydraulic flow induced by the hydraulic gradient, significant errors will be induced for fine-grained soil. This paper will concentrate on the interactions between a hydraulic and an electrical gradient in the driving of a hydraulic flow. Readers interested in the analysis of the coupled flows of heat and mass in porous media should refer to Luikov (1975).

THEORETICAL ANALYSIS

Though the determination of hydraulic conductivity is a routine test in most geotechnical engineering testing laboratories, there are numerous possible variations in the experimental apparatus, technique and procedure (Yeung 1990a). The measured hydraulic conductivity of fine-grained soil can vary over orders of magnitude because of these variations. Basically, different hydraulic gradients are applied across

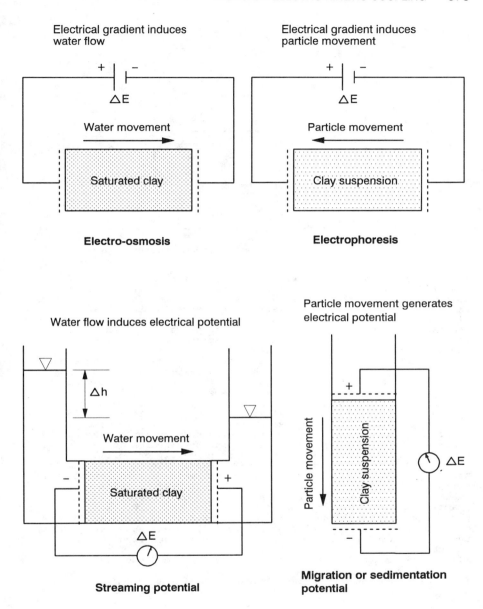

Fig. 1 -- Electro-kinetic phenomena in clay (after Yeung and Mitchell 1993)

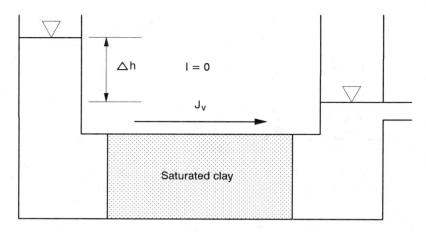

Fig. 2 -- Schematic of hydraulic conductivity measurement

a soil sample, and the corresponding volume flow rates of the fluid are measured. The ratio of the volume flow rate per unit gross total area to the applied hydraulic gradient gives the hydraulic conductivity according to Eq. (1). A schematic of the experimental arrangement is illustrated in Fig. 2. The fundamental assumption for the validity of the experimental results obtained by this procedure is that there exists no driving force other than a hydraulic gradient in the system or the magnitude of such force, if exists, is negligible.

As illustrated in Fig. 1, an electrical gradient is developed when a hydraulic gradient is imposed on a clay sample. It is uncommon to have electrodes installed at both ends of the sample for hydraulic conductivity measurements. Thus the development of a streaming potential in the direction shown in Fig. 1 as a result of the generated hydraulic flow cannot be prevented by short-circuiting such electrodes. Hence the measured hydraulic volume rate is actually induced by a hydraulic gradient and an electrical gradient imposed on the sample simultaneously. The significance of the contribution to the hydraulic flow by the electrical gradient depends primarily on soil type.

As the hydraulic and electrical potential in the soil sample are not homogeneous, the soil sample is a soil-water system out of thermodynamic equilibrium (Groenevelt and Bolt 1969). The formalism of thermodynamics of irreversible processes is employed to provide a theoretical framework for the macroscopic formulation of the irreversible transport processes, to define the problem clearly and to give a quantitative description of the phenomenological coefficients. Application of the theory of thermodynamics of irreversible processes requires the following steps (Fitts 1962):

(1) finding the dissipation function Φ for the flows occurring in the soil-water system;
(2) defining the conjugated fluxes J_i and the driving forces X_i from

$$\Phi = \sum_{i=1}^{n} J_i X_i \qquad (4)$$

(3) formulating the set of phenomenological flow equations in the form of Eq. (3);
(4) applying the Onsager reciprocal relations; and

(5) relating the phenomenological coefficients to measurable quantities.

The dissipation function Φ of any system out of thermodynamic equilibrium can be derived from the local formulations of the conservation laws of mass, energy and momentum on the basis of the consequences of the second law of thermodynamics. Detailed derivation of the dissipation function of a system involves the simultaneous coupled flows of fluid, electricity, cations and anions under the influences of simultaneously imposed hydraulic, electrical and chemical gradients is given by Yeung (1990b). The developed coupled theory has been able to predict the transport of inorganic contaminants in an electro-kinetic barrier (Mitchell and Yeung 1991). Once a proper choice of the fluxes J_i and driving forces X_i is made, the matrix of phenomenological coefficients L_{ij} given by the set of phenomenological flow equations in the form of Eq. (3) is symmetric,

$$L_{ij} = L_{ji} \quad (i,j = 1, 2, \ldots, n) \tag{5}$$

These identities are called the Onsager reciprocal relations. The number of phenomenological coefficients in a system of n conjugated forces and n conjugated fluxes is reduced from n^2 to $n(n+1)/2$.

In the theoretical analysis of coupled hydraulic and electrical flows, the conjugated fluxes are determined to be the volume flow rate of fluid per unit area and electrical current density, and the conjugated driving forces are the hydraulic pressure gradient and electrical gradient, respectively. Hence the coupled flow equations are

$$J_v = L_{11}\nabla(-P) + L_{12}\nabla(-E) \tag{6}$$

$$I = L_{21}\nabla(-P) + L_{22}\nabla(-E) \tag{7}$$

where
- J_v = fluid flux or volume flow rate per unit area (m/s)
- I = electric current density (A/m^2)
- P = hydraulic pressure (N/m^2)
- E = electrical potential (V)
- L_{ij} = phenomenological coefficients

The negative signs in Eqs. (6) and (7) indicate that the fluid and electrical current flow are in the direction of decreasing hydraulic pressure and electrical potential, respectively. As the electrical circuit is always open during a hydraulic conductivity measurement, the electric current must vanish. Putting $I = 0$ in Eq. (7),

$$\nabla(-E) = -\frac{L_{21}}{L_{22}}\nabla(-P) \tag{8}$$

gives the gradient of the streaming potential generated by the applied hydraulic gradient. Substituting Eq. (8) into (6)

$$J_v = \left[L_{11} - \frac{L_{12}L_{21}}{L_{22}}\right]\nabla(-P) = \left[L_{11} - \frac{L_{12}L_{21}}{L_{22}}\right]\gamma_f\nabla(-h) \tag{9}$$

where
- γ_f = unit weight of the fluid (N/m^3)
- h = hydraulic head (m)

The volume flow rate per unit gross total area of a fluid is related to the imposed hydraulic gradient by Darcy's law,

$$J_v = k_h \nabla(-h) \qquad (10)$$

Equating the coefficients of Eqs. (9) and (10),

$$\left[L_{11} - \frac{L_{12}L_{21}}{L_{22}}\right]\gamma_f = k_h \qquad (11)$$

Rearranging

$$L_{11} = \frac{k_h}{\gamma_f} + \frac{L_{12}L_{21}}{L_{22}} \qquad (12)$$

When an electrical gradient is maintained across a wet clay, water is moved from the anode to the cathode as shown in Fig. 1 by electro-osmosis. In practice, the relationship between the volume flow rate and the imposed electrical gradient is given by an expression analogous to Darcy's law,

$$Q = k_e i_e A \qquad (13)$$

where
- Q = volume flow rate (m^3/s)
- k_e = coefficient of electro-osmotic conductivity (m^2/Vs)
- i_e = electrical gradient (V/m)
- A = gross total flow area (m^2)

The phenomenological coefficients L_{12} (L_{21}) and L_{22} can be determined from the measurement of the coefficient of electro-osmotic conductivity and electrical conductivity of a soil sample. Fig. 3 illustrates a schematic for these laboratory measurements. The two compartments housing solutions of identical composition are maintained at equal hydraulic pressure to keep the sample saturated and to eliminate the effects of hydraulic coupling during the experiment. Different electrical gradients are applied across the sample, and the volume flow rates of fluid and electrical currents passing through the sample are measured. The ratio of the volume flow rate per unit gross total area of the fluid to the imposed electrical gradient gives the coefficient of electro-osmotic conductivity as given in Eq. (13), and the ratio of the electrical current density to the imposed electrical gradient yields the electrical conductivity. As $\nabla(-P) = 0$, Eqs. (6) and (7) reduce to

$$J_v = L_{12}\nabla(-E) \qquad (14)$$

$$I = L_{22}\nabla(-E) \qquad (15)$$

Comparing the coefficients of Eqs. (13) and (14)

$$L_{12} = k_e \qquad (16)$$

Eq. (15) is simply the point form of Ohm's law applied to the soil sample. Hence L_{22} is given by

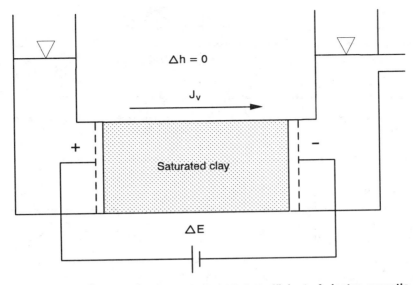

Fig. 3 -- Schematic of the measurement of coefficient of electro-osmotic permeability and electrical conductivity

$$L_{22} = \frac{I}{\nabla(-E)} = \sigma \qquad (17)$$

where
σ = bulk electrical conductivity of the soil (S/m)

As a consequence of Onsager's reciprocity theorem, the equality of L_{12} and L_{21} gives rise to the equivalence between streaming potential and electro-osmosis,

$$\left(\frac{Q}{I}\right)_{\Delta P = 0} = -\left(\frac{\Delta E}{\Delta P}\right)_{I=0} \qquad (18)$$

which is known as Saxén's law. Detailed derivation of the equivalence is presented by Mitchell (1993). The equivalence was first verified by Saxén (1892) using a clay diaphragm and solutions of different concentrations of zinc, cadmium or copper sulfate. His results are tabulated in Table 2. In their study on how the variation in certain key soil parameters influence electro-osmotic behavior, Gray and Mitchell (1967) obtained experimental data for three types of materials: a pure kaolinite (Hydrite UF), an illitic clay (Grundite), and an artificial silty clay comprised of equal parts by weight of kaolinite and silica flour. The clays were made as nearly homoionic to sodium ion as possible. The samples have been consolidated to predetermined water contents prior to the measurements of transport coefficients. Precautions were taken to eliminate or suppress to a negligible level the effects of electrolysis, ion exchange, gas formation at the electrodes, physicochemical alteration of the clays, heating, and the building up of concentration gradients across the clay plug. Saxén's law was tested over a wide range of water contents — from slurry to a consistency obtained from consolidation under pressures as high as 7.7 MPa. The results are

TABLE 2 -- Electro-osmotic flow and streaming potential measurements (after Saxén 1892)

Solution (1)	Q/I (m³/C) (2)	$-\Delta E/\Delta P$ (V/Pa) (3)	Ratio (3)/(2) (4)
0.0174 M $ZnSO_4$	3.60×10^{-8}	3.52×10^{-8}	0.978
0.0261 M $ZnSO_4$	3.82×10^{-8}	3.79×10^{-8}	0.992
0.0348 M $ZnSO_4$	3.46×10^{-8}	3.44×10^{-8}	0.994
0.0195 M $CdSO_4$	5.82×10^{-8}	5.88×10^{-8}	1.010
0.0390 M $CdSO_4$	1.16×10^{-8}	1.15×10^{-8}	0.991
0.0400 M $CuSO_4$	3.85×10^{-8}	3.85×10^{-8}	1.000
0.0800 M $CuSO_4$	2.33×10^{-8}	2.37×10^{-8}	1.017

tabulated in Table 3. The equivalence holds quite well in silty clay and kaolinite but not so well in the illitic clay. However, it should be noted that it is more difficult to measure the small streaming potential induced in the illitic clays because of electrical shielding. Experimental data obtained by Olsen (1969, 1972) on kaolinite saturated with sodium and equilibrated with 0.001 M sodium chloride (NaCl) solution are also in support of the equivalence. Transport coefficients were measured after successive increments of consolidation and rebound under loads ranging from 1.33 MPa to 66.7 MPa. The homoionic system with chloride as the anion was used together with silver-silver chloride electrodes to provide a system in which electric current could only be passed through the system by means of simple electrolytic transport, so that extraneous electrochemical reactions and gas generation at the electrodes could be avoided. The electro-osmotic flow volume per unit electric current Q/I and the streaming potential generated per unit applied hydraulic pressure difference $\Delta E/\Delta P$ were computed by the author using the transport coefficients A_{ij} reported by Olsen (1969, 1972). The results are tabulated in Table 4. As tabulated in the last column of Tables 2, 3 and 4, it appears that Saxén's law is valid in most clay-water-electrolyte systems under a wide range of conditions. Thus the applicability of the formalism of thermodynamics of irreversible processes is verified.

Substituting Eqs. (16) and (17) into (12),

$$L_{11} = \frac{k_h}{\gamma_f} + \frac{k_e^2}{\sigma} \qquad (19)$$

Most recent analyses of the coupled hydraulic and electrical flow phenomena are performed using Eq. (3). The volume flow rate of the fluid induced by a hydraulic gradient and that induced by an electrical gradient are summed algebraically to give the total volume flow rate. However, the phenomenological coefficient L_{11} is generally taken to be k_h/γ_f and the effects of electro-kinetic coupling are not included in the formulation (Esrig 1968; Lewis and Garner 1972; Lewis and Humpheson 1973; Renaud and Probstein 1987; Wan and Mitchell 1976). The approach may be adequate if the hydraulic conductivity used in the coupled flow

TABLE 3 -- Electro-osmotic flow and streaming potential measurements (adapted from Gray and Mitchell 1967)

Material (1)	Solution (2)	Void ratio e (3)	Q/I (m³/C) (4)	$-\Delta E/\Delta P$ (V/Pa) (5)	Ratio (5)/(4) (6)
Kaolinite	0.001 M NaCl	1.72 1.44 1.17 0.90 0.80	4.496×10^{-7} 3.843×10^{-7} 3.451×10^{-7} 2.649×10^{-7} 2.351×10^{-7}	3.974×10^{-7} 3.768×10^{-7} 3.209×10^{-7} 2.547×10^{-7} 2.220×10^{-7}	0.884 0.980 0.930 0.961 0.944
Illitic clay	0.001 M NaCl	2.84 2.34 1.04	7.649×10^{-8} 5.000×10^{-8} 2.239×10^{-8}	6.306×10^{-8} 4.235×10^{-8} 1.940×10^{-8}	0.824 0.847 0.866
Silty clay	0.001 M NaCl	1.33 1.11 0.85 0.63	6.418×10^{-7} 5.839×10^{-7} 4.888×10^{-7} 3.619×10^{-7}	6.287×10^{-7} 5.709×10^{-7} 4.925×10^{-7} 3.657×10^{-7}	0.980 0.978 1.008 1.011
Kaolinite	0.01 M NaCl	1.94 1.62 1.37 1.12 0.88	1.539×10^{-7} 1.429×10^{-7} 1.319×10^{-7} 1.218×10^{-7} 1.060×10^{-7}	1.504×10^{-7} 1.353×10^{-7} 1.280×10^{-7} 1.220×10^{-7} 1.026×10^{-7}	0.977 0.947 0.970 1.002 0.968
Illitic clay	0.01 M NaCl	2.72 2.19 1.86 1.21	3.992×10^{-8} 3.004×10^{-8} 2.705×10^{-8} 1.996×10^{-8}	3.116×10^{-8} 2.873×10^{-8} 2.630×10^{-8} 1.735×10^{-8}	0.781 0.956 0.972 0.869
Silty clay	0.01 M NaCl	1.96 1.35 1.10 0.85 0.69	6.530×10^{-8} 6.063×10^{-8} 5.970×10^{-8} 5.783×10^{-8} 5.690×10^{-8}	6.809×10^{-8} 6.492×10^{-8} 6.194×10^{-8} 6.194×10^{-8} 5.783×10^{-8}	1.043 1.071 1.038 1.071 1.016

TABLE 4 -- Electro-osmotic flow and streaming potential measurements (adapted from Olsen 1969, 1972)

Consolidation Pressure (MPa) (1)	Void ratio e (2)	Q/I (m^3/C) (3)	$-\Delta E/\Delta P$ (V/Pa) (4)	Ratio (4)/(3) (5)
5.3	0.399	5.207×10^{-7}	5.207×10^{-7}	1.000
10.0	0.353	4.438×10^{-7}	4.493×10^{-7}	1.012
20.0	0.279	2.733×10^{-7}	2.723×10^{-7}	0.996
40.0	0.194	1.523×10^{-7}	1.523×10^{-7}	1.000
66.7	0.114	9.492×10^{-8}	9.469×10^{-8}	0.998
6.7	0.180	1.025×10^{-7}	1.038×10^{-7}	1.013
1.3	0.213	1.380×10^{-7}	1.391×10^{-7}	1.008

equation is determined without the interference of an electrical gradient or the coupling effects are negligible. Otherwise, appropriate corrections should be made to the phenomenological coefficient L_{11}.

The significance of the second term in Eq. (19) depends on the relative magnitudes of the parameters. Values of the hydraulic conductivity for different soils range from 1×10^{-4} to 1×10^{-20} m/s, and values of the coefficient of electro-osmotic conductivity range from 1×10^{-9} to 1×10^{-8} m^2/Vs. The electrical conductivities of different soils varies from 0.01 to 0.1 S/m (Yeung 1993). Hence the value of the second term in Eq. (19) will have a lower bound of

$$\frac{(1 \times 10^{-9})^2}{0.1} = 1 \times 10^{-17} \frac{m^4}{Ns} \tag{20}$$

and a upper bound of

$$\frac{(1 \times 10^{-8})^2}{0.01} = 1 \times 10^{-14} \frac{m^4}{Ns} \tag{21}$$

If the fluid is taken to be water of unit weight of 9810 N/m^3, the first term in Eq. (19) will vary between 1.01968×10^{-24} to 1.01968×10^{-8} m^4/Ns. The percentage error induced by taking k_h/γ_f as L_{11} can be computed by

$$\% \ error = \frac{\frac{k_h}{\gamma_f} - L_{11}}{L_{11}} \times 100\% = \frac{-\frac{k_e^2}{\sigma}}{\frac{k_h}{\gamma_f} + \frac{k_e^2}{\sigma}} \times 100\% \tag{22}$$

The results of these sensitivity analyses are tabulated in Table 5 and illustrated in Fig. 4. It can be observed that the error is negligible if the hydraulic conductivity of the soil is greater than 1×10^{-9} m/s as in most coarse-grained soils. However, the percentage error can be as great as 100% in fine-grained soils of low hydraulic conductivity. As electro-kinetics are most effective in the remediation of fine-grained soils, the error should be properly rectified before Eq. (3) can be used to yield correct answers.

TABLE 5 -- Results of sensitivity analyses

k_h (m/s) (1)	k_h/γ_w (m^4/Ns) (2)	$k_e = 1.0\times10^{-9}$ m^2/Vs and $\sigma = 0.1$ S/m		$k_e = 10.0\times10^{-9}$ m^2/Vs and $\sigma = 0.01$ S/m	
		L_{ll} (3)	% error (4)	L_{ll} (5)	% error (6)
1.0×10^{-20}	1.019680×10^{-24}	1.000000×10^{-17}	-99.999990	1.000000×10^{-14}	-100.000000
1.0×10^{-19}	1.019680×10^{-23}	1.000001×10^{-17}	-99.999898	1.000000×10^{-14}	-100.000000
1.0×10^{-18}	1.019680×10^{-22}	1.000010×10^{-17}	-99.998980	1.000000×10^{-14}	-99.999999
1.0×10^{-17}	1.019680×10^{-21}	1.000102×10^{-17}	-99.989804	1.000000×10^{-14}	-99.999990
1.0×10^{-16}	1.019680×10^{-20}	1.001020×10^{-17}	-99.898136	1.000001×10^{-14}	-99.999898
1.0×10^{-15}	1.019680×10^{-19}	1.010197×10^{-17}	-98.990613	1.000010×10^{-14}	-99.998980
1.0×10^{-14}	1.019680×10^{-18}	1.101968×10^{-17}	-90.746738	1.000102×10^{-14}	-99.989804
1.0×10^{-13}	1.019680×10^{-17}	2.019680×10^{-17}	-49.512799	1.001020×10^{-14}	-99.898136
1.0×10^{-12}	1.019680×10^{-16}	1.119680×10^{-16}	-8.931125	1.010197×10^{-14}	-98.990613
1.0×10^{-11}	1.019680×10^{-15}	1.029680×10^{-15}	-0.971176	1.101968×10^{-14}	-90.746738
1.0×10^{-10}	1.019680×10^{-14}	1.020680×10^{-14}	-0.097974	2.019680×10^{-14}	-49.512799
1.0×10^{-9}	1.019680×10^{-13}	1.019780×10^{-13}	-0.009806	1.119680×10^{-13}	-8.931125
1.0×10^{-8}	1.019680×10^{-12}	1.019690×10^{-12}	-0.000981	1.029680×10^{-12}	-0.971176
1.0×10^{-7}	1.019680×10^{-11}	1.019681×10^{-11}	-0.000098	1.020680×10^{-11}	-0.097974
1.0×10^{-6}	1.019680×10^{-10}	1.019680×10^{-10}	-0.000010	1.019780×10^{-10}	-0.009806
1.0×10^{-5}	1.019680×10^{-9}	1.019680×10^{-9}	-0.000001	1.019690×10^{-9}	-0.000981
1.0×10^{-4}	1.019680×10^{-8}	1.019680×10^{-8}	-0.000000	1.019681×10^{-8}	-0.000098

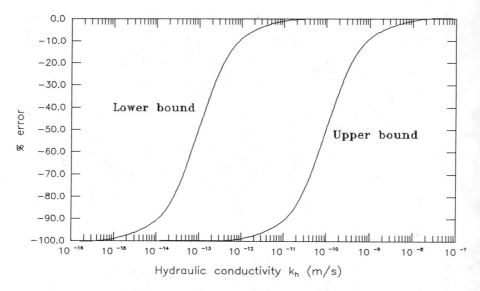

Fig. 4 -- Results of sensitivity analyses

EXPERIMENTAL EVALUATION

Substituting Eqs. (12), (16) and (17) into (6), the relationship between the hydraulic volume flux and the imposed hydraulic and electrical gradients is given by

$$J_v = \left[k_h + \frac{k_e^2}{\sigma}\gamma_f\right]\nabla(-h) + k_e\nabla(-E) \qquad (23)$$

Its validity in a compacted clay has been evaluated experimentally. The soil used is a gray brown silty clay of moderate plasticity (Unified Soil Classification CH). The maximum dry density is 17.3 kN/m³ (110 pcf) and the optimum water content is 17.4 percent as determined by the modified Proctor compaction test. The liquid limit, plastic limit and plasticity index of the soil are 52, 27 and 25 percent, respectively.

Two samples of 35.6 mm in diameter and 100 mm in length were compacted by kneading compaction at moisture contents of 16.6 percent and 18.0 percent to dry densities of 15.5 kN/m³ (98.8 pcf) and 15.6 kN/m³ (99 pcf), respectively. They were then mounted onto fixed wall permeameters made of lucite and equipped with electrodes. Detailed description of the experimental apparatus and laboratory procedures are given by Yeung et al. (1992). The samples were then fully saturated by tap water. The required parameters were measured in the order given below to minimize any possible adverse effects of each measurement on subsequent measurements.

The hydraulic volume flow rates of these samples were measured under a hydraulic gradient of fifty and the hydraulic conductivities k_h were determined to be 7.16×10⁻¹¹ and 6.94×10⁻¹¹ m/s, respectively.

The initial polarization potentials across the samples were measured to be 0.333 and 0.105 V, respectively. As the required electrical potentials to induce comparable volume flow rates induced by hydraulic gradients are very small, failure to take these polarization

potential into account could result in tremendous measurement errors. Electrical potentials were applied across the samples so that the electrically induced flows and the hydraulically induced flows were in the same direction and the resulting hydraulic volume flow rates and the electrical currents passing through the samples were measured.
The volume flow rates induced by the electrical gradients alone were then measured. The coefficients of electro-osmotic conductivity of the two samples were determined to be 3.03×10^{-9} and 3.13×10^{-9} m^2/Vs, respectively. The electrical conductivities of the samples were determined in the same measurements to be 0.1127 and 0.1172 S/m, respectively.
The hydraulic gradient of fifty was then re-applied across the sample. The polarities of the applied electrical gradients were reversed and the volume flow rates induced by the combined gradients were measured.
Finally the hydraulic conductivities of the samples were measured again to ensure no significant adverse effects have been caused by the electrical treatment of the soil.
The theoretical hydraulic volume flow rates induced by combined hydraulic and electrical gradients were calculated by Eq. (23) using the measured hydraulic conductivities, coefficients of electro-osmotic conductivity, electrical conductivities, and imposed hydraulic and electrical gradients. When the hydraulic flows induced by the two gradients were in the same direction, the ratios of the theoretical combined flow rates to the experimentally measured flow rates were determined to be 1.02 and 1.00 for the two samples, respectively. When the flows induced by the two gradients were in opposite directions, the ratios were determined to be 1.04 and 1.02, respectively. The results obtained in this experiment indicate the applicability of Eq. (23) in saturated compacted clay. The hydraulic volume flow rates induced by hydraulic and electrical gradients are algebraically additive provided the correct phenomenological coefficients are used.

CONCLUSIONS

Hydraulic conductivity is the phenomenological transport coefficient relating the hydraulic volume flow rate and the imposed hydraulic gradient. However, no precaution is taken in state-of-the-art apparatus, technique and procedure to eliminate the streaming potential generated by the hydraulic flow through a soil sample. Thus the measured hydraulic conductivity is a function of the hydraulic and electrical properties of the soil. The effects of electro-kinetic coupling are more pronounced in fine-grained soils of low hydraulic conductivity. Theoretical analysis of these coupling effects using the formalism of thermodynamics of irreversible processes is presented in this paper. Experimental results in support of the analysis are also included. If coupled flow equations in the form of Eq. (3) is used to describe flows and contaminant migration during electro-kinetic remediation of contaminated fine-grained soils, proper correction should be taken for the conductivity coefficient or significant errors will result.

REFERENCES

Abd-El-Aziz, M. H., and Taylor, S. A., 1965, "Simultaneous Flow of Water and Salt through Unsaturated Porous Media: I. Rate Equations," Soil Science Society of America Proceedings, Vol. 29, No. 2, pp. 141-143.

Darcy, H., 1856, "Détermination des lois d'écoulement de l'eau à travers le sable," Les fontaines publiques de la ville de Dijon, Victor Dalmont, Paris, pp. 590-594. (Freeze, R. A., 1983, Trans., "Determination of the Laws of the Flow of Water through Sand,"

Physical Hydrogeology, Benchmark Papers in Geology, Vol. 72, Freeze, R. A., and Back, W. Eds., Hutchinson Ross, Stroudsburg, pp. 14-20.)

Esrig, M. I., 1968, "Pore Pressures, Consolidation and Electrokinetics," *Journal of the Soil Mechanics and Foundations Division*, Proceedings of ASCE, Vol. 94, No. SM4, pp. 899-921.

Fitts, D. D., 1962, *Nonequilibrium Thermodynamics*, McGraw-Hill, New York.

Gairon, S., and Swartzendruber, D., 1975, "Water Flux and Electrical Potentials in Water-Saturated Bentonite," *Soil Science Society of America Proceedings*, Vol. 39, No. 5, pp. 811-817.

Gray, D. H., and Mitchell, J. K., 1967, "Fundamental Aspects of Electro-osmosis in Soils," *Journal of the Soil Mechanics and Foundations Division*, Proceedings of ASCE, Vol. 93, No. SM6, pp. 209-236.

Kemper, W. D., and Letey, J., 1968, "Solute and Solvent Flow as Influenced and Coupled by Surface Reactions," *Transactions of the 9th International Congress of Soil Science*, Adelaide, Vol. 1, pp. 233-241.

Letey, J., and Kemper, W. D., 1969, "Movement of Water and Salt through a Clay-Water System: Experimental Verification of Onsager Reciprocal Relation,' *Soil Science Society of America Proceedings*, Vol. 33, No. 1, pp. 25-29.

Lewis, R. W., and Garner, R. W., 1972, "A Finite Element Solution of Coupled Electro-kinetic and Hydrodynamic Flow in Porous Media," *International Journal for Numerical Methods in Engineering*, Vol. 5, No. 1, pp. 41-55.

Lewis, R. W., and Humpheson, C., 1973, "Numerical Analysis of Electro-osmotic Flow in Soil," *Journal of the Soil Mechanics and Foundations Division*, Proceedings of ASCE, Vol. 99, No. SM8, pp. 603-616.

Luikov, A. V., 1975, "Systems of Differential Equations of Heat and Mass Transfer in Capillary-Porous Bodies (Review)," *International Journal of Heat and Mass Transfer*, Vol. 18, No. 1-A, pp. 1-14.

Mitchell, J. K., 1986, "Potential Uses of Electro-kinetics for Hazardous Waste Site Remediation," *Unpublished Proceedings of the Workshop on Electro-kinetic Treatment and Its Application in Environmental-Geotechnical Engineering for Hazardous Waste Site Remediation*, Seattle, Section II, pp. 1-20.

Mitchell, J. K., 1991, "Conduction Phenomena: from Theory to Geotechnical Practice," *Géotechnique*, Vol. 41, No. 3, pp. 299-340.

Mitchell, J. K., 1993, *Fundamentals of Soil Behavior*, 2nd Edition, John Wiley & Sons, New York.

Mitchell, J. K., and Yeung, A. T., 1991, "Electro-kinetic Flow Barriers in Compacted Clay," *Geotechnical Engineering 1990*, Transportation Research Record 1288, Transportation Research Board, National Research Council, Washington, D. C., pp. 1-9.

Olsen, H. W., 1969, "Simultaneous Flows of Liquid and Charge in Saturated Kaolinite," *Soil Science Society of America Proceedings*, Vol. 33, No. 3, pp. 338-344.

Olsen, H. W., 1972, "Liquid Movement through Kaolinite under Hydraulic, Electric and Osmotic Gradients," *American Association of Petroleum Geologists Bulletin*, Vol. 56, No. 10, pp. 2022-2028.

Renaud, P. C., and Probstein, R. F., 1987, "Electroosmotic control of hazardous wastes," *PCH PhysioChemical Hydrodynamics*, Vol. 9, No. 1/2, pp. 345-360.

Saxén, U. (1892). *Wied. Ann.*, Vol. 47, pp. 46.

United States Environmental Protection Agency, 1986, *Unpublished Proceedings of the Workshop on Electro-kinetic Treatment and Its Application in Environmental-Geotechnical Engineering for Hazardous Waste Site Remediation*, Seattle.

Wan, T. Y., and Mitchell, J. K., 1976, "Electro-osmotic Consolidation of Soils," *Journal of the Geotechnical Engineering Division*, Proceedings of ASCE, Vol. 102, No. GT5, pp. 473-491.

Yeung, A. T., 1990a, "On the Laboratory Measurement of Hydraulic Conductivity of Earth Barriers for Waste Containment," *Hong Kong Engineer*, Vol. 18, No. 5, pp. 23-29.

Yeung, A. T., 1990b, "Coupled Flow Equations for Water, Electricity, and Ionic Contaminants through Clayey Soils under Hydraulic, Electrical and Chemical Gradients," *Journal of Non-Equilibrium Thermodynamics*, Vol. 15, No. 3, pp. 247-267.

Yeung, A. T., 1993, "Electro-kinetic Flow Processes in Porous Media and Their Applications," *Advances in Porous Media*, Elsevier, Amsterdam, Vol. 2, in press.

Yeung, A. T., and Mitchell, J. K., 1993, "Coupled Fluid, Electrical and Chemical Flows in Soil," *Géotechnique*, Vol. 43, No. 1, pp. 121-134.

Yeung, A. T., Sadek, S. M., and Mitchell, J. K., 1992, "A New Apparatus for the Evaluation of Electro-kinetic Processes in Hazardous Waste Management," *Geotechnical Testing Journal*, ASTM, Vol. 15, No. 3, pp. 207-216.

Raymond N. Yong,[1] Boon-Kong Tan,[2] and Abdel M.O. Mohamed[3]

EVALUATION OF ATTENUATION CAPABILITY OF A MICACEOUS SOIL AS DETERMINED FROM COLUMN LEACHING TESTS

REFERENCE: Yong, R. N., Tan, B. k., and Mohamed, A. M. O., "Evaluation of Attenuation Capability of a Micaceous Soil as Determined from Column Leaching Test," Hydraulic Conductivity and Waste Contaminant Transport in Soil, ASTM STP 1142, David E. Daniel and Stephen J. Trautwein, Eds., American Society for Testing and Materials, Philadelphia, 1994.

ABSTRACT: The attenuation capability of a hydrobiotite-vermiculite (HV) soil from southern Ontario is studied as a pollutant adsorbent landfill liner. The soil has specific surface areas varying from 90-206 m^2/g, low cation exchange capacities (CEC) of about 10-20 meq/100 g of dry soil, a maximum dry unit weight of about 1.83 Mg/m^3, and low hydraulic conductivities of 10^{-9} m/sec for the compacted samples. Increase in density with a corresponding decrease in hydraulic conductivity can be obtained by mixing the HV soil with kaolinite or Champlain Sea Clay. Heavy metal retention studies of HV soil and soil admixtures indicate that heavy metals retention can exceed the CEC of the soil, implying that other processes or mechanisms besides cation exchange such as precipitation are involved. While the HV soil alone shows satisfactory values for hydraulic conductivity (k= 10^{-9} m/sec), pH = 7-8, and high adsorption/retention of cations, especially heavy metals, the addition of Champlain sea clay in particular enhances or improves its performance.

Key Words: cation exchange capacity, buffering intensity, micaceous soil, heavy metals, adsorption, precipitation, hydraulic conductivity, Champlain sea clay.

Regulations and/or guidelines articulated by many regulatory agencies mandate the use of clay soil barriers as primary and secondary barriers for attenuation of contaminants in transport of leachates in and through the barriers. A very necessary requirement for successful barrier containment against transport of contaminants is the chemical buffering

---

[1] William Scott Professor of Civil Engineering and Applied Mechanics; Director, Geotechnical Research Centre, McGill University, Montreal, P.Q. Canada, H3A 2K6

[2] Lecturer, Department of Geology, Universiti Kebangsaan Malaysia, 36000 UKM Bangi, Selangor, Malaysia

[3] Senior Research Associate, Geotechnical Research Centre, McGill University, Montreal, P.Q., Canada, H3A 2K6

intensity of the clay soil material used for construction of the barrier. The same kinds of scrutiny are required in situations where the soil in-place is to be used as a secondary barrier system. The study of soil chemical buffering intensity not only responds to the above concern, and to the concerns relating to chemical compatibility of soil barrier system [1], but also permits one to establish the means whereby soil conditioning or "an improvement/additive" can be usefully applied to the soil to improve its functionality.

Previous studies [2, 3, 4] of soil buffering intensity are most often obtained experimentally from titration of soil suspensions as well as by computation at individual pH values. To obtain the buffer intensity, the experimentally obtained titration curve should show the pH change in the soil solution tested as a result of the addition of strong acid or base. Hence, the buffering intensity is always positive and is proportional to the reciprocal of the slope of the titration curve. However, in studying transport of contaminants in soils, some recent studies [5, 6, 7] have evaluated the soil buffering intensity using column leaching tests. In these studies, the ability of the soil, in its compacted state, to maintain its natural pH as a function of depth and time was evaluated. The latter approach simulates the actual state of a soil material in a compacted clay barrier system.

In the present study, the chemical buffering intensity of a natural micaceous soil (hydrobiotite-vermiculite) found in eastern Ontario (Canada) is studied using soil column leaching tests. Because of the relative coarseness of the natural micaceous soil, studies were also conducted on soil mixtures where kaolinite (20% by dry weight), and Champlain sea clay (25% by dry weight) were added separately to the micaceous soil. The soil column leaching tests performed used a municipal solid waste leachate spiked with heavy metals and cations in the form of chlorides. Analysis of soils and information used to evaluate the soil column leaching test results include specific surface area, cation exchange capacity (CEC), adsorption isotherms (from separate tests), hydraulic conductivity (from separate tests), pore fluid and soil analysis after leaching, and leaching column adsorption characteristics. By analyzing both the soil and pore fluid at various stages of leaching, information can be obtained with respect to retention characteristics and capability, and also efficiency of the chemical attenuation system.

CHARACTERIZATION

A description of the samples obtained from the chosen area is given in Table 1. Overall, the samples were generally granular, with some fine fraction (silt and clay). The specific surface area shown in Table 1 relates to the fraction passing the No. 200 sieve (< 0.074 mm). This fraction was in general less than 5 percent (by weight) of the total sample.

Basic characterization of the soils included determination of: dominant mineral species in the micaceous soil, specific surface area, cation exchange capacity, compaction characteristics and hydraulic conductivity. The mineralogical studies show that the micaceous soil, which is a hydrobiotite-vermiculite, consists of phlogophite, serpentine and mixed layer clay minerals which expand and exfoliate upon heating. More recent mineralogical analyses have shown that the coarser fraction consists of intergrowths of phlogopite-lizardite, in a finer grained matrix of talc, smectite and aliettite [8].

Table 1 -- Samples Tested

Sample No.	Location	Comments	Specific Surface Area, <0.074 mm fraction, m^2/g
ST1-A	Stockpile	Minus 4 mesh (-4.75mm) fraction from a split composite sample of four 3.7 m vertical channels from four pits in the surface stockpile. Dark brown, loose silty material.	90.1
ST1-B	Stockpile	Fine pebble fraction of ST1-A. Light brown, fine to coarse gravel of "vermiculite" rock, micaceous.	96.1
ST1-C	Stockpile	Coarse pebble fraction of ST1-A. Light brown/green gravel to cobble-sized "vermiculite" rock, micaceous.	105.5
TR1-A	Trench 1	Split 53.5 m horizontal channel sample of good grade "vermiculite" bearing saprolite. Excludes one internal 5 m low grade section. Dark brown loose silty material plus some gravel-sized rock, micaceous.	122.1
TR1-B	Trench 1	Split composite of vertical channel samples from 5 sites in the TR1-A interval, average sample height 2 m. Dark brown loose silty material plus some gravel-sized rock, micaceous.	136.5
TR2-A	Trench 2	Split 21.5 m horizontal channel of high grade "vermiculite"/saprolite. Excludes an internal 20 m lower grade section. Dark brown loose silty material plus some gravel-sized rock, micaceous.	110.8
TR2-B	Trench 2	Split 39 m horizontal channel sample, lower grade "vermiculite"/saprolite. Includes the 20 m section from TR2-A. Dark brown loose silty material plus some gravel-sized rock, micaceous.	130.5
TR4-A	Trench 4	Split 17 horizontal channel sample, high grade "vermiculite"/saprolite. Dark brown loose silty material plus some gravel-sized rock, micaceous.	100.9
CF-1	Open pit	Minus 80 mesh fraction (clay with some silt) from a 21 m horizontal channel sample of weathered vermiculite/saprolite along east wall of pit. Dark brown, loose and silty. Micaceous.	199.1
No. 10	Open pit	Light brown loose silty material plus some gravel-sized rock, micaceous.	206.2

Specific Surface Area

Specific surface areas of the soil samples were determined using the Ethylene Glycol Monoethyl Ether (EGME) adsorption method [9]. The results are given in Table 1. The values of the specific surface areas range from 90 to 206 m^2/g, i.e., moderately high, indicating perhaps mixtures of mainly illitic/micaceous minerals and others, since the surface area of illite is 60-200 m^2/g, and vermiculite is 400-800 m^2/g [1, 10].

Cation Exchange Capacity (CEC)

Exchangeable cations, sodium, potassium, magnesium and calcium (Na^+, K^+, Mg^{2+}, Ca^{2+}) and the CEC of the soil samples were determined using two methods: (a) batch equilibrium test, ASTM D4319, standard test method for distribution ratios by the short-term batch method, and (b) silver-thiourea method [11]. The two techniques were adopted to investigate the difference in the calculated CEC values. The results are shown in Tables 2 and 3 respectively. It should be noted that in order to convert from meq/l to meq/100 g of dry soil, the volume of solution and the amount of dry soil used should be included in the conversion. The results show that the CEC values are rather low, i.e., ranging from 7 to 19 meq/100 g of dry soil (batch equilibrium method), or 3 to 14 meq/100 g soil (silver-thiourea method). The low CEC values once again indicate that the soils are mainly illitic/mica in composition, since the CEC of illite is 15-40 meq/100 g of dry soil, and the CEC of vermiculite is 120-200 meq/100 g of dry soil [1, 12, 13].

Table 2 - Exchangeable Cations and Cation Exchange Capacity Based on Batch Equilibrium Test (ASTM D4319).

	Na^+		K^+		Ca^{2+}		Mg^{2+}		CEC = Σ
	ppm	meq/100 g of dry soil	ppm	meq/100 g of dry soil	ppm	meq/100g of dry soil	ppm	meq/100g of dry soil	
ST1A	0.7	0.03	4.9	0.12	56.53	2.83	73.3	6.08	9.06
ST1B	0.6	0.03	5.3	0.13	55.23	2.76	75.8	6.29	9.21
ST1C	1.2	0.05	13.9	0.35	56.79	2.84	77.7	6.45	9.69
TR1A	0.6	0.03	3.6	0.09	31.59	1.58	122.9	10.20	11.90
TR1B	--	0.16	--	0.27	--	3.00	116.5	10.62	14.05
TR2A	--	--	4.2	0.11	29.05	1.45	152.5	9.67	11.23
TR2B	0.5	0.02	7.3	0.18	45.05	2.25	78.3	12.66	15.11
TR4A	0.4	0.02	5.8	0.15	18.56	0.93	203.4	6.50	7.60
CF1	0.7	0.03	11.3	0.28	28.06	1.40	---	16.88	18.59
No. 10	--	---	--	0.05	--	7.2		7.64	14.89

Compaction

The results of compaction tests performed on samples No. TR1A and No. 10, using the standard Proctor compaction method ASTM D698, test methods for moisture density relations of soil and soil aggregate mistures, are as follows: maximum dry unit weights of 1.84 and 1.83 Mg/m^3, and optimum moisture contents of 14.4 and 16.1% respectively. The compacted soil samples lacked cohesion and disintegrated readily.

Compaction tests were also performed on soil samples mixed with other clay soils (Champlain sea clay from Quebec and kaolinite clay), to determine the effects of the addition of other clay soils on the geotechnical and adsorption characteristics of the hydrobiotite-vermiculite (HV) soil. The compaction tests were performed on: (1) HV (sample no. 10) + kaolinite (20% by dry weight) designated as HVK, and (2) HV soil + Champlain sea clay (25% by dry weight) designated as VHQ.

Table 3 - Cation Exchange Capacity Based on Silver-Thiourea Method

Sample No.	CEC (meq/100 g of dry soil)	Extractable Cations (meq/100 g of dry soil)				
		Na^+	K^+	Ca^{2+}	Mg^{2+}	CEC= Σ
ST1A	2.80	nd	0.39	28.80	9.96	39.15
ST1B	5.60	nd	0.40	31.20	7.30	38.90
ST1C(f)	5.60	nd	0.39	36.40	7.30	44.09
TR1A(f)	4.00	nd	0.49	4.90	15.27	20.66
TR1A(g)	3.20	nd	0.49	4.80	15.94	21.23
TR1B	5.20	nd	0.44	7.30	18.26	26.00
TR2A	5.20	nd	0.46	3.80	14.94	19.20
TR2B	4.00	nd	0.45	6.90	18.26	25.61
TR4A(g)	2.40	nd	0.46	1.90	14.28	16.64
CF1(f)	8.00	nd	0.51	3.40	19.92	23.83
10	13.20	nd	0.30	15.40	17.93	33.63
10(g)	13.60	nd	0.33	16.00	19.26	35.59

(f) - "fine" portion of natural sample (coarse silt/fine sand)
(g) - "ground" sample (fine silt/clay)
nd - not detected

The compaction results are shown in Fig. 1. The results show that the addition of kaolinite or Champlain sea clay results in an increase in the maximum dry unit weight obtained, from 1.83 Mg/m^3 (HV soil) to 1.85 Mg/m^3 (HVK soil) and 1.91 Mg/m^3 (HVQ soil), with corresponding decreases in optimum moisture contents from 16.1% to 15.9% and 14.5% respectively. The increase in maximum dry unit weight obtained with the addition of kaolinite or Champlain sea clay is considered to be advantageous with respect to the proposed use of the soil as a clay liner, inasmuch as it would increase the shear strength of the liner and reduce its compressibility and hydraulic conductivity.

Permeability/ Hydraulic Conductivity

Falling head hydraulic conductivity (permeability) tests were conducted on the HV, HVK and HVQ soils. The samples were compacted at or close to, wet side, the optimum moisture content. From the permeability results shown in Table 4, it is noted that the permeabilities of HV, HVK, and HVQ soils range from 2.3×10^{-9} to 2.9×10^{-10} m/sec. The addition of

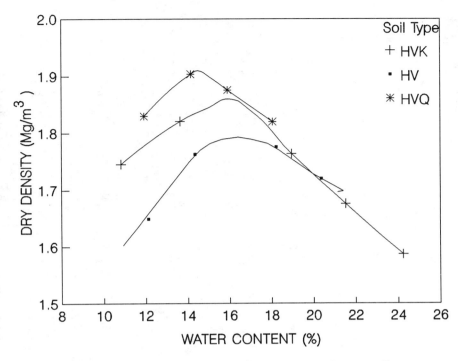

Fig. 1 - Compaction curves for HV, HVK, and HVQ soils.

kaolinite (i.e. the HVK soil) or Champlain Sea Clay (i.e. the HVQ soil) causes a reduction in the permeability. In the case of the HVQ soil, a reduction by one order of magnitude was obtained.

Table 4: Permeability Tests (Falling Head)

Soil Type	Dry Unit Weight γ_d (Mg/m^3)	Initial Water Content, W_o(%)	Permeability K(m/sec)
HV	1.82 (1.83)	16.2 (16.1)	2.3×10^{-9}
HVK	1.84 (1.85)	14.2 (15.9)	1.4×10^{-9}
HVQ	1.90 (1.91)	14.3 (14.5)	2.9×10^{-10}

Note: Values in brackets denote maximum dry unit weight or optimum moisture content

LEACHING COLUMN TESTS

Materials and Methods

Leaching column tests were conducted on the HV, HVK, and HVQ soils to study the adsorption characteristics of the soils, in particular with respect to the retention and migration of heavy metals through the soil columns. The soils were compacted in three layers, at their optimum moisture contents using static compaction (compression machine) into plexiglass cylinders which were later reassembled as leaching cells. The leaching cells used were identical to those used previously for tests reported by Yong et al,[14], and the procedures implemented were also similar to the previously tested. For each soil type, three leaching cells were prepared, e.g., HV1, HV3, and HV5, representing HV leaching cells where 1 pore volume (PV), 3 PVs and 5 PVs of leachate were collected respectively. The information concerning the compacted soil samples used for the leaching tests are shown in Table 5. The internal diameter of the mould used was 115 mm and except for the HVQ series and sample HVK5, the heights of the samples were 121 mm. For the HVQ series, the heights varied between 95 mm and the height of the HVK5 sample was 118 mm.

Table 5   Leaching Test - Dimensions & Compaction
(Compaction by Compression Machine in 3 Layers - Static Compaction)

No.	Pore Volume (PV, cubic cm)	Weight of Compacted Soil (g)	Bulk Unit Weight of Compacted Soil (Mg/ cubic m)	Opt. Moisture Content (%)	γ dry (Mg/ cubic m)	γ_{dmax}^{*} (Mg/cu. m)
HV1	398.1	2646.6	2.14	16.1	1.84	1.83
HV3	398.1	--	--	16.1	--	1.83
HV5	398.1	2654.2	2.15	16.1	1.85	1.83
HVK1	390.4	2676.1	2.17	15.9	1.87	1.85
HVK3	390.4	2666.5	2.16	15.9	1.86	1.85
HVK5	380.7	2615.2	2.17	15.9	1.87	1.85
HVQ1	290.0	2138.1	2.17	14.5	1.90	1.91
HVQ3	285.0	2125.3	2.19	14.5	1.91	1.91
HVQ5	265.6	1973.3	2.18	14.5	1.90	1.91

*: $W_{opt.}$ and γ_{dmax} from standard Proctor compaction (ASTM D698)

The municipal solid waste (MSW) leachate used for the leaching tests was spiked with lead and zinc (Pb^{2+}, Zn^{2+}, respectively) and Na^+, K^+, Mg^{2+}, Ca^{2+} in the form of chloride salts. The pH of the reconstituted leachate was also lowered by adding some concentrated HCl. Due to the lowering of pH and the introduction of chloride salts in the solution, the chloride concentration in the leachate is increased. This in turn results in precipitation of some of the lead chloride due to the increase in lead chloride concentration over its solubility product. The chemical compositions of the reconstituted leachate are as shown in Table 6.

Leaching was performed under a constant applied air pressure equivalent to a water head of 8.4 or 10.6 m depending on the sample height. This resulted in a hydraulic gradient of 87-89 for all samples except sample HVQ5 which had a slightly higher hydraulic gradient of 97.

Using the same test scheme reported previously [14], effluents were collected at every 0.5 PV during the leaching process and analyzed. At the end of the PV 1, PV 3 or PV 5 test, the soil sample (e.g., HV1 for PV 1 test, HV3 for PV 3 test and HV5 for PV 5 test) was extruded, sectioned into 10 mm thick slices, and the soil slices analyzed for soluble ions in the pore fluid, and exchangeable cations (ASTM D4319). Cations and heavy metals were determined using atomic absorption spectrophotometry (AAS), whilst chloride was determined using titration with silver nitrate ($AgNO_3$). The pH and conductivity were determined using a pH meter and the electrophoretic mass-transport analyzer, respectively.

Table 6: Chemical Compositions of the Reconstituted Leachate. (all ionic concentrations are in ppm).

pH @ room temperature	1.33
Specific Conductivity (mS/cm)	16.833
Sodium, Na^+	346
Potassium, K^+	164.8
Magnesium, Mg^{2+}	43.8
Calcium, Ca^{2+}	95.4
Lead, Pb^{2+}	1372.2
Zinc, Zn^{2+}	1141.6
Chloride, Cl^-	5258.4

RESULTS AND DISCUSSIONS

Permeability/ Hydraulic Conductivity

The variations in the coefficient of permeability (k) during the leaching process are shown in Fig. 2. For the HV soil, k values show a slight decrease from the initial condition before leaching to that after passing 1 PV. This appears to be the time period for "stabilizing" the system. Following this period, the k values remain almost constant, tapering off at 1.9×10^{-9} m/sec after the passage of 5 PV. The HVK soil shows an almost constant permeability. Similarly, the HVQ soil shows a slight decrease in permeability at the beginning of leaching. To all intents and purposes, within the experimental constraints and "accuracy" of the measuring technique, one could argue that leaching with the spiked leachate did not result in any significant change in the permeability of the soil samples tested relative to that tested with distilled water.

The initial decrease in the k value for all the samples is required to achieve "equilibrium" since the soils were not initially fully saturated. This is identified as the "stabilizing" period. However, this also could be attributed to the decrease in the initial ion concentrations (sodium, potassium, magnesium, and calcium) in the pore fluid because of leaching, resulting in the tendency for the diffuse double layer (DDL) to expand [1, 13]. Also, note that a decrease in the permeability could be due to dispersion and/or translocation of clay particles -- (void plugging).

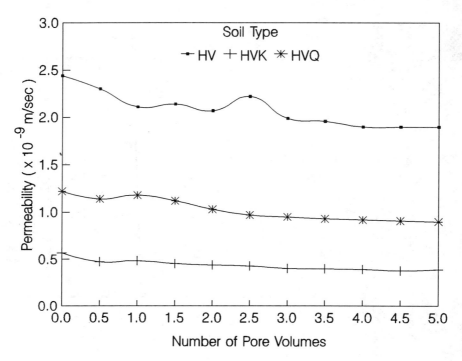

Fig. 2 - Permeability (coefficient of permeability k) variation for HV, HVK, and HVQ soils due to leaching.

pH

The pH of the original MSW leachate was slightly above neutral (7.1). However, for the leaching column tests, the pH of the reconstituted leachate was lowered to 1.3 (highly acidic) by the addition of concentrated HCl. This was done to facilitate the study of the migration of lead and zinc (Pb^{2+}, Zn^{2+}) through the soil columns since it is well known that heavy metals generally would precipitate if the solution pH values are high (e.g., Pb^{2+} precipitates at pH > 5)

Figure 3 shows the effluent pH of the samples HV, HVK, and HVQ after passage of 5 PVs, indicating a general gradual decrease in the pH's of the effluents of the three soil types with increasing permeation by leachate. Whereas the initial pH values showed HVQ > HV > HVK, the performance of the HVK soil mixture appears to be influenced by the initial pH of the kaolinite soil component. As reported previously [4], the low pH value of the kaolinite can initially play a dominant role in soil leaching experiments. As more pore volumes are leached through the soil, this initial pH influence becomes muted -- as witness the results for the HVK mixture at PV 4.5 and above. In the case of the initial pH value shown by the HVQ soil, the Champlain Sea Clay from the Matagami (Quebec) site contains abundant carbonates, up to 30-40% [14]. One would expect that the dissolution of carbonates (calcite, $CaCO_3$, and dolomite, $CaMg(CO_3)_2$) would contribute to higher pH values at the initial stages of leaching. This would decrease with continued leaching.

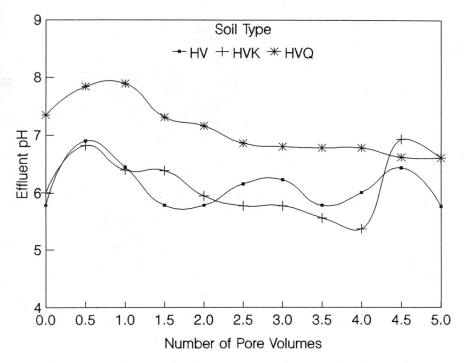

Fig. 3 - Effluent pHs for HV, HVK and HVQ soils (5 p.v.).

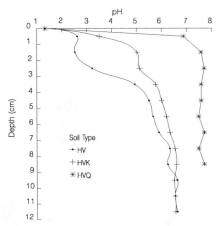

Fig. 4 - pH variation for HV, HVK, and HVQ soils (5 p.v.).

The variation of the pH values of the pore fluids (of the leached soils) with depth of the soil column are shown in Fig. 4. As can be expected, the curves show a general increase in the pH from the top (upper slices) to the bottom (lower slices) of the soil column. This is clearly seen for the HV and HVK soils. The pH increases from about 3 at the uppermost slice (slice no. 1) to about 6.5 at the lowermost slice (slice no. 12). With increasing permeation by the leachate (PV 1 to 5), the pH of each slice decreases progressively as shown by the HV soil. The HVQ soil shows high pH values (pH > 7) throughout the profile. This indicates the high acid neutralization capacity of the HVQ soil. Also, comparing soil types, the acid neutralization capacities of the studied soils are HVQ > HV > HVK.

Specific Conductivity

The specific conductivity, which is a measure of the total amount of dissolved solids in a solution, is shown in Fig. 5 for the 5 PV test effluents, for the various soils (HV, HVQ and HVK). Measurements were made at various pore volumes during the 5 PV leaching test. The results, which are also typical of the PV 1 and PV 3 test series effluents (not shown herein) demonstrate the general trend for all soil types, i.e., the specific conductivities begin with low values (0.2 to 0.7 ms/cm) at PV 0.5, and then rise sharply to reach a maximum value of 3.4 ms/cm at PV 1.5 to PV 2.5. This is in accord with the diffuse double layer model expectation of cation and anion adsorption. With continued leaching, the availability of adsorption sites decreases, and thus less cations will be

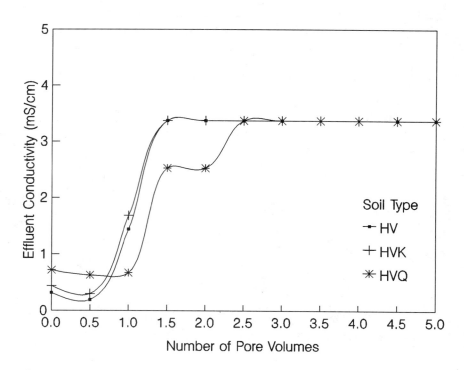

Fig. 5 - Effluent conductivity for HV, HVK, and HVQ soils (5 p.v.).

adsorbed within the soil-water system. Equilibrium is reached when the cation concentration in the effluents show a constant value. A comparison of the adsorption capabilities of the three soil types tested show HVQ > HV > HVK. Note that in the case of the HVQ soil, the initial gradual increase in conductivity from PV 0.5 to PV 1.0 indicates high adsorptions at the initial stages.

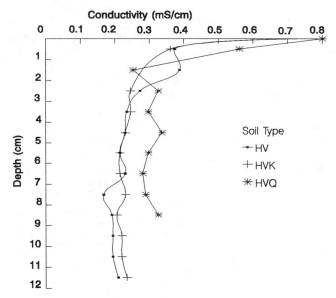

Fig. 6 - Conductivity variation for HV, HVK, and HVQ soils (5 p.v.).

Specific conductivity variations with depth for the soil columns are shown in Fig. 6. As might be expected, the figure shows the general trend of decreasing specific conductivity values with depth, reflecting the greater amounts of dissolved solids present in the pore fluids (adsorbed within the soil-water system) at the upper portions of the soil columns. With increasing permeation, (i.e., greater leaching values, PV 5 instead of PV 3) specific conductivity increases because of the higher concentration of dissolved solids present in the soil-water system. Thus, as shown from the test results, the following order for specific conductivity is obtained: HVQ5 > HVQ3 > HVQ1. Similar results are obtained for HV and HVK soils. Comparing soil types, specific conductivity of HVQ > HV or HVK. This is attributed to the presence of carbonates in the HVQ soil which would react with the acidic leachate to release more cations (Mg^{2+}, Ca^{2+}) into solution.

Breakthrough Curves

The breakthrough curve, which portrays the variation of the concentration of a particular ion (Ce) with respect to the original concentration of the particular ion (Co) in relation to the number of pore volumes of leachate permeated, for the various cations (Na^+, K^+, Mg^{2+}, Ca^{2+}, Pb^{2+} and Zn^{2+}) and the anion Cl^- are shown in Figs. 7a through 7g for the 3 types of soils (HV, HVK, and HVQ) studied. It should be noted that a breakthrough occurs at Ce/Co= 0.5. A brief examination of the results of the relative concentration (Ce/Co) variation of the cations at various PV's can provide further insight into the adsorption and/or buffering intensity of the soils.

Sodium, Na^+ -- The breakthrough curves for Na^+ shown in Fig. 7a indicate Na^+ breakthrough occurring at PV 2.7 for the HVK soil, PV 3.0 for the HV soil and PV 3.7 for the HVQ soil. The earlier breakthrough experienced in the HVK soil, followed by the HV and the HVQ soils is expected, inasmuch as the adsorption capacities of the soils shows that the adsorption capacity of HVQ > HV > HVK. It is noted that with an increasing number of pore volumes permeated, the relative concentration of Na^+ begins to exceed 1.0, indicative of elution or desorption of Na^+ from the soil solids.

Potassium, K^+ -- The breakthrough curves for K^+ (Fig. 7b) show that no breakthrough in any of the soils, i.e., the highest relative concentration (Ce/Co) attained at PV 5.0 was 0.3, i.e., less than 0.5. The inability of K^+ to obtain breakthrough within the PV 5 range is likely due to the fact that K^+ is often adsorbed and incorporated into the interlayer lattice of the hydrobiotite-vermiculite, HV, soil.

Magnesium, Mg^{2+} -- In the results shown in Fig. 7c, all three soil types (HV, HVK, and HVQ) show very high relative concentrations (Ce/Co up to about 30). This would indicate elution of Mg^{2+} from the soil solids during the leaching process, consistent with the information shown in Tables 2 and 3, i.e., the high extractable Mg^{2+} content. Whereas breakthrough is interpreted as occurring at Ce/Co= 0.5 for all soil types, it is not clear whether the conventional method for assessment of breakthrough performance should be rigidly applied to this particular circumstance. Comparing soil types, the similar behaviour of HV and HVK is obvious, while the HVQ soil shows better adsorption capacities for PV < 2.5. For PV > 2.5, the HVQ soil shows the highest Ce/Co values, indicating excessive release of Mg^{2+} from the carbonates.

Calcium, Ca^{2+} -- As in the case of the performance of the Mg^{2+} leaching tests, the relatively high relative concentrations of Ce/Co (10 to 12) for Ca^{2+} (Fig. 7d) indicates high elution of Ca^{2+} from the soil solids during the leaching process. The situation is similar to the case of Mg^{2+}, i.e., the high extractable Ca^{2+} presence in the HV soil. Breakthrough for all soil types occurred at PV 0.5 or 0.6. The similarity in behaviour of HV and HVK soils is again observed. The HVQ soil shows better adsorption capacities for PV < 2.5. The high relative concentrations of divalent Mg^{2+} and Ca^{2+} in the effluents collected can also be attributed in part to cation exchange or replacement by Pb^{2+} and Zn^{2+} since these heavy metals were introduced in high concentrations in the leachate (>1000 ppm).

Lead, Pb^{2+} -- The breakthrough curves for Pb^{2+} are shown in Fig. 7e. Note that the ordinate is expressed in relative concentration (ratio) of 10^{-4}. The relative concentration values are all extremely low, i.e., Ce/Co are in the order of 1×10^{-4}, indicating that most of the Pb^{2+} is retained in the soils in both precipitated and adsorbed forms. No evidence of breakthrough can be observed in this range of PV leaching. The results or interpretations relating to non-specific adsorption of Pb^{2+} are consistent with previous studies in relation to heavy metal transport and adsorption in soils [1, 4, 5]. The retention capabilities of the three soil types with respect to Pb^{2+} are similar, with all three showing high attenuation of Pb^{2+}.

Zinc, Zn^{2+} -- The results shown for Zn^{2+} (Fig. 7f) are somewhat inconsistent for the HVK soil at low PV leaching. The evidence suggests some experimental error, and that a retesting of the soil will need to be

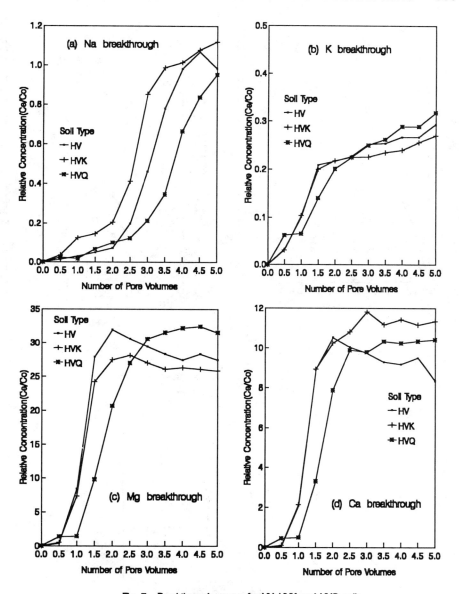

Fig. 7 - Breakthrough curves for HV, HVK and HVQ soils.

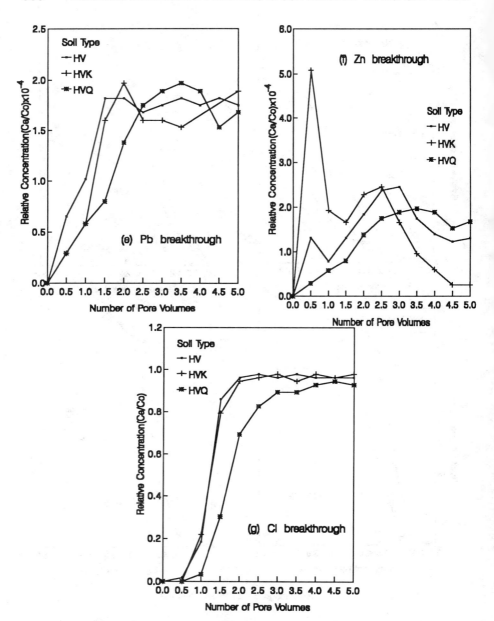

Fig. 7 - Breakthrough curves for HV, HVK and HVQ soils.

performed. Discounting the PV 0.5 spike of the HVK soil, one can observe a consistent performance of the breakthrough curves. Whereas the high retention by all three soil types is indicated. $Ce/Co = 10^{-4}$ as in the case of Pb^{2+}, it is not immediately clear why there is a diminishing value for the Ce/Co ratio at the higher PV's. A possible reason for the performance of these "curves" could be in the resultant effects obtained in speciation (note presence of relatively high Cl^- content), selectivity, and competition for "adsorption" sites between the Pb^{2+} and Zn^{2+} [1, 16, 17]. This problem constitutes the continuing study underway on the overall problem of transport and fate of multi-component heavy metal species.

Chloride, Cl^- -- Breakthrough curves for Cl^- are shown in Fig. 7g. Whereas the chloride ion is generally considered to be a very mobile and non-interacting anion, recent studies [16] on ligand affiliation (speciation with Cl^-) would raise some considerable doubt, especially if chloride concentration is high relative to the heavy metals present within the system. Assuming some degree of "non-interaction" as is commonly assumed, one would observe that attenuation of the Cl^- is low, and that the HV and HVK soils show very similar curves, with breakthrough of Cl^- at PV 1.2, and Ce/Co=1 at PV 2. The HVQ soil shows more attenuation of Cl^-, with breakthrough at PV 1.7, and Ce/Co= 0.9 at PV 3.

Migration Profiles

The migration profiles for the Na^+, K^+, Pb^{2+}, and Zn^{2+} are shown in Fig. 8 for the PV 5 test series. A comparison with the test results obtained for PV 1 and PV 3 series (not shown), indicates increasing concentration of the various ions with increasing PV application of the leachate. The elution of Na^+ ions has been noted previously, and is seen to be larger for the mixed soils (HVQ and HVK). This is not exactly true for the performance of the K^+ ions (Fig. 8b) with resect to the HVQ soil. Further testing and analyses will be needed to form a more complete picture.

In the case of Pb^{2+} and Zn^{2+}, (Figs. 8c, 8d) the results show significant reductions in the concentration of lead and zinc with penetration depth. The curious spike in the results shown for the HVK soil in Fig. 8c is not well understood, and at the present time is considered to be an experimental anomaly which needs to be re-examined. The results (not shown) indicate that with increasing permeation, from PV 1 to PV 5, the Pb^{2+} concentrations increase progressively and the Pb^{2+} "front" penetrated deeper into the soil column. This is consistent with previous observations [14]. Also consistent with previous studies is the order of Pb^{2+} retention capability is HVQ >> HV> HVK -- i.e., relative to CEC and acid neutralization capacities [1, 4, 17]. Because of the presence of carbonates in the natural clay soil used, the HVQ soil shows almost complete attenuation of Pb^{2+} within the uppermost 10 mm of the soil column.

The migration profiles for Zn^{2+} are very similar to those of Pb^{2+} as shown in Fig. 8d. As for the case of Pb^{2+}, the mechanisms involved are similar (in general) and hence the migration profiles for Zn^{2+} also show high attenuations of Zn^{2+} by all the three types of soils (HV, HVK, and HVQ) tested, from 1142 ppm (initial Zn^{2+} concentration in leachate) to zero ppm. The Zn^{2+} concentrations increase with increasing number of pore volumes of leachate permeated, and the depth of migrations also increases. For the HV and HVK soils, total attenuation of Zn^{2+} occurred at depths > 80 mm; for the HVQ soils, total attenuation occurred within the uppermost 20 mm. The order of retention capability for Zn^{2+} is HVQ >> HV > HVK.

A comparison of the migration profiles for Pb^{2+} and Zn^{2+} confirm the higher mobility of Zn^{2+} -- as witness the greater penetration depth. The results indicate up to 80 mm penetration depth for Zn^{2+} compared to 60 mm for Pb^{2+}, and slightly higher Zn^{2+} concentrations compared to Pb^{2+}, in spite

602   HYDRAULIC CONDUCTIVITY AND WASTE CONTAMINANT TRANSPORT IN SOIL

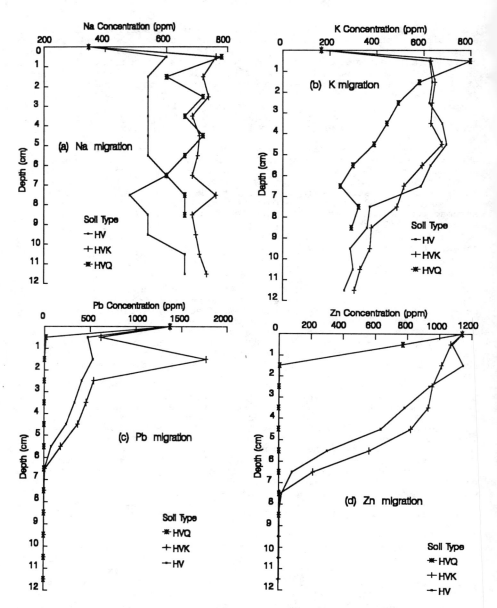

Fig. 8 - 5 p.v. migration profiles for HV, HVK, and HVQ soils.

of the lower initial concentration of Zn^{2+}, i.e., 1142 ppm Zn^{2+} versus 1372 ppm Pb^{2+} in the leachate. This situation confirms the observation with respect to the exchange mechanism of cations of equal charge as being generally inversely proportional to the hydrated radii or proportional to the unhydrated radii of the cations [18]. If one predicts the order of soil retention based on unhydrated radii, one obtains a preferential adsorption for Pb^{2+}, i.e., adsorption of Pb^{2+} (0.12 nm) > Zn^{2+} (0.074 nm), which agrees with the experimental results. If on the other hand, one predicts preferential adsorption in relation to metal ion softness, which is a function of ionization potential, charge of metal ion, and ionic radius [17, 19], the adsorption order is also Pb^{2+} > Zn^{2+}, which again agrees with the experimental results.

Exchangeable Cations

To better illustrate the adsorption or retention of the cations by the soil columns, the variation of exchangeable (adsorbed) cations with sample depth can be examined for the soils tests. In implementing this procedure the exchangeable cations were determined using NH_4OAc in batch equilibrium testing. Whereas all the major cations (Na^+, K^+, Mg^{2+}, and Ca^{2+}) were examined, only the Na^+ and K^+ results are shown in Fig. 9. The results for the cations do not show any real deviation from expectations. By and large, the order of concentration of the cations shows HVQ > HV > HVK, indicating higher adsorption by HVQ soils.

The total exchangeable cations results (Fig. 9d) represents the CEC of the soil. The results show that the CEC's of the soils are in the order of HVQ > HV > HVK, i.e., with HVQ= 40- 50 meq/100 g of dry soil, HV= 10-30 meq/100 g of dry soil, and HVK= 10-20 meq/100 g of dry soil. These results are to be expected since HVQ contains Champlain Sea Clay which is known to have a higher CEC (in the order of 60 meq/100 g of dry soil [1]), whereas HVK contains kaolinite which has a lower CEC (in the order of 5-15 meq/100 g of dry soil [13]) and the hydrobiotite-vermiculite soil samples have CEC's of about 10-37 meq/100 g of dry soil [1].

CONCLUSIONS

The tests performed in this experimental investigation were designed to provide information in regard to the capability of the hydrobiotite-vermiculite soil and its admixtures to function as a clay barrier against a specific type of municipal solid waste leachate. In order to maximize contaminant soil interactions, high concentrations of heavy metals in the leachate were used. The performance of the soils was studied with particular reference to the acid neutralizing capacity and permeability of the soils. From the test results, one could conclude that the soil can function as a natural soil liner material, so long as the finer fractions of the soil are used. If the more granular fractions are included in the soil, the use of another clay soil is recommended. This is best demonstrated by the soil mixture which includes the HV soil and the natural Quebec clay (Champlain sea clay), HVQ soil mixture. The higher CEC and presence of carbonates in the natural Quebec clay soil contributes significantly to the buffering capability of the soil mixture. The addition of a lower CEC soil (kaolinite) without benefit of carbonates, to obtain a mixture of HV and Kaolinite (HVK soil mixture) produces better results than the HV soil because of the higher CEC, but does not perform as well as the HVQ soil mixture. Other observations arising from the test results obtained can be indicated as follows:

(1) The specific surface areas of the hydrobiotite-vermiculite soil ranges from 90 to 206 m^2/g, i.e., moderately high, indicating that the mixtures of mainly illitic/mica minerals are sensitive to source location and size sorting -- as might be expected for the deposit. This testifies to the well established problem of determination of uniformity of soil deposit prior to soil sampling and also to the need for control on

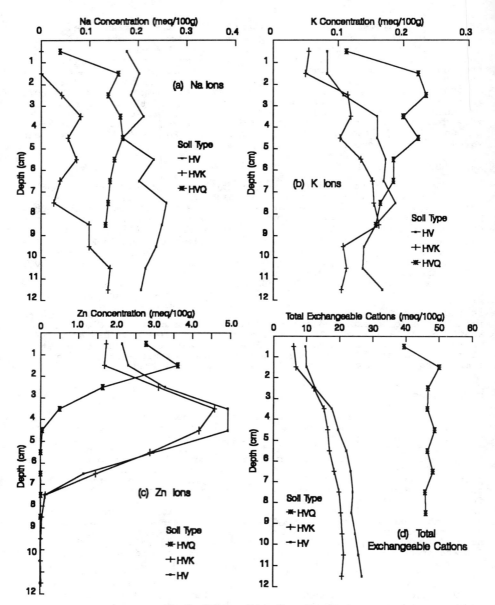

Fig. 9 - Exchangeable cations, 5 p.v. tests.

procurement of samples.

(2) The CEC of the soils are low, ranging from 7 to 19 meq/ 100 g of dry soil (batch equilibrium test), or 3 to 14 meq/100 g of dry soil (silver-thiourea method),-- within the range of illite/ mica. The results are consistent with the type of material obtained.

(3) The results obtained from leaching column tests indicate that a very slight decrease in the permeabilities of the soils at the initial stages of leaching, attesting to the compatibility of the soils to the test leachate. One needs to be careful in how one interprets the initial performance of the leaching penetration front when the soil is initially unsaturated. Interpretations concerning "equilibrating time" or "equilibrating pore volume" may need to be seriously considered. A "pre-saturation" leaching volume which has been used in some reported studies, and which does not ensure full soil saturation could establish steady-state permeation, thereby eliminating the "equilibrating time or pore volume" problem. Conventional wisdom generally ignores or minimizes the impact of results obtained for less than 2 PV's.

(4) The migration profiles indicate high attenuation of Pb^{2+} and Zn^{2+} by all the three soil types (HV, HVK, and HVQ) tested. The order of retention capability is HVQ > HV > HVK consistent with the CEC and buffering intensity of the soils. The differences in the retention, adsorption and buffering intensities of HVQ, HV, and HVK soils are attributed to differences in their composition, notably the presence of carbonates in the HVQ soil.

(5) Whereas the HV soil exhibits satisfactory properties (e.g., k = 10^{-9} m/sec), pH = 7-8, high adsorption/retention of cations, especially Pb^{2+} and Zn^{2+}) the addition of the natural clay (Champlain sea clay, HVQ soil) enhances or improves its performance. The addition of kaolinite to the HV soil (HVK soil) does not produce comparable (to the HVQ soil) adsorption or retention of cations. The CEC of HVQ= 40- 50 meq/100 g of dry soil, HV= 10- 30 meq/100 g of dry soil, and HVK= 10- 20 meq/100 g of dry soil.

(6) The retention of heavy metals is thought to be mainly through precipitation (high pH's of > 5) and cation exchange, with the latter being confined to the upper portions of the soil columns only.

ACKNOWLEDGEMENTS

The authors acknowledge the technical assistance provided by the staff of the Geotechnical Research Centre of McGill University. The study was jointly funded by the National Science and Engineering Research Council of Canada (Grant No. A-882) and the Ministry of the Environment, Ontario, Canada.

REFERENCES

[1] Yong, R.N., Mohamed, A.M.O. and Warkentin, B.P., "Principles of Contaminant Transport in Soils," Elsevier Scientific Publishing Company, Amsterdam, 1992, 327.p.

[2] Phadungchewit, Y., "The Role of pH and Soil Buffering Capacity in Heavy Metal Retention in Clay Soils," Ph.D. Thesis, McGill University, Montreal, Canada, 1990, 176 p.

[3] Van Breeman, N., and Wielemaker, W.G., "Buffer Intensities and Equilibrium pH of Materials and Soils: I. The Contribution of Minerals and Aqueous Carbonate to pH Buffering," Proceedings, Soil Science Society of America, 1974, Vol. 38, pp. 5-60.

[4] Yong, R.N., Warkentin, B.P., Phadungchewit, Y., and Galvez, R., "Buffer Capacity and Lead Retention in Some Clay Materials," Journal, Water, Air, and Soil Pollution, 1990, Vol. 53, pp.53--67.

[5] Mohamed, A.M.O., Yong, R.N., Tan, B.K., Farkas, A., and Curtis, L.W., "Assessment of Chemical Buffering Capability of Ontario Mixed Layer Mica-Vermiculite," 45th Annual Canadian Geotechnical

[6] Conference, Toronto, Oct., 1992.
Mohamed, A.M.O, Yong, R.N., and Tan, B.K., "Mitigation of Acidic Mine Drainage: Engineered Soil Barriers for Reactive Tailings," <u>ASCE Proceedings of the Environmental Engineering Sessions</u>, Water Forum 92, Baltimore, M.D., 1992, pp. 457-462.

[7] Yong, R.N., Tan, B.K., and Mohamed, A.M.O., "Assessment of Chemical Buffering Capability of a Micaceous Soil," <u>International Conference on the Implications of Ground Chemistry/Microbiology for Construction</u>, Bristol, England, 1992, pp. 1-14.

[8] Farkes, A., Curtis, L.W., Yong, R.N., and Mohamed, A.M.O., "Investigation of Ontario Mixed Layer Mica-Vermiculite as Potential Landfill Liner Material and Adsorbent of Organic and Inorganic Pollutant," Environmental Research: 1991, Technology Transfer Conference, Toronto, Canada.

[9] Carter, D., Heilman, T., and Gonzalez, J., "Ethylene Glycol Monoethyl Ether for Determining Surface Area of Silicate Minerals," <u>Journal of Soil Science</u>, 1965, pp. 356-361.

[10] Uehara, G., Gillman, G., "The Mineralogy, Chemistry and Physics of Tropical Soils with Variable Charge Clays," <u>Westview Press</u>, Boulder, Colorado, 1981, 170 p.

[11] Chhabra, R., Pleysier, J., and Cremers, A., "The measurement of the Cation Exchange Capacity and Exchangeable Cations in Soil: A New Method," <u>Proceedings of the International Clay Conference</u>, Illinois, USA, 1975, pp. 439-448.

[12] Morrill, L., Mahilum, B., and Mohinddin, S., "Organic Compounds in Soils: Sorption, Degradation, and Persistence," <u>Ann Arbour</u>, Science Publishers Inc. Ann Arbor, MI, 1982, 326 p.

[13] Yong, R.N., and Warkentin, B.P., "Soil Properties and Behaviour," <u>Elsevier Scientific Publishing Co</u>, Amsterdam, 1975, 449 p.

[14] Yong, R.N., Warith, M.A., and Boonsinsuk, P., "Migration of Leachate Solution Through Clay Liner and Substrate," <u>Hazardous and Industrial Solid Waste Testing and Disposal</u>, ASTM STP 933, pp. 208-225.

[15] Quigley, R.M., Sethi, A.J., Boonsinsuk, P., Sheeran, D.E. and Yong, R.N., "Geologic Control on Soil Composition and Properties, Lake Ojibway Clay Plain, Matagami, Quebec," <u>Proceedings, 36th Canadian Geotechnical Conference</u>, Montreal, Canada, 1982.

[16] Yong, R.N., and Sheremata, T.W., "Effect of Chloride Ions on Adsorption of Cadmium from a Landfill Leachate", <u>Canadian Geotechnical Journal</u>, Vol. 28, pp. 84-91.

[17] Yong, R.N., and Phadungchewit, Y., "pH Influence on Selectivity and Retention of Heavy Metals in Some Clay Soils," Paper submitted to <u>Canadian Geotechnical Journal</u>, July, 1992.

[18] Bohn, H.L., McNeal, B.L., and O'Connor, G.A., "Soil Chemistry," <u>John Wiley and Sons Inc.</u>, 1979, 329. p.

[19] Stumm, W., and Morgan, J.J., "Aquatic Chemistry," 2nd ed, <u>John Wiley and Sons Inc.</u>, 1981, 780 p.

Author Index

B

Baumgartner, N., 375
Benson, C. H., 3, 227
Boldt-Leppin, B., 407
Boutwell, G. P., 184
Bowders, J. J., Jr., 461
Bruner, D. R., 255
Buettner, W. G., 390

C

Chamberlain, E. J., 227
Cheng, S.-H., 266

D

Daniel, D. E., 30, 422
Day, S. R., 284
DeGroot, D. J., 300, 439
Drumm, E. C., 318
Dunn, R. J., 335

E

Edil, T. B., 353
Elrick, D. E., 375
Eppstein, L. K., 546
Evans, J. C., 79

F

Fallow, D. J., 375

G

Glade, M. J., 521
Guven, C., 505

H

Hardinato, F. S., 3
Haug, M. D., 390, 407
Heim, D. P., 353
Herzog, B. L., 95

K

Kiusalaas, N. J., 482

L

Larralde, J. L., 266
Liljestrand, H. M., 422
Lo, I. M.-C., 422
Lutenegger, A. J., 255, 300, 439

M

Maltby, V., 546
Martin, J. P., 266
McClelland, S., 461
Mohamed, A. M. O., 586
Motan, E. S., 3

N

Nelson, K. R., 482

O

Olsen, H. W., 482
Othman, M. A., 227

P

Palmer, B. S., 335
Pamukcu, S., 505
Park, J. K., 353
Parkin, G. W., 375
Phifer, M. A., 318
Poeter, E. P., 482

R

Reynolds, W. D., 375
Rosik, E. E., 559

S

Sacrison, R. R., 559
Shackelford, C. D., 111, 521
Stephens, D. B., 169

T

Tan, B.-K., 586
Topcu, I. B., 505
Trautwein, S. J., 184

W

Wallace, J. F., 559
Willden, A. T., 482
Wilson, G. V., 318
Wong, L. C., 390

Y

Yeung, A. T., 569
Yong, R. N., 586

Z

Zimmie, T. F., 227

Subject Index

A

Acid permeants, 111
Adsorption, 586
Aluminum hydroxides, 422
Aluminosilicates, 111
ASTM standards, 335
Attapulgite, 284

B

Backfill materials, 79
Bailer test, 255
Barrier cap, 505
Beam flexural strength, 266
Bentonite, 284, 407, 439, 521
    cement-bentonite, 79
    sand-bentonite, 111
Block specimen, 3
    collection, 227
Borehole sealant, 439
Borehole test, 184
Bouwer and Rice method, 95
Brooks and Corey, 169
Buffering intensity, 586

C

Cadmium release, 505
Cation exchange capacity, 586
Cement, 505
Cement-bentonite, 79
Champlain sea clay, 586
Chlorobenzenes, 422
Claymax, 422
Clays, 266, 461, 546
    Champlain sea, 586
    compacted, 184, 227, 318, 335
    liners, 3, 184, 353
    organically modified, 422
    particles, 111
    varved, 300
Compacted liners, 3
Compaction, 266, 318
Compressibility, 482
Compressive strength, 266

Concrete, plastic, 79
Consolidation tests, 300
Constant gradient, 521
Constant head technique, 375
Cooper, Bredehoeft, and
    Papadoupulos method, 95
Coupling, electro-kinetic, 569
Crust method, 169
Cutoff walls, 79, 284

D

Deformation moduli, 266
Desiccation, 284, 318
Dilatometer, 300
Dispersion coefficient,
    hydrodynamic, 353
Dissipation tests, 300
Distortion, 266
Durability, 505

E

Electro-kinetic coupling, 569
Electro-osmosis, 569
Evapotranspiration, 546

F

Falling head technique, 375
Flexure, 266
Flocculation, 111
Flow, coupled, 569
Flow model, double porosity, 255
Flow pumps, 482, 521
Flow rate, 390
Flow, unsaturated, 375
Fluid pressure, 390
Fly ash, 521
Freeze-thaw, 227, 461, 505

G

Glacial tills, 255
  slug tests, 95
Gold mine tailings, 559
Granular materials, 30
Grout, 439

H

HELP (Hydrologic Evaluation of Landfill Performance) model, 546
Humic acid, 422
Hvorslev method, 95

I

Immersion, 284
Infiltration, early-time, 375
Infiltrometer
  double-ring, 184, 559
  sealed double-ring, 3
  testing, 335

K

Kaolinite, 318, 461, 482

L

Leachate, 30, 353
Leaching, 505
  column, 586
Lead, 422
Leakage analysis, 390
Liners, 335, 439, 461, 559
  clay, 3, 111, 184, 353, 422
  soil, 3, 375, 390

M

Matric potential, 482
Metals
  cadmium, 505
  heavy, 586
Micaceous soil, 586
Mine tailings, 559
Models
  HELP, 546
  probabilistic, 3
Montmorillonite, 422

N

Nguyen and Pinder method, 95

O

Organic chemicals, 111, 422, 353
  volatile, 353
Ottawa sand, 407
Outflow method, one-step, 169

P

Paper mill sludge, 546
Partition coefficient, 353
Permeability, 111, 184, 461, 482
  constant-gradient tests, 521
  flexible wall constant head, 559
  low, 95
  slow, 375
  testing, triaxial, 407
  water, 505
Permeameters
  flexible-wall, 30, 111, 255, 390, 521
  rigid-wall, 30, 111
  test, 255
Piezocone, 300
Piezometers, 255, 300
Plastic concrete, 79
Plasticity, 266
Polymers, 407
Pore pressure, 300
Porosity, effective, 353
Precipitation, 586
Pressure plate method, 169
Proctor method, 461
Profile method, instantaneous, 169
Pumping test, 255

R

Residual materials, 505
Retardation factor, 353
Rock, slug tests, 95
Runoff, 546

S

Sand-bentonite, 111, 407, 521
Sand, coarse-grained, 95
Sand, silty, 482
Sand, slug tests, 95
Saturated soils, 30
SDRI, 559
Sealants, borehole, 439
Sealed double-ring infiltrometers (SDRIs), 3
Sedimentation, 284
Seepage, 169, 546
Shale, 461, 559
Shrinkage, 111
Sludge, 505
    paper mill, 546
Slug tests, 95, 300
Slurry, 407
    borehole, 439
    cutoff, 284
    wall, 79
Sodium bentonite, 407
Soil-bentonite, 79
Solidification, 505
Sorption, 422
Specimen size, representative, 3
Standards, 335
Steel residue, 505
Streaming potential, 569
Stress state, 227
Suction, 335
Swelling, 335
    test, 111

T

Tailings, gold mine, 559
Tensile strength, 266
Till
    fine-grained, 255
    glacial, 255
    slug tests, 95
Triaxial systems, 482

U

Unsaturated soils, 169

V

Vadose zone, 169
Volatile organic compounds, 353
Volume change, 390, 407
Volume control, 482
Volumetric strain, 318

W

Walls, vertical cutoff, 79
Water content, 318